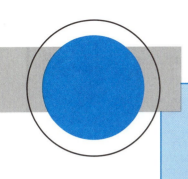

Digital Logic and Microprocessor Design with VHDL

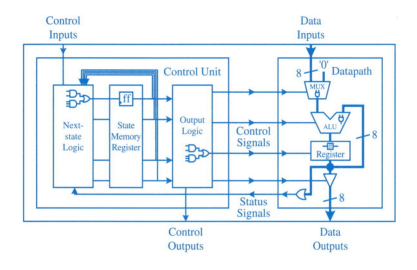

Enoch O. Hwang

La Sierra University
Riverside

Australia • Brazil • Japan • Korea • Mexico • Singapore • Spain • United Kingdom • United States

CENGAGE
Learning™

Digital Logic and Microprocessor Design with VHDL
Enoch O. Hwang

Associate Vice-President and Editorial Director: Evelyn Veitch

Publisher: Bill Stenquist

Sales and Marketing Manager: John More

Developmental Editor: Kamilah Reid Burrell

Permissions Coordinator: Vicki Gould

Production Services: RPK Editorial Services

Copy Editor: Shelly Gerger-Knechtl

Proofreader: Rose Kernan/
Shelly Gerger-Knechtl

Indexer: Enoch O. Hwang

Production Manager: Renate McCloy

Creative Director: Angela Cluer

Interior Design: Carmela Pereira

Cover Design: Andrew Adams

Compositor: PreTEX, Inc.

For product information and technology assistance, contact us at
Cengage Learning Customer & Sales Support, 1-800-354-9706

For permission to use material from this text or product, submit all requests online at **www.cengage.com/permissions**
Further permissions questions can be e-mailed to
permissionrequest@cengage.com

Library of Congress Control Number: 2004117355

ISBN-13: 978-0-534-46593-3

ISBN-10: 0-534-46593-5

Cengage Learning
5191 Natorp Boulevard
Mason, OH 45040
USA

Cengage Learning is a leading provider of customized learning solutions with office locations around the globe, including Singapore, the United Kingdom, Australia, Mexico, Brazil, and Japan. Locate your local office at **www.cengage.com/global**

Cengage Learning products are represented in Canada by Nelson Education, Ltd.

To learn more about Cengage Learning, visit **www.cengage.com**

Purchase any of our products at your local college store or at our preferred online store **www.cengagebrain.com**

Printed and bound in Canada
3 4 5 6 13 12 11 10

To my wife and children, Windy, Jonathan, and Michelle

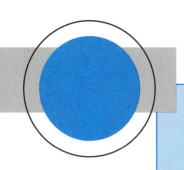

Contents

● ● ● ● ● ● ● ● ● ● ● ● ● ● ● ● ●

5 Implementation Technologies 145

● ● ● ● ● ● ● ● ● ● ● ● ● ● ● ● ··

8 Standard Sequential Components 274

● ● ● ● ● ● ● ● ● ● ● ● ● ● ● ● ● ● ● ●

9 Datapaths 310

● ● ● ● ● ● ● ● ● ● ● ● ● ● ● ● ● ● ● ●

10 Control Units 359

● ● ● ● ● ● ● ● ● ● ● ● ● ● ●

11 Dedicated Microprocessors 417

● ● ● ● ● ● ● ● ● ● ● ● ● ● ●

12 General-Purpose Microprocessors 467

● ● ● ● ● ● ● ● ● ● ● ● ● ● ●

A Schematic Entry—Tutorial 1 511

● ● ● ● ● ● ● ● ● ● ● ● ● ● ●

B VHDL Entry—Tutorial 2 525

● ● ● ● ● ● ● ● ● ● ● ● ● ● ●

C UP2 Programming—Tutorial 3 534

● ● ● ● ● ● ● ● ● ● ● ● ● ● ● ● ●

D VHDL Summary 557

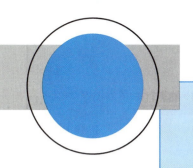

Preface

This book is about the digital logic design of microprocessors. It is intended to provide both an understanding of the basic principles of digital logic design, and how these fundamental principles are applied in the building of complex microprocessor circuits using current technologies. Although the basic principles of digital logic design have not changed, the design process and the implementation of the circuits have changed. With the advances in fully integrated modern computer-aided design (CAD) tools for logic synthesis, simulation, and the implementation of circuits in programmable logic devices (PLDs), such as field programmable gate arrays (FPGAs), it is now possible to design and implement complex digital circuits very easily and quickly.

Many excellent books on digital logic design have followed the traditional approach of introducing the basic principles and theories of logic design and the building of separate combinational and sequential components. However, students are left to wonder about the purpose of these individual components and how they are used in the building of microprocessors—the ultimate in digital circuits. One primary goal of this book is to fill in this gap by going beyond the logic principles and the building of individual components. The use of these principles and the individual components are combined together to create datapaths and control units, and finally, the building of real, dedicated custom microprocessors and general-purpose microprocessors.

Previous logic design and implementation techniques mainly focus on the logic-gate level. At this low level, it is difficult to discuss larger and more complex circuits beyond the standard combinational and sequential circuits. However, with the introduction of the register-transfer technique for designing datapaths and the concept of a finite state machine for control units, we easily can implement an arbitrary algorithm as a dedicated microprocessor in hardware. The construction of a general-purpose microprocessor then comes naturally as a generalization of a dedicated microprocessor.

With the provided CAD tool and the optional FPGA hardware development kit, students actually can implement these microprocessor circuits and see them execute, both in software simulation and in hardware. The book contains many interesting examples with complete circuit schematic diagrams and VHDL codes for both simulation and implementation in hardware. With the hands-on exercises, the student will learn not only the principles of digital logic design but, also in practice, how circuits are implemented using current technologies.

To actually see your own microprocessor come to life in real hardware is an exciting experience. Hopefully, this will help the students to not only remember what they have learned but will also get them interested in the world of digital circuit design.

● ● ● ● ● ● ● ● ● ● ● ● ● ● ● ● ●
Advanced and Historical Topics

Sections that are designated with an asterisk (*) are either advanced topics or topics for a historical perspective. These sections may be skipped without any loss of continuity in learning how to design a microprocessor.

● ● ● ● ● ● ● ● ● ● ● ● ● ● ● ● ●
Summary Checklist

There is a chapter summary checklist at the end of each chapter. These checklists provide a quick way for students to evaluate whether they have understood the materials presented in the chapter. The items in the checklists are divided into two categories. The first set of items deal with new concepts, ideas, and definitions, while the second set deals with practical "how to do something" types.

● ● ● ● ● ● ● ● ● ● ● ● ● ● ● ● ●
Design of Circuits using VHDL

Although this book provides coverage on VHDL for all of the circuits, it can be omitted entirely for the understanding and designing of digital circuits. For an introductory course in digital logic design, learning the basic principles is more important than learning how to use a hardware description language. In fact, instructors may find that students may get lost in learning the principles while trying to learn the language at the same time. With this in mind, the VHDL code in the text is totally independent of the presentation of each topic and may be skipped without any loss of continuity.

On the other hand, by studying the VHDL codes, the student can not only learn the use of a hardware description language but also learn how digital circuits can be designed automatically using a synthesizer. This book provides a basic introduction to VHDL and uses the "learn-by-examples" approach. In writing VHDL code at the dataflow and behavioral levels, the student will see the power and usefulness of a state-of-the-art CAD synthesis tool.

● ● ● ● ● ● ● ● ● ● ● ● ● ● ● ●

Using this Book

This book can be used in either an introductory or a more advanced course in digital logic design. For an introductory course with no previous background in logic, Chapters 1 through 4 are intended to provide the fundamental concepts in designing combinational circuits, and Chapters 6 through 8 cover the basic sequential circuits. Chapters 9 through 12 on microprocessor design can be introduced and covered lightly. For an advanced course, where students already have an exposure to logic gates and simple digital circuits, Chapters 1 through 4 will serve as a review. The focus should be on the register-transfer design of datapaths and control units and the building of dedicated and general-purpose microprocessors, as covered in Chapters 9 through 12. A lab component should complement the course where students can have a hands-on experience in implementing the circuits presented using the included CAD software and the optional hardware development kit.

> **Chapter 1—Designing a Microprocessor** gives an overview of the various components of a microprocessor circuit and the different abstraction levels in which a circuit can be designed.

> **Chapter 2—Digital Circuits** provides the basic principles and theories for designing digital logic circuits by introducing the use of truth tables and Boolean algebra and how the theories get translated into logic gates and circuit diagrams. A brief introduction to VHDL also is given.

> **Chapter 3—Combinational Circuits** shows how combinational circuits are analyzed, synthesized, and reduced.

> **Chapter 4—Combinational Components** discusses the standard combinational components that are used as building blocks for larger digital circuits. These components include the adder, subtractor, arithmetic logic unit, decoder, encoder, multiplexer, tri-state buffer, comparator, shifter, and multiplier. In an hierarchical design, these components will be used to build larger circuits such as the microprocessor.

> **Chapter 5—Implementation Technologies** digresses a little by looking at how logic gates are implemented at the transistor level and the various programmable logic devices available for implementing digital circuits.

> **Chapter 6—Latches and Flip-Flops** introduces the basic storage elements: specifically, the latch and the flip-flop.

> **Chapter 7—Sequential Circuits** shows how sequential circuits in the form of finite state machines are analyzed and synthesized. This chapter also shows how the operation of sequential circuits precisely can be described using state diagrams.

> **Chapter 8—Sequential Components** discusses the standard sequential components that are used as building blocks for larger digital circuits. These components include the register, shift register, counter, register file, and memory. Similar to the combinational components, these sequential components will be used in an hierarchical fashion to build larger circuits.

Chapter 9—Datapaths introduces the register-transfer design methodology and shows how an arbitrary algorithm can be performed by a datapath.

Chapter 10—Control Units shows how a finite state machine (introduced in Chapter 7) is used to control the operations of a datapath so that the algorithm can be executed automatically.

Chapter 11—Dedicated Microprocessors ties the separate datapath and control unit together to form one coherent circuit—the custom dedicated microprocessor. Several complete dedicated microprocessor examples are provided.

Chapter 12—General-Purpose Microprocessors continues on from Chapter 11 to suggest that a general-purpose microprocessor is really a dedicated microprocessor that is dedicated to only read, decode, and execute instructions. A simple general-purpose microprocessor is designed and implemented, and programs written in machine language can be executed on it.

• • • • • • • • • • • • • • • ⋯

Software and Hardware Packages

The newest student edition of Altera's MAX+plus II CAD software is included with this book on the accompanying CD-ROM. The optional Altera UP2 hardware development kit is available from Altera at a special student price. An order form for the kit can be found on the accompanying CD-ROM.

Source files for all of the circuit drawings and VHDL codes presented in this book also can be found on the accompanying CD-ROM.

• • • • • • • • • • • • • ⋯

Website for the Book

The website for this book is located at the following URL:

www.cs.lasierra.edu/~ehwang/digitaldesign/toc.html

The website provides many resources for both faculty and students.

ENOCH O. HWANG
Riverside, California

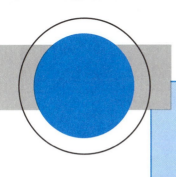

About the Author

Enoch Hwang has a Ph.D. in Computer Science from the University of California, Riverside. He is currently an Associate Professor of Computer Science at La Sierra University in Southern California and a Lecturer in the Departments of Electrical Engineering, and Computer Science and Engineering at the University of California, Riverside, teaching digital logic design.

Even from his childhood days, he has been fascinated with electronic circuits. In one of his first experiments, he attempted to connect a microphone to the speaker inside a portable radio through the earphone plug. Instead of hearing sound from the microphone through the speaker, smoke was seen coming out of the radio. Thus ended that experiment and his family's only radio. He now continues on his interest in digital circuits with research in embedded microprocessor systems, controller automation, power optimization, and robotics.

Designing Microprocessors

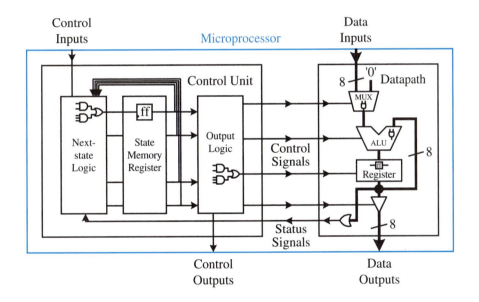

Being a computer science or electrical engineering student, you probably have assembled a PC before. You may have gone out to purchase the motherboard, central processing unit (CPU), memory, disk drive, video card, sound card, and other necessary parts, assembled them together, and have made yourself a state-of-the-art working computer. But have you ever wondered how the circuits inside those integrated circuit (IC) chips are designed? You know how the PC works at the system level by installing the operating system and seeing your machine come to life. But have you thought about how your PC works at the circuit level, how the memory is designed, or how the CPU circuit is designed?

In this book, I will show you from the ground up, how to design the digital circuits for **microprocessors**, also known as CPUs. When we hear the word "microprocessor," the first thing that probably comes to many of our minds is the Intel Pentium® CPU, which is found in most PCs. However, there are many more microprocessors that are not Pentiums, and many more microprocessors that are used in areas other than the PCs.

Microprocessors are the heart of all "smart'" devices, whether they be electronic devices or otherwise. Their smartness comes as a direct result of the decisions and controls that microprocessors make. For example, we usually do not consider a car to be an electronic device. However, it certainly has many complex, smart electronic systems, such as the anti-lock brakes and the fuel-injection system. Each of these systems is controlled by a microprocessor. Yes, even the black, hardened blob that looks like a dried-up and pressed-down piece of gum inside a musical greeting card is a microprocessor.

There are generally two types of microprocessors: **general-purpose microprocessors** and **dedicated microprocessors**. General-purpose microprocessors, such as the Pentium CPU, can perform different tasks under the control of software instructions. General-purpose microprocessors are used in all personal computers.

Dedicated microprocessors, also known as **application-specific integrated circuits** (**ASIC**s), on the other hand, are designed to perform just one specific task. For example, inside your cell phone, there is a dedicated microprocessor that controls its entire operation. The embedded microprocessor inside the cell phone does nothing else but control the operation of the phone. Dedicated microprocessors are, therefore, usually much smaller and not as complex as general-purpose microprocessors. However, they are used in every smart electronic device, such as the musical greeting cards, electronic toys, TVs, cell phones, microwave ovens, and anti-lock brake systems in your car. From this short list, I'm sure that you can think of many more devices that have a dedicated microprocessor inside them. Although the small dedicated microprocessors are not as powerful as the general-purpose microprocessors, they are being sold and used in a lot more places than the powerful general-purpose microprocessors that are used in personal computers.

Designing and building microprocessors may sound very complicated, but don't let that scare you, because it is not really all that difficult to understand the basic principles of how microprocessors are designed. We are not trying to design a Pentium microprocessor here, but after you have learned the material presented in this book, you will have the basic knowledge to understand how it is designed.

This book will show you in an easily understandable approach, starting with the basics and leading you through to the building of larger components, such as the arithmetic logic unit (ALU), register, datapath, control unit, and finally to the building of the microprocessor—first dedicated microprocessors, and then general-purpose microprocessors. Along the way, there will be many sample circuits that you can try out and actually implement in hardware using the optional Altera UP2 development board. These circuits, forming the various components found inside a microprocessor, will be combined together at the end to produce real, working microprocessors. Yes, the exciting part is that at the end, you actually can implement your microprocessor in a real IC and see that it really can execute software programs or make lights flash!

1.1   Overview of a Microprocessor

The Von Neumann model of a computer, shown in Figure 1.1, consists of four main components: the input, the output, the memory, and the microprocessor (or CPU). The parts that you purchased for your computer can all be categorized into one of these four groups. The keyboard and mouse are examples of input devices. The CRT (cathode ray tube) and speakers are examples of output devices. The different types of memory (cache, read-only memory (ROM), random-access memory (RAM), and disk drive) are all considered part of the memory box in the model. In this book, the focus is not on the mechanical aspects of the input, output, and storage devices. Rather, the focus is on the design of the digital circuitry of the microprocessor, the memory, and other supporting digital logic circuits.

The logic circuit for the microprocessor can be divided into two parts: the **data-path** and the **control unit**, as shown in Figure 1.1. Figure 1.2 shows the details inside the control unit and the datapath. The datapath is responsible for the actual execution of all data operations performed by the microprocessor, such as the addition of two numbers inside the arithmetic logic unit (ALU). The datapath also includes registers for the temporary storage of your data. The functional units inside the datapath, which in our example includes the ALU and the register, are connected together with multiplexers and data signal lines. The data signal lines are for transferring data between two functional units. Data signal lines in the circuit diagram are represented by lines connecting two functional units. Sometimes, several data signal lines are

Figure 1.1
Von Neumann model
of a computer.

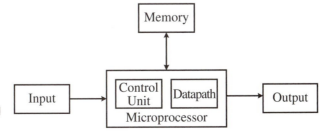

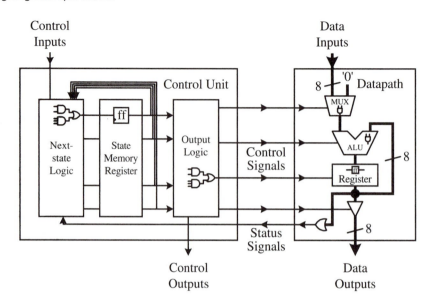

Figure 1.2
Internal parts of a
microprocessor.

grouped together to form a **bus**. The width of the bus (that is, the number of data signal lines in the group) is annotated next to the bus line. In the example, the bus lines are thicker and are 8-bits wide. Multiplexers, also known as MUXs, are for selecting data from two or more sources to go to one destination. In the sample circuit, a 2-to-1 multiplexer is used to select between the input data and the constant '0' to go to the left operand of the ALU. The output of the ALU is connected to the input of the register. The output of the register is connected to three different destinations: (1) the right operand of the ALU, (2) an OR gate used as a comparator for the test "not equal to 0," and (3) a tri-state buffer. The tri-state buffer is used to control the output of the data from the register.

Even though the datapath is capable of performing all of the data operations of the microprocessor, it cannot, however, do it on its own. In order for the datapath to execute the operations automatically, the control unit is required. The control unit, also known as the controller, controls all of the operations of the datapath and therefore, the operations of the entire microprocessor. The control unit is a **finite state machine** (FSM) because it is a machine that executes by going from one state to another and there are only a finite number of states for the machine to go to. The control unit is made up of three parts: (1) the **next-state logic**, (2) the **state memory**, and (3) the **output logic**. The purpose of the state memory is to remember the current state that the FSM is in. The next-state logic is the circuit for determining what the next state should be for the machine. The output logic is the circuit for generating the actual control signals for controlling the datapath.

Every digital logic circuit, regardless of whether it is part of the control unit or the datapath, is categorized as either a **combinational circuit** or a **sequential circuit**. A combinational circuit is one where the output of the circuit is dependent only on the current inputs to the circuit. For example, an adder circuit is a combinational circuit. It takes two numbers as inputs. The adder evaluates the sum of these two numbers and outputs the result.

A sequential circuit, on the other hand, is dependent not only on the current inputs, but also on all the previous inputs. In other words, a sequential circuit has to remember its past history. For example, the up-channel button on a TV remote is part of a sequential circuit. Pressing the up-channel button is the input to the circuit. However, just having this input is not enough for the circuit to determine what TV channel to display next. In addition to the up-channel button input, the circuit must also know the current channel that is being displayed, which is the history. If the current channel is channel 3, then pressing the up-channel button will change the channel to channel 4.

Since sequential circuits are dependent on the history, they must therefore contain memory elements for remembering the history; whereas combinational circuits do not have memory elements. Examples of combinational circuits inside the microprocessor include the next-state logic and output logic in the control unit, and the ALU, multiplexers, tri-state buffers, and comparators in the datapath. Examples of sequential circuits include the register for the state memory in the controller and the registers in the datapath. The memory in the Von Neuman computer model is also a sequential circuit.

Regardless of whether a circuit is combinational or sequential, they are all made up of the three basic logic gates: AND, OR, and NOT gates. From these three basic gates, the most powerful computer can be made. Furthermore, these basic gates are built using transistors—the fundamental building blocks for all digital logic circuits. Transistors are just electronic binary switches that can be turned on or off. The on and off states of a transistor are used to represent the two binary values: 1 and 0.

Figure 1.3 summarizes how the different parts and components fit together to form the microprocessor. From transistors, the basic logic gates are built. Logic gates are combined together to form either combinational circuits or sequential circuits. The difference between these two types of circuits is only in the way the logic gates are connected together. Latches and flip-flops are the simplest forms of sequential circuits, and they provide the basic building blocks for more complex sequential circuits. Certain combinational circuits and sequential circuits are used as standard building blocks for larger circuits, such as the microprocessor. These standard combinational and sequential components usually are found in standard libraries and serve as larger building blocks for the microprocessor. Different combinational components and sequential components are connected together to form either the datapath or the control unit of a microprocessor. Finally, combining the datapath and the control unit together will produce the circuit for either a dedicated or a general-purpose microprocessor.

● ● ● ● ● ● ● ● ● ● ● ● ● ● ● ⋯⋯

1.2   Design Abstraction Levels

Digital circuits can be designed at any one of several abstraction levels. When designing a circuit at the **transistor level**, which is the lowest level, you are dealing with

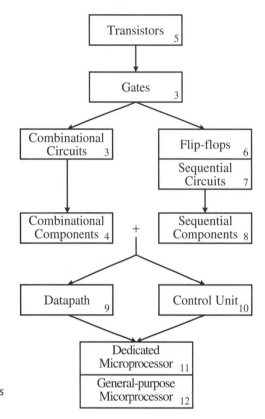

Figure 1.3
Summary of how the parts of a microprocessor fit together. The numbers in each box denote the chapter number in which the topic is discussed.

discrete transistors and connecting them together to form the circuit. The next level up in the abstraction is the **gate level**. At this level, you are working with logic gates to build the circuit. At the gate level, you also can specify the circuit using either a truth table or a Boolean equation. In using logic gates, a designer usually creates standard combinational and sequential components for building larger circuits. In this way, a very large circuit, such as a microprocessor, can be built in a hierarchical fashion. Design methodologies have shown that solving a problem hierarchically is always easier than trying to solve the entire problem as a whole from the ground up. These combinational and sequential components are used at the **register-transfer level** in building the datapath and the control unit in the microprocessor. At the register-transfer level, we are concerned with how the data is transferred between the various registers and functional units to realize or solve the problem at hand. Finally, at the highest level, which is the **behavioral level**, we construct the circuit by describing the behavior or operation of the circuit using a hardware description language. This is very similar to writing a computer program using a programming language.

1.3 **Examples of a 2-to-1 Multiplexer**

As an example, let us look at the design of the 2-to-1 multiplexer from the different abstraction levels. At this point, don't worry too much if you don't understand the details of how all of these circuits are built. This is intended just to give you an idea of what the description of the circuits look like at the different abstraction levels. We will get to the details in the rest of the book.

An important point to gain from these examples is to see that there are many different ways to create the same functional circuit. Although they are all functionally equivalent, they are different in other respects such as size (how big the circuit is or how many transistors it uses), speed (how long it takes for the output result to be valid), cost (how much it costs to manufacture), and power usage (how much power it uses). Hence, when designing a circuit, besides being functionally correct, there will always be economic versus performance tradeoffs that we need to consider.

The multiplexer is a component that is used a lot in the datapath. An analogy for the operation of the 2-to-1 multiplexer is similar in principle to a railroad switch in which two railroad tracks are to be merged onto one track. The switch controls which one of the two trains on the two separate tracks will move onto the one track. Similarly, the 2-to-1 multiplexer has two data inputs, d_0 and d_1, and a select input, s. The select input determines which data from the two data inputs will pass to the output, y.

Figure 1.4 shows the graphical symbol also referred to as the **logic symbol** for the 2-to-1 multiplexer. From looking at the logic symbol, you can tell how many signal lines the 2-to-1 multiplexer has, and the name or function designated for each line. For the 2-to-1 multiplexer, there are two data input signals, d_1 and d_0, a select input signal, s, and an output signal, y.

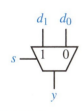

Figure 1.4 Logic symbol for the 2-to-1 multiplexer.

1.3.1 **Behavioral Level**

We can describe the operation of the 2-to-1 multiplexer simply, using the same names as in the logic symbol, by saying that

$$\text{if } s = 0 \text{ then } d_0 \text{ passes to } y,$$

otherwise

$$d_1 \text{ passes to } y$$

Or more precisely, the value that is at d_0 passes to y if $s = 0$, and the value that is at d_1 passes to y if $s = 1$.

We use a hardware description language (HDL) to describe a circuit at the **behavioral** level. When describing a circuit at this level, you would write basically the same thing as in the description, except that you have to use the correct syntax required by the hardware description language. Figure 1.5 shows the description of the 2-to-1 multiplexer using the hardware description language called VHDL.

```
LIBRARY IEEE;
USE IEEE.STD_LOGIC_1164.ALL;

ENTITY multiplexer IS PORT (
   d0, d1, s: IN STD_LOGIC;
   y: OUT STD_LOGIC);
END multiplexer;

ARCHITECTURE Behavioral OF multiplexer IS
BEGIN
   PROCESS(s, d0, d1)
   BEGIN
     IF (s = '0') THEN
       y <= d0;
     ELSE
       y <= d1;
     END IF;
   END PROCESS;
END Behavioral;
```

Figure 1.5
Behavioral level
VHDL description of
the 2-to-1
multiplexer.

The LIBRARY and USE statements are similar to the "#include" preprocessor command in C. The IEEE library contains the definition for the STD_LOGIC type used in the declaration of signals. The ENTITY section declares the interface for the circuit by specifying the input and output signals of the circuit. In this example, there are three input signals of type STD_LOGIC and one output signal also of type STD_LOGIC. The ARCHITECTURE section defines the actual operation of the circuit. The operation of the multiplexer is defined in the conditional IF-THEN-ELSE statement:

```
IF (s = '0') THEN
  y <= d0;
ELSE
  y <= d1;
END IF;
```

The two signal assignment statements, which uses the symbol <= to denote the signal assignment, in conjunction with the IF-THEN-ELSE statement, says that the signal y gets the value of d_0 if s is equal to 0; otherwise, y gets the value of d_1.

As you can see, when designing circuits at the behavioral level, we do not need to know what logic gates are needed or how they are connected together. We only need to know their interface and operation.

1.3.2   Gate Level

At the gate level, you can draw a **schematic diagram**, which is a diagram showing how the logic gates are connected together. Two schematic diagrams of a circuit are shown in Figure 1.6(a) and (b). In Figure 1.6(a), the circuit uses three INVERTERS ($\rightarrow$), four

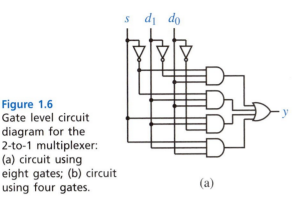

Figure 1.6
Gate level circuit diagram for the 2-to-1 multiplexer: (a) circuit using eight gates; (b) circuit using four gates.

(a)

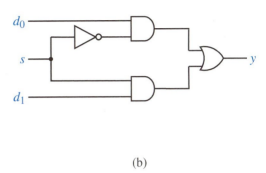

(b)

3-input AND gates ($\equiv$D-), and one 4-input OR gate ($\equiv$D-). In Figure 1.6(b), only one INVERTER, two 2-input AND gates, and one 2-input OR gate are needed. Although one circuit is larger (in terms of the number of gates needed) than the other, both of these circuits realize the same 2-to-1 multiplexer function. Therefore, when we want to actually implement a 2-to-1 multiplexer circuit, we will want to use the second, smaller circuit rather than the first.

At the gate level, you can also describe the 2-to-1 multiplexer using a **truth table** or with a **Boolean equation**, as shown in Figure 1.7(a) and (b), respectively. For the truth table, we list all possible combinations of the binary values for the three inputs, s, d_0, and d_1, and then determine what the output value y should be based on the functional description of the circuit. We see that for the first four rows of the table when $s = 0$, y has the same values as d_0; whereas, in the last four rows when $s = 1$, y has the same values as d_1.

The Boolean equation in Figure 1.7(b) can be derived from either the schematic diagram or the truth table. The first equality in Figure 1.7(b) matches the truth table in (a) and also the schematic diagram in Figure 1.6(a). The second equality in Figure 1.7(b) matches the schematic diagram in Figure 1.6(b). To derive the equation from the truth table, we look at all of the rows where the output y is a 1. Each of these

Figure 1.7
Gate level description of the 2-to-1 multiplexer: (a) using a truth table; (b) using a Boolean equation.

s	d_1	d_0	y
0	0	0	0
0	0	1	1
0	1	0	0
0	1	1	1
1	0	0	0
1	0	1	0
1	1	0	1
1	1	1	1

(a)

$$y = s'd_1'd_0 + s'd_1d_0 + sd_1d_0' + sd_1d_0$$
$$= s'd_0 + sd_1$$

(b)

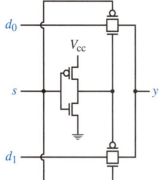

Figure 1.8
Transistor circuit for
the 2-to-1 multiplexer.

rows results in a term in the equation. For each term, the variable is primed ($'$) when
the value of the variable is a 0, and unprimed when the value of the variable is a 1.

1.3.3   Transistor Level

The 2-to-1 multiplexer circuit at the transistor level is shown in Figure 1.8. It contains
six transistors, three of which are PMOS ($\triangle$), and three are NMOS ($\perp$). The pair
of transistors on the left forms an inverter for the signal s, while the two pairs of
transistors on the right form two transmission gates. The transmission gate allows or
disallows the data signal d_0 or d_1 to pass through, depending on the control signal s.
The top transmission gate is turned on when s is a 0, and the bottom transmission
gate is turned on when s is a 1. Hence, when s is 0, the value at d_0 is passed to y,
and when s is 1, the value at d_1 is passed to y.

1.4   Introduction to VHDL

The popularity of using hardware description languages (HDL) for designing digital
circuits began in the mid-1990s when commercial synthesis tools became available.
Two popular HDLs used by many engineers today are VHDL and Verilog. VHDL,
which stands for VHSIC Hardware Description Language (VHSIC, in turn, stands
for Very High Speed Integrated Circuit), was sponsored and developed jointly by the
U.S. Department of Defense and the IEEE in the mid-1980s. It was standardized by
the IEEE in 1987 (VHDL-87), and later extended in 1993 (VHDL-93). Verilog, on
the other hand, was first introduced in 1984, and again later in 1988, as a proprietary
hardware description language by two companies: Synopsys and Cadence Design
Systems. In this book, we will use VHDL.

VHDL, in many respects, is similar to a regular computer programming language, such as C. For example, it has constructs for variable assignments, conditional
statements, loops, and functions (just to name a few). In a computer programming
language, a compiler is used to translate the high-level source code to machine code.
In VHDL, however, a synthesizer is used to translate the source code to a description of the actual hardware circuit that implements the code. From this description,

```
LIBRARY IEEE;
USE IEEE.STD_LOGIC_1164.ALL;

ENTITY multiplexer IS PORT(
   d0, d1, s: IN STD_LOGIC;
   y: OUT STD_LOGIC);
END multiplexer;

ARCHITECTURE Dataflow OF multiplexer IS
BEGIN
   y <= ((NOT s) AND d0) OR (s AND d1);
END Dataflow;
```

Figure 1.9
Dataflow level VHDL description of the 2-to-1 multiplexer.

which we call a **netlist**, the actual, physical digital device that realizes the source code can be made automatically. Accurate functional and timing simulation of the code is also possible in order to test the correctness of the circuit.

You saw in Section 1.3.1 how we used VHDL to describe the 2-to-1 multiplexer at the behavioral level. VHDL can also be used to describe a circuit at other levels. Figure 1.9 shows the VHDL code for the multiplexer written at the **dataflow** level. The main difference between the behavioral VHDL code shown in Figure 1.5 and the dataflow VHDL code is that, in the behavioral code, there is a PROCESS block statement; whereas, in the dataflow code, there is no PROCESS statement. Statements within a PROCESS block are executed sequentially like in a computer program, while statements outside a PROCESS block (including the PROCESS block itself) are executed concurrently or in parallel. The signal assignment statement, using the symbol <=, is derived directly from the Boolean equation for the multiplexer, as shown in Figure 1.7(b), using the built-in VHDL operators: AND, OR, and NOT.

In addition to the behavioral and dataflow levels, we can also write VHDL code at the **structural** level. Figure 1.11 shows the VHDL code for the multiplexer written at the structural level; the code is based on the circuit shown in Figure 1.10. The three different gates (*and2gate*, *or2gate*, and *notgate*) used in the circuit first are declared and defined using the ENTITY and ARCHITECTURE statements, respectively. After this, the multiplexer is declared (also with the ENTITY statement). The actual, structural definition of the multiplexer is in the ARCHITECTURE section for *multiplexer2*. First of all, the COMPONENT statements specify what components are used in the circuit.

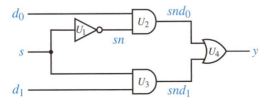

Figure 1.10
2-to-1 multiplexer circuit.

The SIGNAL statement declares three internal signals (*sn*, *snd0*, and *sd1*) that will be used in the connection of the circuit. Finally, the PORT MAP statements declare the instances of the gates used in the circuit and also specify how they are connected using the external and internal signals.

```
---------------- NOT gate ----------------------
LIBRARY IEEE;
USE IEEE.STD_LOGIC_1164.ALL;

ENTITY notgate IS PORT(
  i: IN STD_LOGIC;
  o: OUT STD_LOGIC);
END notgate;

ARCHITECTURE Dataflow OF notgate IS
BEGIN
  o <= not i;
END Dataflow;


---------------- 2-input AND gate ---------------
LIBRARY IEEE;
USE IEEE.STD_LOGIC_1164.ALL;

ENTITY and2gate IS PORT(
  i1, i2: IN STD_LOGIC;
  o: OUT STD_LOGIC);
END and2gate;

ARCHITECTURE Dataflow OF and2gate IS
BEGIN
  o <= i1 AND i2;
END Dataflow;


---------------- 2-input OR gate ----------------
LIBRARY IEEE;
USE IEEE.STD_LOGIC_1164.ALL;

ENTITY or2gate IS PORT(
  i1, i2: IN STD_LOGIC;
  o: OUT STD_LOGIC);
END or2gate;

ARCHITECTURE Dataflow OF or2gate IS
BEGIN
  o <= i1 OR i2;
END Dataflow;
```

Figure 1.11
Structural level VHDL description of the 2-to-1 multiplexer.
(continued on next page)

```
---------------- 2-to-1 multiplexer ------------
LIBRARY IEEE;
USE IEEE.STD_LOGIC_1164.ALL;

ENTITY multiplexer IS PORT(
  d0, d1, s: IN STD_LOGIC;
  y: OUT STD_LOGIC);
END multiplexer;

ARCHITECTURE Structural OF multiplexer IS
  COMPONENT notgate PORT(
    i: IN STD_LOGIC;
    o: OUT STD_LOGIC);
  END COMPONENT;
  COMPONENT and2gate PORT(
    i1, i2: IN STD_LOGIC;
    o: OUT STD_LOGIC);
  END COMPONENT;
  COMPONENT or2gate PORT(
    i1, i2: IN STD_LOGIC;
    o: OUT STD_LOGIC);
  END COMPONENT;

  SIGNAL sn, snd0, sd1: STD_LOGIC;

BEGIN
  U1: notgate PORT MAP(s,sn);
  U2: and2gate PORT MAP(d0, sn, snd0);
  U3: and2gate PORT MAP(d1, s, sd1);
  U4: or2gate PORT MAP(snd0, sd1, y);
END Structural;
```

Figure 1.11
Structural level VHDL
description of the
2-to-1 multiplexer.

● ● ● ● ● ● ● ● ● ● ● ● ● ●

1.5 Synthesis

Given a gate level circuit diagram, such as the one shown in Figure 1.6, you actually can get some discrete logic gates and manually connect them together with wires on a breadboard. Traditionally, this is how electronic engineers actually designed and implemented digital logic circuits. However, this is not how electronic engineers design circuits anymore. They write programs, such as the one in Figure 1.5, just like computer programmers do. The question then is how does the program that describes the operation of the circuit actually gets converted to the physical circuit?

The problem here is similar to translating a computer program written in a high-level language to machine language for a particular computer to execute. For a computer program, we use a compiler to do the translation. For translating a digital logic circuit, we use a **synthesizer**. Instead of using a high-level computer language to describe a computer program, we use a hardware description language (HDL) to describe the operations of a digital logic circuit. Writing a description of a digital logic circuit is similar to writing a computer program; the only difference is that a different language is used. A synthesizer is then used to translate the HDL program into the circuit **netlist**. A netlist is a description of how a circuit actually is realized or connected using basic gates. This translation process from an HDL description of a circuit to its netlist is referred to as **synthesis**.

Furthermore, the netlist from the output of the synthesizer can be used directly to implement the actual circuit in a programmable logic device (PLD) chip, such as a field programmable gate array (FPGA). With this final step, the creation of a digital circuit that is implemented fully in an integrated circuit (IC) chip can be done easily. The Appendices give tutorials of the complete process: from writing the VHDL code to synthesizing the circuit and uploading the netlist to the FPGA chip using Altera's development system.

● ● ● ● ● ● ● ● ● ● ● ● ● ● ● ●

1.6   Going Forward

We will now embark upon a journey that will take you from a simple transistor to the building of a microprocessor. Figure 1.2 will serve as our guide and map. If you get lost on the way, and do not know where a particular component fits in the overall picture, just refer to this map. At the beginning of each chapter, I will refresh your memory with this map by highlighting the components in the map that the chapter will cover.

Figure 1.12 is an actual picture of the circuitry inside an Intel Pentium 4 CPU. When you reach the end of this book, you still may not be able to design the circuit for the P4, but you certainly will have the knowledge of how a microprocessor is designed, because you actually will have designed and implemented a working microprocessor yourself.

● ● ● ● ● ● ● ● ● ● ● ● ● ● ● ●

1.7   Summary Checklist

- Microprocessor
- General-purpose microprocessor
- Dedicated microprocessor
- ASIC
- Datapath

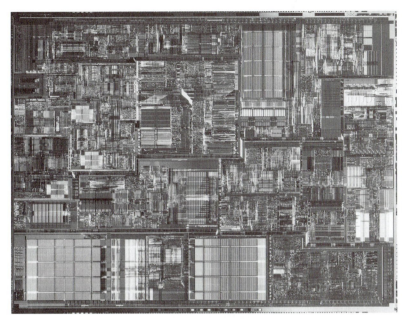

Figure 1.12   The internal circuitry of the Intel P4 CPU.

- Control unit
- Finite state machine (FSM)
- Next-state logic
- State memory
- Output logic
- Combinational circuit
- Sequential circuit
- Transistor level design
- Gate level design
- Register-transfer level design
- Behavioral level design
- Logic symbol
- VHDL
- Synthesis
- Netlist

● ● ● ● ● ● ● ● ● ● ● ● ● ● ● ● ● ● ●

1.8 **Problems**

P1.1. Find out the approximate number of general-purpose microprocessors sold in the U.S. in a year versus the number of dedicated microprocessors sold.

P1.2. Compile a list of devices that you use during one regular day that are controlled by a microprocessor.

P1.3. Describe what your regular daily routine will be like if there is no electrical power (including battery power) available.

P1.4. Apply the Von Neumann model of a computer system, as shown in Figure 1.1, to the following systems. Determine what parts of the system correspond to the different parts of the model.
(a) Traffic light
(b) Heart pacemaker
(c) Microwave oven
(d) Musical greeting card
(e) Hard disk drive (not the entire personal computer)

P1.5. The speed of a microprocessor is often measured by its clock frequency. What is the clock frequency of the fastest general-purpose microprocessor available?

P1.6. Compare some typical clock speeds between general-purpose microprocessors versus dedicated microprocessors.

P1.7. Summarize the mainstream generations of the Intel general-purpose microprocessors used in personal computers starting with the 8086 CPU. List the year introduced, the clock speed, and the number of transistors in each.

P1.8. Using Figure 1.9 as a template, write the dataflow VHDL code for the 2-to-1 multiplexer circuit shown in Figure 1.6(a).

P1.9. Using Figure 1.11 as a template, write the structural VHDL code for the 2-to-1 multiplexer circuit shown in Figure 1.6(a).

P1.10. Do Tutorial 1 in Appendix A.

P1.11. Do Tutorial 2 in Appendix B.

P1.12. Do Tutorial 3 in Appendix C.

CHAPTER 2

Digital Circuits

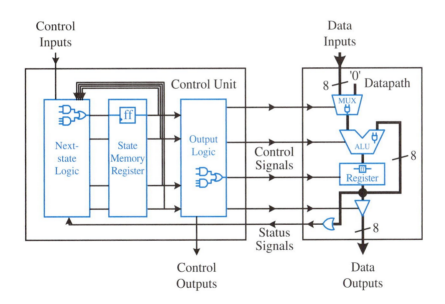

Our world is an analog world. Measurements that we make of the physical objects around us are never in discrete units, but rather in a continuous range. We talk about physical constants such as 2.718281828... or 3.141592.... To build analog devices that can process these values accurately is next to impossible. Even building a simple analog radio requires very accurate adjustments of frequencies, voltages, and currents at each part of the circuit. If we were to use voltages to represent the constant 3.14, we would have to build a component that will give us exactly 3.14 volts every time. This is again impossible; due to the imperfect manufacturing process, each component produced is slightly different from the others. Even if the manufacturing process can be made as perfect as perfect can get, we still would not be able to get 3.14 volts from this component every time we use it. The reason being that the physical elements used in producing the component behave differently in different environments, such as temperature, pressure, and gravitational force, just to name a few. Therefore, even if the manufacturing process is perfect, using this component in different environments will not give us exactly 3.14 volts every time.

To make things simpler, we work with a digital abstraction of our analog world. Instead of working with an infinite continuous range of values, we use just two values! Yes, just two values: 1 and 0, on and off, high and low, true and false, black and white, or however you want to call it. It certainly is much easier to control and work with two values rather than an infinite range. We call these two values a binary value for the reason that there are only two of them. A single 0 or a single 1 is then a **binary digit** or **bit**. This sounds great, but we have to remember that the underlying building block for our digital circuits still is based on an analog world.

This chapter provides the theoretical foundations for building digital logic circuits using logic gates, the basic building blocks for all digital circuits. In order to understand how logic gates are used to implement digital circuits, we need to have a good understanding of the basic theory of Boolean algebra, Boolean functions, and how to use and manipulate them. Most people may find Sections 2.5 and 2.6 on these theories to be boring, but let me encourage you to grind through it patiently, because if you do not understand it now, you quickly will get lost in the later chapters. The good news is that these two sections are the only sections in this book on theory, and I will try to keep them as short and simple as possible. You also will find that many of the Boolean theorems are very familiar, because they are similar to the algebra theorems that you have learned from your high school math class. As you can see from the microprocessor road map, this chapter affects all of the parts for building a microprocessor.

● ● ● ● ● ● ● ● ● ● ● ● ● ● ● ● ●

2.1 Binary Numbers

Since digital circuits deal with binary values, we will begin with a quick introduction to binary numbers. A bit, having either the value of 0 or 1, can represent only two things or two pieces of information. It is, therefore, necessary to group many bits together to represent more pieces of information. A string of n bits can represent 2^n

different pieces of information. For example, a string of two bits results in the four combinations: 00, 01, 10, and 11. By using different encoding techniques, a group of bits can be used to represent different information, such as a number, a letter of the alphabet, a character symbol, or a command for the microprocessor to execute.

The use of decimal numbers is quite familiar to us. However, since the binary digit is used to represent information within the computer, we also need to be familiar with **binary numbers**. Note that the use of binary numbers is just a form of representation for a string of bits. We can just as well use octal, decimal, or hexadecimal numbers to represent the string of bits. In fact, you will find that hexadecimal numbers are often used as a shorthand notation for binary numbers.

The decimal number system is a positional system. In other words, the value of the digit is dependent on the position of the digit within the number. For example, in the decimal number 48, the decimal digit 4 has a greater value than the decimal digit 8 because it is in the tenth position, whereas the digit 8 is in the unit position. The value of the number is calculated as $(4 \times 10^1) + (8 \times 10^0)$.

Like the decimal number system, the binary number system is also a positional system. The only difference between the two is that the binary system is a base-2 system, and so, it uses only two digits, 0 and 1, instead of ten. The binary numbers from 0 to 15 (decimal) are shown in Figure 2.1. The range from 0 to 15 has 16 different combinations. Since $2^4 = 16$, therefore, we need a 4-bit binary number (i.e., a string of four bits) to represent this range.

Decimal	Binary	Octal	Hexadecimal
0	0000	0	0
1	0001	1	1
2	0010	2	2
3	0011	3	3
4	0100	4	4
5	0101	5	5
6	0110	6	6
7	0111	7	7
8	1000	10	8
9	1001	11	9
10	1010	12	A
11	1011	13	B
12	1100	14	C
13	1101	15	D
14	1110	16	E
15	1111	17	F

Figure 2.1
Numbers from 0 to 15 in binary, octal, and hexidecimal number systems.

When we count in decimals, we count from 0 to 9. After 9, we go back to 0 and have a carry of a 1 to the next digit. When we count in binary numbers, we do the same thing, except that we only count from 0 to 1. After 1, we go back to 0 and have a carry of a 1 to the next bit.

The decimal value of a binary number can be found just like that for a decimal number, except that we raise the base number 2 to a power rather than the base number 10 to a power. For example, the value for the decimal number 658 is

$$658_{10} = (6 \times 10^2) + (5 \times 10^1) + (8 \times 10^0) = 600 + 50 + 8 = 658_{10}$$

Similarly, the decimal value for the binary number 1011011_2 is

$$1011011_2 = (1 \times 2^6) + (0 \times 2^5) + (1 \times 2^4) + (1 \times 2^3) + (0 \times 2^2) + (1 \times 2^1) + (1 \times 2^0)$$
$$= 64 + 16 + 8 + 2 + 1 = 91_{10}$$

To get the decimal value, the least significant bit (in this case, the rightmost 1) is multiplied with 2^0. The next bit to the left is multiplied with 2^1, and so on. Finally, they are all added together to give the value 91_{10}.

Notice the subscript 10 in the decimal number 658_{10}, and the 2 in the binary number 1011011_2. This subscript is used to denote the base of the number whenever there might be confusion as to what base the number is in.

Converting a decimal number to its binary equivalent can be done by successively dividing the decimal number by 2 and keeping track of the remainder at each step. Combining the remainders together (starting with the last one) forms the equivalent binary number. For example, using the decimal number 91, we divide it by 2 to get 45 with a remainder of 1. Then we divide 45 by 2 to get 22 with a remainder of 1. We continue in this fashion until the end as shown here.

Concatenating the remainders together, starting with the last one (most significant bit) results in the binary number 1011011_2.

Binary numbers usually consist of a long string of bits. A shorthand notation for writing out this lengthy string of bits is to use either octal or hexadecimal numbers. Since the octal system is base-8 and the hexadecimal system is base-16 (both of which are a power of 2), a binary number can be converted easily to an octal or hexadecimal number, or vice versa.

Octal numbers only use the digits from 0 to 7 for the eight different combinations. When counting in octals, the number after 7 is 10, as shown in Figure 2.1. To convert a binary number to octal, we simply group the bits into groups of threes, starting from the right (least significant bit). The reason for this is because $8 = 2^3$. For each group of three bits, we write the equivalent octal digit for it. For example, the conversion of the binary number 1110011_2 to the octal number 163_8 is shown here.

$$\frac{001}{1} \qquad \frac{110}{6} \qquad \frac{011}{3}$$

Since the original binary number has seven bits, we need to extend it with two leading 0's to get three bits for the leftmost group. Note that when we are dealing with negative numbers, we may require extending the number with leading ones instead of 0's.

Converting an octal number to its binary equivalent is just as easy. For each octal number, we write down the equivalent three bits. These groups of three bits are concatenated together to form the final binary number. For example, the conversion of the octal number 5724_8 to the binary number 101111010100_2 is shown here.

$$\begin{array}{cccc} 5 & 7 & 2 & 4 \\ 101 & 111 & 010 & 100 \end{array}$$

The decimal value of an octal number can be found just like that for a binary or decimal number, except that we raise the base number 8 to a power instead of the base number 2 or 10 to a power. For example, the octal number 5724_8 has the value:

$$5724_8 = (5 \times 8^3) + (7 \times 8^2) + (2 \times 8^1) + (4 \times 8^0) = 2560 + 448 + 16 + 4 = 3028_{10}$$

Hexadecimal numbers are treated basically the same way as octal numbers except with the appropriate changes to the base. Hexadecimal (or hex for short) numbers use base-16, and thus require 16 different digit symbols, as shown in Figure 2.1. Converting binary numbers to hexadecimal numbers involves grouping the bits into groups of fours since $16 = 2^4$. For example, the conversion of the binary number 11011011011_2 to the hexadecimal number $6DB_{16}$ is shown below. Again, we need to extend it with a leading 0 to get four bits for the leftmost group.

$$\frac{0110}{6} \qquad \frac{1101}{D} \qquad \frac{1011}{B}$$

To convert a hex number to a binary number, we write down the equivalent four bits for each hex digit, and then concatenate them together to form the final binary number. For example, the conversion of the hexadecimal number $5C4A_{16}$ to the binary number 0101110001001010_2 is shown here.

$$\begin{array}{cccc} 5 & C & 4 & A \\ 0101 & 1100 & 0100 & 1010 \end{array}$$

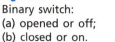

Figure 2.2
Binary switch:
(a) opened or off;
(b) closed or on.

The following example shows how the decimal value of the hexadecimal number $C4A_{16}$ is evaluated.

$$C4A_{16} = (C \times 16^2) + (4 \times 16^1) + (A \times 16^0)$$
$$= (12 \times 16^2) + (4 \times 16^1) + (10 \times 16^0)$$
$$= 3072 + 64 + 10 = 3146_{10}$$

• • • • • • • • • • • • • • • • •

2.2   **Binary Switch**

Besides the fact that we are working only with binary values, digital circuits are easy to understand because they are based on one simple idea of turning a switch on or off to obtain either one of the two binary values. Since the switch can be in either one of two states (on or off), we call it a **binary switch**, or just a **switch** for short. The switch has three connections: an input, an output, and a control for turning the switch on or off, as shown in Figure 2.2. When the switch is opened, as in Figure 2.2(a), it is turned off, and nothing gets through from the input to the output. When the switch is closed, as in Figure 2.2(b), it is turned on, and whatever is presented at the input is allowed to pass through to the output.

Uses of the binary switch idea can be found in many real-world devices. For example, the switch can be an electrical switch with the input connected to a power source and the output connected to a siren S, as shown in Figure 2.3.

When the switch is closed, the siren turns on. The usual convention is to use a 1 to mean "on" and a 0 to mean "off." Therefore, when the switch is closed, the output is 1 and the siren will turn on. We can also use a variable, x, to denote the state of the switch. We can let $x = 1$ to mean the switch is closed and $x = 0$ to mean the switch is opened. Using this convention, we can describe the state of the siren S in terms of the variable x using a simple logic expression. Since $S = 1$ if $x = 1$ and $S = 0$ if $x = 0$, we can write

$$S = x$$

This logic expression describes the output S in terms of the input variable x.

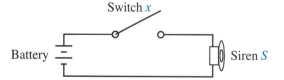

Figure 2.3
A siren controlled
by a switch.

2.3 Basic Logic Operators and Logic Expressions

Two binary switches can be connected together either in series or in parallel, as shown in Figure 2.4.

If two switches are connected in series, as in Figure 2.4(a), then both switches have to be on in order for the output F to be 1. In other words, $F = 1$ if $x = 1$ and $y = 1$. If either x or y is off, or both are off, then $F = 0$. Translating this into a logic expression, we get

$$F = x \text{ AND } y$$

Hence, two switches connected in series give rise to the logical **AND** operator. In a Boolean function (which we will explain in more detail in Section 2.5), the AND operator is denoted either with a dot ($\cdot$) or no symbol at all. Thus, we can rewrite the above expression as

$$F = x \cdot y$$

or simply

$$F = xy$$

If we connect two switches in parallel, as in Figure 2.4(b), then only one switch needs to be on in order for the output F to be 1. In other words, $F = 1$ if either $x = 1$, or $y = 1$, or both x and y are 1's. This means that $F = 0$ only if both x and y are 0's. Translating this into a logic expression, we get

$$F = x \text{ OR } y$$

and this gives rise to the logical **OR** operator. In a Boolean function, the OR operator is denoted with a "plus" symbol ($+$). Thus, we can rewrite the above expression as

$$F = x + y$$

In addition to the AND and OR operators, there is another basic logic operator—the **NOT** operator, also known as the **INVERTER**. Whereas, the AND and OR operators

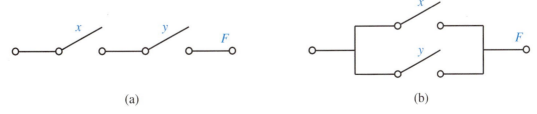

(a)   (b)

Figure 2.4   Connection of two binary switches: (a) in series; (b) in parallel.

have multiple inputs; the NOT operator has only one input and one output. The NOT operator simply inverts its input, so a 0 input will produce a 1 output, and a 1 becomes a 0. In a Boolean function, the NOT operator is denoted either with an "apostrophe" symbol ($'$) or a "bar" on top ($^-$) as in

$$F = x'$$

or

$$F = \bar{x}$$

When several operators are used in the same expression, the precedence given to the operators are (from highest to lowest) NOT, AND, and OR. The order of evaluation can be changed by means of using parenthesis. For example, the expression

$$F = xy + z'$$

means (x AND y) OR (NOT z), and the expression

$$F = x(y + z)'$$

means x AND (NOT (y OR z)).

● ● ● ● ● ● ● ● ● ● ● ● ● ● ● ● ● ● ●

2.4 **Truth Tables**

The operation of the AND, OR, and NOT logic operators can be described formally by using a **truth table**, as shown in Figure 2.5. A truth table is a two-dimensional array where there is one column for each input and one column for each output (a circuit may have more than one output). Since we are dealing with binary values, each input can be either a 0 or a 1. We simply enumerate all possible combinations of 0's and 1's for all of the inputs. Usually, we want to write these input values in the normal binary counting order. With two inputs, there are 2^2 combinations giving us the four rows in the table. The values in the output column are determined from applying the corresponding input values to the functional operator. For the AND truth table in Figure 2.5(a), $F = 1$ only when x and y are both 1, otherwise, $F = 0$. For the

Figure 2.5
Truth tables for the three basic logical operators: (a) AND; (b) OR; (c) NOT.

x	y	F
0	0	0
0	1	0
1	0	0
1	1	1

(a)

x	y	F
0	0	0
0	1	1
1	0	1
1	1	1

(b)

x	F
0	1
1	0

(c)

OR truth table in Figure 2.5(b), $F = 1$ when either x or y, or both is a 1, otherwise $F = 0$. For the NOT truth table, the output F is just the inverted value of the input x.

Using a truth table is one method to formally describe the operation of a circuit or function. The truth table for any given logic expression (no matter how complex it is) can always be derived. Examples on the use of truth tables to describe digital circuits are given in the following sections. Another method to formally describe the operation of a circuit is by using Boolean expressions or Boolean functions.

2.5 Boolean Algebra and Boolean Functions

2.5.1 Boolean Algebra

George Boole, in 1854, developed a system of mathematical logic, which we now call **Boolean algebra**. Based on Boole's idea, Claude Shannon, in 1938, showed that circuits built with binary switches can easily be described using Boolean algebra. The abstraction from switches being on and off to the use of Boolean algebra is as follows. Let $B = \{0, 1\}$ be the Boolean algebra whose elements are one of the two values, 0 and 1. We define the operations AND ($\cdot$), OR ($+$), and NOT ($'$) for the elements of B by the axioms in Figure 2.6(a). These axioms are simply the definitions for the AND, OR, and NOT operators.

A variable x is called a **Boolean variable** if x takes on only values in B (i.e., either 0 or 1). Consequently, we obtain the theorems in Figure 2.6(b) for single variables and Figure 2.6(c) for two and three variables.

The theorems in Figure 2.6(b) can be proven easily by substituting the binary values into the expressions and using the axioms. For example, to show that Theorem 6a is true, we substitute 0 into x to get Axiom 3a, and substitute 1 into x to get Axiom 2a.

To prove the theorems in Figure 2.6(c), we can use either one of two methods: (1) use a truth table, or (2) use axioms and theorems that have already been proven. We show these two methods in the following two examples.

Example 2.1   **Proof of a theorem using a truth table**

Prove Theorem 12a from Figure 2.6(c), using a truth table.

Theorem 12a states that $x \cdot (y + z) = (x \cdot y) + (x \cdot z)$. To prove that Theorem 12a is true using a truth table, we need to show that, for every combination of values for the three variables x, y, and z, the left-hand side of the expression is equal to the right-hand side.

We start with the first three columns labeled x, y, and z; and we enumerate all possible combinations of values for these three variables giving us the eight rows as shown next.

1a.	$0 \cdot 0 = 0$	1b.	$1 + 1 = 1$
2a.	$1 \cdot 1 = 1$	2b.	$0 + 0 = 0$
3a.	$0 \cdot 1 = 1 \cdot 0 = 0$	3b.	$1 + 0 = 0 + 1 = 1$
4a.	$0' = 1$	4b.	$1' = 0$

(a)

5a.	$x \cdot 0 = 0$	5b.	$x + 1 = 1$	Null Element
6a.	$x \cdot 1 = 1 \cdot x = x$	6b.	$x + 0 = 0 + x = x$	Identity
7a.	$x \cdot x = x$	7b.	$x + x = x$	Idempotent
8a.	$(x')' = x$			Double Complement
9a.	$x \cdot x' = 0$	9b.	$x + x' = 1$	Inverse

(b)

10a.	$x \cdot y = y \cdot x$	10b.	$x + y = y + x$	Commutative
11a.	$(x \cdot y) \cdot z = x \cdot (y \cdot z)$	11b.	$(x + y) + z = x + (y + z)$	Associative
12a.	$x \cdot (y + z) = (x \cdot y) + (x \cdot z)$	12b.	$x + (y \cdot z) = (x + y) \cdot (x + z)$	Distributive
13a.	$x \cdot (x + y) = x$	13b.	$x + (x \cdot y) = x$	Absorption
14a.	$(x \cdot y) + (x \cdot y') = x$	14b.	$(x + y) \cdot (x + y') = x$	Combining
15a.	$(x \cdot y)' = x' + y'$	15b.	$(x + y)' = x' \cdot y'$	DeMorgan's

(c)

Figure 2.6 Boolean algebra axioms and theorems: (a) axioms; (b) single-variable theorems; (c) two- and three-variable theorems.

x	y	z	$(y + z)$	$(x \cdot y)$	$(x \cdot z)$	$x \cdot (y + z)$	$(x \cdot y) + (x \cdot z)$
0	0	0	0	0	0	0	0
0	0	1	1	0	0	0	0
0	1	0	1	0	0	0	0
0	1	1	1	0	0	0	0
1	0	0	0	0	0	0	0
1	0	1	1	0	1	1	1
1	1	0	1	1	0	1	1
1	1	1	1	1	1	1	1

For each combination (row), we evaluate the intermediate expressions $(y + z)$, $(x \cdot y)$, and $(x \cdot z)$ by substituting the values of x, y, and z into the expression. Finally, we obtain the values for the last two columns, which correspond to the left-hand side and right-hand side of Theorem 12a. The values in these two columns are identical for every combination of x, y, and z; therefore, we can say that Theorem 12a is true.

Example 2.2	**Proof of a theorem using axioms and theorems**

Prove Theorem 13b from Figure 2.6(c) using other axioms and theorems from the figure.

Theorem 13b states that $x + (x \cdot y) = x$. To prove that Theorem 13b is true using other axioms and theorems, we can argue as follows:

$$x + (x \cdot y) = (x \cdot 1) + (x \cdot y) \qquad \text{(by Identity Theorem 6a)}$$

$$= x \cdot (1 + y) \qquad \text{(by Distributive Theorem 12a)}$$

$$= x \cdot (1) \qquad \text{(by Null Element Theorem 5b)}$$

$$= x \qquad \text{(by Identity Theorem 6a)}$$

Example 2.2 shows that some theorems can be derived from others that already have been proven with the truth table. Full treatment of Boolean algebra is beyond the scope of this book and can be found in the references. For our purposes, we simply assume that all of the theorems are true and will use them just to show that two circuits are equivalent, as depicted in the next two examples.

Example 2.3	**Using Boolean algebra to reduce an equation**

Use Boolean algebra to reduce the equation

$$F(x, y, z) = (x' + y' + x'y' + xy)(x' + yz)$$

as much as possible.

$$F = (x' + y' + x'y' + xy)(x' + yz)$$

$$= (x' \cdot 1 + y' \cdot 1 + x'y' + xy)(x' + yz) \qquad \text{(by Identity Theorem 6a)}$$

$$= (x'(y + y') + y'(x + x') + x'y' + xy)(x' + yz) \qquad \text{(by Inverse Theorem 9b)}$$

$$= (x'y + x'y' + y'x + y'x' + x'y' + xy)(x' + yz) \qquad \text{(by Distributive Theorem 12a)}$$

$$= (x'y + x'y' + y'x + \cancel{y'x'} + \cancel{x'y'} + xy)(x' + yz) \qquad \text{(by Idempotent Theorem 7b)}$$

$$= (x'(y + y') + x(y + y'))(x' + yz) \qquad \text{(by Distributive Theorem 12a)}$$

$$= (x' \cdot 1 + x \cdot 1)(x' + yz) \qquad \text{(by Inverse Theorem 9b)}$$

$$= (x' + x)(x' + yz) \qquad \text{(by Identity Theorem 6a)}$$

$$= 1(x' + yz) \qquad \text{(by Inverse Theorem 9b)}$$

$$= (x' + yz) \qquad \text{(by Identity Theorem 6a)}$$

Since the expression $(x' + y' + x'y' + xy)(x' + yz)$ reduces down to $(x' + yz)$, therefore, we do want to implement the circuit for the latter expression rather than the former because the circuit size for the latter is much smaller.

Example 2.4   **Using Boolean algebra to show that two equations are equivalent**

Show, using Boolean algebra, that the two equations

$$F_1 = (xy' + x'y + x' + y' + z')(x + y' + z)$$

and

$$F_2 = y' + x'z + xz'$$

are equivalent.

$$F_1 = (xy' + x'y + x' + y' + z')(x + y' + z)$$

$$= xy'x + xy'y' + xy'z + x'yx + x'yy' + x'yz + x'x + x'y' + x'z + y'x + y'y' + y'z + z'x + z'y' + z'z$$

$$= xy' + xy' + xy'z + 0 + 0 + x'yz + 0 + x'y' + x'z + xy' + y' + y'z + xz' + y'z' + 0$$

$$= xy' + xy'z + x'yz + x'y' + x'z + y' + y'z + xz' + y'z'$$

$$= y'(x + xz + x' + 1 + z + z') + x'z(y + 1) + xz'$$

$$= y' + x'z + xz'$$

$$= F_2$$

---

*2.5.2 Duality Principle

Notice in Figure 2.6 that we have listed the axioms and theorems in pairs. Specifically, we define the **dual** of a logic expression as one that is obtained by exchanging all $+$ operators with $\cdot$ operators, and vice versa, and by exchanging all 0's with 1's, and vice versa. For example, the dual of the logic expression:

$$(x \cdot y' \cdot z) + (x \cdot y \cdot z') + (y \cdot z) + 0$$

is

$$(x + y' + z) \cdot (x + y + z') \cdot (y + z) \cdot 1$$

The **duality principle** states that if a Boolean expression is true, then its dual is also true. Be careful in that it does not say that a Boolean expression is equivalent to its dual. For example, Theorem 5a in Figure 2.6(b) says that $x \cdot 0 = 0$ is true, thus by the duality principle, its dual, $x + 1 = 1$ is also true. However, $x \cdot 0 = 0$ is not equal to $x + 1 = 1$, since 0 definitely is not equal to 1.

We will see in Section 2.5.3 that the inverse of a Boolean expression easily can be obtained by first taking the dual of that expression and then complementing each Boolean variable in the resulting dual expression. In this respect, the duality principle is often used in digital logic design. Whereas an expression might be complex to implement, its inverse might be simpler, thus resulting in a smaller circuit; inverting the final output of this circuit will produce the same result as from the original expression.

2.5.3   Boolean Functions and their Inverses

As we have seen, any digital circuit can be described by a logical expression, also known as a **Boolean function**. Any Boolean functions can be formed from binary variables and the Boolean operators $\cdot$, $+$, and $'$ (for AND, OR, and NOT, respectively). For example, the following Boolean function uses the three variables (or literals) x, y, and z. It has three AND **terms** (also referred to as **product terms**), and these AND terms are ORed (summed) together. The first two AND terms each contain all three variables, while the last AND term contains only two variables. By definition, an AND (or product) term is either a single variable or two or more variables ANDed together. Quite often, we refer to functions that are in this format as a **sum-of-products** or **or-of-ands**.

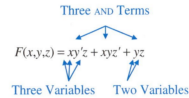

Three AND Terms

$$F(x,y,z) = xy'z + xyz' + yz$$

Three Variables   Two Variables

The value of a function evaluates to either 0 or 1, depending on the given set of values for the variables. For example, the function in the above figure evaluates to 1 when any one of the three AND terms evaluate to 1, since 1 OR x is 1. The first AND term, $xy'z$, equals 1 if

$$x = 1, y = 0, \text{ and } z = 1$$

because, if we substitute these values for x, y, and z into the first AND term $xy'z$, we get a 1. Similarly, the second AND term, xyz', equals 1 if

$$x = 1, y = 1, \text{ and } z = 0$$

The last AND term, yz, has only two variables. What this means is that the value of this term is not dependent on the missing variable x. In other words, x can be either 0 or 1, but as long as $y = 1$ and $z = 1$, this term will be equal to 1.

Thus, we can summarize by saying that F evaluates to 1 if

$$x = 1, y = 0, \text{ and } z = 1$$

or

$$x = 1, y = 1, \text{ and } z = 0$$

or

$$x = 0, y = 1, \text{ and } z = 1$$

or

$$x = 1, y = 1, \text{ and } z = 1$$

Otherwise, F evaluates to 0.

x	y	z	F	F'
0	0	0	0	1
0	0	1	0	1
0	1	0	0	1
0	1	1	1	0
1	0	0	0	1
1	0	1	1	0
1	1	0	1	0
1	1	1	1	0

Figure 2.7
Truth table for the function
$F = xy'z + xyz' + yz$.

It is often more convenient to summarize this verbal description of a function with a truth table, as shown in Figure 2.7 under the column labeled F. Notice that the four rows in the table, where $F = 1$, match the four cases in the description.

The inverse of a function, denoted by F', can be obtained easily from the truth tables for F by simply changing all of the 0's to 1's and 1's to 0's, as shown in the truth table in Figure 2.7 under the column labeled F'. Therefore, we can write the Boolean function for F' in the sum-of-products format, where the AND terms are obtained from those rows where $F' = 1$. Thus, we get

$$F' = x'y'z' + x'y'z + x'yz' + xy'z'$$

To deduce F' algebraically from F requires the use of DeMorgan's theorem (Theorem 15a in Figure 2.6(c)) twice. For example, using the same function

$$F = xy'z + xyz' + yz$$

we obtain F' as follows

$$F' = (xy'z + xyz' + yz)'$$
$$= (xy'z)' \cdot (xyz')' \cdot (yz)'$$
$$= (x' + y + z') \cdot (x' + y' + z) \cdot (y' + z')$$

There are three things to notice about this equation for F'. First, F' is just the dual of F (as defined in Section 2.5.2) with all the variables inverted. Second, instead of being in a sum-of-products format, it is in a **product-of-sums** (**and-of-ors**) format where three OR terms (also referred to as "sum" terms) are ANDed together. Third, from the same original function F, we obtained two different equations for F'. From the truth table in Figure 2.7, we obtained

$$F' = x'y'z' + x'y'z + x'yz' + xy'z'$$

and from applying DeMorgan's theorem to F, we obtained

$$F' = (x' + y + z') \cdot (x' + y' + z) \cdot (y' + z')$$

Hence, we must conclude that these two expressions for F', where one is in the sum-of-products format and the other is in the product-of-sums format, are equivalent. In general, all functions can be expressed in either the sum-of-products or product-of-sums format.

Thus, we should also be able to express the function, $F = xy'z + xyz' + yz$, in the product-of-sums format. We can derive it using one of two methods. For method one, we can start with F' and apply DeMorgan's theorem to it just like how we obtained F' from F.

$$F = (F')'$$
$$= (x'y'z' + x'y'z + x'yz' + xy'z')'$$
$$= (x'y'z')' \cdot (x'y'z)' \cdot (x'yz')' \cdot (xy'z')'$$
$$= (x + y + z) \cdot (x + y + z') \cdot (x + y' + z) \cdot (x' + y + z)$$

For the second method, we start with the original F and convert it to the product-of-sums format using the Boolean theorems from Figure 2.6.

$F = xy'z + xyz' + yz$

$= (x + x + y) \cdot (x + x + z) \cdot (x + y + y) \cdot (x + y + z) \cdot (x + z' + y) \cdot$ (Step 1)

$\quad (x + z' + z) \cdot (y' + x + y) \cdot (y' + x + z) \cdot (y' + y + y) \cdot$

$\quad (y' + y + z) \cdot (y' + z' + y) \cdot (y' + z' + z) \cdot (z + x + y) \cdot$

$\quad (z + x + z) \cdot (z + y + y) \cdot (z + y + z) \cdot (z + z' + y) \cdot (z + z' + z)$

$= (x + y) \cdot (x + z) \cdot (x + y) \cdot (x + y + z) \cdot (x + z' + y) \cdot$ (Step 2)

$\quad (y' + x + z) \cdot (z + x + y) \cdot (z + x) \cdot (z + y) \cdot (z + y)$

$= (x + y) \cdot (x + z) \cdot (x + y + z) \cdot (x + y + z') \cdot (x + y' + z) \cdot (y + z)$ (Step 3)

$= (x + y + zz') \cdot (x + yy' + z) \cdot (x + y + z) \cdot (x + y + z') \cdot$ (Step 4)

$\quad (x + y' + z) \cdot (xx' + y + z)$

$= (x + y + z) \cdot (x + y + z') \cdot (x + y + z) \cdot (x + y' + z) \cdot (x + y + z) \cdot$ (Step 5)

$\quad (x + y + z') \cdot (x + y' + z) \cdot (x + y + z) \cdot (x' + y + z)$

$= (x + y + z) \cdot (x + y + z') \cdot (x + y' + z) \cdot (x' + y + z)$

In Step 1, we apply Theorem 12b (Distributive) to get every possible combination of the sum terms. For example, the first sum term $(x + x + y)$ is obtained from getting the first x from $xy'z$, the second x from xyz', and the y from yz. The second sum term $(x + x + z)$ is obtained from getting the first x from $xy'z$, the second x from xyz', and the z from yz. This is repeated for all combinations. In this step, the sum terms, such as $(x + z' + z)$, where it contains variables of the form $v + v'$, can be eliminated, since $v + v' = 1$, and $1 \cdot x = x$.

In Step 2 and Step 3, duplicate variables and terms are eliminated. For example, the term $(x+x+y)$ is equal to $(x+y+y)$, which is just $(x+y)$. The term $(x+z'+z)$ is equal to $(x+1)$, which is equal to just 1, and therefore, can be eliminated completely from the expression.

In Step 4, every sum term with a missing variable will have that variable added back in by using Theorems 6b and 9a, which say that $x + 0 = x$ and $yy' = 0$, therefore, $x + yy' = x$.

Step 5 uses the Distributive theorem, and the resulting duplicate terms are again eliminated to give us the format that we want.

Functions that are in the product-of-sums format (such as the one shown below) are more difficult to deduce when they evaluate to 1. For example, using

$$F' = (x' + y + z') \cdot (x' + y' + z) \cdot (y' + z')$$

F' evaluates to 1 when all three terms evaluate to 1. For the first term to evaluate to 1, x can be 0, y can be 1, or z can be 0. For the second term to evaluate to 1, x can be 0, y can be 0, or z can be 1. Finally, for the last term, y can be 0, z can be 0, or x can be either 0 or 1. As a result, we end up with many more combinations to consider, even though many of the combinations are duplicates.

However, it is easier to determine when a product-of-sums format expression evaluates to 0. For example, using the same expression:

$$F' = (x' + y + z') \cdot (x' + y' + z) \cdot (y' + z')$$

F' evaluates to 0 when any one of the three OR terms is 0, since 0 AND x is 0; and this happens when

$$x = 1, y = 0, \text{ and } z = 1 \text{ for the first OR term,}$$

or

$$x = 1, y = 1, \text{ and } z = 0 \text{ for the second OR term,}$$

or

$$y = 1, z = 1, \text{ and } x \text{ can be either 0 or 1 for the last OR term.}$$

Similarly, for a sum-of-products format expression, it is easy to evaluate when it is a 1 but difficult to evaluate when it is a 0.

These four conditions in which F' evaluates to 0 match exactly the four rows in the truth table shown in Figure 2.7 where $F' = 0$. Therefore, we see that, in general, the unique algebraic expression for any Boolean function can be specified by either (1) selecting the rows from the truth table where the function is a 1 and using the sum-of-products format, or (2) selecting the rows from the truth table where the function is a 0 and using the product-of-sums format. Whatever format we decide to use, the one thing to remember is that we are always interested in only when the function (or its inverse) is equal to a 1.

Figure 2.8 summarizes these two formats for the function $F = xy'z + xyz' + yz$ and its inverse F'. Notice that the sum-of-products format for F is the dual with its variables inverted from the product-of-sums format for F'. Similarly, the product-of-sums format for F is the dual with its variables inverted from the sum-of-products format for F'.

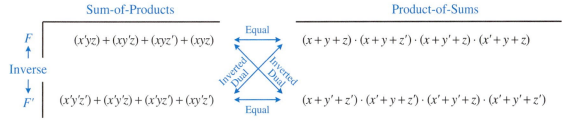

Figure 2.8   Relationships between the function $F = xy'z + xyz' + yz$ and its inverse F', and the sum-of-products and product-of-sums formats. The label "Inverted Dual" means applying the duality principle and then inverting the variables.

• • • • • • • • • • • • • • ⋯

2.6 Minterms and Maxterms

As you recall, a product term is a term with either a single variable or two or more variables ANDed together, and a sum term is a term with either a single variable or two or more variables ORed together. To differentiate between a term that contains any number of variables with a term that contains *all* of the variables used in the function, we use the words minterm and maxterm. We are not introducing new ideas here; rather, we are just introducing two new words and notations for defining what we have already learned.

2.6.1 Minterms

A **minterm** is a product term that contains all of the variables used in a function. For a function with n variables, the notation m_i, where $0 \leq i < 2^n$, is used to denote the minterm whose index i is the binary value of the n variables, such that the variable is complemented if the value assigned to it is a 0 and uncomplemented if it is a 1.

For example, for a function with three variables x, y, and z, the notation m_3 is used to represent the term in which the values for the variables xyz are 011 (for the subscript 3). Since we want to complement the variable whose value is a 0 and uncomplement it if it is a 1, therefore, m_3 is for the minterm $x'yz$. Figure 2.9(a) shows the eight minterms and their notations for $n = 3$ using the three variables x, y, and z.

When specifying a function, we usually start with product terms that contain all of the variables used in the function. In other words, we want the **sum of minterms**, and, more specifically, the sum of the one-minterms (i.e., the minterms for which the function is 1) as opposed to the zero-minterms (i.e., that is the minterms for which the function is 0). We use the notation **1-minterm** to denote one-minterm, and **0-minterm** to denote zero-minterm.

The function from the previous section:

$$F = xy'z + xyz' + yz$$

$$= x'yz + xy'z + xyz' + xyz$$

x	y	z	Minterm	Notation
0	0	0	$x'y'z'$	m_0
0	0	1	$x'y'z$	m_1
0	1	0	$x'yz'$	m_2
0	1	1	$x'yz$	m_3
1	0	0	$xy'z'$	m_4
1	0	1	$xy'z$	m_5
1	1	0	xyz'	m_6
1	1	1	xyz	m_7

x	y	z	Maxterm	Notation
0	0	0	$x+y+z$	M_0
0	0	1	$x+y+z'$	M_1
0	1	0	$x+y'+z$	M_2
0	1	1	$x+y'+z'$	M_3
1	0	0	$x'+y+z$	M_4
1	0	1	$x'+y+z'$	M_5
1	1	0	$x'+y'+z$	M_6
1	1	1	$x'+y'+z'$	M_7

(a)   (b)

Figure 2.9 (a) Minterms for three variables; (b) Maxterms for three variables.

and repeated in the following truth table has the 1-minterms: m_3, m_5, m_6, and m_7.

x	y	z	F	F'	Minterm	Notation
0	0	0	0	1	$x'y'z'$	m_0
0	0	1	0	1	$x'y'z$	m_1
0	1	0	0	1	$x'yz'$	m_2
0	1	1	1	0	$x'yz$	m_3
1	0	0	0	1	$xy'z'$	m_4
1	0	1	1	0	$xy'z$	m_5
1	1	0	1	0	xyz'	m_6
1	1	1	1	0	xyz	m_7

Thus, a shorthand notation for the function is

$$F(x, y, z) = m_3 + m_5 + m_6 + m_7$$

By using just the minterm notations, we do not know how many variables are in the original function. Consequently, we need to explicitly specify the variables used by the function, as in $F(x, y, z)$. We can further simplify the notation by using the standard algebraic symbol Σ for summation and listing out the minterm index numbers. Therefore, we have

$$F(x, y, z) = \Sigma(3, 5, 6, 7)$$

These are just different ways of expressing the same function.

Since a function is obtained from the sum of the 1-minterms, the inverse of the function, therefore, must be the sum of the 0-minterms. This can be obtained easily by replacing the set of indices with those that were excluded from the original set. Therefore,

$$F'(x, y, z) = \Sigma(0, 1, 2, 4).$$

Example 2.5 **Converting a function to the sum-of-minterms format using Boolean algebra**

Given the Boolean function $F(x, y, z) = y + x'z$, use Boolean algebra to convert the function to the sum-of-minterms format.

This function has three variables. In a sum-of-minterms format, all product terms must have all variables. To do so, we need to expand each product term by ANDing it with $(v + v')$ for every missing variable v in that term. Since $(v + v') = 1$, ANDing a product term with $(v + v')$ does not change the value of the term.

$$F = y + x'z$$

$$= y(x + x')(z + z') + x'z(y + y')$$

(expand the first term by ANDing it with $(x + x')$ $(z + z')$ and the second term with $(y + y')$)

$$= (xyz) + (xyz') + (x'yz) + (x'yz') + \cancel{(x'yz)} + (x'y'z)$$

(use the Distributive theorem to change to sum-of-products format)

$$= m_7 + m_6 + m_3 + m_2 + m_1$$

$$= \Sigma(1, 2, 3, 6, 7)$$

(sum of 1-minterms)

Example 2.6 **Converting the inverse of a function to the sum-of-minterms format**

Given the Boolean function $F(x, y, z) = y + x'z$, use Boolean algebra to convert the inverse of the function to the sum-of-minterms format.

$$F' = (y + x'z)'$$

(inverse)

$$= y' \cdot (x'z)'$$

(use DeMorgan's theorem)

$$= y' \cdot (x + z')$$

(use DeMorgan's theorem)

$$= y'x + y'z'$$

(use the Distributive theorem to change to sum-of-products format)

$$= y'x(z + z') + y'z'(x + x')$$

(expand the first term by ANDing it with $(z + z')$ and the second term with $(x + x')$)

$$= (xy'z) + (xy'z') + \cancel{(xy'z')} + (x'y'z')$$

$$= m_5 + m_4 + m_0$$

$$= \Sigma(0, 4, 5)$$

(sum of 0-minterms)

*2.6.2 Maxterms

Analogous to a minterm, a **maxterm** is a sum term that contains all of the variables used in the function. For a function with n variables, the notation M_i, where $0 \leq i < 2^n$, is used to denote the maxterm whose index i is the binary value of the n variables, such that the variable is complemented if the value assigned to it is a 1 and uncomplemented if it is a 0.

For example, for a function with three variables x, y, and z, the notation M_3 is used to represent the term in which the values for the variables xyz are 011 (for the subscript 3). For maxterms, we want to complement the variable whose value is a 1 and uncomplement it if it is a 0, hence, M_3 is for the maxterm $x + y' + z'$. Figure 2.9(b) shows the eight maxterms and their notations for $n = 3$ using the three variables x, y, and z.

We have seen that a function can also be specified as a product-of-sums, or more specifically, a **product of 0-maxterms** (i.e., the maxterms for which the function is a 0). Just like the minterms, we use the notation **1-maxterm** to denote one-maxterm, and **0-maxterm** to denote zero-maxterm. Thus, the function:

$$F(x, y, z) = (xy'z) + (xyz') + (yz)$$

$$= (x + y + z) \cdot (x + y + z') \cdot (x + y' + z) \cdot (x' + y + z)$$

which is shown in the following table:

x	y	z	F	F'	Maxterm	Notation
0	0	0	0	1	$x + y + z$	M_0
0	0	1	0	1	$x + y + z'$	M_1
0	1	0	0	1	$x + y' + z$	M_2
0	1	1	1	0	$x + y' + z'$	M_3
1	0	0	0	1	$x' + y + z$	M_4
1	0	1	1	0	$x' + y + z'$	M_5
1	1	0	1	0	$x' + y' + z$	M_6
1	1	1	1	0	$x' + y' + z'$	M_7

can be specified as the product of the 0-maxterms M_0, M_1, M_2, and M_4. The shorthand notation for the function is

$$F(x, y, z) = M_0 \cdot M_1 \cdot M_2 \cdot M_4$$

By using the standard algebraic symbol Π for product and listing out the maxterm index numbers, the notation is further simplified to

$$F(x, y, z) = \Pi(0, 1, 2, 4)$$

The following summarizes these relationships for the function $F = xy'z + xyz' + yz$ and its inverse. Comparing these equations with those in Figure 2.8, we see that they are identical.

$F(x,y,z) = (x'yz) + (xy'z) + (xyz') + (xyz)$

$\qquad = m_3 + m_5 + m_6 + m_7$

$\qquad = \Sigma(3, 5, 6, 7)$ Σ 1-minterms

$\qquad = (x + y + z) \cdot (x + y + z') \cdot (x + y' + z) \cdot (x' + y + z)$

$\qquad = M_0 \cdot M_1 \cdot M_2 \cdot M_4$

$\qquad = \Pi(0, 1, 2, 4)$ Π 0-maxterms

$F'(x,y,z) = (x'y'z') + (x'y'z) + (x'yz') + (xy'z')$

$\qquad = m_0 + m_1 + m_2 + m_4$

$\qquad = \Sigma(0, 1, 2, 4)$ Σ 0-minterms

$\qquad = (x + y' + z') \cdot (x' + y + z') \cdot (x' + y' + z) \cdot (x' + y' + z')$

$\qquad = M_3 \cdot M_5 \cdot M_6 \cdot M_7$

$\qquad = \Pi(3, 5, 6, 7)$ Π 1-maxterms

Equivalent — Inverted Duals — Inverse — Equivalent

Notice that it is always the Σ of minterms and Π of maxterms; you never have Σ of maxterms or Π of minterms.

Example 2.7

Converting a function to the product-of-maxterms format

Given the Boolean function $F(x, y, z) = y + x'z$, use Boolean algebra to convert the function to the product-of-maxterms format.

To change a sum term to a maxterm, we expand each term by ORing it with (vv') for every missing variable v in that term. Since $(vv') = 0$, therefore, ORing a sum term with (vv') does not change the value of the term.

$F = y + x'z$

$\qquad = y + (x'z)$

$\qquad = (y + x') \cdot (y + z)$ (use the Distributive theorem to change to product-of-sums format)

$\qquad = (y + x' + zz') \cdot (y + z + xx')$ (expand the first term by ORing it with zz', and the second term with xx')

$\qquad = (x' + y + z) \cdot (x' + y + z') \cdot (x + y + z) \cdot \cancel{(x' + y + z)}$

$\qquad = M_4 \cdot M_5 \cdot M_0$

$\qquad = \Pi(0, 4, 5)$ (product of 0-maxterms)

Example 2.8 **Converting the inverse of a function to the product-of-maxterms format**

Given the Boolean function $F(x, y, z) = y + x'z$, use Boolean algebra to convert the inverse of the function to the product-of-maxterms format.

$F' = (y + x'z)'$	(inverse)
$\quad = y' \cdot (x'z)'$	(use DeMorgan's theorem)
$\quad = y' \cdot (x + z')$	(use DeMorgan's theorem)
$\quad = (y' + xx' + zz') \cdot (x + z' + yy')$	(expand the first term by ORing it with $xx' + zz'$, and the second term with yy')
$\quad = (x + y' + z) \cdot (x + y' + z') \cdot (x' + y' + z) \cdot$	(use the Distributive theorem to change to product-of-sums format)
$\quad \quad (x' + y' + z') \cdot (x + y + z') \cdot \cancel{(x + y' + z')}$	
$\quad = M_2 \cdot M_3 \cdot M_6 \cdot M_7 \cdot M_1$	
$\quad = \Pi(1, 2, 3, 6, 7)$	(product of 1-maxterms)

• • • • • • • • • • • • • • • •

2.7 Canonical, Standard, and Non-Standard Forms

Any Boolean function that is expressed as a sum-of-minterms, or as a product-of-maxterms is said to be in its **canonical form**. For example, the following two expressions are in their canonical forms

$$F = (x'yz) + (xy'z) + (xyz') + (xyz)$$
$$F' = (x + y' + z') \cdot (x' + y + z') \cdot (x' + y' + z) \cdot (x' + y' + z')$$

As noted from the previous section, to convert a Boolean function from one canonical form to its other equivalent canonical form, simply interchange the symbols Σ with Π and list the index numbers that were excluded from the original form. For example, the following two expressions are equivalent

$$F_1(x, y, z) = \Sigma(3, 5, 6, 7)$$
$$F_2(x, y, z) = \Pi(0, 1, 2, 4)$$

To convert a Boolean function from one canonical form to its inverse, simply interchange the symbols Σ with Π and list the same index numbers from the original form. For example, the following two expressions are inverses

$$F_1(x, y, z) = \Sigma(3, 5, 6, 7)$$

$$F_2(x, y, z) = \Pi(3, 5, 6, 7)$$

A Boolean function is said to be in a **standard form** if a sum-of-products expression or a product-of-sums expression has at least one term that is not a minterm or a maxterm, respectively. In other words, at least one term in the expression is missing at least one variable. For example, the following expression is in a standard form because the last term is missing the variable x

$$F = xy'z + xyz' + yz$$

Sometimes, common variables in a standard form expression can be factored out. The resulting expression is no longer in a sum-of-products or product-of-sums format. These expressions are in a **non-standard form**. For example, starting with the previous expression, if we factor out the common variable x from the first two terms, we get the following expression, which is in a nonstandard form

$$F = x(y'z + yz') + yz$$

● ● ● ● ● ● ● ● ● ● ● ● ● ● ● ●

2.8 **Logic Gates and Circuit Diagrams**

Logic gates are the actual physical implementations of the logical operators discussed in the previous sections. Transistors, acting as tiny electronic binary switches, are connected together to form these gates. Thus, we have the AND gate, the OR gate, and the NOT gate (also called the INVERTER) for the corresponding AND, OR, and NOT logical operators. These gates form the basic building blocks for all digital logic circuits. The name "gate" comes from the fact that these devices operate like a door or gate to let or not to let things (in our case, current) through.

In drawing digital circuit diagrams (also called **schematic diagrams** or just **schematics**), we use special **logic symbols** to denote these gates, as shown in Figure 2.10. The AND gate (specifically, the 2-input AND gate) in Figure 2.10(a) has two input connections coming in from the left and one output connection going out on the right. Similarly, the 2-input OR gate in Figure 2.10(b) has two input connections and one

Figure 2.10
Logic symbols for the three basic logic gates: (a) 2-input AND; (b) 2-input OR; (c) NOT.

(a)   (b)   (c)

Figure 2.11
Logic symbols for:
(a) 3-input AND;
(b) 4-input AND;
(c) 3-input OR;
(d) 4-input OR;
(e) 2-input NAND;
(f) 2-input NOR;
(g) 3-input NAND;
(h) 3-input NOR;
(i) 2-input XOR;
(j) 2-input XNOR.

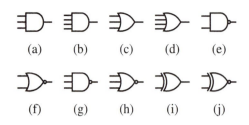

(a)   (b)   (c)   (d)   (e)

(f)   (g)   (h)   (i)   (j)

output connection. The INVERTER in Figure 2.10(c) has one input coming from the left and one output going to the right. The outputs from these gates, of course, are dependent on their inputs and are defined by their logical functions.

Sometimes, an AND gate or an OR gate with more than two inputs are needed. Hence, in addition to the 2-input AND and OR gates, there are 3-input, 4-input, or as many inputs as are needed of the AND and OR gates. In practice, however, the number of inputs is limited to a small number, such as five. The logic symbols for some of these gates are shown in Figure 2.11(a) through (d).

There are several other gates that are variants of the three basic gates that also are used often in digital circuits. They are the NAND gate, the NOR gate, the XOR gate, and the XNOR gate. The NAND gate is derived from an AND gate and the INVERTER connected in series, so that the output of the AND gate is inverted. The name "NAND" comes from the description "NOT AND." Similarly, the NOR gate is the OR gate with its output inverted. The XOR (or eXclusive OR) gate is like the OR gate except that when both inputs are 1, the output is a 0 instead. The XNOR (or eXclusive NOR) gate is just the inverse of the XOR gate for when there are an even number of inputs (like two inputs). When there are an odd number of inputs (like three inputs), the XOR is the same as the XNOR. The logic symbols and their truth tables for some of these gates are shown in Figure 2.11(e) through (j) and Figure 2.12, respectively.

Notice, in Figure 2.11, the use of the little circle or bubble at the output of some of the logic symbols. This bubble is used to denote the inverted value of a signal. For example, the NAND gate is the inverse of the AND gate. Thus, the NAND gate logic symbol is the same as the AND gate logic symbol, except that it has the extra bubble at the output.

The notations used for these gates in a logical expression are $(xy)'$ for the 2-input NAND gate, $(x+y)'$ for the 2-input NOR gate, $x \oplus y$ for the 2-input XOR gate, and $x \odot y$ for the 2-input XNOR gate.

Looking at the truth table in Figure 2.12 for the 2-XOR gate, we can derive the equation for the 2-input XOR gate as

$$x \oplus y = x'y + xy'$$

Similarly, the equation for the 2-input XNOR gate as derived from the 2-XNOR truth table in Figure 2.12 is

$$x \odot y = x'y' + xy$$

x	y	2-NAND $(x \cdot y)'$	2-NOR $(x + y)'$	2-XOR $x \oplus y$	2-XNOR $x \odot y$
0	0	1	1	0	1
0	1	1	0	1	0
1	0	1	0	1	0
1	1	0	0	0	1

x	y	z	3-AND $x \cdot y \cdot z$	3-OR $x + y + z$	3-NAND $(x \cdot y \cdot z)'$	3-NOR $(x+y+z)'$	3-XOR $x \oplus y \oplus z$	3-XNOR $x \odot y \odot z$
0	0	0	0	0	1	1	0	0
0	0	1	0	1	1	0	1	1
0	1	0	0	1	1	0	1	1
0	1	1	0	1	1	0	0	0
1	0	0	0	1	1	0	1	1
1	0	1	0	1	1	0	0	0
1	1	0	0	1	1	0	0	0
1	1	1	1	1	0	0	1	1

Figure 2.12   Truth table for: 2-input NAND; 2-input NOR; 2-input XOR; 2-input XNOR; 3-input AND; 3-input OR; 3-input NAND; 3-input NOR; 3-input XOR; 3-input XNOR.

The equation for the 3-input XOR gate is derived as follows

$$x \oplus y \oplus z = (x \oplus y) \oplus z$$
$$= (x'y + xy') \oplus z$$
$$= (x'y + xy')z' + (x'y + xy')'z$$
$$= x'yz' + xy'z' + (x'y)'(xy')'z$$
$$= x'yz' + xy'z' + (x + y')(x' + y)z$$
$$= x'yz' + xy'z' + \cancel{xx'z} + xyz + x'y'z + \cancel{y'yz}$$
$$= x'y'z + x'yz' + xy'z' + xyz$$

The last four product terms in this derivation are the four 1-minterms in the 3-input XOR truth table in Figure 2.12. For three or more inputs, the XOR gate has a value of 1 when there is an odd number of 1's in the inputs; otherwise, it is a 0.

Notice also that the truth tables in Figure 2.12 for the 3-input XOR and XNOR gates are identical. It turns out that for an even number of inputs, XOR is the inverse of XNOR, but for an odd number of inputs, XOR is equal to XNOR.

All of these gates can be interconnected together to form large, complex circuits which we call **networks**. These networks can be described graphically either using **circuit diagrams**, with Boolean expressions, or with truth tables.

Example 2.9 **Drawing a circuit diagram**

Draw the circuit diagram for the equation $F(x, y, z) = y + x'z$.

In the equation, we need to first invert x and then AND it with z. Finally, we need to OR y with the output of the AND. The resulting circuit is shown here. For easy reference, the internal nodes in the circuit are annotated with the two intermediate values x' and $x'z$.

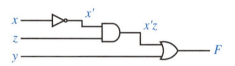

Example 2.10 **Drawing a circuit diagram**

Draw the circuit diagram for the equation $F(x, y, z) = xyz + xyz' + x'yz + x'yz' + x'y'z$.

The equation consists of five AND terms that are ORed together. Each AND term requires three inputs for the three variables. Hence, the circuit shown below has five 3-input AND gates, whose outputs are connected to a 5-input OR gate. The inputs to the AND gates come directly from the three variables x, y, and z (or their inverted values). Notice that in the equation, there are six inverted variables. However, in the circuit, we do not need six INVERTERS. Rather, only three INVERTERS are used; one for each variable.

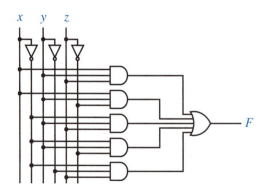

2.9 Designing a Car Security System

In a car security system, we usually want to connect the siren in such a way that the siren will activate when it is triggered by one or more sensors. In addition, there will be a master switch to turn the system on or off. Let us assume that there is a car door switch D, a vibration detector switch V, and a master switch M. We will use the convention that when the door is opened, $D = 1$, otherwise, $D = 0$. Similarly, when the car is being shaken, $V = 1$, otherwise, $V = 0$. Thus, we want the siren S to turn on (that is, set $S = 1$) when either $D = 1$ or $V = 1$ or when both $D = 1$ and $V = 1$, but only for when the system is turned on (that is, when $M = 1$). However, when we turn off the system and either enter or drive the car, we do not want the siren to turn on. Hence, when $M = 0$, it does not matter what values D and V have, the siren should remain off.

Given the above description of a car security system, we can build a digital circuit that meets our specifications. We start by constructing a truth table, which is basically a precise way of stating the operations for the device. The table will have three input columns M, D, and V, and an output column S, as shown in Figure 2.13(a).

Under the three input columns, we enumerate all possible binary values for the three inputs. The values under the S column are obtained from interpreting the description of when we want the siren to turn on. When $M = 0$, we don't want the siren to come on, regardless of what the values for D and V are. When $M = 1$, we want the siren to come on when either D or V is a 1, or both D and V are 1's.

M	D	V	S
0	0	0	0
0	0	1	0
0	1	0	0
0	1	1	0
1	0	0	0
1	0	1	1
1	1	0	1
1	1	1	1

(a)

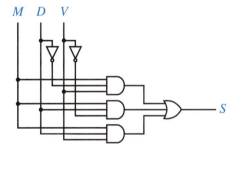

(b)

Figure 2.13
Car security system:
(a) truth table;
(b) circuit diagram derived from the truth table;
(c) simplified circuit diagram.

(c)

The truth table in Fig. 2.13(a) table can be described formally with a logic expression written in words as

$$S = (M \text{ AND } (\text{NOT } D) \text{ AND } V) \text{ OR } (M \text{ AND } D \text{ AND } (\text{NOT } V)) \text{ OR } (M \text{ AND } D \text{ AND } V)$$

or preferably, using the simpler notation of a Boolean function

$$S = (MD'V) + (MDV') + (MDV)$$

Again, what this equation is saying is that we want the siren to activate (i.e., $S = 1$) when:

- the master switch is on and the door is not opened and the vibration switch is on, or
- the master switch is on and the door is opened and the vibration switch is not on, or
- the master switch is on and the door is opened and the vibration switch is on.

Notice that we are only interested in the situations when $S = 1$. We ignore the rows when $S = 0$. When we construct circuits from truth tables, we always use only the rows where the output is a 1.

Finally, we can translate this equation into a circuit diagram. The translation is a simple one-to-one mapping of changing the AND operator into the AND gate, the OR operator into the OR gate, and the NOT operator into the INVERTER. Thus, we get the circuit diagram shown in Figure 2.13(b) for our car security system.

A careful reader might notice that the Boolean equation shown above for specifying when the siren is to be turned on can be simplified to

$$S = M(D + V)$$

This simplified equation says that the siren is to be turned on only when the master switch is on and either the door switch or vibration switch is on. The corresponding simplified circuit diagram is shown in Fig. 2.13(c). Just by using simple reasoning, we can see that this simplified circuit will do exactly what the circuit in Fig. 2.13(b) does. In other words, both circuits are functionally equivalent.

More formally, we can use the Boolean theorems from Section 2.5.1 to show that these two equations (and therefore, the two circuits) are indeed equivalent as follows:

$$\begin{aligned}
S &= (MD'V) + (MDV') + (MDV) \\
&= M(D'V + DV' + DV) && \text{(by Distributive Theorem 12a)} \\
&= M(D'V + DV' + DV + DV) && \text{(by Idempotent Theorem 7b)} \\
&= M(D(V' + V) + V(D' + D)) && \text{(by Distributive Theorem 12a)} \\
&= M(D(1) + V(1)) && \text{(by Inverse Theorem 9b)} \\
&= M(D + V) && \text{(by Identity Theorem 6a)}
\end{aligned}$$

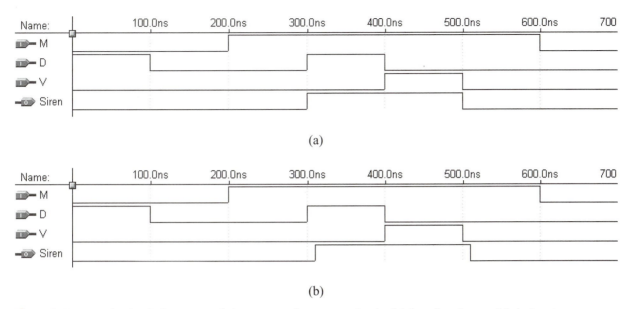

Figure 2.14   Sample simulation trace of the car security system circuit: (a) functional trace (b) timing trace.

Figure 2.14(a) shows a sample simulation trace of the car security system circuit. Between times 0 and 200 ns, the master switch M is a 0, so regardless of the values of D and V, the siren is off ($Siren = 0$). Between times 200 ns and 600 ns, $M = 1$. During this time, whenever either $D = 1$ or $V = 1$, the siren is on.

Figure 2.14(a) is a **functional** trace of the circuit, so all of the signal edges line up exactly, i.e., the output signal edge changes at exactly the same time (with no delay) as the input signal edge that caused it to change. For a **timing** trace, on the other hand, the output signal edge will be delayed slightly after the causing input edge, as shown in Figure 2.14(b).

When building a circuit, besides having a functionally correct circuit, we also want to optimize it in terms of its size, speed, heat dissipation, and power consumption. We will see in later sections how circuits are optimized.

· · · · · · · · · · · · · · · ·

2.10   VHDL for Digital Circuits

A digital circuit that is described with a Boolean function can easily be converted to VHDL code using the dataflow model. At the **dataflow** level, a circuit is defined using built-in VHDL logic operators, such as AND, OR, and NOT. These operators are applied to signals using concurrent signal assignment statements.

2.10.1 VHDL Code for a 2-Input NAND Gate

Figure 2.15 shows the VHDL code for a 2-input NAND gate. It also serves as a basic template for all VHDL codes. Lines starting with two hyphens are comments. The LIBRARY and USE statements specify that the IEEE library is needed and that all of the components in that library package can be used. These two statements are equivalent to the "#include" preprocessor line in C.

Every component defined in VHDL, whether it is a simple NAND gate or a complex microprocessor, has two parts: an ENTITY section and an ARCHITECTURE section. The ENTITY section is similar to a function declaration in C and serves as the interface between the component and the outside. It declares all of the input and output signals for a circuit. Every entity must have a unique name; in this example, the name *NAND2gate* is used. The entity contains a PORT list, which, like a parameter list, specifies the data to be passed in and out of the component. In this example, there are two input signals called x and y of type STD_LOGIC and an output signal called f of the same type. The STD_LOGIC type is like the BIT type, except that it contains additional values besides just 0's and 1's.

The ARCHITECTURE section defines the operation of the entity; it contains the code that realizes the operation of the component. For every architecture, you need to specify its name and which entity it is for. In this example, the name is *Dataflow*, and it is for the entity *NAND2gate*. It is possible for one entity to have more than one architecture, since an entity can be implemented in more than one way. Within the body of the architecture, we can have one or more concurrent statements. Unlike statements in C where they are executed in sequential order, concurrent statements in the architecture body are executed in parallel. Thus, the ordering of these statements is irrelevant. The symbol "<=" is used for a signal assignment statement. The expression on the right-hand side of the <= symbol is evaluated when either x or y changes values (either from 0 to 1 or from 1 to 0), and the result is assigned to the signal on the left-hand side. The NAND operator is a built-in VHDL operator.

Figure 2.15
VHDL code for a
2-input NAND gate.

```
-- this is a dataflow model of a 2-input NAND gate
LIBRARY IEEE;
USE IEEE.STD_LOGIC_1164.ALL;
ENTITY NAND2gate IS PORT (
  x: IN STD_LOGIC;
  y: IN STD_LOGIC;
  f: OUT STD_LOGIC);
END NAND2gate;

ARCHITECTURE Dataflow OF NAND2gate IS
BEGIN
  f <= x NAND y;          -- signal assignment
END Dataflow;
```

2.10.2   VHDL Code for a 3-Input NOR Gate

Figure 2.16(a) shows the VHDL code for a 3-input NOR gate. In addition to the three input signals x, y, and z, and one output signal f declared in the entity section, this example has two internal signals, *xory* and *xoryorz*, both of which are of the STD_LOGIC type. The keyword SIGNAL in the architecture section is used to declare these two internal signals. Internal signals are used for naming connection points (or nodes) within a circuit. Three concurrent signal assignment statements are used. All of the signal assignment statements are executed concurrently, so the ordering of

```
LIBRARY IEEE;
USE IEEE.STD_LOGIC_1164.ALL;

ENTITY NOR3gate IS PORT (
  x: IN STD_LOGIC;
  y: IN STD_LOGIC;
  z: IN STD_LOGIC;
  f: OUT STD_LOGIC);
END NOR3gate;

ARCHITECTURE Dataflow OF NOR3gate IS
  SIGNAL xory, xoryorz : STD_LOGIC;
BEGIN
  xory <= x OR y;            -- three concurrent signal assignments
  xoryorz <= xory OR z;
  f <= NOT xoryorz;
END Dataflow;
```

(a)

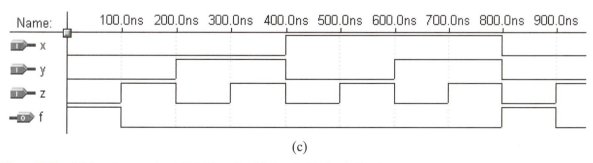

(b)

(c)

Figure 2.16   A 3-input NOR gate: (a) VHDL code; (b) circuit; (c) simulation trace.

the statements is irrelevant. The coding of these three signal assignment statements is based on the 3-input NOR gate circuit shown in Figure 2.16(b).

Figure 2.16(c) shows a sample simulation trace of the circuit. In the trace, we see that the output signal f is 1 only when all three inputs are 0's. This occurs twice: the first time between 0 and 100 ns, and the second time between 800 ns and 900 ns. For all of the other times, f is 0, since not all three inputs are 0's. Hence, the simulation trace shows the correct operation of this circuit for the 3-input NOR gate.

2.10.3　VHDL Code for a Function

Figure 2.17 shows the VHDL code and the simulation trace for the car security system circuit discussed in Section 2.9. The function implemented is $S = (MD'V) + (MDV') + (MDV)$.

```
LIBRARY IEEE;
USE IEEE.STD_LOGIC_1164.ALL;

ENTITY Siren IS PORT (
  M: IN STD_LOGIC;
  D: IN STD_LOGIC;
  V: IN STD_LOGIC;
  S: OUT STD_LOGIC);
END Siren;

ARCHITECTURE Dataflow OF Siren IS
  SIGNAL term_1, term_2, term_3: STD_LOGIC;
BEGIN
  term_1 <= M AND (NOT D) AND V;
  term_2 <= M AND D AND (NOT V);
  term_3 <= M AND D AND V;
  S <= term_1 OR term_2 OR term_3;
END Dataflow;
```

(a)

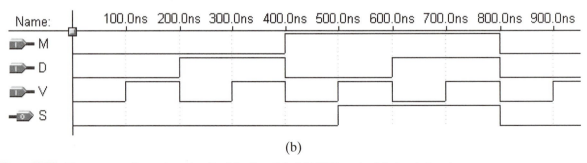

(b)

Figure 2.17　The car security system circuit of Section 2.9: (a) VHDL code; (b) simulation trace.

This VHDL code (as well as the ones from the two previous sections) is written at the dataflow level. This is not because the name of the architecture is "Dataflow". Dataflow level coding uses logic equations to describe a circuit, and this is done by using the built-in VHDL operators such as AND, OR, and NOT in concurrent signal assignment statements.

● ● ● ● ● ● ● ● ● ● ● ● ● · · · · ·
2.11 Summary Checklist

- Binary number
- Hexadecimal number
- Binary switch
- AND, OR, and NOT
- Truth table
- Boolean algebra axioms and theorems
- Duality principle
- Boolean function and the inverse
- Product term
- Sum term
- Sum-of-products (or-of-ands)
- Product-of-sums (and-of-ors)
- Minterm and maxterm
- Sum-of-minterms
- Product-of-maxterms
- Canonical, standard, and non-standard form
- Logic gate, logic symbol
- Circuit diagram
- NAND, NOR, XOR, XNOR
- Network
- VHDL
- Be able to derive the Boolean equation from a truth table (or vice versa)
- Be able to derive the circuit diagram from a Boolean equation (or vice versa)
- Be able to derive the circuit diagram from a truth table (or vice versa)
- Be able to use Boolean algebra to reduce a Boolean equation

• • • • • • • • • • • • • • • •

2.12 **Problems**

P2.1. Convert the following decimal numbers to binary numbers.
  (a) 66
  (b) 49
  (c) 513
  (d) 864
  (e) 1897
  (f) 2004

P2.2. Convert the following unsigned binary numbers to decimal, hexadecimal and octal numbers.
  (a) 11110
  (b) 11010
  (c) 100100011
  (d) 1011011
  (e) 1101101110
  (f) 101111010100

P2.3. Convert the following hexadecimal numbers to binary numbers.
  (a) 66
  (b) E3
  (c) 2FE8
  (d) 7C2
  (e) 5A2D
  (f) E08B

P2.4. Derive the truth table for the following Boolean functions.
  (a) $F(x, y, z) = x'y'z' + x'yz + xy'z' + xyz$
  (b) $F(x, y, z) = xy'z + x'yz' + xyz + xyz'$
  (c) $F(w, x, y, z) = w'xy'z + w'xyz + wxy'z + wxyz$
  (d) $F(w, x, y, z) = wxy'z + w'yz' + wxz + xyz'$
  (e) $F(x, y, z) = xy' + x'y'z + xyz'$
  (f) $F(w, x, y, z) = w'z' + w'xy + wx'z + wxyz$
  (g) $F(x, y, z) = [(x + y')(yz)'](xy' + x'y)$
  (h) $F(N_3, N_2, N_1, N_0) = N_3'N_2'N_1N_0' + N_3'N_2'N_1N_0 + N_3N_2'N_1N_0' + N_3N_2'N_1N_0 + N_3N_2N_1'N_0' + N_3N_2N_1N_0$

P2.5. Derive the Boolean function for the following truth tables

(a)

a	b	c	F
0	0	0	0
0	0	1	0
0	1	0	1
0	1	1	1
1	0	0	0
1	0	1	0
1	1	0	1
1	1	1	0

(b)

w	x	y	z	F
0	0	0	0	0
0	0	0	1	0
0	0	1	0	1
0	0	1	1	0
0	1	0	0	1
0	1	0	1	1
0	1	1	0	0
0	1	1	1	1
1	0	0	0	0
1	0	0	1	1
1	0	1	0	1
1	0	1	1	0
1	1	0	0	1
1	1	0	1	1
1	1	1	0	0
1	1	1	1	1

(c)

w	x	y	z	F_1	F_2
0	0	0	0	1	1
0	0	0	1	0	1
0	0	1	0	0	1
0	0	1	1	1	1
0	1	0	0	0	0
0	1	0	1	1	1
0	1	1	0	1	0
0	1	1	1	0	0
1	0	0	0	0	1
1	0	0	1	1	1
1	0	1	0	1	0
1	0	1	1	0	0
1	1	0	0	1	1
1	1	0	1	0	1
1	1	1	0	0	1
1	1	1	1	1	1

(d)

N_3	N_2	N_1	N_0	F
0	0	0	0	0
0	0	0	1	0
0	0	1	0	1
0	0	1	1	1
0	1	0	0	0
0	1	0	1	0
0	1	1	0	1
0	1	1	1	0
1	0	0	0	0
1	0	0	1	0
1	0	1	0	1
1	0	1	1	1
1	1	0	0	1
1	1	0	1	0
1	1	1	0	0
1	1	1	1	1

P2.6. Use a truth table to show that the following expressions are true.

(a) $w'z' + w'xy + wx'z + wxyz = w'z' + xyz + wx'y'z + wyz$

(b) $z + y' + yz' = 1$

(c) $xy'z' + x' + xyz' = x' + z'$

(d) $xy + x'z + yz = xy + x'z$

(e) $w'x'yz' + w'x'yz + wx'yz' + wx'yz + wxyz = y(x' + wz)$

(f) $w'xy'z + w'xyz + wxy'z + wxyz = xz$

(g) $x_i y_i + c_i(x_i + y_i) = x_i y_i c_i + x_i y_i c_i' + x_i y_i' c_i + x_i' y_i c_i$

(h) $x_i y_i + c_i(x_i + y_i) = x_i y_i + c_i(x_i \oplus y_i)$

P2.7. Use Boolean algebra to show that the expressions in Problem P2.6 are true.

P2.8. Use Boolean algebra to reduce the functions in Problem P2.4 as much as possible.

P2.9. Any function can be implemented directly either as specified or as its inverted form with a NOT gate added at the final output. Assume that the circuit size is proportional to only the number of AND gates and OR gates (i.e., ignore the number of NOT gates in determining the circuit size). Determine which form of the function (the inverted or non-inverted) will result in a smaller circuit size for the following function. Give your reason, and specify how many AND and OR gates are needed to implement the smaller circuit.

$$F(x, y, z) = x'y'z' + x'y'z + xy'z + xy'z' + xyz$$

P2.10. Derive the truth table for the following logic gates.

(a) 4-input AND gate

(b) 4-input NAND gate

(c) 4-input NOR gate

(d) 4-input XOR gate

(e) 4-input XNOR gate

(f) 5-input XOR gate

(g) 5-input XNOR gate

P2.11. Derive the truth table for the following Boolean functions.

(a) $F(w, x, y, z) = [(x \odot y)' + (xyz)'](w' + x + z)$

(b) $F(x, y, z) = x \oplus y \oplus z$

(c) $F(w, x, y, z) = [w'xy'z + w'z(y \oplus x)]'$

P2.12. Use Boolean algebra to convert the functions in Problem P2.11 to:

(a) The sum-of-minterms format

(b) The product-of-maxterms format

P2.13. Use Boolean algebra to reduce the functions in Problem P2.11 as much as possible.

P2.14. Use a truth table to show that the following expressions are true.
(a) $(x \oplus y) = (x \odot y)'$
(b) $x \oplus y' = x \odot y$
(c) $(w \oplus x) \odot (y \oplus z) = (w \odot x) \odot (y \odot z) = (((w \odot x) \odot y) \odot z)$
(d) $[((xy)'x)'((xy)'y)']' = x \oplus y$

P2.15. Use Boolean algebra to show that the expressions in Problem P2.14 are true.

P2.16. Use Boolean algebra to show that XOR = XNOR for three inputs.

P2.17. Express the Boolean functions in Problem P2.4 using:
(a) The Σ notation
(b) The Π notation

P2.18. Write the following expression as a Boolean function in the canonical form.
(a) $F(x, y, z) = \Sigma(1, 3, 7)$
(b) $F(w, x, y, z) = \Sigma(1, 3, 7)$
(c) $F(x, y, z) = \Pi(1, 3, 7)$
(d) $F(w, x, y, z) = \Pi(1, 3, 7)$
(e) $F'(x, y, z) = \Sigma(1, 3, 7)$
(f) $F'(x, y, z) = \Pi(1, 3, 7)$

P2.19. Given $F'(x, y, z) = \Sigma(1, 3, 7)$, express the function F using a truth table.

P2.20. Use Boolean algebra to convert the function $F(x, y, z) = \Sigma(3, 4, 5)$ to its equivalent product-of-sums canonical form.

P2.21. Given $F = xy'z' + xy'z + xyz' + xyz$, write the expression for F' using:
(a) The product-of-sums format
(b) The sum-of-products format

P2.22. Use Boolean algebra to convert the equation $F = w \odot x \odot y \odot z$ to:
(a) The sum-of-minterms format
(b) The product-of-maxterms format

P2.23. Write the complete dataflow VHDL code for the Boolean functions in Problem P2.4.

P2.24. Write the complete dataflow VHDL code for the logic gates in Problem P2.10.

P2.25. Write the complete dataflow VHDL code for the Boolean functions in Problem P2.11.

Combinational Circuits

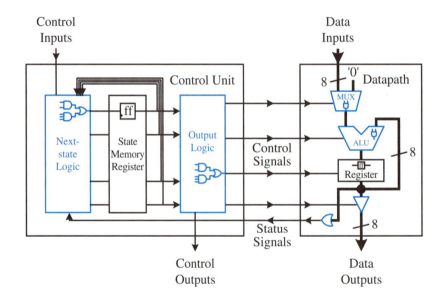

Digital circuits, regardless of whether they are part of the control unit or the datapath, are classified as either one of two types: combinational or sequential. **Combinational circuits** are the class of digital circuits where the outputs of the circuit are dependent only on the current inputs. In other words, a combinational circuit is able to produce an output simply from knowing what the current input values are. **Sequential circuits**, on the other hand, are circuits whose outputs are dependent on not only the current inputs, but also on all of the past inputs. Therefore, in order for a sequential circuit to produce an output, it must know the current input and all past inputs. Because of their dependency on past inputs, sequential circuits must contain memory elements in order to remember the history of past input values. Combinational circuits do not need to know the history of past inputs, and therefore, do not require any memory elements. A "large" digital circuit may contain both combinational circuits and sequential circuits. However, regardless of whether it is a combinational circuit or a sequential circuit, it is nevertheless a digital circuit, and so they use the same basic building blocks—the AND, OR, and NOT gates. What makes them different is in the way the gates are connected.

The car security system from Section 2.9 is an example of a combinational circuit. In the example, the siren is turned on when the master switch is on and someone opens the door. If you close the door then the siren will turn off immediately. With this setup, the output, which is the siren, is dependent only on the inputs, which are the master and door switches. For the security system to be more useful, the siren should remain on even after closing the door after it is first triggered. In order to add this new feature to the security system, we need to modify it so that the output is not only dependent on the master and door switches, but also dependent on whether the door has been opened previously or not. A memory element is needed in order to remember whether the door previously was opened or not, and this results in a sequential circuit.

In this and the next chapter, we will look at the design of combinational circuits. In this chapter, we will look at the analysis and design of general combinational circuits. Chapter 4 will look at the design of specific combinational components. Some sample combinational circuits in our microprocessor road map include the next-state logic and output logic in the control unit, and the multiplexer, ALU, comparator, and tri-state buffer in the datapath. We will leave the design of sequential circuits for a later chapter.

In addition to being able to design a functionally correct circuit, we would also like to be able to optimize the circuit in terms of size, speed, and power consumption. Usually, reducing the circuit size will also increase the speed and reduce the power usage. In this chapter, we will look only at reducing the circuit size. Optimizing the circuit for speed and power usage is beyond the scope of this book.

● ● ● ● ● ● ● ● ● ● ● ● ● ● ●

3.1   Analysis of Combinational Circuits

Very often, we are given a digital logic circuit, and we would like to know the operation of the circuit. The analysis of combinational circuits is the process in which

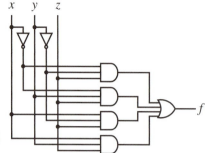

Figure 3.1
Sample combinational circuit.

we are given a combinational circuit, and we want to derive a precise description of the operation of the circuit. In general, a combinational circuit can be described precisely either with a truth table or with a Boolean function.

3.1.1 Using a Truth Table

For example, given the combinational circuit of Figure 3.1, we want to derive the truth table that describes the circuit. We create the truth table by first listing all of the inputs found in the circuit, one input per column, followed by all of the outputs found in the circuit. Hence, we start with a table with four columns: three columns (x, y, z) for the inputs, and one column (f) for the output, as shown in Figure 3.2(a).

The next step is to enumerate all possible combinations of 0's and 1's for all of the input variables. In general, for a circuit with n inputs, there are 2^n combinations, from 0 to $2^n - 1$. Continuing on with the example, the table in Figure 3.2(b) lists the eight combinations for the three variables in order.

Now, for each row in the table (that is, for each combination of input values), we need to determine what the output value is. This is done by substituting the values for the input variables and tracing through the circuit to the output. For example, using $xyz = 000$, the outputs for all of the AND gates are 0, and ORing all of the zeros gives a zero. Therefore, $f = 0$ for this set of values for x, y, and z. This is shown in the annotated circuit in Figure 3.2(c).

For $xyz = 001$, the output of the top AND gate gives a 1, and 1 ORed with anything gives a 1; therefore, $f = 1$, as shown in the annotated circuit in Figure 3.2(d).

Continuing in this fashion for all of the input combinations, we can complete the final truth table for the circuit, as shown in Figure 3.2(e).

A faster method for evaluating the values for the output signals is to work backwards, that is, to trace the circuit from the output back to the inputs. You want to ask the question: When is the output a 1 (or a 0)? Then trace back to the inputs to see what the input values ought to be in order to get the 1 output. For example, using the circuit in Figure 3.1, f is a 1 when any one of the four OR-gate inputs is a 1. For the first input of the OR gate to be a 1, the inputs to the top AND gate must be all 1's. This means that the values for x, y, and z must be 0, 0, and 1, respectively. Repeat this analysis with the remaining three inputs to the OR gate. What you will end up with are the four input combinations for which f is a 1. The remaining input combinations, of course, will produce a 0 for f.

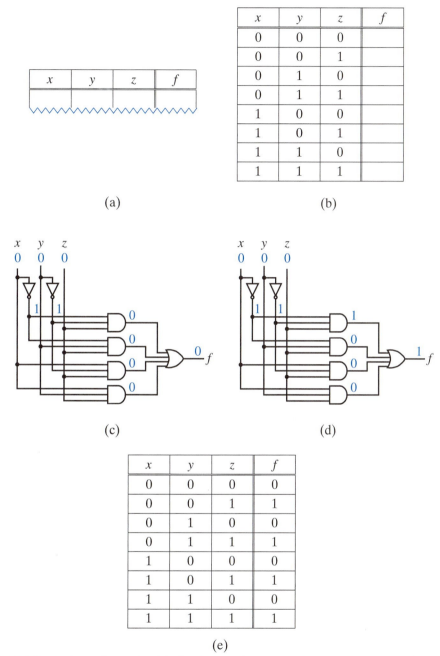

x	y	z	f
0	0	0	
0	0	1	
0	1	0	
0	1	1	
1	0	0	
1	0	1	
1	1	0	
1	1	1	

(a)   (b)

(c)   (d)

x	y	z	f
0	0	0	0
0	0	1	1
0	1	0	0
0	1	1	1
1	0	0	0
1	0	1	1
1	1	0	0
1	1	1	1

(e)

Figure 3.2   Deriving the truth table for the sample circuit in Figure 3.1: (a) listing the input and output columns; (b) enumerating all possible combinations of the three input values; (c) circuit annotated with the input values $xyz = 000$; (d) circuit annotated with the input values $xyz = 001$; (e) complete truth table for the circuit.

Example 3.1   **Deriving a truth table from a circuit diagram**

Derive the truth table for the following circuit with three inputs, A, B, and C, and two outputs, P and Q:

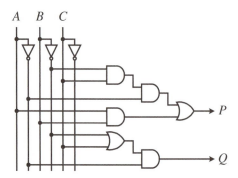

The truth table will have three columns for the three inputs and two columns for the two outputs. Enumerating all possible combinations of the three input values gives eight rows in the table. For each combination of input values, we need to evaluate the output values for both P and Q. For P to be a 1, either of the OR-gate inputs must be a 1. The first input to this OR gate is a 1 if $ABC = 001$. The second input to this OR gate is a 1 if $AB = 11$. Since C is not specified in this case, it means that C can be either a 0 or a 1. Hence, we get the three input combinations for which P is a 1, as shown in the following truth table under the P column. The rest of the input combinations will produce a 0 for P. For Q to be a 1, both inputs of the AND gate must be a 1. Hence, A must be a 0, and either B is a 0 or C is a 1. This gives three input combinations for which Q is a 1, as shown in the truth table under the Q column.

A	B	C	P	Q
0	0	0	0	1
0	0	1	1	1
0	1	0	0	0
0	1	1	0	1
1	0	0	0	0
1	0	1	0	0
1	1	0	1	0
1	1	1	1	0

3.1.2 **Using a Boolean Function**

To derive a Boolean function that describes a combinational circuit, we simply write down the Boolean logical expression at the output of each gate (instead of substituting actual values of 0's and 1's for the inputs) as we trace through the circuit from the primary input to the primary output.   Using the sample combinational circuit of Figure 3.1, we note that the logical expression for the output of the top AND gate is $x'y'z$. The logical expressions for the remaining AND gates are, respectively, $x'yz$, $xy'z$, and xyz. Finally, the outputs from these AND gates are all ORed together. Hence, we get the final expression

$$f = x'y'z + x'yz + xy'z + xyz$$

To help keep track of the expressions at the output of each logic gate, we can annotate the outputs of each logic gate with the resulting logical expression as shown here.

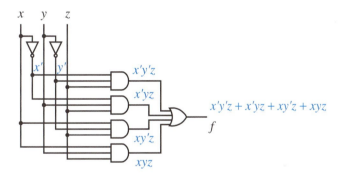

If we substitute all possible combinations of values for all of the variables in the final equation, we should obtain the same truth table as before.

Example 3.2   **Deriving a Boolean function from a circuit diagram**

Derive the Boolean function for the following circuit with three inputs, x, y, and z, and one output, f.

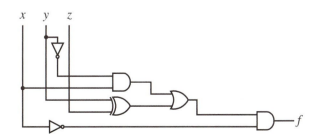

Starting from the primary inputs x, y, and z, we annotate the outputs of each logic gate with the resulting logical expression. Hence, we obtain the annotated circuit:

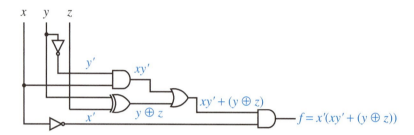

The Boolean function for the circuit is the final equation $f = x'(xy' + (y \oplus z))$ at the output of the circuit.

If a circuit has two or more outputs, then there must be one equation for each of the outputs. The equations are derived totally independent of each other.

• • • • • • • • • • • • • • • •

3.2 Synthesis of Combinational Circuits

Synthesis of combinational circuits is just the reverse procedure of the analysis of combinational circuits. In synthesis, we start with a description of the operation of the circuit. From this description, we derive either the truth table or the Boolean logical function that precisely describes the operation of the circuit. Once we have either the truth table or the logical function, we easily can translate that into a circuit diagram.

For example, let us construct a 3-bit comparator circuit that outputs a 1 if the number is greater than or equal to 5 and outputs a 0 otherwise. In other words, construct a circuit that outputs a 0 if the input is a number between 0 and 4 (inclusive) and outputs a 1 if the input is a number between 5 and 7 (inclusive). The reason why the maximum number is 7 is because the range for an unsigned 3-bit binary number is from 0 to 7. Hence, we can use the three bits, x_2, x_1, and x_0, to represent the 3-bit input value to the comparator. From the description, we obtain the following truth table.

Decimal	Binary Number			Output
Number	x_2	x_1	x_0	f
0	0	0	0	0
1	0	0	1	0
2	0	1	0	0
3	0	1	1	0
4	1	0	0	0
5	1	0	1	1
6	1	1	0	1
7	1	1	1	1

In constructing the circuit, we are interested only in when the output is a 1 (i.e., when the function f is a 1). Thus, we only need to consider the rows where the output function $f = 1$. From the previous truth table, we see that there are three rows where $f = 1$, which give the three AND terms $x_2 x_1' x_0$, $x_2 x_1 x_0'$, and $x_2 x_1 x_0$. Notice that the variables in the AND terms are such that it is inverted if its value is a 0, and not inverted if its value is a 1. In the case of the first AND term, we want $f = 1$ when $x_2 = 1$, $x_1 = 0$, and $x_0 = 1$; and this is satisfied in the expression $x_2 x_1' x_0$. Similarly, the second and third AND terms are satisfied in the expressions $x_2 x_1 x_0'$ and $x_2 x_1 x_0$, respectively. Finally, we want $f = 1$ when either one of these three AND terms is equal to 1. So we ORed the three AND terms together, giving us our final expression:

$$f = x_2 x_1' x_0 + x_2 x_1 x_0' + x_2 x_1 x_0 \tag{3.1}$$

In drawing the schematic diagram, we simply convert the AND operators to AND gates, OR operators to OR gates, and primes to NOT gates. The equation is in the sum-of-products format, meaning that it is summing (ORing) the product (AND) terms. A sum-of-products equation translates to a two-level circuit with the first level being made up of AND gates and the second level made up of OR gates. Each of the three AND terms contains three variables, so we use a 3-input AND gate for each of the three AND terms. The three AND terms are ORed together, so we use a 3-input OR gate to connect the output of the three AND gates. For each inverted variable, we need an inverter. The schematic diagram derived from Equation 3.1 is shown here.

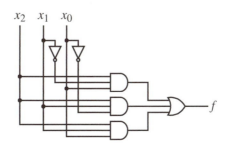

From this discussion we see that any combinational circuit can be constructed using only AND, OR, and NOT gates from either a truth table or a Boolean equation.

Example 3.3 | **Synthesizing a combinational circuit from a truth table**

Synthesize a combinational circuit from the following truth table. The three variables, a, b, and c, are input signals, and the two variables, x and y, are output signals.

a	b	c	x	y
0	0	0	1	0
0	0	1	0	0
0	1	0	1	1
0	1	1	1	0
1	0	0	0	1
1	0	1	1	1
1	1	0	1	0
1	1	1	0	0

We either can derive the Boolean equation from the truth table and then derive the circuit from the equation, or we can derive the circuit directly from the truth table. For this example, we first will derive the Boolean equation. Since there are two output signals, there will be two equations; one for each output signal.

From Section 2.6, we saw that a function is formed by summing its 1-minterms. For output x, there are five 1-minterms: m_0, m_2, m_3, m_5, and m_6. These five minterms represent the five AND terms, $a'b'c'$, $a'bc'$, $a'bc$, $ab'c$, and abc'. Hence, the equation for x is

$$x = a'b'c' + a'bc' + a'bc + ab'c + abc'$$

Similarly, the output signal y has three 1-minterms, and they are $a'bc'$, $ab'c'$, and $ab'c$. Hence, the equation for y is

$$y = a'bc' + ab'c' + ab'c$$

The combinational circuit constructed from these two equations is shown in Figure 3.3(a). Each 3-variable AND term is replaced by a 3-input AND gate. The three inputs to these AND gates are connected to the three input variables a, b, and c, either directly if the variable is not primed or through a NOT gate if the variable is primed. For output x, a 5-input OR gate is used to connect the outputs of the five AND gates for the corresponding five AND terms. For output y, a 3-input OR gate is used to connect the outputs of the three AND gates.

Notice that the two AND terms, $a'bc'$, and $ab'c$, appear in both the x and y equations. As a result, we do not need to generate these two signals twice. Hence, we can reduce the size of the circuit by not duplicating these two AND gates, as shown in Figure3.3(b).

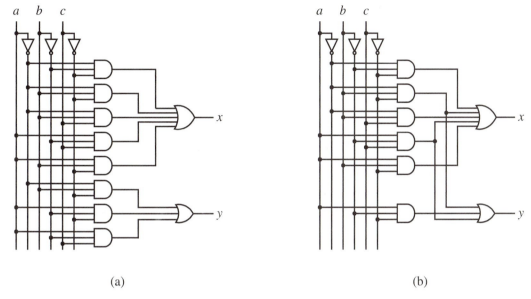

Figure 3.3   Combinational circuit for Example 3.3: (a) no reduction; (b) with reduction.

*3.3 Technology Mapping

To reduce implementation cost and turnaround time to produce a digital circuit on an IC, designers often make use of off-the-shelf semi-custom gate arrays. Many gate arrays are ICs that have only NAND gates or NOR gates built in them, but their input and output connections are not yet connected. To use these gate arrays, a designer simply has to specify where to make these connections between the gates. The problem here is that, when we use these gate arrays to implement a circuit, we need to convert all AND, OR, and NOT gates in the circuit to use only NAND or NOR gates, depending on what is available in the gate array. In addition, these NAND and NOR gates usually have the same number of fixed inputs, for example, only three inputs.

In Section 3.2, we saw that any combinational circuit can be constructed with only AND, OR, and NOT gates. It turns out that any combinational circuit can also be constructed with either only NAND gates or only NOR gates. The reason why we want to use only NAND or NOR gates will be made clear when we look at how these gates are built at the transistor level in Chapter 5. We will now look at how a circuit with AND, OR, and NOT gates is converted to one with only NAND or only NOR gates.

The conversion of any given circuit to use only 2-input NAND or 2-input NOR gates is possible by observing the following equalities. These equalities, in fact, are obtained from the Boolean algebra theorems from Chapter 2.

Rule 1: $x'' = x$ (double NOT)

Rule 2: $x' = (x \cdot x)' = (x \cdot 1)'$ (NOT to NAND)

Rule 3: $x' = (x + x)' = (x + 0)'$ (NOT to NOR)

Rule 4: $xy = ((xy)')'$ (AND to NAND)

Rule 5: $x + y = ((x + y)')' = (x'y')'$ (OR to NAND)

Rule 6: $xy = ((xy)')' = (x' + y')'$ (AND to NOR)

Rule 7: $x + y = ((x + y)')'$ (OR to NOR)

Rule 1 simply says that a double inverter can be eliminated altogether. Rules 2 and 3 convert a NOT gate to a NAND gate or a NOR gate, respectively. For both Rules 2 and 3, there are two ways to convert a NOT gate to either a NAND gate or a NOR gate. For the first method, the two inputs are connected in common. For the second method, one input is connected to the logic 1 for the NAND gate and to 0 for the NOR gate. Rule 4 applies Rule 1 to the AND gate. The resulting expression gives us a NAND gate followed by a NOT gate. We can then use Rule 2 to change the NOT gate to a NAND gate. Rule 5 changes an OR gate to use two NOT gates and a NAND gate by first applying Rule 1 and then De Morgan's theorem. Again, the two NOT gates can be changed to two NAND gates using Rule 2. Similarly, Rule 6 converts an AND gate to use two NOT gates and a NOR gate, and Rule 7 converts an OR gate to a NOR gate followed by a NOT gate.

In a circuit diagram, these rules are translated to the equivalent circuits, as shown in Figure 3.4. Rules 2, 4, and 5 are used if we want to convert a circuit to use only 2-input NAND gates; whereas, Rules 3, 6, and 7 are used if we want to use only 2-input NOR gates.

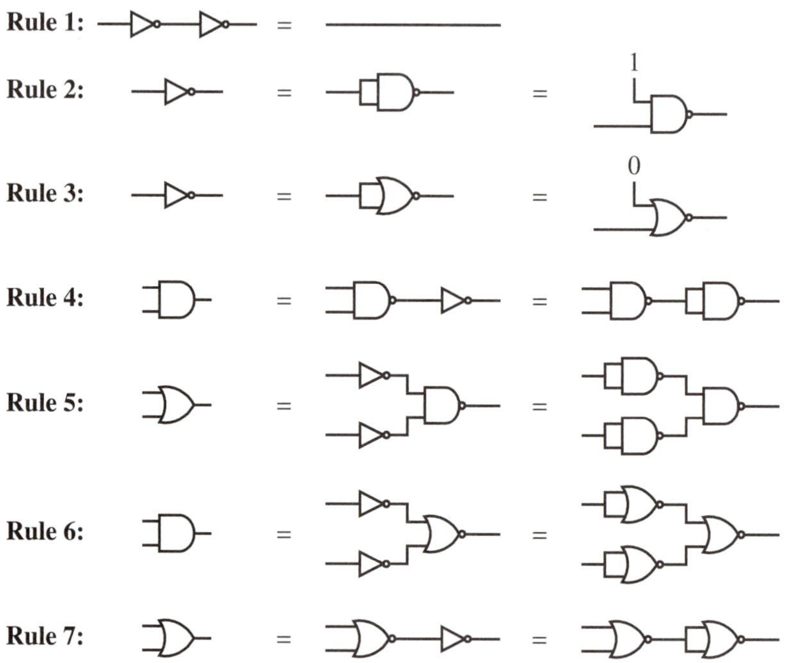

Figure 3.4
Circuits for converting from AND, OR, and NOT gates to NAND or NOR gates.

Rule 8: (gate diagram) = (gate diagram)

Rule 9: (gate diagram) = (gate diagram)

Rule 10: (gate diagram) = (gate diagram) = (gate diagram)

Rule 11: (gate diagram) = (gate diagram) = (gate diagram)

Figure 3.5
Circuits for converting 2-input to 3-input NAND or NOR gate and vice versa.

Another thing that we might want is to get the functionality of a 2-input NAND or 2-input NOR gate from a 3-input NAND or 3-input NOR gate, respectively. In other words, we want to use a 3-input NAND or NOR gate to work like a 2-input NAND or NOR gate, respectively. On the other hand, we might also want to get the reverse of that (that is, to get the functionality of a 3-input NAND or 3-input NOR gate from a 2-input NAND or 2-input NOR gate, respectively). These equalities are shown in the following rules and their corresponding circuits in Figure 3.5.

Rule 8: $(x \cdot y)' = (x \cdot y \cdot y)'$ (2-input to 3-input NAND)

Rule 9: $(x + y)' = (x + y + y)'$ (2-input to 3-input NOR)

Rule 10: $(abc)' = ((ab)c)' = ((ab)''c)'$ (3-input to 2-input NAND)

Rule 11: $(a + b + c)' = ((a + b) + c)' = ((a + b)'' + c)'$ (3-input to 2-input NOR)

Rule 8 converts from a 2-input NAND gate to a 3-input NAND gate. Rule 9 converts from a 2-input NOR gate to a 3-input NOR gate. Rule 10 converts from a 3-input NAND gate to using only 2-input NAND gates. Rule 11 converts from a 3-input NOR gate to using only 2-input NOR gates. Notice that for Rules 10 and 11, an extra NOT gate is needed in between the two gates.

Example 3.4   **Converting a circuit to use only 3-input NAND gates**

Convert the following circuit to use only 3-input NAND gates.

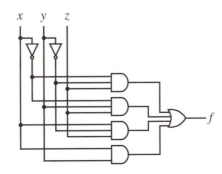

First, we need to change the 4-input OR gate to a 3- and 2-input OR gates.

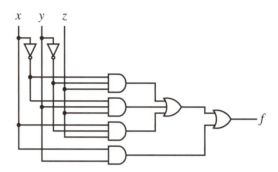

Then we will use Rule 4 to change all of the AND gates to 3-input NAND gates with inverters and Rule 5 to change all of the OR gates to 3-input NAND gates with inverters. The 2-input NAND gates are replaced with 3-input NAND gates with two of its inputs connected together.

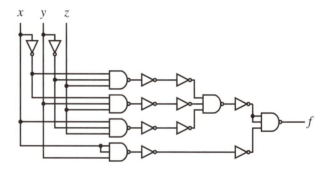

Finally, we eliminate all the double inverters and replace the remaining inverters with NAND gates with their inputs connected together.

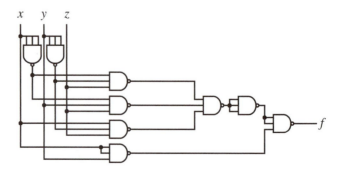

3.4 Minimization of Combinational Circuits

When constructing digital circuits, in addition to obtaining a functionally correct circuit, we like to optimize them in terms of circuit size, speed, and power consumption. In this section, we will focus on the reduction of circuit size. Usually, by reducing the circuit size, we will also improve on speed and power consumption. We have seen in the previous sections that any combinational circuit can be represented using a Boolean function. The size of the circuit is proportional directly to the size or complexity of the functional expression. In fact, it is a one-to-one correspondence between the functional expression and the circuit size. In Section 2.5.1, we saw how we can transform a Boolean function to another equivalent function by using the Boolean algebra theorems. If the resulting function is simpler than the original, then we want to implement the circuit based on the simpler function, since that will give us a smaller circuit size.

Using Boolean algebra to transform a function to one that is simpler is not an easy task, especially for the computer. There is no formula that says which is the next theorem to use. Luckily, there are easier methods for reducing Boolean functions. The **Karnaugh map** method is an easy way for reducing an equation manually and is discussed in Section 3.4.1. The **Quine-McCluskey** or **tabulation** method for reducing an equation is ideal for programming the computer and is discussed in Section 3.4.3.

3.4.1 Karnaugh Maps

To minimize a Boolean equation in the sum-of-products form, we need to reduce the number of product terms by applying the Combining Boolean theorem (Theorem 14) from Section 2.5.1. In so doing, we also will have reduced the number of variables used in the product terms. For example, given the following 3-variable function:

$$F = xy'z' + xyz'$$

we can factor out the two common variables xz' and reduce it to

$$F = xz'(y' + y)$$

$$= xz'1$$

$$= xz'$$

In other words, two product terms that differ by only one variable, whose value is a 0 (primed) in one term and a 1 (unprimed) in the other term, can be combined together to form just one term with that variable omitted, as shown in the previous equations. Thus, we have reduced the number of product terms, and the resulting product term has one less variable. By reducing the number of product terms, we reduce the number of OR operators required, and by reducing the number of variables in a product term, we reduce the number of AND operators required.

Looking at a logic function's truth table, sometimes it is difficult to see how the product terms can be combined and minimized. A **Karnaugh map** (**K-map** for short) provides a simple and straightforward procedure for combining these product terms. A K-map is just a graphical representation of a logic function's truth table, where the minterms are grouped in such a way that it allows one to easily see which of the minterms can be combined. The K-map is a two-dimensional array of squares, each of which represents one minterm in the Boolean function. Thus, the map for an n-variable function is an array with 2^n squares.

Figure 3.6 shows the K-maps for functions with 2, 3, 4, and 5 variables. Notice the labeling of the columns and rows are such that any two adjacent columns or rows differ in only one bit change. This condition is required because we want minterms in adjacent squares to differ in the value of only one variable or one bit, and so these minterms can be combined together. This is why the labeling for the third and fourth columns and for the third and fourth rows are always interchanged. When we read K-maps, we need to visualize them as such that the two end columns or rows wrap around, so that the first and last columns and the first and last rows are really adjacent to each other, because they also differ in only one bit.

In Figure 3.6, the K-map squares are annotated with their minterms and minterm numbers for easy reference only. For example, in Figure 3.6(a) for a 2-variable K-map, the entry in the first row and second column is labeled $x'y$ and annotated

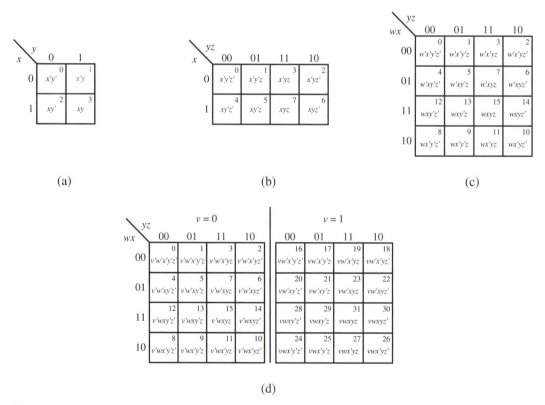

Figure 3.6   Karnaugh maps for: (a) 2-variables; (b) 3-variables; (c) 4-variables; (d) 5-variables.

with the number 1. This is because the first row is when the variable x is a 0, and the second column is when the variable y is a 1. Since, for minterms, we need to prime a variable whose value is a 0 and not prime it if its value is a 1, this entry represents the minterm $x'y$, which is minterm number 1. Be careful that, if we label the rows and columns differently, the minterms and the minterm numbers will be in different locations. When we use K-maps to minimize an equation, we will not write these in the squares. Instead, we will be putting 0's and 1's in the squares.

For a 5-variable K-map, as shown in Figure 3.6(d), we need to visualize the right half of the array (where $v = 1$) to be on top of the left half (where $v = 0$). In other words, we need to view the map as three-dimensional. Hence, although the squares for minterms 2 and 16 are located next to each other, they are not considered to be adjacent to each other. On the other hand, minterms 0 and 16 are adjacent to each other, because one is on top of the other.

Given a Boolean function, we set the value for each K-map square to either a 0 or a 1, depending on whether that minterm for the function is a 0-minterm or a 1-minterm, respectively. However, since we are only interested in using the 1-minterms for a function, the 0's are sometimes not written in the 0-minterm squares.

For example, the K-map for the 2-variable function:

$$F = x'y' + x'y + xy$$

is

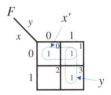

The 1-minterms, m_0 $(x'y')$ and m_1 $(x'y)$, are adjacent to each other, which means that they differ in the value of only one variable. In this case, x is 0 for both minterms, but for y, it is a 0 for one minterm and a 1 for the other minterm. Thus, variable y can be dropped, and the two terms are combined together giving just x'. The prime in x' is because x is 0 for both minterms. This reasoning corresponds with the expression:

$$x'y' + x'y = x'(y' + y) = x'(1) = x'$$

Similarly, the 1-minterms m_1 $(x'y)$ and m_3 (xy) are also adjacent, and y is the variable having the same value for both minterms, and so they can be combined to give

$$x'y + xy = (x' + x)y = (1)y = y$$

We use the term **subcube** to refer to a rectangle of adjacent 1-minterms. These subcubes must be rectangular in shape and can only have sizes that are powers of two. Formally, for an n-variable K-map, an m-*subcube* is defined as that set of 2^m minterms in which $n - m$ of the variables will have the same value in every minterm, while the remaining variables will take on the 2^m possible combinations of 0's and 1's. Thus, a 1-minterm all by itself is called a 0-subcube, two adjacent 1-minterms is called a 1-subcube, and so on. In the previous 2-variable K-map, there are two 1-subcubes: one labeled with x' and one labeled with y.

A 2-subcube will have four adjacent 1-minterms and can be in the shape of any one of those shown in Figure 3.7(a) through (e). Notice that Figure 3.7(d) and (e) also form 2-subcubes, even though the four 1-minterms are not physically adjacent to each other. They are considered to be adjacent because the first and last rows and the first and last columns wraparound in a K-map. In Figure 3.7(f), the four 1-minterms cannot form a 2-subcube, because even though they are physically adjacent to each other, they do not form a rectangle. However, they can form three 1-subcubes—$y'z$, $x'y'$, and $x'z$.

We say that a subcube is *characterized* by the variables having the same values for all of the minterms in that subcube. In general, an m-subcube for an n-variable K-map will be characterized by $n - m$ variables. If the value that is similar for all of the variables is a 1, that variable is unprimed; whereas, if the value that is similar for all of the variables is a 0, that variable is primed. In an expression, this is equivalent

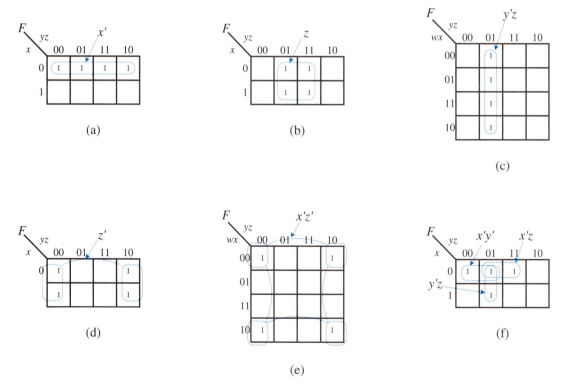

Figure 3.7 Examples of K-maps with 2-subcubes: (a) and (b) 3-variable; (c) 4-variable; (d) 3-variable with wraparound subcube; (e) 4-variable with wraparound subcube; (f) four adjacent minterms that cannot form a 2-subcube.

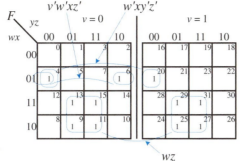

Figure 3.8
A 5-variable
K-map with three
wraparound subcubes.

to the resulting smaller product term when the minterms are combined together. For example, the 2-subcube in Figure 3.7(d) is characterized by z', since the value of z is 0 for all of the minterms, whereas the values for x and y are not all the same for all of the minterms. Similarly, the 2-subcube in Figure 3.7(e) is characterized by $x'z'$.

For a 5-variable K-map, as shown in Figure 3.8, we need to visualize the right half of the array (where $v = 1$) to be on top of the left half (where $v = 0$). Thus, for example, minterm 20 is adjacent to minterm 4 since one is on top of the other, and they form the 1-subcube $w'xy'z'$. Even though minterm 6 is physically adjacent to minterm 20 on the map, they cannot be combined together, because when you visualize the right half as being on top of the left half, then they really are not on top of each other. Instead, minterm 6 is adjacent to minterm 4 because the columns wrap around, and they form the subcube $v'w'xz'$. Minterms 9, 11, 13, 15, 25, 27, 29, and 31 all are adjacent, and together they form the subcube wz. Now that we are viewing this 5-variable K-map in three dimensions, we also need to change the condition of the subcube shape to be a three-dimensional rectangle.

You can see that this visualization becomes almost impossible to work with very quickly as we increase the number of variables. In more realistic designs with many more variables, tabular methods instead of K-maps are used for reducing the size of equations.

The K-map method reduces a Boolean function from its canonical form to its standard form. The goal for the K-map method is to find as few subcubes as possible to cover all of the 1-minterms in the given function. This naturally implies that the size of the subcube should be as big as possible. The reasoning for this is that each subcube corresponds to a product term, and all of the subcubes (or product terms) must be ORed together to get the function. Larger subcubes require fewer AND gates because of fewer variables in the product term, and fewer subcubes will require fewer inputs to the OR gate.

The procedure for using the K-map method is as follows:

1. Draw the appropriate K-map for the given function and place a 1 in the squares that correspond to the function's 1-minterms.

2. For each 1-minterm, find the largest subcube that covers this 1-minterm. This largest subcube is known as a prime implicant (PI). By definition, a **prime**

implicant is a subcube that is not contained within any other subcube. If there is more than one subcube that is of the same size as the largest subcube, then they are all prime implicants.

3. Look for 1-minterms that are covered by only one prime implicant. Since this prime implicant is the only subcube that covers this particular 1-minterm, this prime implicant must be in the final solution. This prime implicant is referred to as an *essential* prime implicant (EPI). By definition, an **essential prime implicant** is a prime implicant that includes a 1-minterm that is not included in any other prime implicant.

4. Create a minimal cover list by selecting the smallest possible number of prime implicants such that every 1-minterm is contained in at least one prime implicant. This cover list must include all of the essential prime implicants plus zero or more of the remaining prime implicants. It is acceptable that a particular 1-minterm is covered in more than one prime implicant, but all 1-minterms must be covered.

5. The final minimized function is obtained by ORing all of the prime implicants from the minimal cover list.

Note that the final minimized function obtained by the K-map method may not be in its most reduced form. It is only in its most reduced *standard* form. Sometimes, it is possible to reduce the standard form further into a nonstandard form.

| **Example 3.5** | **Using K-map to minimize a 4-variable function** |

Use the K-map method to minimize a 4-variable (w, x, y, and z) function F with the 1-minterms: m_0, m_2, m_5, m_7, m_{10}, m_{13}, m_{14}, and m_{15}.

We start with the following 4-variable K-map with a 1 placed in each of the eight minterms squares:

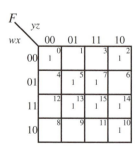

The prime implicants for each of the 1-minterms are shown in the following K-map and table.

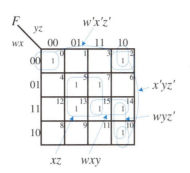

1-minterm	Prime Implicant
m_0	$w'x'z'$
m_2	$w'x'z'$, $x'yz'$
m_5	xz
m_7	xz
m_{10}	$x'yz'$, wyz'
m_{13}	xz
m_{14}	wyz', wxy
m_{15}	xz

For minterm m_0, there is only one prime implicant $w'x'z'$. For minterm m_2, there are two 1-subcubes that cover it, and they are the largest. Therefore, m_2 has two prime implicants, $w'x'z'$ and $x'yz'$. When we consider m_{14}, again, there are two 1-subcubes that cover it, and they are the largest. So m_{14} also has two prime implicants. Minterm m_{15}, however, has only one prime implicant, xz. Although the 1-subcube wxy also covers m_{15}, it is not a prime implicant for m_{15} because it is smaller than the 2-subcube, xz.

From the K-map, we see that there are five prime implicants: $w'x'z'$, $x'yz'$, xz, wyz', and wxy. Of these five prime implicants, $w'x'z'$ and xz are essential prime implicants, since m_0 is covered only by $w'x'z'$, and m_5, m_7, and m_{13} are covered only by xz.

We start the cover list by including the two essential prime implicants $w'x'z'$ and xz. These two subcubes will have covered the minterms m_0, m_2, m_5, m_7, m_{13}, and m_{15}. To cover the remaining two uncovered minterms, m_{10} and m_{14}, we want to use as few prime implicants as possible. Hence, we select the prime implicant wyz', which covers both of them.

Finally, our reduced standard-form equation is obtained by ORing the two essential prime implicants and one prime implicant in the cover list.

$$F = w'x'z' + xz + wyz'$$

Notice that we can reduce this standard-form equation even further by factoring out the z' from the first and last term to get the nonstandard-form equation:

$$F = z'(w'x' + wy) + xz$$

Example 3.6   **Using K-map to minimize a 5-variable function**

Use the K-map method to minimize a 5-variable function $F(v, w, x, y$ and $z)$ with the 1-minterms: $v'w'x'yz'$, $v'w'x'yz$, $v'w'xy'z$, $v'w'xyz$, $vw'x'yz'$, $vw'x'yz$, $vw'xyz'$, $vw'xyz$, $vwx'y'z$, $vwx'yz$, $vwxy'z$, and $vwxyz$.

First, we obtain the following K-maps.

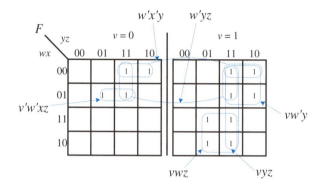

The list of prime implicants is: $v'w'xz$, $w'x'y$, $w'yz$, $vw'y$, vyz, and vwz. From this list of prime implicants, $w'yz$ and vyz are not essential. The four remaining essential prime implicants are able to cover all of the 1-minterms. Hence, the solution in standard form is

$$F = v'w'xz + w'x'y + vw'y + vwz$$

3.4.2 Don't-Cares

There are times when a function is not specified fully. In other words, there are some minterms for the function where we do not care whether their values are a 0 or a 1. When drawing the K-map for these "**don't-care**" minterms, we assign an "$\times$" in that square instead of a 0 or a 1. Usually, a function can be reduced even further if we remember that these $\times$'s can be either a 0 or a 1. As you recall when drawing K-maps, enlarging a subcube reduces the number of variables for that term. Thus, in drawing subcubes, some of them may be enlarged if we treat some of these $\times$'s as 1's. On the other hand, if some of these $\times$'s will not enlarge a subcube, then we want to treat them as 0's so that we do not need to cover them. It is not necessary to treat all $\times$'s to be all 1's or all 0's. We can assign some $\times$'s to be 0's and some to be 1's.

For example, given a function having the following 1-minterms and don't-care minterms:

$$\text{1-minterms}: \quad m_0, m_1, m_2, m_3, m_4, m_7, m_8, \text{ and } m_9$$

$$\times\text{-minterms}: \quad m_{10}, m_{11}, m_{12}, m_{13}, m_{14}, \text{ and } m_{15}$$

we obtain the following K-map with the prime implicants x', yz, and $y'z'$.

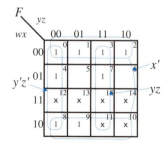

Notice that, in order to get the 4-subcube characterized by x', the two don't-care minterms, m_{10} and m_{11}, are taken to have the value 1. Similarly, the don't-care minterms, m_{12} and m_{15}, are assigned a 1 for the subcubes $y'z'$ and yz, respectively. On the other hand, the don't-care minterms, m_{13} and m_{14}, are taken to have the value 0, so that they do not need to be covered in the solution. The reduced standard-form function as obtained from the K-map is, therefore,

$$F = x' + yz + y'z'$$

Again, this equation can be reduced further by recognizing that $yz + y'z' = y \odot z$. Thus,

$$F = x' + (y \odot z)$$

*3.4.3   Tabulation Method

K-maps are useful for manually obtaining the minimized standard-form Boolean function for maybe up to, at most, five variables. However, for functions with more than five variables, it becomes very difficult to visualize how the minterms should be combined into subcubes. Moreover, the K-map algorithm is not as straightforward for converting to a computer program. There are **tabulation methods** that are better suited for programming the computer, and thus, can solve any function given in a canonical form having any number of variables. One tabulation method is known as the **Quine-McCluskey** method.

Example 3.7   **Illustrating the Quine-McClusky algorithm**

We now illustrate the Quine-McCluskey algorithm using the same four-variable function from Example 3.5 and repeated here

$$F(w, x, y, z) = \Sigma(0, 2, 5, 7, 10, 13, 14, 15)$$

To construct the initial table, the minterms are grouped according to the number of 1's in that minterm number's binary representation. For example, m_0 (0000) has no 1's; m_2 (0010) has one 1; m_5 (0101) has two 1's; etc.. Thus, the initial table of 0-subcubes (i.e., subcubes having only one minterm) as obtained from the function stated above is

Group	Subcube Minterms	Subcube Value				Subcube Covered
		w	x	y	z	
G_0	m_0	0	0	0	0	✓
G_1	m_2	0	0	1	0	✓
G_2	m_5	0	1	0	1	✓
	m_{10}	1	0	1	0	✓
G_3	m_7	0	1	1	1	✓
	m_{13}	1	1	0	1	✓
	m_{14}	1	1	1	0	✓
G_4	m_{15}	1	1	1	1	✓

The "Subcube Covered" column is filled in from the next step.

In Step 2, we construct a second table by combining those minterms in adjacent groups from the first table that differ in only one bit position, as shown next. For example, m_0 and m_2 differ in only the y bit. Therefore, in the second table, we have an entry for the 1-subcube containing the two minterms, m_0 and m_2. A dash (–) is used in the bit position that is different in the two minterms. Since this 1-subcube covers the two individual minterms, m_0 and m_2, we make a note of it by checking these two minterms in the "Subcube Covered" column in the previous table. This process is equivalent to saying that the two minterms, m_0 ($w'x'y'z'$) and m_2 ($w'x'yz'$), can be combined together and are reduced to the one term, $w'x'z'$. The dash under the y column simply means that y can be either a 0 or a 1, and therefore, y can be discarded. Thus, this second table simply lists all of the 1-subcubes. Again, the "Subcube Covered" column in this second table will be filled in from the third step.

Group	Subcube Minterms	Subcube Value				Subcube Covered
		w	x	y	z	
G_0	m_0, m_2	0	0	–	0	
G_1	m_2, m_{10}	–	0	1	0	
G_2	m_5, m_7	0	1	–	1	✓
	m_5, m_{13}	–	1	0	1	✓
	m_{10}, m_{14}	1	–	1	0	
G_3	m_7, m_{15}	–	1	1	1	✓
	m_{13}, m_{15}	1	1	–	1	✓
	m_{14}, m_{15}	1	1	1	–	

In Step 3, we perform the same matching process as before. We look for subcubes in adjacent groups that differ in only one bit position. In the matching, the dash must also match. These subcubes are combined to create the next subcube table. The resulting table, however, is a table containing 2-subcubes. From the above 1-subcube table, we get the following 2-subcube table:

Group	Subcube Minterms	Subcube Value				Subcube Covered
		w	x	y	z	
G_2	m_5, m_7, m_{13}, m_{15}	–	1	–	1	

From the 1-subcube table, subcubes $m_5 m_7$ and $m_{13} m_{15}$ can be combined together to form the subcube $m_5 m_7 m_{13} m_{15}$ in the 2-subcube table, since they differ in only the w bit. Similarly, subcubes $m_5 m_{13}$ and $m_7 m_{15}$ from the 1-subcube table can also be combined together to form the subcube, $m_5 m_7 m_{13} m_{15}$, because they differ in only the y bit. From both of these combinations, the resulting subcube is the same. Therefore, we have the four checks in the 1-subcube table, but only one resulting subcube in the 2-subcube table. Notice that in the subcube $m_5 m_7 m_{13} m_{15}$, there are two dashes;

one that is carried over from Step 2, and one for where the bit is different from the current step.

We continue to repeat the matching step as long as there are adjacent subcubes that differ in only one bit position. We stop when there are no more subcubes that can be combined. The prime implicants are those subcubes that are not covered (i.e., those without a check mark in the "Subcube Covered" column). The only subcube in the 2-subcube table does not have a check mark, and it has the value $x = 1$ and $z = 1$; thus, we get the prime implicant, xz. The 1-subcube table has four subcubes that do not have a checkmark; they are the four prime implicants: $w'x'z'$, $x'yz'$, wyz', and wxy. Note that these prime implicants may not be necessarily all in the last table. These five prime implicants (xz, $w'x'z'$, $x'yz'$, wyz', and wxy) are exactly the same as those obtained in Example 3.5.

- - - - - - - - - - - - - - - - -

*3.5 Timing Hazards and Glitches

As you probably know, things in practice don't always work according to what you learn in school. Hazards and glitches in circuits are such examples of things that may go awry. In our analysis of combinational circuits, we have been performing only functional analysis. A functional analysis assumes that there is no delay for signals to pass from the input to the output of a gate. In other words, we look at a circuit only with respect to its logical operation as defined by the Boolean theorems. We have not considered the timing of the circuit. When a circuit is actually implemented, the timing of the circuit (that is, the time for the signals to pass from the input of a logic gate to the output) is very critical and must be treated with care. Otherwise, an actual implementation of the circuit may not work according to the functional analysis of the same circuit. **Timing hazards** are problems in a circuit as a result of timing issues. These problems can be observed only from a timing analysis of the circuit or from an actual implementation of the circuit. A functional analysis of the circuit will not reveal timing hazard problems.

A **glitch** is when a signal is expected to be stable (from a functional analysis), but it changes value for a brief moment and then goes back to what it is expected to be. For example, if a signal is expected to be at a stable 0, but instead, it goes up to a 1 and then drops back to a 0 very quickly. This sudden, unexpected transition of the signal is a glitch, and the circuit having this behavior contains a hazard.

Take, for example, the simple 2-to-1 multiplexer circuit shown in Figure 3.9(a). Let us assume that both d_0 and d_1 are at a constant 1 and that s goes from a 1 to a 0. For a functional analysis of the circuit, the output y should remain at a constant 1. However, if we perform a timing analysis of the circuit, we will see something different in the timing diagram. Let us assume that all of the logic gates in the circuit have a delay of one time unit. The resulting timing trace is shown in Figure 3.9(b). At time t_0, s drops to a 0. Since it takes one time unit for s to be inverted through

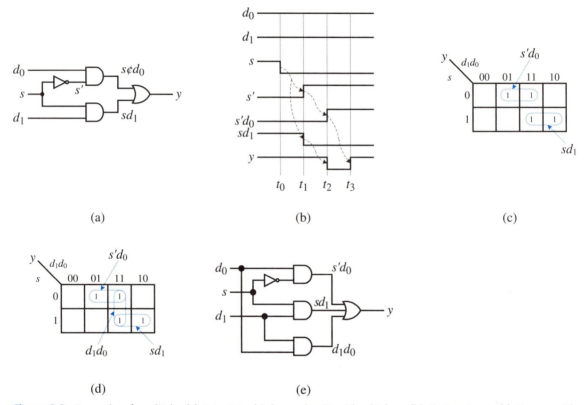

(a)   (b)   (c)

(d)   (e)

Figure 3.9   Example of a glitch: (a) 2-to-1 multiplexer circuit with glitches; (b) timing trace; (c) K-map with glitches; (d) K-map without glitches; (e) 2-to-1 multiplexer circuit without glitches.

the inverter, s' changes to a 1 after one time unit at time t_1. At the same time, it takes the bottom AND gate one time unit for the output sd_1 to change to a 0 at time t_1. However, the top AND gate will not see any input change until time t_1, and when it does, it takes another one time unit for its output $s'd_0$ to rise to a 1 at time t_2. Starting at time t_1, both inputs of the OR gate are 0, so after one time unit, the OR gate outputs a 0 at time t_2. At time t_2, when the top AND gate outputs a 1, the OR gate will take this 1 input and outputs a 1 after one time unit at t_3. So between times t_2 and t_3, output y unexpectedly drops to a 0 for one time unit and then rises back to a 1. Hence, the output signal y has a glitch, and the circuit has a hazard.

As you may have noticed, glitches in a signal are caused by multiple sources having paths of different delays driving that signal. These types of simple glitches can be solved easily using K-maps. A glitch generally occurs if, by simply changing one input, we have to go out of one prime implicant in a K-map and into an adjacent one (i.e., moving from one subcube to another). The glitch can be eliminated by adding an extra prime implicant, so that when going from one prime implicant to the adjacent one, we remain inside the third prime implicant.

Figure 3.9(c) shows the K-map with the two original prime implicants, $s'd_0$ and sd_1, that correspond to the circuit in Figure 3.9(a). When we change s from a 1 to a 0, we have to go out of the prime implicant sd_1 and into the prime implicant $s'd_0$. Figure 3.9(d) shows the addition of the extra prime implicant d_1d_0. This time, when moving from the prime implicant sd_1 to the prime implicant $s'd_0$, we remain inside the new prime implicant d_1d_0. The 2-to-1 multiplexer circuit with the extra prime implicant d_1d_0 added, as shown in Figure 3.9(e), will prevent the glitch from happening.

3.5.1   Using Glitches

Sometimes, we can use glitches to our advantage, as shown in the following example.

Example 3.8   **A one-shot circuit using glitches**

A circuit that outputs a single, short pulse when given an input of arbitrary time length is known as a *one-shot*. A one-shot circuit is used, for example, for generating a single, short 1 pulse when a key is pressed. Sometimes, when a key is pressed, we do not want to generate a continuous 1 signal for as long as the key is pressed. Instead, we want the output signal to be just a single, short pulse, even if the key is still being pressed.

Since logic gates have an inherent signal delay, we can use this delay to determine the duration of the short pulse that we want. This short pulse, of course, is really just a glitch in the circuit. Figure 3.10(a) shows a sample one-shot circuit using signal delays through three inverters. Figure 3.10(b) shows a sample timing trace for it.

Initially, assume that the values for *Input* is a 0, and point *A* is a 1; therefore, the output of the AND gate is 0. When we set *Input* to a 1 momentarily, both inputs to the AND gate will be 1's, and so after a delay through the AND gate, *Output* will be a 1. After a delay through the three inverters, with *Input* still at 1, point *A* will go to a 0, and *Output* will change back to a 0. When we set *Input* back to a 0, *Output* will continue to be a 0. After the delay through the inverters when point *A* goes back to a 1, *Output* remains at 0.

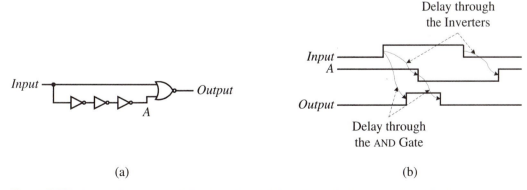

(a)   (b)

Figure 3.10   A one-shot circuit: (a) using signal delay through three inverters; (b) timing trace.

As a result, a glitch is created by the signal delay through the three inverters. This glitch, however, is the short 1 pulse that we want, and the length of this pulse is determined by the delay through the inverters. With this one-shot circuit, it does not matter how long the input key is being pressed, the output signal will always be the same 1 pulse each time that the key is pressed.

• • • • • • • • • • • • • • •
3.6 BCD to 7-Segment Decoder

We will now synthesize the circuit for a BCD to 7-segment decoder for driving a 7-segment LED display. The decoder converts a 4-bit binary coded decimal (BCD) input to seven output signals for turning on the seven lights in a 7-segment LED display. The 4-bit input encodes the binary representation of a decimal digit. Given the decimal digit input, the seven output lines are turned on in such a way so that the LED displays the corresponding digit. The 7-segment LED display schematic with the names of each segment labeled is shown here.

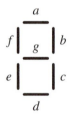

The operation of the BCD to 7-segment decoder is specified in the truth table in Figure 3.11. The four inputs to the decoder are i_3, i_2, i_1, and i_0, and the seven outputs for each of the seven LEDs are labeled a, b, c, d, e, f, and g. For each input combination, the corresponding digit to display on the 7-segment LED is shown in the "Display" column. The segments that need to be turned on for that digit will have a 1, while the segments that need to be turned off for that digit will have a 0. For example, for the 4-bit input 0000, which corresponds to the digit 0, segments a, b, c, d, e, and f need to be turned on, while segment g needs to be turned off.

Notice that the input combinations 1010 to 1111 are not used, and so don't-care values are assigned to all of the segments for these six combinations.

From the truth table in Figure 3.11, we are able to specify seven equations that are dependent on the four inputs for each of the seven segments. For example, the canonical form equation for segment a is

$$a = i_3'i_2'i_1'i_0' + i_3'i_2'i_1i_0' + i_3'i_2'i_1i_0 + i_3'i_2i_1'i_0 + i_3'i_2i_1i_0' + i_3'i_2i_1i_0 + i_3i_2'i_1'i_0' + i_3i_2'i_1'i_0$$

Before implementing this equation directly in a circuit, we want to simplify it first using the K-map method. The K-map for the equation for segment a is

Inputs				Decimal Digit	Display	a	b	c	d	e	f	g
i_3	i_2	i_1	i_0									
0	0	0	0	0		1	1	1	1	1	1	0
0	0	0	1	1		0	1	1	0	0	0	0
0	0	1	0	2		1	1	0	1	1	0	1
0	0	1	1	3		1	1	1	1	0	0	1
0	1	0	0	4		0	1	1	0	0	1	1
0	1	0	1	5		1	0	1	1	0	1	1
0	1	1	0	6		1	0	1	1	1	1	1
0	1	1	1	7		1	1	1	0	0	0	0
1	0	0	0	8		1	1	1	1	1	1	1
1	0	0	1	9		1	1	1	0	0	1	1
Rest of the Combinations						×	×	×	×	×	×	×

Figure 3.11   Truth table for the BCD to 7-segment decoder.

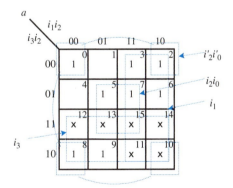

From evaluating the K-map, we derive the simpler equation for segment a as

$$a = i_3 + i_1 + i_2' i_0' + i_2 i_0 = i_3 + i_1 + (i_2 \odot i_0)$$

Proceeding in a similar manner, we get the following remaining six equations:

$$b = i_2' + (i_1 \odot i_0)$$

$$c = i_2 + i_1' + i_0$$

$$d = i_1 i_0' + i_2' i_0' + i_2' i_1 + i_2 i_1' i_0$$

$$e = i_1 i_0' + i_2' i_0'$$

$$f = i_3 + i_2 i_1' + i_2 i_0' + i_1' i_0'$$

$$g = i_3 + (i_2 \oplus i_1) + i_1 i_0'$$

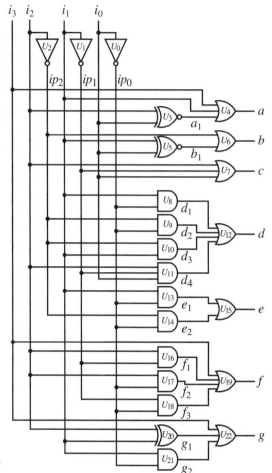

Figure 3.12
Circuit for the BCD
to 7-segment decoder.

From these seven simplified equations, we can now implement the circuit, as shown in Figure 3.12. The labeling of the nodes and gates in the drawing will be explained and used in Section 3.7.1.

3.7 VHDL for Combinational Circuits

Writing VHDL code to describe a digital circuit can be done using any one of three models or levels of abstraction: **structural**, **dataflow**, or **behavioral**. The choice of which model to use usually depends on what is known about the circuit. At the structural level, which is the lowest level, you first have to manually design the circuit. Having drawn the circuit, you use VHDL to specify the components and gates that are needed by the circuit and how they are connected together by following your circuit exactly. Synthesizing a structural VHDL description of a circuit will produce a netlist that is like your original circuit. The advantage of working at the structural level is that you have full control as to what components are used and how they are connected together. The disadvantage, of course, is that you need to manually come up with the circuit, and so the full capabilities of the synthesizer are not utilized. A simple example of a structural VHDL code for a 2-input multiplexer was shown in Figure 1.11.

At the dataflow level, the circuit is defined using built-in VHDL logic operators (such as the AND, OR, and NOT) that are applied to input signals. In order to work at this level, you need to have the Boolean equations for the circuit. Hence, the dataflow level is best suited for describing a circuit that is already expressed as a Boolean function. The equations easily are converted to the required VHDL syntax using signal-assignment statements. A simple example of a dataflow VHDL code for a 2-input multiplexer was shown in Figure 1.9.

All of the statements used in the structural and dataflow levels are executed concurrently, as opposed to statements in a computer program, which usually are executed in a sequential manner. In other words, the ordering of the VHDL statements written in the structural or dataflow level does not matter—the results would be exactly the same.

Describing a circuit at the behavioral level is very similar to writing a computer program. You have all of the standard high-level programming constructs—such as the FOR LOOP, WHILE LOOP, IF THEN ELSE, CASE, and variable assignments. The statements are enclosed in a PROCESS block, and are executed sequentially. A simple example of a behavioral VHDL code for a 2-input multiplexer was shown in Figure 1.5.

3.7.1 Structural BCD to 7-Segment Decoder

Figure 3.13 shows the structural VHDL code for the BCD to 7-segment decoder based on the circuit shown in Figure 3.12. The code starts with declaring and defining all of the components needed in the circuit. For this decoder circuit, only basic gates (such as the NOT gate, 2-input AND, 3-input AND, etc.) are used. The ENTITY statement is used to declare all of these components, and the ARCHITECTURE statement is used to define the operation of these components. Since we are using only simple gates, defining these components using the dataflow model is the simplest. For more complex components (as we will see in later chapters), we want to choose the model that is best suited for the information that we have available for the circuit. The reason why the code shown in Figure 3.13 is structural is not because of how these components are defined, but rather on how these components are connected together to form the enclosing entity; in this case, the *bcd* entity. Notice that the LIBRARY and USE statements need to be repeated for every ENTITY declaration.

The actual structural code begins with the *bcd* ENTITY declaration. The *bcd* circuit shown in Figure 3.12 has four input signals: i_3, i_2, i_1, and i_0 and seven output signals: a, b, c, d, e, f, and g. These signals are declared in the PORT list using the keyword IN for the input signals, and OUT for the output signals; both of which are of type STD_LOGIC.

```
----------------- NOT gate ----------------------
LIBRARY IEEE;
USE IEEE.STD_LOGIC_1164.ALL;
ENTITY notgate IS PORT(
  i: IN STD_LOGIC;
  o: OUT STD_LOGIC);
END notgate;
ARCHITECTURE Dataflow OF notgate IS
BEGIN
 o <= NOT i;
END Dataflow;


----------------- 2-input AND gate ---------------
LIBRARY IEEE;
USE IEEE.STD_LOGIC_1164.ALL;
ENTITY and2gate IS PORT(
  i1, i2: IN STD_LOGIC;
  o: OUT STD_LOGIC);
END and2gate;
ARCHITECTURE Dataflow OF and2gate IS
BEGIN
  o <= i1 AND i2;
END Dataflow;


----------------- 3-input AND gate ---------------
LIBRARY IEEE;
USE IEEE.STD_LOGIC_1164.ALL;
ENTITY and3gate IS PORT(
  i1, i2, i3: IN STD_LOGIC;
  o: OUT STD_LOGIC);
END and3gate;
ARCHITECTURE Dataflow OF and3gate IS
BEGIN
  o <= (i1 AND i2 AND i3);
END Dataflow;


----------------- 2-input OR gate ----------------
LIBRARY IEEE;
USE IEEE.STD_LOGIC_1164.ALL;
ENTITY or2gate IS PORT(
  i1, i2: IN STD_LOGIC;
  o: OUT STD_LOGIC);
END or2gate;
```

Figure 3.13
Structural VHDL
code of the BCD to
7-segment decoder.
(continued on next page)

```
ARCHITECTURE Dataflow OF or2gate IS
BEGIN
  o <= i1 OR i2;
END Dataflow;


---------------- 3-input OR gate ----------------
LIBRARY IEEE;
USE IEEE.STD_LOGIC_1164.ALL;
ENTITY or3gate IS PORT(
  i1, i2, i3: IN STD_LOGIC;
  o: OUT STD_LOGIC);
END or3gate;
ARCHITECTURE Dataflow OF or3gate IS
BEGIN
  o <= i1 OR i2 OR i3;
END Dataflow;


---------------- 4-input OR gate ----------------
LIBRARY IEEE;
USE IEEE.STD_LOGIC_1164.ALL;
ENTITY or4gate IS PORT(
  i1, i2, i3, i4: IN STD_LOGIC;
  o: OUT STD_LOGIC);
END or4gate;
ARCHITECTURE Dataflow OF or4gate IS
BEGIN
  o <= i1 OR i2 OR i3 OR i4;
END Dataflow;


---------------- 2-input XOR gate ----------------
LIBRARY IEEE;
USE IEEE.STD_LOGIC_1164.ALL;
ENTITY xor2gate IS PORT(
  i1, i2: IN STD_LOGIC;
  o: OUT STD_LOGIC);
END xor2gate;
ARCHITECTURE Dataflow OF xor2gate IS
BEGIN
  o <= i1 XOR i2;
END Dataflow;
```

Figure 3.13
Structural VHDL
code of the BCD to
7-segment decoder.
(continued on next page)

```
---------------- 2-input XNOR gate --------------
LIBRARY IEEE;
USE IEEE.STD_LOGIC_1164.ALL;
ENTITY xnor2gate IS PORT(
  i1, i2: IN STD_LOGIC;
  o: OUT STD_LOGIC);
END xnor2gate;
ARCHITECTURE Dataflow OF xnor2gate IS
BEGIN
  o <= NOT(i1 XOR i2);
END Dataflow;


---------------- bcd entity --------------------
LIBRARY IEEE;
USE IEEE.STD_LOGIC_1164.ALL;

ENTITY bcd IS PORT(
  i0, i1, i2, i3: IN STD_LOGIC;
  a, b, c, d, e, f, g: OUT STD_LOGIC);
END bcd;
ARCHITECTURE Structural OF bcd IS
COMPONENT notgate PORT(
  i: IN STD_LOGIC;
  o: OUT STD_LOGIC);
END COMPONENT;
COMPONENT and2gate PORT(
  i1, i2: IN STD_LOGIC;
  o: OUT STD_LOGIC);
END COMPONENT;
COMPONENT and3gate PORT(
  i1, i2, i3: IN STD_LOGIC;
  o: OUT STD_LOGIC);
END COMPONENT;
COMPONENT or2gate PORT(
  i1, i2: IN STD_LOGIC;
  o: OUT STD_LOGIC);
END COMPONENT;
COMPONENT or3gate PORT(
  i1, i2, i3: IN STD_LOGIC;
  o: OUT STD_LOGIC);
END COMPONENT;
```

Figure 3.13
Structural VHDL
code of the BCD to
7-segment decoder.
(continued on next page)

```
COMPONENT or4gate PORT(
  i1, i2, i3, i4: IN STD_LOGIC;
  o: OUT STD_LOGIC);
END COMPONENT;
COMPONENT xor2gate PORT(
  i1, i2: IN STD_LOGIC;
  o: OUT STD_LOGIC);
END COMPONENT;
COMPONENT xnor2gate PORT(
  i1, i2: IN STD_LOGIC;
  o: OUT STD_LOGIC);
END COMPONENT;

SIGNAL ip0,ip1,ip2,a1,b1,d1,d2,d3,d4,e1,e2,f1,f2,f3,g1,g2: STD_LOGIC;
BEGIN
  U0: notgate PORT MAP(i0,ip0);
  U1: notgate PORT MAP(i1,ip1);
  U2: notgate PORT MAP(i2,ip2);
  U3: xnor2gate PORT MAP(i2, i0, a1);
  U4: or3gate PORT MAP(i3, i1, a1, a);
  U5: xnor2gate PORT MAP(i1, i0, b1);
  U6: or2gate PORT MAP(ip2, b1, b);
  U7: or3gate PORT MAP(i2, ip1, i0, c);
  U8: and2gate PORT MAP(i1, ip0, d1);
  U9: and2gate PORT MAP(ip2, ip0, d2);
  U10: and2gate PORT MAP(ip2, i1, d3);
  U11: and3gate PORT MAP(i2, ip1, i0, d4);
  U12: or4gate PORT MAP(d1, d2, d3, d4, d);
  U13: and2gate PORT MAP(i1, ip0, e1);
  U14: and2gate PORT MAP(ip2, ip0, e2);
  U15: or2gate PORT MAP(e1, e2, e);
  U16: and2gate PORT MAP(i2, ip1, f1);
  U17: and2gate PORT MAP(i2, ip0, f2);
  U18: and2gate PORT MAP(ip1, ip0, f3);
  U19: or4gate PORT MAP(i3, f1, f2, f3, f);
  U20: xor2gate PORT MAP(i2, i1, g1);
  U21: and2gate PORT MAP(i1, ip0, g2);
  U22: or3gate PORT MAP(i3, g1, g2, g);
END Structural;
```

Figure 3.13
Structural VHDL
code of the BCD to
7-segment decoder.

The ARCHITECTURE section begins by specifying the components needed in the circuit using the COMPONENT statement. The port list in the COMPONENT statements must match exactly the port list in the entity declarations of the components. They must match not only in the number, direction, and type of the signals but also in the names given to the signals. Note also that names in the component port list can be the same as the names in the *bcd* entity port list, but they are not the same signals. For example, the *and2gate* component port list and the *bcd* entity port list both have two signals called i_1 and i_2. References to these two signals in the body of the *bcd* architecture are for the signals declared in the *bcd* entity.

After the COMPONENT statements, the internal node signals are declared using the SIGNAL statement. The names listed are the same as the internal node names used in the circuit in Figure 3.12 for easy reference.

Following all of the declarations, the body of the architecture starts with the keyword BEGIN. For each gate used in the circuit, there is a corresponding PORT MAP statement. Each PORT MAP statement begins with an optional label (i.e., U_1, U_2, and so on) followed by the name of the component (as previously declared with the COMPONENT statements) to use. Again, the labels used in the PORT MAP statements correspond to the labels on the gates in the circuit in Figure 3.12. The parameter list in the PORT MAP statement matches the port list in the component declaration. For example, U_0 is instantiated with the component *notgate*. The first parameter in the PORT MAP statement is the input signal i_0, and the second parameter is the output signal ip_0. U_4 is instantiated with the 3-input OR gate. The three inputs are i_3, i_1, and a_1, and the output is a. Here, a_1 is the output from the 2-input XNOR gate of U_3. The rest of the PORT MAP statements in the program are obtained in a similar manner.

All of the PORT MAP statements are executed concurrently, and therefore, the ordering of these statements is irrelevant. In other words, changing the ordering of these statements will still produce the same result. Any time when a signal in a PORT MAP statement changes value (i.e., from a 0 to a 1 or vice versa) that PORT MAP statement is executed.

3.7.2 Dataflow BCD to 7-Segment Decoder

Figure 3.14 shows the dataflow VHDL code for the BCD to 7-segment decoder based on the Boolean equations derived in Section 3.6. The ENTITY declaration for this dataflow code is exactly the same as that for the structural code, since the interface for the decoder remains the same.

In the ARCHITECTURE section, seven concurrent signal assignment statements are used: one for each of the seven Boolean equations, which corresponds to the seven LED segments. For example, the equation for segment a is

$$a = i_3 + i_1 + (i_2 \odot i_0)$$

```
LIBRARY IEEE;
USE IEEE.STD_LOGIC_1164.ALL;
ENTITY bcd IS PORT(
   i0, i1, i2, i3: IN STD_LOGIC;
   a, b, c, d, e, f, g: OUT STD_LOGIC);
END bcd;
ARCHITECTURE Dataflow OF bcd IS
BEGIN
   a <= i3 OR i1 OR (i2 XNOR i0);                -- seg a
   b <= (NOT i2) OR NOT (i1 XOR i0);             -- seg b
   c <= i2 OR (NOT i1) OR i0;                    -- seg c
   d <= (i1 AND NOT i0) OR (NOT i2 AND NOT i0)   -- seg d
        OR (NOT i2 AND i1) OR (i2 AND NOT i1 AND i0);
   e <= (i1 AND NOT i0) OR (NOT i2 AND NOT i0);  -- seg e
   f <= i3 OR (i2 AND NOT i1)                    -- seg f
        OR (i2 AND NOT i0) OR (NOT i1 AND NOT i0);
   g <= i3 OR (i2 XOR i1) OR (i1 AND NOT i0);    -- seg g
END Dataflow;
```

Figure 3.14
Dataflow VHDL code
of the BCD to
7-segment decoder.

This is converted to the signal assignment statement:

$$a <= i3 \text{ OR } i1 \text{ OR } (i2 \text{ XNOR } i0);$$

Proceeding in a similar manner, we obtain the signal assignment statements in the dataflow code for the remaining six equations.

All of the signal assignment statements are executed concurrently, and therefore, the ordering of these statements is irrelevant. In other words, changing the ordering of these statements will still produce the same result. Any time when a signal on the right-hand side of an assignment statement changes value (i.e., from a 0 to a 1 or vice versa) that assignment statement is executed.

3.7.3 Behavioral BCD to 7-Segment Decoder

The behavioral VHDL code for the BCD to 7-segment decoder is shown in Figure 3.15. The port list for this entity is slightly different from the two entities in the previous sections. Instead of having the four separate input signals, i_0, i_1, i_2, and i_3, we have declared a vector, I, of length four. This vector, I, is declared with the type keyword STD_LOGIC_VECTOR, that is, a vector of type STD_LOGIC. The length of the vector is specified by the range (3 DOWNTO 0). The first number (3) in the range denotes the index of the most significant bit of the vector, and the second number (0) in the range denotes the index of the least significant bit of the vector. Likewise, the seven output signals, a to g, are replaced with the STD_LOGIC_VECTOR *Segs* of length 7. This time, however, the keyword TO is used in the range to mean that the

```
LIBRARY IEEE;
USE IEEE.STD_LOGIC_1164.ALL;

ENTITY bcd IS PORT (
   I: IN STD_LOGIC_VECTOR (3 DOWNTO 0);
   Segs: OUT STD_LOGIC_VECTOR (1 TO 7));
END bcd;

ARCHITECTURE Behavioral OF bcd IS
BEGIN
   PROCESS(I)
     BEGIN
       CASE I IS
       WHEN "0000" => Segs <= "1111110";
       WHEN "0001" => Segs <= "0110000";
       WHEN "0010" => Segs <= "1101101";
       WHEN "0011" => Segs <= "1111001";
       WHEN "0100" => Segs <= "0110011";
       WHEN "0101" => Segs <= "1011011";
       WHEN "0110" => Segs <= "1011111";
       WHEN "0111" => Segs <= "1110000";
       WHEN "1000" => Segs <= "1111111";
       WHEN "1001" => Segs <= "1110011";
       WHEN OTHERS => Segs <= "0000000";
     END CASE;
   END PROCESS;
END Behavioral;
```

Figure 3.15
Behavioral VHDL
code of the BCD to
7-segment decoder.

most significant bit in the vector is index 1 and the least significant bit in the vector is index 7.

In the architecture section, a PROCESS statement is used. All of the statements inside the process block are executed sequentially. The process block itself, however, is treated as a single, concurrent statement. Thus, the architecture section can have two or more process blocks together with other concurrent statements, and these will all execute concurrently.

The parenthesized list of signals after the PROCESS keyword is referred to as the **sensitivity list**. The purpose of the sensitivity list is that, when a value for any of the listed signals changes, the entire process block is executed from the beginning to the end.

In the code, there is a CASE statement inside the process block. Depending on the value of I, one of the WHEN parts will be executed. A WHEN part consists of the keyword WHEN followed by a constant value for the variable I to match, followed by the symbol "=>." The statement or statements after the symbol "=>" is executed when I matches that corresponding constant. In the code, all of the WHEN parts contain

Name:		100.0ns	200.0ns	300.0ns	400.0ns	500.0ns	600.0ns	700.0ns	800.0ns	900.0ns	1.0
I	0	1	2	3	4	5	6	7	8	9	
Segs	1111110	0110000	1101101	1111001	0110011	1011011	1011111	1110000	1111111	1110011	

Figure 3.16 A sample simulation trace of the behavioral 7-segment decoder code.

one signal assignment statement. All of the signal assignment statements assign a string of seven bits to the output signal *Segs*. This string of seven bits corresponds to the on–off values of the seven segments, *a* to *g*, as shown in the 7-segment decoder truth table of Figure 3.11. For example, looking at the truth table, we see that when $I = $ "0000" (that is, for the decimal digit 0) we want all of the segments to be on except for segment *g*. Recall that in the declaration of the *Segs* vector, the most significant bit, which is the leftmost bit in the bit string, is index 1, and the least significant bit, which is the rightmost bit, is index 7. In VHDL, the notation $Segs(n)$ is used to denote the index *n* of the *Segs* vector. In the code, we have designated $Segs(1)$ for segment *a*, $Segs(2)$ for segment *b*, and so on to $Segs(7)$ for segment *g*. So, in order to display the decimal digit 0, we need to assign the bit string "1111110" to *Segs*.

If the value of *I* does not match any of the WHEN parts, then the WHEN OTHERS part will be chosen. In this case, all of the segments will be turned off. Notice that for both the structural and the dataflow code, the segments are not all turned off when *I* is one of these values. Instead, a certain combination of LEDs are turned on because the K-maps assigned some of the don't-cares to 1's. If we assign all the don't-cares to 0, then all the LEDs will be turned off. An alternative to turning all of the segments off for the remaining six cases is to display the six alphabets, *A*, *b*, *C*, *d*, *E*, and *F*, for the six hexadecimal digits. The two letters, *b*, and *d*, have to be displayed in lowercase, because otherwise, it will be the same as the numbers 8 and 0, respectively.

A sample simulation trace of the behavioral 7-segment decoder code is shown in Figure 3.16.

3.8 Summary Checklist

- Combinational circuit
- Analysis of combinational circuit
- Synthesis of combinational circuit
- Technology mapping
- Using K-maps to minimize a Boolean function
- The use of "don't-cares"
- Using "don't-cares" in a K-map
- Using the Quine-McCluskey method to minimize a Boolean function
- Timing hazards and glitches

- How to eliminate simple glitches
- Writing structural, dataflow, and behavioral VHDL code
- Be able to analyze any combinational circuit by deriving its truth table or Boolean function
- Be able to synthesize a combinational circuit from a truth table or Boolean function
- Be able to reduce any combinational circuit to its smallest size

3.9 Problems

P3.1. Derive the truth table for the following circuits.

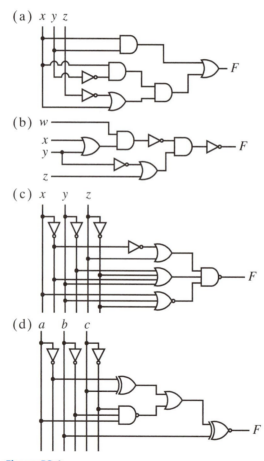

Figure P3.1

P3.2. Derive the Boolean function directly from the circuits in Problem P3.1.

P3.3. Draw the circuit diagram that implements the following truth tables.

(a)

a	b	c	F
0	0	0	0
0	0	1	0
0	1	0	1
0	1	1	1
1	0	0	0
1	0	1	0
1	1	0	1
1	1	1	0

(b)

w	x	y	z	F
0	0	0	0	0
0	0	0	1	0
0	0	1	0	1
0	0	1	1	0
0	1	0	0	1
0	1	0	1	1
0	1	1	0	0
0	1	1	1	1
1	0	0	0	0
1	0	0	1	1
1	0	1	0	1
1	0	1	1	0
1	1	0	0	1
1	1	0	1	1
1	1	1	0	0
1	1	1	1	1

(c)

w	x	y	z	F_1	F_2
0	0	0	0	1	1
0	0	0	1	0	1
0	0	1	0	0	1
0	0	1	1	1	1
0	1	0	0	0	0
0	1	0	1	1	1
0	1	1	0	1	0
0	1	1	1	0	0
1	0	0	0	0	1
1	0	0	1	1	1
1	0	1	0	1	0
1	0	1	1	0	0
1	1	0	0	1	1
1	1	0	1	0	1
1	1	1	0	0	1
1	1	1	1	1	1

(d)

N_3	N_2	N_1	N_0	F
0	0	0	0	0
0	0	0	1	0
0	0	1	0	1
0	0	1	1	1
0	1	0	0	0
0	1	0	1	0
0	1	1	0	1
0	1	1	1	0
1	0	0	0	0
1	0	0	1	0
1	0	1	0	1
1	0	1	1	1
1	1	0	0	1
1	1	0	1	0
1	1	1	0	0
1	1	1	1	1

P3.4. Draw the circuit diagram that implements the following expressions.

(a) $F(x, y, z) = \Sigma(0, 1, 6)$

(b) $F(w, x, y, z) = \Sigma(0, 1, 6)$

(c) $F(w, x, y, z) = \Sigma(2, 6, 10, 11, 14, 15)$

(d) $F(x, y, z) = \Pi(0, 1, 6)$

(e) $F(w, x, y, z) = \Pi(0, 1, 6)$

(f) $F(w, x, y, z) = \Pi(2, 6, 10, 11, 14, 15)$

P3.5. Draw the circuit diagram that implements the following Boolean functions using as few basic gates as possible, but without modifying the equation.

(a) $F = xy' + x'y'z + xyz'$

(b) $F = w'z' + w'xy + wx'z + wxyz$

(c) $F = w'xy'z + w'xyz + wxy'z + wxyz$

(d) $F = N_3'N_2'N_1N_0' + N_3'N_2'N_1N_0 + N_3N_2'N_1N_0' + N_3N_2'N_1N_0 + N_3N_2N_1'N_0' + N_3N_2N_1N_0$

(e) $F = [(x \odot y)' + (xyz)'](w' + x + z)$

(f) $F = x \oplus y \oplus z$

(g) $F = [w'xy'z + w'z(y \oplus x)]'$

P3.6. Draw the circuit diagram that implements the Boolean functions in Problem P3.5 using only 2-input AND, 2-input OR, and NOT gates.

P3.7. Design a circuit that inputs a 4-bit number. The circuit outputs a 1 if the input number is any one of the following numbers: 2, 3, 10, 11, 12, and 15. Otherwise, it outputs a 0.

P3.8. Design a circuit that inputs a 4-bit number. The circuit outputs a 1 if the input number is greater than or equal to 5. Otherwise, it outputs a 0.

P3.9. Design a circuit that inputs a 4-bit number. The circuit outputs a 1 if the input number has an even number of zeros. Otherwise, it outputs a 0.

P3.10. Construct the following circuit. The circuit has five input signals and one output signal. The five input lines are labeled W, X, Y, Z, and E, and the output line is labeled F. E is used to enable (turn on) or disable (turn off) the circuit; thus, when $E = 0$, the circuit is disabled and F is always 0. When $E = 1$, the circuit is enabled, and F is determined by the value of the four input signals W, X, Y, and Z, where W is the most significant bit. If the value is odd, then $F = 1$; otherwise, $F = 0$.

P3.11. Draw the smallest circuit that inputs two 2-bit numbers. The circuit outputs a 2-bit number that represents the count of the number of even numbers in the inputs. The number 0 is taken as an even number. For example, if the two input numbers are 0 and 3, then the circuit outputs the number 1 in binary. If the two input numbers are 0 and 2, then the circuit outputs the number 2 in binary. Show your work by deriving the truth table, the equation, and finally the circuit. You need to minimize all of the equations to standard forms.

P3.12. Derive and draw the circuit that inputs two 2-bit unsigned numbers. The circuit outputs a 3-bit signed number that represents the difference between the two input numbers (i.e., it is the result of the first number minus the second number). Derive the truth table and equations in canonical form.

P3.13. Use Boolean algebra to show that the following circuit is equivalent to the NOT gate.

Figure P3.13

P3.14. Construct a 4-input NAND gate circuit using only 2-input NAND gates.

P3.15. Implement the following circuit using as few NAND gates (with any number of inputs) as possible.

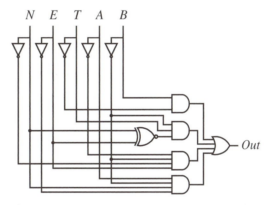

Figure P3.15

P3.16. Draw the circuit diagram that implements the Boolean functions in Problem P3.5, using only 2-input NAND gates.

P3.17. Draw the circuit diagram that implements the Boolean functions in Problem P3.5, using only 3-input NAND gates.

P3.18. Draw the circuit diagram that implements the Boolean functions in Problem P3.5, using only 3-input NOR gates.

P3.19. Convert the following circuit as is (i.e., do not reduce it first) to use only 2-input NOR gates.

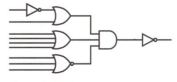

Figure P3.19

P3.20. Convert the following full-adder circuit to use only eleven 2-input NAND gates.

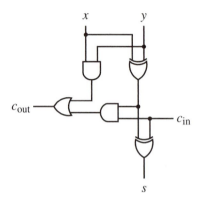

Figure P3.20

P3.21. Perform a timing analysis of the circuit shown in Figure 3.9(e) to see that the circuit does not produce any glitches.

P3.22. Derive a circuit for the 2-input XOR gate that uses only 2-input NAND gates.

P3.23. Use K-maps to reduce the Boolean functions represented by the truth tables in Problem P3.3 to standard form.

P3.24. Use K-maps to reduce the Boolean functions in Problem P3.4 to standard form.

P3.25. Use K-maps to reduce the Boolean functions in Problem P3.5 to standard form.

P3.26. List all of the PIs, EPIs, and all of the minimized standard-form solutions for the following equation.

$$F(v, w, x, y, z) = \Pi(2, 3, 4, 5, 6, 7, 8, 9, 11, 13, 15, 18, 19, 20, 21, 22, 29, 30, 31)$$

P3.27. Use K-maps to reduce the following 4-variable Boolean functions $F(w, x, y, z)$ to standard form.
   (a) 1-minterms: m_2, m_3, m_4, m_5
       Don't-care minterms: $m_{10}, m_{11}, m_{12}, m_{13}, m_{14}, m_{15}$
   (b) 1-minterms: 1, 3, 4, 7, 9
       Don't-care minterms: 0, 2, 13, 14, 15
   (c) 1-minterms: 2, 3, 8, 9
       Don't-care minterms: 1, 5, 6, 7, 13, 15

P3.28. Use K-maps to reduce the following 5-variable Boolean functions $F(v, w, x, y, z)$ to standard form.
   (a) 1-minterms: 1, 3, 4, 7, 9
       Don't-care minterms: 0, 2, 13, 14, 15
   (b) 1-minterms: 2, 4, 10, 15, 16, 21, 26, 29
       Don't-care minterms: 5, 7, 13, 18, 23, 24, 31

P3.29. Use the Quine-McCluskey method to simplify the function:

$$f(w, x, y, z) = \Sigma(0, 2, 5, 7, 13, 15)$$

List all the PIs, EPIs, cover lists, and solutions.

P3.30. Use the Quine-McCluskey method to reduce the Boolean functions in Problem P3.4 to standard form.

P3.31. Write the function that eliminates the static hazard(s) in the function:

$$F = w'z + xyz' + wx'y$$

P3.32. Write the function that eliminates the static hazard(s) in the function:

$$F = y'z' + wz + w'x'y$$

P3.33. Write the complete structural VHDL code for the Boolean functions in Problem P3.4.

P3.34. Write the complete dataflow VHDL code for the Boolean functions in Problem P3.4.

P3.35. Write the complete behavioral VHDL code for the Boolean functions in Problem P3.4.

P3.36. Write the complete dataflow VHDL code for the Boolean functions in Problem P3.5.

P3.37. Write the behavioral VHDL code for converting an 8-bit unsigned binary number to two 4-bit BCD numbers. These two BCD numbers represent the tenth and unit digits of a decimal number. Also, turn on the decimal point LED for the unit digit if the 8-bit binary number is in the one hundreds range, and turn on the decimal point LED for the tenth digit if the 8-bit binary number is in the two hundreds range. This circuit is used as the output circuit for many designs in later chapters.

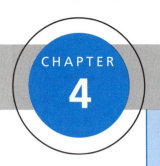

CHAPTER 4

Standard Combinational Components

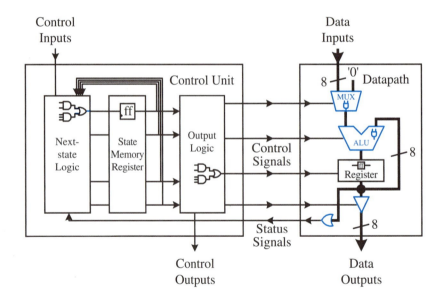

As with many construction projects, it is often easier to build in a hierarchical fashion. Initially, we use the very basic building blocks to build slightly larger building blocks, and then from these larger building blocks, we build yet larger building blocks, and so on. Similarly, in constructing large digital circuits, instead of starting with the basic logic gates as building blocks each time, we often start with larger building blocks. Many of these larger building blocks are often used over and over again in different digital circuits, and therefore, are considered as standard components for large digital circuits. In order to reduce the design time, these standard components are often made available in standard libraries so that they do not have to be redesigned each time that they are needed. For example, many digital circuits require the addition of two numbers; therefore, an adder circuit is considered a standard component and is available in most standard libraries.

Standard **combinational components** are combinational circuits that are available in standard libraries. These combinational components are used mainly in the construction of datapaths. For example, in our microprocessor road map, the standard combinational components are the multiplexer, ALU, comparator, and tri-state buffer. Other standard combinational components include adders, subtractors, decoders, encoders, shifters, rotators, and multipliers. Although the next-state logic and output logic circuits in the control unit are combinational circuits, they are not considered as standard combinational components because they are designed uniquely for a particular control unit to solve a specific problem and usually are never reused in another design.

In this chapter, we will design some standard combinational components. These components will be used in later chapters for the building of the datapath in the microprocessor. When we use these components to build the datapath, we do not need to know the detailed construction of these components. Instead, we only need to know how these components operate, and how they connect to other components. Nevertheless, in order to see the whole picture, we should understand how these individual components are designed.

● ● ● ● ● ● ● ● ● ● ● ● ● ● ●

4.1 Signal Naming Conventions

So far in our discussion, we have always used the words "high" and "low" to mean 1 or 0, or "on" or "off," respectively. However, this is somewhat arbitrary, and there is no reason why we can't say a 0 is a high or a 1 is off. In fact, many standard off-the-shelf components use what we call **negative logic** where 0 is for on and 1 is for off. Using negative logic usually is more difficult to understand because we are used to **positive logic**, where 1 is for on and 0 is for off. In all of our discussions, we will use the more natural, positive logic that we are familiar with.

Nevertheless, in order to prevent any confusion as to whether we are using positive logic or negative logic, we often use the words "assert," "de-assert," "active-high," and "active-low." Regardless of whether we are using positive or negative logic, **active-high** always means that a 1 (i.e., a high) will cause the signal to be active or enabled and that a 0 will cause the signal to be inactive or disabled. For example, if there is an active-high signal called *add* and we want to enable it (i.e., to make it do what it is intended for, which in this case is to add something), then we need to set this signal line to a 1. Setting this signal to a 0 will cause this signal to be disabled or inactive. An **active-low** signal, on the other hand, means that a 0 will cause the signal to be active or enabled, and that a 1 will cause the signal to be inactive or disabled. So if the signal *add* is an active-low signal, then we need to set it to a 0 to make it add something.

We also use the word "**assert**" to mean: to make a signal active or to enable the signal. To **de-assert** a signal is to disable the signal or to make it inactive. For example, to assert the active-high *add* signal line means to set the *add* signal to a 1. To de-assert an active-low line also means to set the line to a 1—since a 0 will enable the line (active-low)—and we want to disable (de-assert) it.

● ● ● ● ● ● ● ● ● ● ● ● ● ● ● ● ● ●

4.2   Adder

4.2.1   Full Adder

To construct an adder for adding two n-bit binary numbers, $X = x_{n-1} \ldots x_0$ and $Y = y_{n-1} \ldots y_0$, we need to first consider the addition of a single bit slice, x_i with y_i, together with the carry-in bit, c_i, from the previous bit position on the right. The result from this addition is a sum bit, s_i, and a carry-out bit, c_{i+1}, for the next bit position. In other words, $s_i = x_i + y_i + c_i$, and $c_{i+1} = 1$ if there is a carry from the addition to the next bit on the left. Note that the + operator in this equation is addition and not the logical OR.

For example, consider the following addition of the two 4-bit binary numbers, $X = 1001$ and $Y = 0011$.

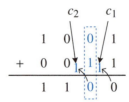

The result of the addition is 1100. The addition is performed just like that for decimal numbers, except that there is a carry whenever the sum is either a 2 or a 3 in decimals, since 2 is 10 in binary and 3 is 11. The most significant bit in the 10 or the 11 is the carry-out bit. Looking at the bit slice that is highlighted in blue where $x_1 = 0$, $y_1 = 1$, and $c_1 = 1$, the addition for this bit slice is $x_1 + y_1 + c_1 = 0 + 1 + 1 = 10$. Therefore, the sum bit is $s_1 = 0$, and the carry-out bit is $c_2 = 1$.

x_i	y_i	c_i	c_{i+1}	s_i
0	0	0	0	0
0	0	1	0	1
0	1	0	0	1
0	1	1	1	0
1	0	0	0	1
1	0	1	1	0
1	1	0	1	0
1	1	1	1	1

(a)

$$
\begin{aligned}
s_i &= x_i'y_i'c_i + x_i'y_ic_i' + x_iy_i'c_i' + x_iy_ic_i \\
&= (x_i'y_i + x_iy_i')c_i' + (x_i'y_i' + x_iy_i)c_i \\
&= (x_i \oplus y_i)c_i' + (x_i \oplus y_i)'c_i \\
&= x_i \oplus y_i \oplus c_i
\end{aligned}
$$

$$
\begin{aligned}
c_{i+1} &= x_i'y_ic_i + x_iy_i'c_i + x_iy_ic_i' + x_iy_ic_i \\
&= x_iy_i(c_i' + c_i) + c_i(x_i'y_i + x_iy_i') \\
&= x_iy_i + c_i(x_i \oplus y_i)
\end{aligned}
$$

(b)

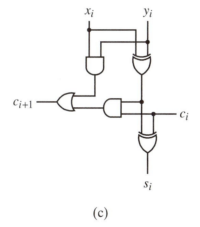

(c)

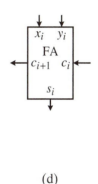

(d)

Figure 4.1
Full adder: (a) truth table; (b) equations for s_i and c_{i+1}; (c) circuit; (d) logic symbol.

The circuit for the addition of a single bit slice is known as a **full adder** (**FA**), and its truth table is shown in Figure 4.1(a). The derivation of the equations for s_i and c_{i+1} are shown in Figure 4.1(b). From these two equations, we get the circuit for the full adder, as shown in Figure 4.1(c). Figure 4.1(d) shows the logic symbol for it. The dataflow VHDL code for the full adder is shown in Figure 4.2.

4.2.2 Ripple-Carry Adder

The full adder is for adding two operands that are only one bit wide. To add two operands that are, say, four bits wide, we connect four full adders together in series. The resulting circuit (shown in Figure 4.3) is called a **ripple-carry adder** for adding two 4-bit operands.

```
LIBRARY IEEE;
USE IEEE.STD_LOGIC_1164.ALL;

ENTITY fa IS PORT (
  Ci, Xi, Yi: IN STD_LOGIC;
  Ci1, Si: OUT STD_LOGIC);
END fa;

ARCHITECTURE Dataflow OF fa IS
BEGIN
  Ci1 <= (Xi AND Yi) OR (Ci AND (Xi XOR Yi));
  Si <= Xi XOR Yi XOR Ci;
END Dataflow;
```

Figure 4.2
Dataflow VHDL code
for a 1-bit full adder.

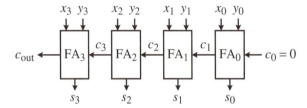

Figure 4.3
Ripple-carry adder.

Since a full adder adds the three bits, x_i, y_i and c_i, together, we need to set the first carry-in bit, c_0, to 0 in order to perform the addition correctly. Moreover, the output signal, c_{out}, is a 1 whenever there is an overflow in the addition.

The structural VHDL code for the 4-bit ripple-carry adder is shown in Figure 4.4. Since we need to duplicate the full-adder component four times, we can use either the PORT MAP statement four times or the FOR-GENERATE statement, as shown in the code, to automatically generate the four components. The statement FOR k IN 3 DOWNTO 0 GENERATE determines how many times to repeat the PORT MAP statement that is in the body of the GENERATE statement and the values used for k. The vector signal *Carryv* is used to propagate the carry bit from one FA to the next.

*4.2.3 Carry-Lookahead Adder

The ripple-carry adder is slow because the carry-in for each full adder is dependent on the carry-out signal from the previous FA. So before FA_i can output valid data, it must wait for FA_{i-1} to have valid data. In the **carry-lookahead adder**, each bit slice eliminates this dependency on the previous carry-out signal and instead uses the values of the two input operands, X and Y, directly to deduce the needed signals. This is possible from the following observations regarding the carry-out signal. For

```
LIBRARY IEEE;
USE IEEE.STD_LOGIC_1164.ALL;

ENTITY Adder4 IS PORT (
  A, B: IN STD_LOGIC_VECTOR(3 DOWNTO 0);
  Cout: OUT STD_LOGIC;
  SUM: OUT STD_LOGIC_VECTOR(3 DOWNTO 0));
END Adder4;

ARCHITECTURE Structural OF Adder4 IS
  COMPONENT FA PORT (
    ci, xi, yi: IN STD_LOGIC;
    co, si: OUT STD_LOGIC);
  END COMPONENT;

  SIGNAL Carryv: STD_LOGIC_VECTOR(4 DOWNTO 0);

BEGIN
  Carryv(0) <= '0';

  Adder: FOR k IN 3 DOWNTO 0 GENERATE
    FullAdder: FA PORT MAP (Carryv(k), A(k), B(k), Carryv(k+1), SUM(k));
  END GENERATE Adder;

  Cout <= Carryv(4);
END Structural;
```

Figure 4.4
VHDL code for a
4-bit ripple-carry
adder using a
FOR-GENERATE
statement.

each FA_i, the carry-out signal, c_{i+1}, is set to a 1 if either one of the following two conditions is true:

$$x_i = 1 \quad \text{and} \quad y_i = 1$$

or

$$(x_i = 1 \quad \text{or} \quad y_i = 1) \quad \text{and} \quad c_i = 1$$

In other words,

$$c_{i+1} = x_i y_i + c_i(x_i + y_i) \tag{4.1}$$

At first glance, this carry-out equation looks completely different from the carry-out equation deduced in Figure 4.1(b). However, they are equivalent (see Problem P2.6(h)).

If we let

$$g_i = x_i y_i$$

and

$$p_i = x_i + y_i$$

then Equation (4.1) can be rewritten as

$$c_{i+1} = g_i + p_i c_i \tag{4.2}$$

Using Equation (4.2) for c_{i+1}, we can recursively expand it to get the carry-out equations for any bit slice, c_i, that is dependent only on the two input operands, X and Y, and the initial carry-in bit, c_0. Using this technique, we get the following carry-out equations for the first four bit slices

$$c_1 = g_0 + p_0 c_0 \tag{4.3}$$

$$
\begin{aligned}
c_2 &= g_1 + p_1 c_1 \\
&= g_1 + p_1(g_0 + p_0 c_0) \\
&= g_1 + p_1 g_0 + p_1 p_0 c_0
\end{aligned} \tag{4.4}
$$

$$
\begin{aligned}
c_3 &= g_2 + p_2 c_2 \\
&= g_2 + p_2(g_1 + p_1 g_0 + p_1 p_0 c_0) \\
&= g_2 + p_2 g_1 + p_2 p_1 g_0 + p_2 p_1 p_0 c_0
\end{aligned} \tag{4.5}
$$

$$
\begin{aligned}
c_4 &= g_3 + p_3 c_3 \\
&= g_3 + p_3(g_2 + p_2 g_1 + p_2 p_1 g_0 + p_2 p_1 p_0 c_0) \\
&= g_3 + p_3 g_2 + p_3 p_2 g_1 + p_3 p_2 p_1 g_0 + p_3 p_2 p_1 p_0 c_0
\end{aligned} \tag{4.6}
$$

Using Equations (4.3) to (4.6), we obtain the circuit for generating the carry-lookahead signals for c_1 to c_4, as shown in Figure 4.5(a). Note that each equation is translated to a three-level combinational logic—one level for generating the g_i and p_i, and two levels (for the sum-of-products format) for generating the c_i expression. This carry-lookahead circuit can be reduced even further because we want c_0 to be a 0 when performing additions and this 0 will cancel the rightmost product term in each equation.

The full adder for the carry-lookahead adder also can be made simpler, since it is no longer required to generate the carry-out signal for the next bit slice. In other words, the carry-in signal for the full adder now comes from the new carry-lookahead circuit rather than from the carry-out signal of the previous bit slice. Thus, this full adder only needs to generate the sum_i signal. Figure 4.5(b) shows one bit slice of the carry-lookahead adder. For an n-bit carry-lookahead adder, we use n bit slices. These n bit slices are not connected in series as with the ripple-carry adder; otherwise, it defeats the purpose of having the more complicated carry-lookahead circuit.

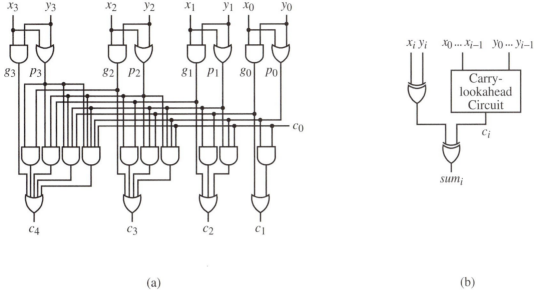

Figure 4.5 (a) Circuit for generating the carry-lookahead signals, c_1 to c_4; (b) one bit slice of the carry-lookahead adder.

- - - - - - - - - - - - - -

4.3   Two's Complement Binary Numbers

Before introducing subtraction circuits, we need to review how negative numbers are encoded using **two's complement** representation. Binary encoding of numbers can be interpreted as either signed or unsigned. Unsigned numbers include only positive numbers and zero, whereas signed numbers include positive, negative, and zero. For signed numbers, the most significant bit (MSB) tells whether the number is positive or negative. If the most significant bit is a 1, then the number is negative; otherwise, the number is positive. The value of a positive signed number is obtained exactly as for unsigned numbers described in Section 2.1. For example, the value for the positive signed number 01101001_2 is just $1 \times 2^6 + 1 \times 2^5 + 1 \times 2^3 + 1 \times 2^0 = 105$ in decimal format.

However, to determine the value of a negative signed number, we need to perform a two-step process: (1) flip all the 1 bits to 0's and all the 0 bits to 1's, and (2) add a 1 to the result obtained from Step 1. The number obtained from applying this two-step process is evaluated as an unsigned number for its value. The negative of this resulting value is the value of the original negative signed number.

Example 4.1   **Finding the value for a signed number**

Given the 8-bit signed number 11101001_2, we know that it is a negative number because of the leading 1. To find out the value of this negative number, we perform the two-step process as follows.

$$11101001 \quad \text{(original number)}$$

$$00010110 \quad \text{(flip bits)}$$

$$00010111 \quad \text{(add a 1 to the previous number)}$$

The value for the resulting number 00010111 is $1 \times 2^4 + 1 \times 2^2 + 1 \times 2^1 + 1 \times 2^0 = 23$. Therefore, the value of the original number 11101001 is negative 23 (-23).

Example 4.2

Finding the value for a signed number

To find the value for the 4-bit signed number 1000, we apply the two-step process to the number as follows.

$$1000 \quad \text{(original number)}$$

$$0111 \quad \text{(flip bits)}$$

$$1000 \quad \text{(add a 1 to the previous number)}$$

The resulting number 1000 is exactly the same as the original number! This, however, should not confuse us if we follow exactly the instructions for the conversion process. We need to interpret the resulting number as an unsigned number to determine the value. Interpreting the resulting number 1000 as an unsigned number gives us the value of 8. Therefore, the original number, which is also 1000, is negative 8 (-8).

Figure 4.6 shows the two's complement numbers for four bits. The range goes from -8 to 7. In general, for an n-bit two's complement number, the range is from -2^{n-1} to $2^{n-1} - 1$.

The nice thing about using two's complement to represent negative numbers is that when we add a number with the negative of the same number, the result is zero, as expected. This is shown in the next example.

Example 4.3

Adding 4-bit signed numbers

Use 4-bit signed arithmetic to perform the following addition.

$$3 \quad = \quad 0011$$

$$\underline{+ \ (-3) \quad = \ + \ 1101}$$

$$0 \quad = \quad \cancel{1}0000$$

The result 10000 has five bits. But since we are using 4-bit arithmetic (that is, the two operands are 4-bits wide), the result must also be in 4-bits. The leading 1 in the result is, therefore, an overflow bit. By dropping the leading one, the remaining result 0000 is the correct answer for the problem. Although this addition resulted in an overflow bit, by dropping this extra bit, we obtained the correct answer.

4-bit Binary	Two's Complement	4-bit Binary	Two's Complement
0000	0	1000	−8
0001	1	1001	−7
0010	2	1010	−6
0011	3	1011	−5
0100	4	1100	−4
0101	5	1101	−3
0110	6	1110	−2
0111	7	1111	−1

Figure 4.6
4-bit two's complement numbers.

Example 4.4   **Adding 4-bit signed numbers**

Use 4-bit signed arithmetic to perform the following addition.

$$
\begin{aligned}
6 &= 0110 \\
+\,3 &= +\,0011 \\
\hline
9 &\neq 1001
\end{aligned}
$$

The result 1001 is a 9 if we interpret it as an unsigned number. However, since we are using signed numbers, we need to interpret the result as a signed number. Interpreting 1001 as a signed number gives −7, which of course is incorrect. The problem here is that the range for a 4-bit signed number is from −8 to +7, and +9 is outside of this range.

Although the addition in this example did not result in an overflow bit, the final answer is incorrect. In order to correct this problem, we need to add (at least) one extra bit by sign extending the number. The corrected arithmetic is shown in Example 4.5.

Example 4.5   **Adding 5-bit signed numbers**

Use 5-bit signed arithmetic to perform the following addition.

$$
\begin{aligned}
6 &= 00110 \\
+3 &= +\,00011 \\
\hline
9 &= 01001
\end{aligned}
$$

The result 01001, when interpreted as a signed number, is 9.

To extend a signed number, we need to add leading 0's or 1's, depending on whether the original most significant bit is a 0 or a 1. If the most significant bit is a 0, we sign extend the number by adding leading 0's. If the most significant bit is a 1, we sign extend the number by adding leading 1's. By performing this sign extension, the value of the number is not changed, as shown in Example 4.6.

Example 4.6   **Performing sign extensions**

Sign extend the numbers 10010 and 0101 to 8 bits.

For the number 10010, since the most significant bit is a 1, therefore, we need to add leading 1's to make the number 8-bits long. The resulting number is 11110010. For the number 0101, since the most significant bit is a 0, therefore, we need to add leading 0's to make the number 8-bits long. The resulting number is 00000101. The following shows that the two resulting numbers have the same value as the two original numbers. Since the first number is negative (because of the leading 1 bit) we need to perform the two-step process to evaluate its value. The second number is positive, so we can evaluate its value directly.

	Original Number	Sign Extended		Original Number	Sign Extended
	10010	11110010		0101	00000101
Flip bits	01101	00001101			
Add 1	01110	00001110			
Value	-14	-14		5	5

● ● ● ● ● ● ● ● ● ● ● ● ● ● ●

4.4 Subtractor

We can construct a one-bit subtractor circuit similar to the method used for constructing the full adder. However, instead of the sum bit, s_i, for the addition, we have a difference bit, d_i, for the subtraction, and instead of having carry-in and carry-out signals, we have borrow-in (b_i) and borrow-out (b_{i+1}) signals. So, when we subtract the ith bit of the two operands, x_i and y_i, we get the difference of $d_i = x_i - y_i$. If, however, the previous bit on the right has to borrow from this i^{th} bit, then input b_i will be set to a 1, and the equation for the difference will be $d_i = x_i - b_i - y_i$. On the other hand, if the ith bit has to borrow from the next bit on the left for the subtraction, then the output b_{i+1} will be set to a 1. The value borrowed is a 2, and so the resulting equation for the difference will be $d_i = x_i - b_i + 2b_{i+1} - y_i$. Note that the symbols $+$ and $-$ used in this equation are for addition and subtraction, and not for logical operations. The term $2b_{i+1}$ is "2 multiply by b_{i+1}." Since b_{i+1} is a 1 when we have to borrow and we borrow a 2 each time, the equation just adds a 2 when there is a borrow. When there is no borrow, b_{i+1} is 0, and so the term b_{i+1} cancels out to 0.

For example, consider the following subtraction of the two 4-bit binary numbers, $X = 0100$ and $Y = 0011$:

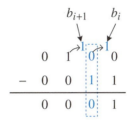

Consider the bit position that is highlighted in blue. Since the subtraction for the previous bit on the right has to borrow, b_i is a 1. Moreover, b_{i+1} is also a 1, because the current bit has to borrow from the next bit on the left. When it borrows, it gets a 2. Therefore, $d_i = x_i - b_i + 2b_{i+1} - y_i = 0 - 1 + 2(1) - 1 = 0$.

The truth table for the 1-bit subtractor is shown in Figure 4.7(a), from which the equations for d_i and b_{i+1}, as shown in Figure 4.7(b), are derived. From these two equations, we get the circuit for the subtractor, as shown in Figure 4.7(c). Figure 4.7(d) shows the logic symbol for the subtractor.

Building a subtractor circuit for subtracting an n-bit operand can be done by daisy-chaining n 1-bit subtractor circuits together, similar to the adder circuit shown in Figure 4.3. However, there is a much better subtractor circuit, as shown in the next section.

● ● ● ● ● ● ● ● ● ● ● ● ● ● ● ● ● ●

4.5 Adder–Subtractor Combination

It turns out that, instead of having to build separate adder and subtractor units, we can modify the ripple-carry adder (or the carry-lookahead adder) slightly to perform both operations. The modified circuit performs subtraction by adding the negated value of the second operand. In other words, instead of performing the subtraction $A - B$, the addition operation $A + (-B)$ is performed.

Recall that in two's complement representation, to negate a value involves inverting all the 0 bits to 1's and 1's to 0's, and then adding a 1. Hence, we need to modify the adder circuit so that we selectively can do either one of two things: (1) flip the bits of the B operand and then add an extra 1 for the subtraction operation, or (2) not flip the bits and not add an extra 1 for the addition operation.

For this adder–subtractor combination circuit (in addition to the two input operands A and B), a select signal, s, is needed to select which operation to perform.

x_i	y_i	b_i	b_{i+1}	d_i
0	0	0	0	0
0	0	1	1	1
0	1	0	1	1
0	1	1	1	0
1	0	0	0	1
1	0	1	0	0
1	1	0	0	0
1	1	1	1	1

(a)

$$d_i = x_i' y_i' b_i + x_i' y_i b_i' + x_i y_i' b_i' + x_i y_i b_i$$
$$= (x_i' y_i + x_i y_i') b_i' + (x_i' y_i' + x_i y_i) b_i$$
$$= (x_i \oplus y_i) b_i' + (x_i \oplus y_i)' b_i$$
$$= x_i \oplus y_i \oplus b_i$$

$$b_{i+1} = x_i' y_i' b_i + x_i' y_i b_i' + x_i' y_i b_i + x_i y_i b_i$$
$$= x_i' b_i (y_i' + y_i) + x_i' y_i (b_i' + b_i) + y_i b_i (x_i' + x_i)$$
$$= x_i' b_i + x_i' y_i + y_i b_i$$

(b)

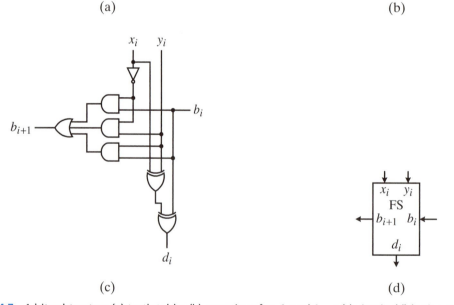

(c)

(d)

Figure 4.7   1-bit subtractor: (a) truth table; (b) equations for d_i and b_{i+1}; (c) circuit; (d) logic symbol.

The assignment of the two operations to the select signal s is shown in Figure 4.8(a). When $s = 0$, we want to perform an addition, and when $s = 1$, we want to perform a subtraction. When $s = 0$, B does not need to be modified, and like the adder circuit from Section 4.2.2, the initial carry-in signal c_0 needs to be set to a 0. On the other hand, when $s = 1$, we need to invert the bits in B and add a 1. The addition of a 1 is accomplished by setting the initial carry-in signal c_0 to a 1. Two circuits are needed for handling the above situations: one for inverting the bits in B and one for setting c_0. Both of these circuits are dependent on s.

s	Function	Operation
0	Add	$F = A + B$
1	Subtract	$F = A + B' + 1$

(a)

s	b_i	y_i	c_0
0	0	0	0
0	1	1	0
1	0	1	1
1	1	0	1

(b)

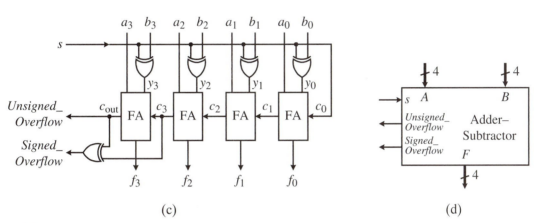

(c)

(d)

Figure 4.8 Adder–subtractor combination: (a) operation table; (b) truth table for y_i and c_0; (c) circuit; (d) logic symbol.

The truth table for these two circuits is shown in Figure 4.8(b). The input variable b_i is the ith bit of the B operand. The output variable y_i is the output from the circuit that either inverts or does not invert the bits in B. From this truth table, we can conclude that the circuit for y_i is just a 2-input XOR gate, while the circuit for c_0 is just a direct connection from s. Putting everything together, we obtain the adder–subtractor combination circuit (for four bits), as shown in Figure 4.8(c). The logic symbol for the circuit is shown in Figure 4.8(d).

Notice the adder–subtractor circuit in Figure 4.8(c) has two different overflow signals, *Unsigned_Overflow* and *Signed_Overflow*. This is because the circuit can deal with both signed and unsigned numbers. When working with unsigned numbers only, the output signal *Unsigned_Overflow* is sufficient to determine whether there is an overflow or not. However, for signed numbers, we need to perform the XOR of *Unsigned_Overflow* with c_3, producing the *Signed_Overflow* signal in order to determine whether there is an overflow or not.

For example, the valid range for a 4-bit *signed* number goes from -2^3 to $2^3 - 1$ (i.e., from -8 to 7). Adding the two signed numbers $4 + 5 = 9$ should result in a signed number overflow, since 9 is outside the range. However, the valid range for a

4-bit *unsigned* number goes from 0 to $2^4 - 1$ (i.e., 0 to 15). If we treat the two numbers 4 and 5 as unsigned numbers, then the result of adding these two unsigned numbers, 9, is inside the range. So when adding the two numbers 4 and 5, the *Unsigned_Overflow* signal should be de-asserted, while the *Signed_Overflow* signal should be asserted. Performing the addition of 4 + 5 in binary as shown here:

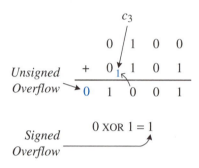

we get 0100 + 0101 = 1001, which produces a 0 for the *Unsigned_Overflow* signal. However, the addition produces a 1 for c_3, and XORing these two values, 0 for the *Unsigned_Overflow* and 1 for c_3, results in a 1 for the *Signed_Overflow* signal.

In another example, adding the two 4-bit signed numbers, $-4 + (-3) = -7$ should not result in a signed overflow. Performing the arithmetic in binary, $-4 = 1100$ and $-3 = 1101$, as shown here:

$$c_3$$

$$
\begin{array}{ccccc}
 & 1 & 1 & 0 & 0 \\
+ & 1 & 1 & 0 & 1 \\
\hline
1 & 1 & 0 & 0 & 1 \\
\end{array}
$$

Unsigned Overflow

$$1 \text{ XOR } 1 = 0$$

Signed Overflow

we get $1100 + 1101 = 11001$, which produces a 1 for both *Unsigned_Overflow* and c_3. XORing these two values together gives a 0 for the *Signed_Overflow* signal. On the other hand, if we treat the two binary numbers, 1100 and 1101, as unsigned numbers, then we are adding $12 + 13 = 25$. Then 25 is outside the unsigned number range, and so the *Unsigned_Overflow* signal should be asserted.

The behavioral VHDL code for the 4-bit adder–subtractor combination circuit is shown in Figure 4.9. The GENERIC keyword declares a read-only constant identifier,

```vhdl
LIBRARY IEEE;
USE IEEE.STD_LOGIC_1164.ALL;
USE IEEE.STD_LOGIC_UNSIGNED.ALL;

ENTITY AddSub IS
GENERIC(n: INTEGER :=4); -- default number of bits = 4
PORT(S: IN STD_LOGIC; -- select subtract signal
  A: IN STD_LOGIC_VECTOR(n-1 DOWNTO 0);
  B: IN STD_LOGIC_VECTOR(n-1 DOWNTO 0);
  F: OUT STD_LOGIC_VECTOR(n-1 DOWNTO 0);
  unsigned_overflow: OUT STD_LOGIC;
  signed_overflow: OUT STD_LOGIC);
END AddSub;

ARCHITECTURE Behavioral OF AddSub IS
  -- temporary result for extracting the unsigned overflow bit
  SIGNAL result: STD_LOGIC_VECTOR(n DOWNTO 0);
  -- temporary result for extracting the c3 bit
  SIGNAL c3: STD_LOGIC_VECTOR(n-1 DOWNTO 0);
BEGIN
  PROCESS(S, A, B)
  BEGIN
    IF (S = '0') THEN -- addition
      -- the two operands are zero extended one extra bit before adding
      -- the & is for string concatination
      result = ('0' & A) + ('0' & B);
      c3 <= ('0' & A(n-2 DOWNTO 0)) + ('0' & B(n-2 DOWNTO 0));
      F <= result(n-1 DOWNTO 0);              -- extract the n-bit result
      unsigned_overflow <= result(n);         -- get the unsigned overflow bit
      signed_overflow <= result(n) XOR c3(n-1); -- get signed overflow bit
    ELSE -- subtraction
      -- the two operands are zero extended one extra bit before subtracting
      -- the & is for string concatination
      result <= ('0' & A) - ('0' & B);
      c3 <= ('0' & A(n-2 DOWNTO 0)) - ('0' & B(n-2 DOWNTO 0));
      F <= result(n-1 DOWNTO 0); -- extract the n-bit result
      unsigned_overflow <= result(n); -- get the unsigned overflow bit
      signed_overflow <= result(n) XOR c3(n-1); -- get signed overflow bit
    END IF;
  END PROCESS;
END Behavioral;
```

Figure 4.9
Behavioral VHDL code for a 4-bit adder–subtractor combination component.

n, of type INTEGER having a default value of 4. This constant identifier then is used in the declaration of the STD_LOGIC_VECTOR size for the three vectors: A, B, and F.

The *Unsigned_Overflow* bit is obtained by performing the addition or subtraction operation using $n + 1$ bits. The two operands are zero extended using the & symbol for concatenation before the operation is performed. The result of the operation is stored in the $n + 1$ bit vector, *result*. The most significant bit of this vector, *result*(n), is the *Unsigned_Overflow* bit.

To get the *Signed_Overflow* bit, we need to XOR the *Unsigned_Overflow* bit with the carry bit, c_3, from the second-to-last bit slice. The c_3 bit is obtained just like how the *Unsigned_Overflow* bit is obtained, except that the operation is performed on only the first $n - 1$ bits of the two operands. The vector *c3* of length n is used for storing the result of the operation. The *Signed_Overflow* signal is the XOR of *result*(n) with *c3*($n - 1$).

4.6 Arithmetic Logic Unit

The **arithmetic logic unit (ALU)** is one of the main components inside a microprocessor. It is responsible for performing arithmetic and logic operations, such as addition, subtraction, logical AND, and logical OR. The ALU, however, is not used to perform multiplications or divisions. It turns out that, in constructing the circuit for the ALU, we can use the same idea as for constructing the adder–subtractor combination circuit, as discussed in the previous section. Again, we will use the ripple-carry adder as the building block and then insert some combinational logic circuitry in front of the two input operands to each full adder. This way, the primary inputs will be modified accordingly, depending on the operations being performed before being passed to the full adder. The general overall circuit for a 4-bit ALU is shown in Figure 4.10(a) and its logical symbol in Figure 4.10(b).

As we can see in Figure 4.10(a), the two combinational circuits in front of the full adder (FA) are labeled LE and AE. The logic extender (LE) is for manipulating all logical operations; whereas, the arithmetic extender (AE) is for manipulating all arithmetic operations. The LE performs the actual logical operations on the two primary operands, a_i and b_i, before passing the result to the first operand, x_i, of the FA. On the other hand, the AE only modifies the second operand, b_i, and passes it to the second operand, y_i, of the FA where the actual arithmetic operation is performed.

We saw from the adder–subtractor circuit that, to perform additions and subtractions, we only need to modify y_i (the second operand to the FA) so that all operations can be done with additions. Thus, the AE only takes the second operand of the primary input, b_i, as its input and modifies the value depending on the operation being performed. Its output is y_i, and it is connected to the second operand input of the FA. As in the adder–subtractor circuit, the addition is performed in the FA. When arithmetic operations are being performed, the LE must pass the first operand unchanged from the primary input a_i to the output x_i for the FA.

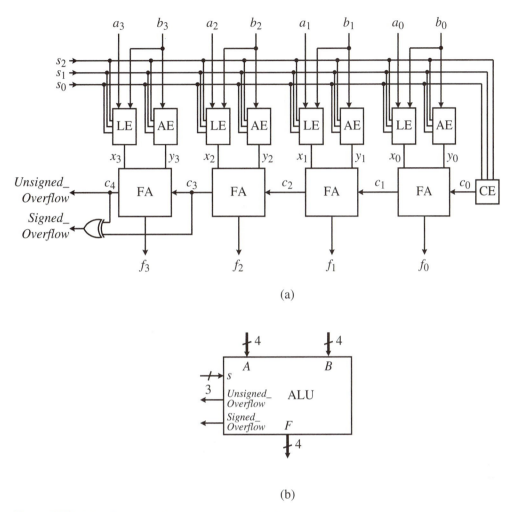

Figure 4.10 4-bit ALU: (a) circuit; (b) logic symbol.

Unlike the AE (where it only modifies the operand), the LE performs the actual logical operations. Thus, for example, if we want to perform the operation A OR B, the LE for each bit slice will take the corresponding bits, a_i and b_i, and OR them together. Hence, one bit from both operands, a_i and b_i, are inputs to the LE. The output of the LE is passed to the first operand, x_i, of the FA. Since this value is already the result of the logical operation, we do not want the FA to modify it but to simply pass it on to the primary output, f_i. This is accomplished by setting both the second operand, y_i, of the FA and c_0 to 0, since adding a 0 will not change the resulting value.

The combinational circuit labeled CE (for carry extender) is for modifying the primary carry-in signal, c_0, so that arithmetic operations are performed correctly. Logical operations do not use the carry signal, so c_0 is set to 0 for all logical operations.

In the circuit shown in Figure 4.10, three select lines, s_2, s_1, and s_0, are used to select the operations of the ALU. With these three select lines, the ALU circuit can implement up to eight different operations. Suppose that the operations that we want to implement in our ALU are as defined in Figure 4.11(a). The x_i column shows the values that the LE must generate for the different operations. The y_i column shows the values that the AE must generate. The c_0 column shows the carry signals that the CE must generate.

For example, for the pass-through operation, the value of a_i is passed through without any modifications to x_i. For the AND operation, x_i gets the result of a_i

s_2	s_1	s_0	Operation Name	Operation	x_i (LE)	y_i (AE)	c_0 (CE)
0	0	0	Pass	Pass A to output	a_i	0	0
0	0	1	AND	A AND B	a_i AND b_i	0	0
0	1	0	OR	A OR B	a_i OR b_i	0	0
0	1	1	NOT	A'	a_i'	0	0
1	0	0	Addition	$A + B$	a_i	b_i	0
1	0	1	Subtraction	$A - B$	a_i	b_i'	1
1	1	0	Increment	$A + 1$	a_i	0	1
1	1	1	Decrement	$A - 1$	a_i	1	0

(a)

s_2	s_1	s_0	x_i
0	0	0	a_i
0	0	1	$a_i b_i$
0	1	0	$a_i + b_i$
0	1	1	a_i'
1	$\times$	$\times$	a_i

(b)

s_2	s_1	s_0	b_i	y_i
0	$\times$	$\times$	$\times$	0
1	0	0	0	0
1	0	0	1	1
1	0	1	0	1
1	0	1	1	0
1	1	0	0	0
1	1	0	1	0
1	1	1	0	1
1	1	1	1	1

(c)

s_2	s_1	s_0	c_0
0	$\times$	$\times$	0
1	0	0	0
1	0	1	1
1	1	0	1
1	1	1	0

(d)

Figure 4.11 ALU operations: (a) function table; (b) LE truth table; (c) AE Truth table; (d) CE truth table.

AND b_i. As mentioned before, both y_i and c_0 are set to 0 for all of the logical operations, because we do not want the FA to change the results. The FA is used only to pass the results from the LE straight through to the output, F. For the subtraction operation, instead of subtracting B, we want to add $-B$. Changing B to $-B$ in two's complement format requires flipping the bits of B and then adding a 1. Thus, y_i gets the inverse of b_i, and the 1 is added through the carry-in, c_0. To increment A, we set y_i to all 0's, and add the 1 through the carry-in, c_0. To decrement A, we add a -1 instead. Negative one in two's complement format is a bit string with all 1's. Hence, we set y_i to all 1's and the carry-in c_0 to 0. For all the arithmetic operations, we need the first operand, A, unchanged for the FA. Thus, x_i gets the value of a_i for all arithmetic operations.

Figure 4.11(b), (c) and (d) shows the truth tables for the LE, AE, and CE, respectively. The LE circuit is derived from the x_i column of Figure 4.11(b); the AE circuit is derived from the y_i column of Figure 4.11(c); and the CE circuit is derived from the c_0 column of Figure 4.11(d). Notice that x_i is dependent on five variables, s_2, s_1, s_0, a_i, and b_i; whereas, y_i is dependent on only four variables, s_2, s_1, s_0, and b_i; and c_0 is dependent on only the three select lines, s_2, s_1, and s_0. The K-maps, equations, and schematics for these three circuits are shown in Figure 4.12.

The behavioral VHDL code for the ALU is shown in Figure 4.13, and a sample simulation trace for all of the operations using the two inputs 5 and 3 is shown in Figure 4.14.

• • • • • • • • • • • • • • • •

4.7 Decoder

A **decoder**, also known as a **demultiplexer**, asserts one out of n output lines, depending on the value of an m-bit binary input data. In general, an m-to-n decoder has m input lines, $A_{m-1}, \ldots, A_0$, and n output lines, $Y_{n-1}, \ldots, Y_0$, where $n = 2^m$. In addition, it has an enable line, E, for enabling the decoder. When the decoder is disabled with E set to 0, all of the output lines are de-asserted. When the decoder is enabled, then the output line whose index is equal to the value of the input binary data is asserted. For example, for a 3-to-8 decoder, if the input address is 101, then the output line Y_5 is asserted (set to 1 for active-high), while the rest of the output lines are de-asserted (set to 0 for active-high).

A decoder is used in a system having multiple components, and we want only one component to be selected or enabled at any one time. For example, in a large memory system with multiple memory chips, only one memory chip is enabled at a time. One output line from the decoder is connected to the enable input on each memory chip. Thus, an address presented to the decoder will enable that corresponding memory chip. The truth table, circuit, and logic symbol for a 3-to-8 decoder are shown in Figure 4.15.

A larger-sized decoder can be implemented using several smaller decoders. For example, Figure 4.16 uses seven 1-to-2 decoders to implement a 3-to-8 decoder. The correct operation of this circuit is left as an exercise for the reader.

The behavioral VHDL code for the 3-to-8 decoder is shown in Figure 4.17.

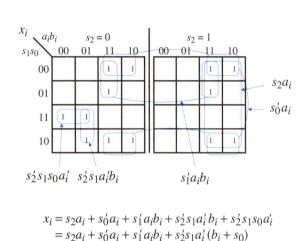

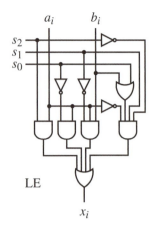

$$x_i = s_2a_i + s_0'a_i + s_1'a_ib_i + s_2's_1a_i'b_i + s_2's_1s_0a_i'$$
$$= s_2a_i + s_0'a_i + s_1'a_ib_i + s_2's_1a_i'(b_i + s_0)$$

(a)

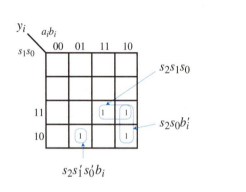

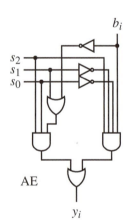

$$y_i = s_2s_1s_0 + s_2s_0b_i' + s_2s_1's_0'b_i$$
$$= s_2s_0(s_1 + b_i') + s_2s_1's_0'b_i$$

(b)

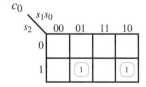

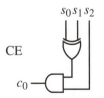

$$c_0 = s_2s_1's_0 + s_2s_1s_0'$$
$$= s_2(s_1 \oplus s_0)$$

(c)

Figure 4.12 K-map, equations, and schematics for: (a) LE; (b) AE; and (c) CE.

```vhdl
LIBRARY IEEE;
USE IEEE.STD_LOGIC_1164.ALL;
-- The following package is needed so that the STD_LOGIC_VECTOR signals
-- A and B can be used in unsigned arithmetic operations.
USE IEEE.STD_LOGIC_UNSIGNED.ALL;

ENTITY alu IS PORT (
  S: IN STD_LOGIC_VECTOR(2 DOWNTO 0); -- select for operations
  A, B: IN STD_LOGIC_VECTOR(3 DOWNTO 0); -- input operands
  F: OUT STD_LOGIC_VECTOR(3 DOWNTO 0)); -- output
END alu;

ARCHITECTURE Behavior OF alu IS
BEGIN
  PROCESS(S, A, B)
    BEGIN
      CASE S IS
      WHEN "000" =>  -- pass A through
        F <= A;
      WHEN "001" =>  -- AND
        F <= A AND B;
      WHEN "010" =>  -- OR
        F <= A OR B;
      WHEN "011" =>  -- NOT A
        F <= NOT A;
      WHEN "100" =>  -- add
        F <= A + B;
      WHEN "101" =>  -- subtract
        F <= A - B;
      WHEN "110" =>  -- increment
        F <= A + 1;
      WHEN OTHERS => -- decrement
        F <= A - 1;
      END CASE;
  END PROCESS;
END Behavior;
```

Figure 4.13   Behavioral VHDL code for an ALU.

Name:	Pass A	AND	OR	NOT A	Add	Subtract	Increment	Decrement
S	0	1	2	3	4	5	6	7
A	5							
B	3							
F	5	1	7	A	8	2	6	4

Figure 4.14 Sample simulation trace with the two input operands, 5 and 3, for all of the eight operations.

E	A_2	A_1	A_0	Y_7	Y_6	Y_5	Y_4	Y_3	Y_2	Y_1	Y_0
0	×	×	×	0	0	0	0	0	0	0	0
1	0	0	0	0	0	0	0	0	0	0	1
1	0	0	1	0	0	0	0	0	0	1	0
1	0	1	0	0	0	0	0	0	1	0	0
1	0	1	1	0	0	0	0	1	0	0	0
1	1	0	0	0	0	0	1	0	0	0	0
1	1	0	1	0	0	1	0	0	0	0	0
1	1	1	0	0	1	0	0	0	0	0	0
1	1	1	1	1	0	0	0	0	0	0	0

(a)

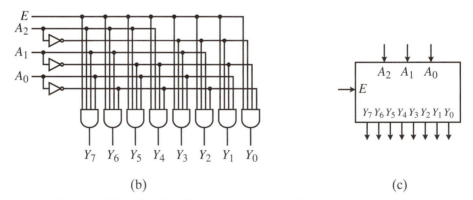

(b)

(c)

Figure 4.15 A 3-to-8 decoder: (a) truth table; (b) circuit; (c) logic symbol.

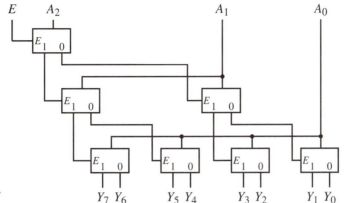

Figure 4.16
A 3-to-8 decoder implemented with seven 1-to-2 decoders.

```
-- A 3-to-8 decoder
LIBRARY IEEE;
USE IEEE.STD_LOGIC_1164.ALL;

ENTITY Decoder IS PORT(
  E: IN STD_LOGIC;                       -- enable
  A: IN STD_LOGIC_VECTOR(2 DOWNTO 0);    -- 3 bit address
  Y: OUT STD_LOGIC_VECTOR(7 DOWNTO 0));  -- data bus output
END Decoder;

ARCHITECTURE Behavioral OF Decoder IS
BEGIN
  PROCESS (E, A)
  BEGIN
    IF (E = '0') THEN                    -- disabled
      Y <= (OTHERS => '0');              -- 8-bit vector of 0
    ELSE
      CASE A IS                          -- enabled
        WHEN "000" => Y <= "00000001";
        WHEN "001" => Y <= "00000010";
        WHEN "010" => Y <= "00000100";
        WHEN "011" => Y <= "00001000";
        WHEN "100" => Y <= "00010000";
        WHEN "101" => Y <= "00100000";
        WHEN "110" => Y <= "01000000";
        WHEN "111" => Y <= "10000000";
        WHEN OTHERS => NULL;
      END CASE;
    END IF;
  END PROCESS;
END Behavioral;
```

Figure 4.17
Behavioral VHDL code for a 3-to-8 decoder.

4.8 Encoder

An **encoder** is almost like the inverse of a decoder where it encodes a 2^n-bit input data into an n-bit code. The encoder has 2^n input lines and n output lines, as shown by the logic symbol in Figure 4.18(d) for $n = 3$. The operation of the encoder is such that exactly one of the input lines should have a 1 while the remaining input lines should have 0's. The output is the binary value of the index of the input line that has the 1. The truth table for an 8-to-3 encoder is shown in Figure 4.18(a). For example, when input I_3 is a 1, the three output bits Y_2, Y_1, and Y_0, are set to 011, which is the binary number for the index 3. Entries having multiple 1's in the truth-table inputs are ignored, since we are assuming that only one input line can be a 1.

I_7	I_6	I_5	I_4	I_3	I_2	I_1	I_0	Y_2	Y_1	Y_0
0	0	0	0	0	0	0	1	0	0	0
0	0	0	0	0	0	1	0	0	0	1
0	0	0	0	0	1	0	0	0	1	0
0	0	0	0	1	0	0	0	0	1	1
0	0	0	1	0	0	0	0	1	0	0
0	0	1	0	0	0	0	0	1	0	1
0	1	0	0	0	0	0	0	1	1	0
1	0	0	0	0	0	0	0	1	1	1

(a)

$$Y_0 = I_1 + I_3 + I_5 + I_7$$
$$Y_1 = I_2 + I_3 + I_6 + I_7$$
$$Y_2 = I_4 + I_5 + I_6 + I_7$$

(b)

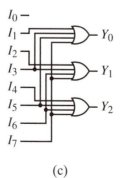

(c)

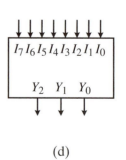

(d)

Figure 4.18   An 8-to-3 encoder: (a) truth table; (b) equations; (c) circuit; (d) logic symbol.

Looking at the three output columns in the truth table, we obtain the three equations shown in Figure 4.18(b) and the resulting circuit in Figure 4.18(c). The logic symbol is shown in Figure 4.18(d).

Encoders are used to reduce the number of bits needed to represent some given data either in data storage or in data transmission. Encoders are also used in a system with 2^n input devices, each of which may need to request for service. One input line is connected to one input device. The input device requesting for service will assert the input line that is connected to it. The corresponding n-bit output value will indicate to the system which of the 2^n devices is requesting for service. For example, if device 5 requests for service, it will assert the I_5 input line. The system will know that device 5 is requesting for service, since the output will be $101 = 5$. However, this only works correctly if it is guaranteed that only one of the 2^n devices will request for service at any one time.

If two or more devices request for service at the same time, then the output will be incorrect. For example, if devices 1 and 4 of the 8-to-3 encoder request for service at the same time, then the output will also be 101, because I_4 will assert the Y_2 signal, and I_1 will assert the Y_0 signal. To resolve this problem, a priority is assigned to each of the input lines so that when multiple requests are made, the encoder outputs the index value of the input line with the highest priority. This modified encoder is known as a **priority encoder**.

*4.8.1 Priority Encoder

The truth table for an active-high 8-to-3 priority encoder is shown in Figure 4.19. The table assumes that input I_7 has the highest priority, and I_0 has the lowest priority. For example, if the highest priority input asserted is I_3, then it doesn't matter whether the lower priority input lines, I_2, I_1 and I_0, are asserted or not; the output will be for that of I_3, which is 011. Since it is possible that no inputs are asserted, there is an extra output, Z, that is needed to differentiate between when no inputs are asserted and when one or more inputs are asserted. Z is set to a 1 when one or more inputs are asserted; otherwise, Z is set to 0. When Z is 0, all of the Y outputs are meaningless.

An easy way to derive the equations for the 8-to-3 priority encoder is to define a set of eight intermediate variables, $v_0, \ldots, v_7$, such that v_k is a 1 if I_k is the highest priority 1 input. Thus, the equations for v_0 to v_7 are:

$$v_0 = I_7' I_6' I_5' I_4' I_3' I_2' I_1' I_0$$

$$v_1 = I_7' I_6' I_5' I_4' I_3' I_2' I_1$$

$$v_2 = I_7' I_6' I_5' I_4' I_3' I_2$$

$$v_3 = I_7' I_6' I_5' I_4' I_3$$

$$v_4 = I_7' I_6' I_5' I_4$$

$$v_5 = I_7' I_6' I_5$$

$$v_6 = I_7' I_6$$

$$v_7 = I_7$$

I_7	I_6	I_5	I_4	I_3	I_2	I_1	I_0	Y_2	Y_1	Y_0	Z
0	0	0	0	0	0	0	0	$\times$	$\times$	$\times$	0
0	0	0	0	0	0	0	1	0	0	0	1
0	0	0	0	0	0	1	$\times$	0	0	1	1
0	0	0	0	0	1	$\times$	$\times$	0	1	0	1
0	0	0	0	1	$\times$	$\times$	$\times$	0	1	1	1
0	0	0	1	$\times$	$\times$	$\times$	$\times$	1	0	0	1
0	0	1	$\times$	$\times$	$\times$	$\times$	$\times$	1	0	1	1
0	1	$\times$	$\times$	$\times$	$\times$	$\times$	$\times$	1	1	0	1
1	$\times$	$\times$	$\times$	$\times$	$\times$	$\times$	$\times$	1	1	1	1

Figure 4.19 An 8-to-3 priority encoder truth table.

Using these eight intermediate variables, the final equations for the priority encoder are similar to the ones for the regular encoder, namely:

$$Y_0 = v_1 + v_3 + v_5 + v_7$$

$$Y_1 = v_2 + v_3 + v_6 + v_7$$

$$Y_2 = v_4 + v_5 + v_6 + v_7$$

Finally, the equation for Z is simply

$$Z = I_7 + I_6 + I_5 + I_4 + I_3 + I_2 + I_1 + I_0$$

4.9 Multiplexer

The **multiplexer**, or **MUX** for short, allows the selection of one input signal among n signals, where $n > 1$ and is a power of two. Select lines connected to the multiplexer determine which input signal is selected and passed to the output of the multiplexer. In general, an n-to-1 multiplexer has n data input lines, m select lines where $m = log_2 n$ (i.e., $2^m = n$), and one output line. For a 2-to-1 multiplexer, there is one select line, s, to select between the two inputs, d_0 and d_1. When $s = 0$, the input line, d_0, is selected, and the data present on d_0 is passed to the output, y. When $s = 1$, the

s	d_1	d_0	y
0	0	0	0
0	0	1	1
0	1	0	0
0	1	1	1
1	0	0	0
1	0	1	0
1	1	0	1
1	1	1	1

(a)

$$y = s'd_1'd_0 + s'd_1d_0 + sd_1d_0' + sd_1d_0$$
$$= s'd_0(d_1' + d_1) + sd_1(d_0' + d_0)$$
$$= s'd_0 + sd_1$$

(b)

Figure 4.20
A 2-to-1 multiplexer:
(a) truth table;
(b) equation;
(c) circuit; (d) logic
symbol.

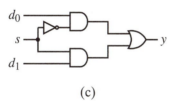

(c)

(d)

input line, d_1, is selected and the data on d_1 is passed to y. The truth table, equation, circuit, and logic symbol for a 2-to-1 multiplexer are shown in Figure 4.20.

Constructing a larger-sized multiplexer, such as the 8-to-1 multiplexer, can be done similarly. In addition to having the eight data input lines, d_0 to d_7, the 8-to-1 multiplexer has three ($2^3 = 8$) select lines, s_0, s_1, and s_2. Depending on the value of the three select lines, one of the eight input lines will be selected and the data on that input line will be passed to the output. For example, if the value of the select lines is 101, then the input line d_5 is selected, and the data that is present on d_5 will be passed to the output.

The truth table, circuit, and logic symbol for the 8-to-1 multiplexer are shown in Figure 4.21. The truth table is written in a slightly different format. Instead of including the d's in the input columns and enumerating all $2^{11} = 2048$ rows (the eleven variables come from the eight d's and the three s's), the d's are written in the entry under the output column. For example, when the select line value is 101, the entry under the output column is d_5, which means that y takes on the value of the input line d_5.

To understand the circuit in Figure 4.21(b), notice that each AND gate acts as a switch and is turned on by one combination of the three select lines. When a particular AND gate is turned on, the data at the corresponding d input is passed through that AND gate. The outputs of the remaining AND gates are all 0's.

s_2	s_1	s_0	y
0	0	0	d_0
0	0	1	d_1
0	1	0	d_2
0	1	1	d_3
1	0	0	d_4
1	0	1	d_5
1	1	0	d_6
1	1	1	d_7

(a)

(b)

(c)

Figure 4.21   An 8-to-1 multiplexer: (a) truth table; (b) circuit; (c) logic symbol.

Instead of using 4-input AND gates (where three of its inputs are used by the three select lines to turn it on) we can use 2-input AND gates, as shown in Figure 4.22(a). This way the AND gate is turned on with just one line. The eight 2-input AND gates can be turned on individually from the eight outputs of a 3-to-8 decoder. Recall from Section 4.7 that the decoder asserts only one output line at any time.

Larger multiplexers can also be constructed from smaller multiplexers. For example, an 8-to-1 multiplexer can be constructed using seven 2-to-1 multiplexers, as shown in Figure 4.22(b). The four top-level 2-to-1 multiplexers provide the eight data inputs and all are switched by the same least significant select line, s_0. This top level selects one from each group of two data inputs. The middle level then groups the four outputs from the top level again into groups of two, and selects one from

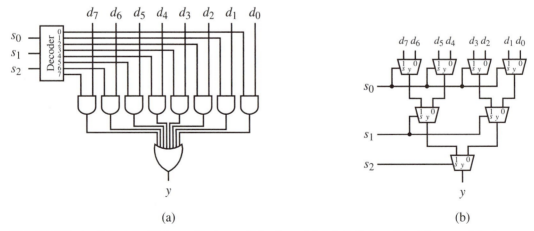

(a)

(b)

Figure 4.22   An 8-to-1 multiplexer implemented using: (a) a 3-to-8 decoder; (b) seven 2-to-1 multiplexers.

each group using the select line s_1. Finally, the multiplexer at the bottom level uses the most significant select line s_2 to select one of the two outputs from the middle level multiplexers.

The VHDL code for an 8-bit wide 4-to-1 multiplexer is shown in Figure 4.23. Two different implementations of the same multiplexer are shown. Figure 4.23(a)

```
-- A 4-to-1 8-bit wide multiplexer
LIBRARY IEEE;
USE IEEE.STD_LOGIC_1164.ALL;

ENTITY Multiplexer IS PORT(
    S: IN STD_LOGIC_VECTOR(1 DOWNTO 0);                 -- select lines
    D0, D1, D2, D3: IN STD_LOGIC_VECTOR(7 DOWNTO 0);    -- data bus input
    Y: OUT STD_LOGIC_VECTOR(7 DOWNTO 0));               -- data bus output
END Multiplexer;

-- Behavioral level code
ARCHITECTURE Behavioral OF Multiplexer IS
BEGIN
  PROCESS (S,D0,D1,D2,D3)
  BEGIN
    CASE S IS
      WHEN "00" => Y <= D0;
      WHEN "01" => Y <= D1;
      WHEN "10" => Y <= D2;
      WHEN "11" => Y <= D3;
      WHEN OTHERS => Y <= (OTHERS => 'U');              -- 8-bit vector of U
    END CASE;
  END PROCESS;
END Behavioral;
```
                                        (a)

```
-- Dataflow level code
ARCHITECTURE Dataflow OF Multiplexer IS
BEGIN
  WITH S SELECT Y <=
    D0 WHEN "00",
    D1 WHEN "01",
    D2 WHEN "10",
    D3 WHEN "11",
    (OTHERS => 'U') WHEN OTHERS;                        -- 8-bit vector of U
END Dataflow;
```
                                        (b)

Figure 4.23
VHDL code for an 8-bit wide 4-1 multiplexer:
(a) behavioral level;
(b) dataflow level.

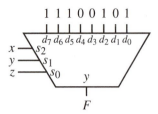

Figure 4.24
Using an 8-to-1
multiplexer to
implement the
function $F(x, y, z) = x'y'z' + x'yz' + xy'z + xyz' + xyz$.

shows the architecture code written at the behavioral level, since it uses a PROCESS statement. Inside the PROCESS block, a CASE statement is used to select between the four choices for s. Figure 4.23(b) shows a dataflow level architecture code using a concurrent selected signal assignment statement using the keyword WITH . . . SELECT. In the first choice, if s is equal to 00, then the value $D0$ is assigned to Y. If s does not match any one of the four choices, 00, 01, 10, and 11, then the WHEN OTHERS clause is selected. The syntax (others => 'U') means to fill the entire vector with the value "U".

*4.9.1 Using Multiplexers to Implement a Function

Multiplexers can be used to implement a Boolean function very easily. In general, for an n-variable function, a 2^n-to-1 multiplexer (that is, a multiplexer with n select lines) is needed. An n-variable function has 2^n minterms, and each minterm corresponds to one of the 2^n multiplexer inputs. The n input variables are connected to the n select lines of the multiplexer. Depending on the values of the n variables, one data input line will be selected, and the value on that input line is passed to the output. Therefore, all we need to do is to connect all of the data input lines to either a 1 or a 0, depending on whether we want that corresponding minterm to be a 1-minterm or a 0-minterm, respectively.

Figure 4.24 shows the implementation of the 3-variable function, $F(x, y, z) = x'y'z' + x'yz' + xy'z + xyz' + xyz$. The 1-minterms for this function are m_0, m_2, m_5, m_6, and m_7, so the corresponding data input lines d_0, d_2, d_5, d_6, and d_7 are connected to a 1, while the remaining data input lines are connected to a 0. For example, the 0-minterm $x'yz$ has the value 011, which will select the d_3 input, so a 0 passes to the output. On the other hand, the 1-minterm $xy'z$ has the value 101, which will select the d_5 input, so a 1 passes to the output.

4.10 Tri-State Buffer

A **tri-state** buffer, as the name suggests, has three states: 0, 1, and a third state denoted by Z. The value Z represents a high-impedance state, which for all practical purposes acts like a switch that is opened or a wire that is cut. Tri-state buffers are used to connect several devices to the same bus. A *bus* is one or more wires for transferring signals. If two or more devices are connected directly to a bus without using tri-state buffers, signals will get corrupted on the bus because the devices are always outputting either a 0 or a 1. However, with a tri-state buffer in between, devices that are not using the bus can disable the tri-state buffer so that it acts as if those devices are physically disconnected from the bus. At any one time, only one active device will have its tri-state buffers enabled, and thus, use the bus.

The truth table and symbol for the tri-state buffer is shown in Figure 4.25(a) and (b). The active-high enable line E turns the buffer on or off. When E is de-asserted with a 0, the tri-state buffer is disabled, and the output y is in its high-impedance Z state. When E is asserted with a 1, the buffer is enabled, and the output y follows the input d.

A circuit consisting of only logic gates cannot produce the high-impedance state required by the tri-state buffer, since logic gates can only output a 0 or a 1. To provide the high-impedance state, the tri-state buffer circuit uses two transistors in conjunction with logic gates, as shown in Figure 4.25(c). Section 5.3 will discuss the operations of these two transistors in detail. For now, we will keep it simple. The top PMOS transistor is enabled with a 0 at the node labeled A, and when it is enabled, a 1 signal from Vcc passes down through the transistor to y. The bottom NMOS transistor is enabled with a 1 at the node labeled B, and when it is enabled, a 0 signal from ground passes up through the transistor to y. When the two transistors are disabled (with $A = 1$ and $B = 0$) they will both output a high impedance Z value; so y will have a Z value.

Having the two transistors, we need a circuit that will control these two transistors so that together they realize the tri-state buffer function. The truth table for this control circuit is shown in Figure 4.25(d). The truth table is derived as follows. When $E = 0$ (it does not matter what the input d is) we want both transistors to be disabled so that the output y has the Z value. The PMOS transistor is disabled when the input $A = 1$; whereas, the NMOS transistor is disabled when the input $B = 0$. When $E = 1$ and $d = 0$, we want the output y to be a 0. To get a 0 on y, we need to enable the bottom NMOS transistor and disable the top PMOS transistor so that a 0 will pass through the NMOS transistor to y. To get a 1 on y for when $E = 1$ and $d = 1$, we need to do the reverse by enabling the top PMOS transistor and disabling the bottom NMOS transistor.

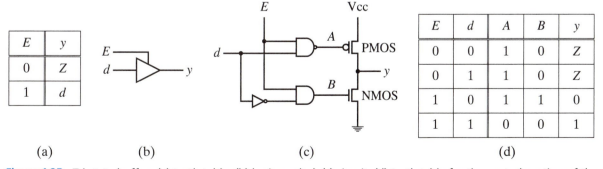

E	d	A	B	y
0	0	1	0	Z
0	1	1	0	Z
1	0	1	1	0
1	1	0	0	1

E	y
0	Z
1	d

(a)   (b)   (c)   (d)

Figure 4.25   Tri-state buffer: (a) truth table; (b) logic symbol; (c) circuit; (d) truth table for the control portion of the tri-state buffer circuit.

```
LIBRARY IEEE;
USE IEEE.STD_LOGIC_1164.ALL;

ENTITY TriState_Buffer IS PORT (
  E: IN STD_LOGIC;
  d: IN STD_LOGIC_VECTOR(7 DOWNTO 0);
  y: OUT STD_LOGIC_VECTOR(7 DOWNTO 0));
END TriState_Buffer;

ARCHITECTURE Behavioral OF TriState_Buffer IS
BEGIN
  PROCESS (E, d)
  BEGIN
    IF (E = '1') THEN
      y <= d;
    ELSE
      y <= (OTHERS => 'Z');      -- to get 8 Z values
    END IF;
  END PROCESS;
END Behavioral;
```

Figure 4.26
VHDL code for an
8-bit wide tri-state
buffer.

The resulting circuit is shown in Figure 4.25(c). When $E = 0$, the output of the NAND gate is a 1 regardless of what the other input is, and so the top PMOS transistor is turned off. Similarly, the output of the AND gate is a 0, and so the bottom NMOS transistor is also turned off. Thus, when $E = 0$, both transistors are off, so the output y is in the Z state.

When $E = 1$, the outputs of both the NAND and AND gates are equal to d'. So if $d = 0$, the output of the two gates are both 1, so the bottom transistor is turned on while the top transistor is turned off. Thus, y will have the value 0, which is equal to d. On the other hand, if $d = 1$, the top transistor is turned on while the bottom transistor is turned off, and y will have the value 1.

The behavioral VHDL code for an 8-bit wide tri-state buffer is shown in Figure 4.26.

• • • • • • • • • • • • • ••

4.11 Comparator

Quite often, we need to compare two values for their arithmetic relationship (equal, greater, less than, etc.). A **comparator** is a circuit that compares two binary values and indicates whether the relationship is true or not. To compare whether a value is

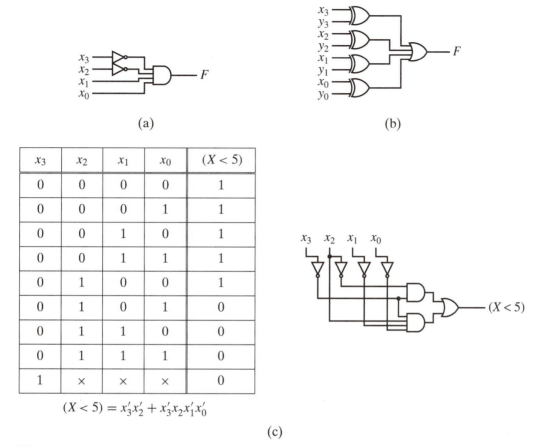

(a)

(b)

$$(X < 5) = x_3'x_2' + x_3'x_2x_1'x_0'$$

(c)

Figure 4.27   Simple 4-bit comparators for: (a) $X = 3$; (b) $X \neq Y$; (c) $X < 5$.

equal or not equal to a constant value, a simple AND gate can be used. For example, to compare a 4-bit variable x with the constant 3, the circuit in Figure 4.27(a) can be used. The AND gate outputs a 1 when the input is equal to the value 3. Since 3 is 0011 in binary, therefore, x_3 and x_2 must be inverted.

The XOR and XNOR gates can be used for comparing inequality and equality, respectively, between two values. The XOR gate outputs a 1 when its two input values are different. Hence, we can use one XOR gate for comparing each bit pair of the two operands. A 4-bit inequality comparator is shown in Figure 4.27(b). Four XOR gates are used, with each one comparing the same bit from the two operands. The outputs of the XOR gates are ORed together so that if any bit pair is different then the two operands are different, and the resulting output is a 1. Similarly, an equality comparator can be constructed using XNOR gates instead, since the XNOR gate outputs a 1 when its two input values are the same.

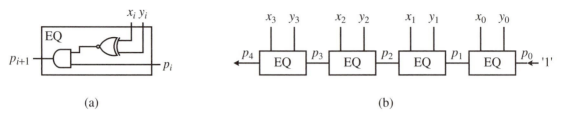

(a) (b)

Figure 4.28 Iterative comparators: (a) 1-bit scale for $x_i = y_i$; (b) 4-bit $X = Y$.

To compare the greater-than or less-than relationships, we can construct a truth table and build the circuit from it. For example, to compare whether a 4-bit value X is less than five, we get the truth table, equation, and circuit shown in Figure 4.27(c).

Instead of constructing a comparator for a fixed number of bits for the input values, we often prefer to construct an **iterative circuit** by constructing a 1-bit slice comparator and then daisy-chaining n of them together to make an n-bit comparator. The 1-bit slice comparator will have (in addition to the two input operand bits, x_i and y_i) a p_i bit that keeps track of whether all the previous bit pairs compared so far are true or not for that particular relationship. The circuit outputs a 1 if $p_i = 1$, and the relationship is true for the current bit pair, x_i and y_i. Figure 4.28(a) shows a 1-bit slice comparator for the equal relationship. If the current bit pair, x_i and y_i, is equal, the XNOR gate will output a 1. Hence, $p_{i+1} = 1$ if the current bit pair is equal and the previous bit pair, p_i, is a 1. To obtain a 4-bit iterative equality comparator, we connect four 1-bit equality comparators in series, as shown in Figure 4.28(b). The initial p_0 bit must be set to a 1. Thus, if all four bit pairs are equal, then the last bit, p_4, will be a 1; otherwise, p_4 will be a 0.

Building an iterative comparator circuit for the greater-than relationship is slightly more difficult. The 1-bit slice comparator circuit for the condition $x_i > y_i$ is constructed as follows. In addition to the two operand input bits, x_i and y_i, there are also two status input bits, g_{in} and e_{in}. Here, g_{in} is a 1 if the condition $x_i > y_i$ is true for the previous bit slice; otherwise, g_{in} is a 0. Furthermore, e_{in} is a 1 if the condition $x_i = y_i$ is true; otherwise, e_{in} is a 0. The circuit also has two status output bits, g_{out} and e_{out}, having the same meaning as the g_{in} and e_{in} signals. These two input and two output status bits allow the bit slices to be daisy-chained together. Following the above description of the 1-bit slice, we obtain the truth table shown in Figure 4.29(a). The equations for e_{out} and g_{out} are shown in Figure 4.29(b), and the 1-bit slice circuit in Figure 4.29(c).

In order for the bit slices to operate correctly, we need to perform the comparisons from the most significant bit to the least significant bit. The complete 4-bit iterative comparator circuit for the condition $x > y$ is shown in Figure 4.29(d). The initial values for g_{in} and e_{in} must be set to $g_{\text{in}} = 0$ and $e_{\text{in}} = 1$.

If $x = y$, then the last e_{out} is a 1; otherwise, e_{out} is a 0. If the last e_{out} is a 0, then the last g_{out} can be either a 1 or a 0. If $x > y$ then g_{out} is a 1; otherwise, g_{out} is a 0. Notice that both e_{out} and g_{out} cannot be both 1's. The operation of this comparator circuit is summarized in Figure 4.29(e).

g_{in}	e_{in}	x_i	y_i	Meaning	g_{out}	e_{out}
0	0	×	×	<	0	0
0	1	0	0	=	0	1
0	1	0	1	<	0	0
0	1	1	0	>	1	0
0	1	1	1	=	0	1
1	0	×	×	>	1	0
1	1	×	×	Invalid	1	1

(a)

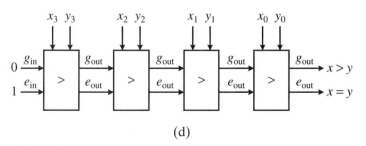

$$g_{out} = g_{in} + e_{in}x_iy_i'$$

(b)

$$e_{out} = g_{in}e_{in} + e_{in}x_i'y_i' + e_{in}x_iy_i$$

(c)

Condition	e_{out}	g_{out}
Invalid	1	1
$x = y$	1	0
$x > y$	0	1
$x < y$	0	0

(d)

(e)

Figure 4.29 Comparator for $x > y$: (a) truth table for a 1-bit slice; (b) K-maps and equations for g_{out} and e_{out}; (c) circuit for 1-bit slice; (d) 4-bit $x > y$ comparator circuit; (e) operational table.

Operation	Comment	Example
Shift left with 0	Shift bits to the left one position. The leftmost bit is discarded and the rightmost bit is filled with a 0.	10110100 X01101000←
Shift left with 1	Same as above, except that the rightmost bit is filled with a 1.	10110100 X01101001←
Shift right with 0	Shift bits to the right one position. The rightmost bit is discarded and the leftmost bit is filled with a 0.	10110100 →01011010X
Shift right with 1	Same as above, except that the leftmost bit is filled with a 1.	10110100 →11011010X
Rotate left	Shift bits to the left one position. The leftmost bit is moved to the rightmost bit position.	10110100 01101001
Rotate right	Shift bits to the right one position. The rightmost bit is moved to the leftmost bit position.	10110100 01011010

Figure 4.30
Shifter and rotator operations.

4.12 Shifter

The **shifter** is used for shifting bits in a binary word one position either to the left or to the right. The operations for the shifter are referred to either as **shifting** or **rotating**, depending on how the end bits are shifted in or out. For a shift operation, the two end bits do not wrap around; whereas for a rotate operation, the two end bits wrap around. Figure 4.30 shows six different shift and rotate operations.

For example, for the "Shift left with 0" operation, all of the bits are shifted one position to the left. The original leftmost bit is shifted out (i.e., discarded) and the rightmost bit is filled with a 0. For the "Rotate left" operation, all of the bits are shifted one position to the left. However, instead of discarding the leftmost bit, it is shifted in as the rightmost bit (i.e., it rotates around).

For each bit position, a multiplexer is used to move a bit from either the left or right to the current bit position. The size of the multiplexer will determine the number of operations that can be implemented. For example, we can use a 4-to-1 multiplexer to implement the four operations, as specified by the table in Figure 4.31(a). Two select lines, s_1 and s_0, are needed to select between the four different operations. For a 4-bit operand, we will need to use four 4-to-1 multiplexers, as shown in Figure 4.31(b). How the inputs to the multiplexers are connected will depend on the given operations.

s_1	s_0	Operation
0	0	Pass through
0	1	Shift left and fill with 0
1	0	Shift right and fill with 0
1	1	Rotate right

(a)

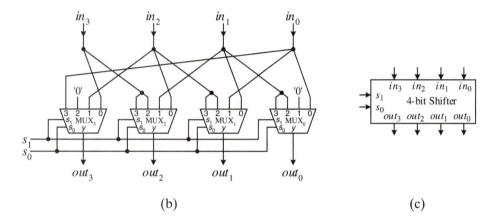

Figure 4.31
A 4-bit shifter:
(a) operation table;
(b) circuit; (c) logic
symbol.

(b)   (c)

In this example, when $s_1 = s_0 = 0$, we want to pass the bit straight through without shifting (i.e., we want the value from in_i to pass to out_i). Given $s_1 = s_0 = 0$, d_0 of the multiplexer is selected, hence, in_i is connected to d_0 of MUX_i, which outputs to out_i. For $s_1 = 0$ and $s_0 = 1$, we want to shift left (i.e., we want the value from in_i to pass to out_{i+1}). With $s_1 = 0$ and $s_0 = 1$, d_1 of the multiplexer is selected, hence, in_i is connected to d_1 of MUX_{i+1}, which outputs to out_{i+1}. For this selection, we also want to shift in a 0 bit, so d_1 of MUX_0 is connected directly to a 0.

The behavioral VHDL code for an 8-bit shifter having the functions as defined in Figure 4.31(a) is shown in Figure 4.32.

*4.12.1   Barrel Shifter

A **barrel shifter** is a shifter that can shift or rotate the data by any number of bits in a single operation. The select lines for a barrel shifter are used, not to determine what kind of operations (shift or rotate) to perform as for the general shifter, but rather, to determine how many bits to move. Hence, only one particular operation can be implemented in a barrel shifter circuit. In general, an n-bit barrel shifter can shift the data bits by as much as $n - 1$ bit distance away in one operation.

```
LIBRARY IEEE;
USE IEEE.STD_LOGIC_1164.ALL;
USE IEEE.STD_LOGIC_UNSIGNED.ALL;

ENTITY shifter IS PORT (
    S: IN STD_LOGIC_VECTOR(1 DOWNTO 0);            -- select for operations
    input: IN STD_LOGIC_VECTOR(7 DOWNTO 0);        -- input
    output: OUT STD_LOGIC_VECTOR(7 DOWNTO 0));     -- output
END shifter;

ARCHITECTURE Behavior OF shifter IS
BEGIN
  PROCESS(S, input)
  BEGIN
    CASE S IS
    WHEN "00" =>                                   -- pass through
      output <= input;
    WHEN "01" =>                                   -- shift left with 0
      output <= input(6 DOWNTO 0) & '0';
    WHEN "10" =>                                   -- shift right with 0
      output <= '0' & input(7 DOWNTO 1);
    WHEN OTHERS =>                                 -- rotate right
      output <= input(0) & input(7 DOWNTO 1);
    END CASE;
  END PROCESS;
END Behavior;
```

Figure 4.32
Behavioral VHDL
code for an 8-bit
shifter having the
operations as
defined in
Figure 4.31(a).

Figure 4.33(a) shows the operation table of a 4-bit barrel shifter implementing the rotate left operation. When $s_1 s_0 = 00$, no rotation is performed (i.e., a pass through). When $s_1 s_0 = 01$, the data bits are rotated one position to the left. When $s_1 s_0 = 10$, the data bits are rotated two positions to the left. The corresponding circuit is shown in Figure 4.33(b).

● ● ● ● ● ● ● ● ● ● ● ● ● ● ●

*4.13 Multiplier

In grade school, we were taught to multiply two numbers using a shift-and-add procedure. Regardless of whether the two numbers are in decimal or binary, we use the same shift-and-add procedure for multiplying them. In fact, multiplying with binary numbers is even easier, because you are always multiplying with either a 0 or a 1.

Select $s_1\ s_0$	Operation	Output $out_3\ out_2\ out_1\ out_0$
00	No rotation	$in_3\ in_2\ in_1\ in_0$
01	Rotate left by 1 bit position	$in_2\ in_1\ in_0\ in_3$
10	Rotate left by 2 bit positions	$in_1\ in_0\ in_3\ in_2$
11	Rotate left by 3 bit positions	$in_0\ in_3\ in_2\ in_1$

(a)

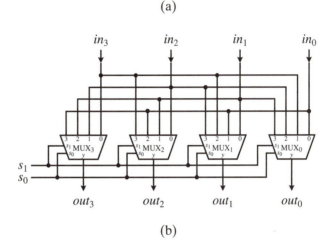

(b)

Figure 4.33
A 4-bit barrel shifter for the rotate left operation:
(a) operation table;
(b) circuit.

Figure 4.34(a) shows the multiplication of two 4-bit unsigned binary numbers—the multiplicand M ($m_3m_2m_1m_0$) with the multiplier Q ($q_3q_2q_1q_0$)—to produce the resulting product P ($p_7p_6p_5p_4p_3p_2p_1p_0$). Notice that the intermediate products are always either the same as the multiplicand (if the multiplier bit is a 1) or it is zero (if the multiplier bit is a 0).

We can derive a combinational multiplication circuit based on this shift-and-add procedure, as shown in Figure 4.34(b). Each intermediate product is obtained by ANDing the multiplicand M with one bit of the multiplier q_i. Since q_i is always a 1 or a 0, the output of the AND gates is always either m_i or 0. For example, bit zero of the first intermediate product is obtained by ANDing m_0 with q_0; bit one is obtained by ANDing m_1 with q_0; and so on. Hence, the four bits for the first intermediate product are m_3q_0, m_2q_0, m_1q_0, and m_0q_0; the four bits for the second intermediate product are m_3q_1, m_2q_1, m_1q_1, and m_0q_1; and so on.

Multiple adders are used to sum all of the intermediate products together to give the final product. Each intermediate product is shifted over to the correct bit position for the addition. For example, p_0 is just m_0q_0; p_1 is the sum of m_1q_0 and m_0q_1; p_2 is the sum of m_2q_0, m_1q_1, and m_0q_2; and so on. The four full adders (1-bit adders) in each row are connected, as in the ripple-carry adder with each carry-out signal

Multiplicand (M)	1 1 0 1		
Multiplier (Q)	× 1 0 1 1		
	1 1 0 1		
Intermediate products	1 1 0 1		
	0 0 0 0		
	+ 1 1 0 1		
Product (P)	1 0 0 0 1 1 1 1		

m_3	m_2	m_1	m_0
× q_3	q_2	q_1	q_0

$$
\begin{array}{cccccccc}
 & & & & m_3q_0 & m_2q_0 & m_1q_0 & m_0q_0 \\
 & & & m_3q_1 & m_2q_1 & m_1q_1 & m_0q_1 & \\
 & & m_3q_2 & m_2q_2 & m_1q_2 & m_0q_2 & & \\
+ & m_3q_3 & m_2q_3 & m_1q_3 & m_0q_3 & & & \\
\hline
P_7 & P_6 & P_5 & P_4 & P_3 & P_2 & P_1 & P_0
\end{array}
$$

(a)

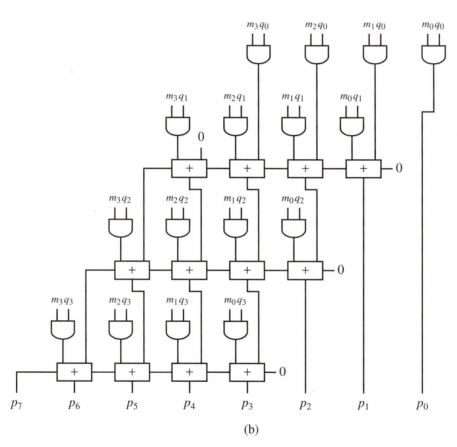

(b)

Figure 4.34 Multiplication: (a) method; (b) circuit.

connected to the carry-in of the next full adder. The carry-out of the last full adder is connected to the input of the last full adder in the row below. The last carry-out from the last row of adders is the value for p_7 of the final product. As in the ripple-carry adder, all of the initial carry-ins, c_0, are set to a 0.

4.14 Summary Checklist

- Full adder
- Ripple-carry adder
- Carry-lookahead adder
- Two's complement
- Sign extension
- Subtractor
- Arithmetic logic unit (ALU)
- Arithmetic extender (AE)
- Logic extender (LE)
- Carry extender (CE)
- Decoder
- Encoder
- Priority encoder
- Multiplexer (MUX)
- Tri-state buffer
- Z value
- Comparator
- Shifter
- Barrel Shifter
- Multiplier

4.15 Problems

P4.1. Convert the following numbers to 12-bit binary numbers using two's complement representation.

(a) 234_{10}

(b) -234_{10}

(c) 234_8

(d) $BC4_{16}$

(e) -472_{10}

P4.2. Convert the following two's complement binary numbers to decimal, octal, and hexadecimal formats.

(a) 1001011

(b) 011110

(c) 101101

(d) 1101011001

(e) 0110101100

P4.3. Write the complete structural VHDL code for the full-adder circuit shown in Figure 4.1(c).

P4.4. Draw the smallest possible complete circuit for a 2-bit carry-lookahead adder.

P4.5. Draw the complete circuit for a 4-bit carry-lookahead adder.

P4.6. Derive the carry-lookahead equation and circuit for c_5.

P4.7. Show that when adding two n-bit signed numbers, $A_{n-1} \ldots A_0$ and $B_{n-1} \ldots B_0$, producing the result $S_{n-1} \ldots S_0$, the *Signed_Overflow* flag can be deduced by the equation:

$$Signed\_Overflow = A_{n-1} \text{ XOR } B_{n-1} \text{ XOR } S_{n-1} \text{ XOR } S_n$$

P4.8. Draw the complete 4-bit ALU circuit having the following operations. Use K-maps to reduce all of the equations to standard form.

s_2	s_1	s_0	Operations
0	0	0	$B - 1$
0	0	1	A NOR B
0	1	0	$A - B$
0	1	1	A XNOR B
1	0	0	1
1	0	1	A NAND B
1	1	0	$A + B$
1	1	1	A'

P4.9. Draw the complete 4-bit ALU circuit having the following operations. Don't-care values are assigned to unused select combinations. Use K-maps to reduce all of the equations to standard form.

s_2	s_1	s_0	Operations
0	0	0	Pass A through the LE
0	0	1	Pass B through the LE
0	1	0	NOT A
0	1	1	NOT B
1	0	0	$A - B$
1	0	1	$B - A$
1	1	0	$B + 1$

P4.10. Draw the complete 4-bit ALU circuit having the following operations. Use K-maps to reduce all of the equations to standard form.

s_2	s_1	s_0	Operations
0	0	0	A plus B
0	0	1	Increment A
0	1	0	Increment B
0	1	1	Pass A
1	0	0	$A - B$
1	0	1	A XOR B
1	1	0	A AND B

P4.11. Draw the complete 4-bit ALU circuit having the following operations. Use K-maps to reduce all of the equations to standard form.

s_2	s_1	s_0	Operations
0	0	0	Pass A
0	0	1	Pass B through the AE
0	1	0	A plus B
0	1	1	A'
1	0	0	A XOR B
1	0	1	A NAND B
1	1	0	$A - 1$
1	1	1	$A - B$

P4.12. Given the following K-maps for the LE, AE, and CE of an ALU, determine the ALU operations assigned to each of the select line combinations.

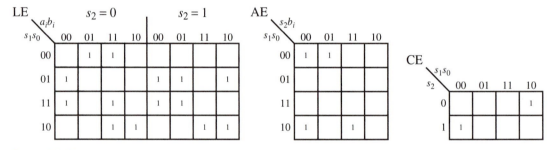

Figure P4.12

P4.13. A four-function ALU has the following equations for its LE, AE, and CE:

$$x_i = a_i + s_1's_0b_i$$
$$y_i = s_1's_0' + s_1s_0b_i'$$
$$c_0 = s_1s_0$$

Determine the four functions in the correct order that are implemented in this ALU. Show all of your work.

P4.14. Draw the circuit for the 2-to-4 decoder.

P4.15. Derive the truth table for a 3-to-8 decoder using negative logic.

P4.16. Draw the circuit for the 4-to-16 decoder using only 2-to-4 decoders.

P4.17. Draw the circuit for the 4-to-2 priority encoder using only 2-input AND, 2-input OR, and NOT gates.

P4.18. Draw the circuit for an 8-to-3 priority encoder.

P4.19. Draw the circuit for the 4-to-2 priority encoder using only 2-to-1 priority encoders and 2-to-1 multiplexers.

P4.20. Write the behavioral VHDL code for the 8-to-3 priority encoder.

P4.21. Draw the circuit for a 16-to-1 multiplexer using only 4-to-1 multiplexers.

P4.22. Draw the circuit for a 16-to-1 multiplexer using only 2-to-1 multiplexers.

P4.23. Use only 2-to-1 multiplexers to implement the function:

$$f(w, x, y, z) = \Sigma(0, 2, 5, 7, 13, 15)$$

P4.24. Use only 2-to-1 multiplexers (as many as you need) to implement the function:

$$F(x, y, z) = \Pi(0, 3, 4, 5, 7)$$

P4.25. Use one 8-to-1 multiplexer to implement the function:

$$F(x, y, z) = \Sigma(0, 3, 4, 6, 7)$$

P4.26. Use 2-to-1 multiplexers to implement the function:

$$F(x, y, z) = \Sigma(0, 2, 4, 5)$$

P4.27. Derive the truth table for comparing two unsigned 2-bit operands for the less-than-or-equal-to relationship. Derive the equation and circuit from this truth table.

P4.28. Construct the circuit for one bit slice of an n-bit magnitude comparator that compares $x_i \geq y_i$.

P4.29. Draw the circuit for a 4-bit iterative comparator that tests for the greater-than-or-equal-to relationship.

P4.30. Draw the circuit for a 4-bit shifter that realizes the following operation table.

s_2	s_1	s_0	Operation
0	0	0	Pass through
0	0	1	Rotate left
0	1	0	Shift right and fill with 1
0	1	1	Not used
1	0	0	Shift left and fill with 0
1	0	1	Pass through
1	1	0	Rotate right
1	1	1	Shift right and fill with 0

P4.31. Draw a 4-bit shifter circuit for the following operational table. Use only the basic gates AND, OR, and NOT. (*Note*: Do not use multiplexers.)

s_1	s_0	Operation
0	0	Shift left fill with 0
0	1	Shift right fill with 0
1	0	Rotate left
1	1	Rotate right

P4.32. Draw a 4-bit shifter circuit for the following operation table using only six 2-to-1 multiplexers.

Operation
Shift left fill with 0
Shift right fill with 0
Rotate left
Rotate right

P4.33. Derive the truth table for the following combinational circuit. Write also the operation name for each row in the table.

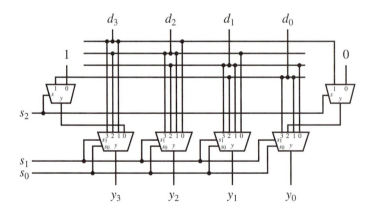

Figure P4.33

P4.34. Draw a 4-bit barrel shifter circuit for the rotate right operation.

P4.35. Draw the complete detail circuit diagram for the 4-bit multiplier based on the circuit shown in Figure 4.34(b).

Implementation Technologies

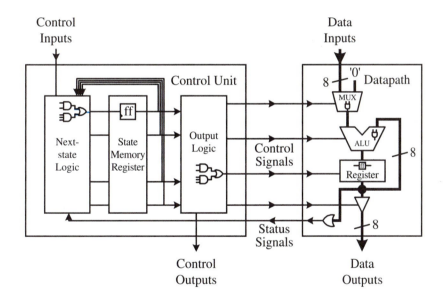

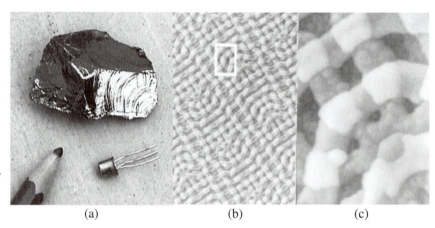

(a)   (b)   (c)

This chapter discusses how digital circuits are implemented at the physical level. As you know, transistors are the fundamental building blocks for all digital circuits. They are the actual physical devices that implement the binary switch and, therefore, also for the logic gates.

There are many different transistor technologies for creating a digital circuit. Some of these technologies are the diode-transistor logic (DTL), transistor–transistor logic (TTL), bipolar logic, and complementary metal-oxide-semiconductor (CMOS) logic. Among them, the most widely used is the CMOS technology.

Figure 5.1(a) shows a single, discrete transistor with its three connections for signal input, output, and control. Above the transistor in the figure is a lump of silicon, which, of course is the main ingredient for the transistor. Figure 5.1(b) is a picture of transistors inside an IC taken with an electron microscope. Figure 5.1(c) is a higher magnification of the rectangular area in Figure 5.1(b).

In this chapter, we will look at how CMOS transistors work and how they are used to build the basic logic gates. Next, we will look at how digital circuits are actually implemented in various programmable logic devices (PLDs), such as read-only memories (ROMs), programmable logic arrays (PLAs), programmable array logic (PAL®) devices, complex programmable logic devices (CPLDs), and field programmable gate arrays (FPGAs). The optional Altera UP2 development board contains both a CPLD and a FPGA chip for implementing your circuits. The information presented in this chapter, however, is not needed to understand how microprocessors are designed at the logic circuit level.

● ● ● ● ● ● ● ● ● ● ● ● ● ● ● ●

5.1 Physical Abstraction

Physical circuits deal with physical properties, such as voltages and currents. Digital circuits use the abstractions 0 and 1 to represent the presence or absence of these physical properties. In fact, a range of voltages is interpreted as the logic 0, and

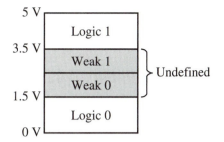

Figure 5.2
Voltage levels for
logic 0 and 1.

another, non-overlapping range is interpreted as the logic 1. Traditionally, digital circuits operate with a 5-volt power supply. In such a case, it is customary to interpret the voltages in the range 0–1.5 V as logic 0, with voltages in the range 3.5–5 V as logic 1. This is shown in Figure 5.2. Voltages in the middle range (from 1.5–3.5 V) are undefined and should not occur in the circuit except during transitions from one state to the other. However, they may be interpreted as a "weak" logic 0 or a "weak" logic 1.

In our discussion of transistors, we will not get into their electrical characteristics of voltages and currents, but we will simply use the abstraction of 0 and 1 to describe their operations.

5.2 Metal-Oxide-Semiconductor Field-Effect Transistor (MOSFET)

The metal-oxide-semiconductor field-effect transistor (MOSFET) acts as a voltage-controlled switch with three terminals: **source**, **drain**, and **gate**. The gate controls whether current can pass from the source to the drain or not. When the gate is asserted or activated, the transistor is turned on and current flows from the source to the drain. When looking at the transistor by itself, there is no physical difference between the source and the drain terminals. They are distinguished only when connected with the rest of the circuit by the differences in the voltage levels.

There are two variations of the MOSFET: the n-channel and the p-channel. The physical structures of these two transistors are shown in Figure 5.3(a) and (b), respectively. The name "metal-oxide-semiconductor" comes from the three layers of material that make up the transistor. The "n" stands for negative and represents the electrons, while "p" stands for positive and represents the holes that flow through a channel in the semiconductor material between the source and the drain.

For the n-channel MOSFET, shown in Figure 5.3(a), a p-type silicon semiconductor material (called the substrate) is doped with n-type impurities at the two ends. These two n-type regions form the source and the drain of the transistor. An insulating oxide layer is laid on top of the two n regions and the p substrate, except for two openings leading to the two n regions. Finally, metal is laid in the two openings in the oxide to form connections to the source and the drain. Another deposit of metal

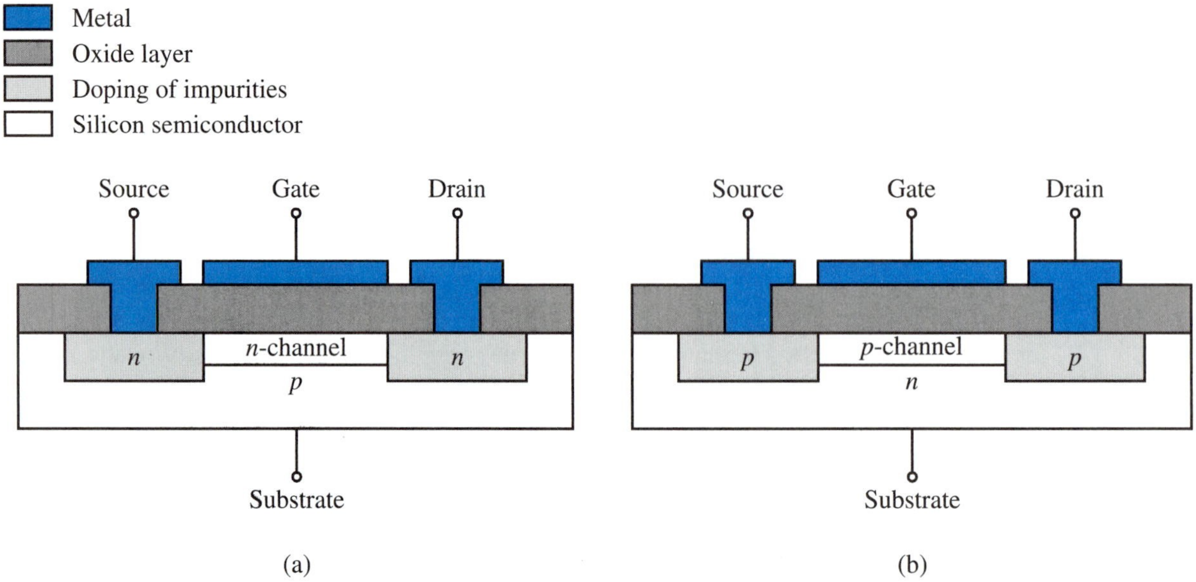

■ Metal
■ Oxide layer
■ Doping of impurities
□ Silicon semiconductor

Figure 5.3 Physical structure of the MOSFET: (a) n-channel (NMOS); (b) p-channel (PMOS).

is laid on top of the oxide between the source and the drain to form the connection to the gate.

The structure of the p-channel MOSFET shown in Figure 5.3(b) is similar to that in Figure 5.3(a), except that the substrate is of n-type material, and the doping for the source and drain is of p-type impurities.

The n-channel and p-channel MOSFETs work opposite of each other. For the n-channel MOSFET, only an n-channel between the source and the drain is created under the control of the gate. This n-channel (n for negative) only allows negative charged electrons (0's) to move from the source to the drain. On the other hand, the p-channel MOSFET can only create a p-channel between the source and the drain under the control of the gate, and this p-channel (p for positive) only allows positive charged holes (1's) to move from the source to the drain.

● ● ● ● ● ● ● ● ● ● ● ● ● ● ●

5.3 CMOS Logic

In CMOS (complementary MOS) logic, only the two complementary MOSFET transistors, (n-channel also known as NMOS and p-channel also known as PMOS),[1] are used to create the circuit. The logic symbols for the NMOS and PMOS transistors are shown in Figures 5.4(a) and 5.5(a), respectively. In designing CMOS circuits, we are

---

[1] For bipolar transistors, these two transistors are referred to as NPN and PNP, respectively.

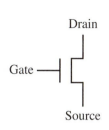

Gate		Switch	Input Signal	Output Signal
1		On	0	0
		(Closed)	1	Weak 1
0	(Any value other than a 1)	Off	×	Z
		(Open)		

(a)   (b)

Figure 5.4 NMOS transistor; (a) logic symbol; (b) truth table.

interested only in the three connections—source, drain, and gate—of the transistor. The substrate for the NMOS is always connected to ground, while the substrate for the PMOS is always connected to V_{CC},[2] so it is ignored in the diagrams for simplicity. Notice that the only difference between these two logic symbols is that one has a circle at the gate input, while the other does not. Using the convention that the circle denotes active-low (i.e., a 0 activates the signal), the NMOS gate input (with no circle) is, therefore, active-high. The PMOS gate input (with a circle) is active-low.

For the NMOS transistor, the source is the terminal with the lower voltage with respect to the drain. You intuitively can think of the source as the terminal that is supplying the 0 value, while the drain consumes the 0 value. The operation of the NMOS transistor is shown in Figure 5.4(b). When the gate is a 1 (asserted), the NMOS transistor is turned on or enabled, and the source input that is supplying the 0 can pass through to the drain output through the connecting n-channel. However,

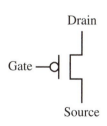

Gate		Switch	Input Signal	Output Signal
0		On	0	Weak 0
		(Closed)	1	1
1	(Any value other than a 0)	Off	×	Z
		(Open)		

(a)   (b)

Figure 5.5 PMOS transistor; (a) logic symbol; (b) truth table.

---

[2] V_{CC} is power or 5 volts in a 5 V circuit, while ground is 0 V.

if the source has a 1, the 1 will not pass through to the drain even if the transistor is turned on, because the NMOS does not create a p-channel. Instead, only a weak 1 will pass through to the drain. On the other hand, when the gate is a 0 (or any value other than a 1), the transistor is turned off, and the connection between the source and the drain is disconnected. In this case, the drain will always have a high-impedance Z value independent of the source value. The $\times$ (don't-care) in the "Input Signal" column means it does not matter what the input value is, the output will be Z. The **high-impedance** value, denoted by Z, means no value or no output. This is like having an insulator with an infinite resistance or a break in a wire, therefore whatever the input is, it will not pass over to the output.

The PMOS transistor works exactly the opposite of the NMOS transistor. For the PMOS transistor, the source is the terminal with the higher voltage with respect to the drain. You intuitively can think of the source as the terminal that is supplying the 1 value, while the drain consumes the 1 value. The operation of the PMOS transistor is shown in Figure 5.5(b). When the gate is a 0 (asserted), the PMOS transistor is turned on or enabled, and the source input that is supplying the 1 can pass through to the drain output through the connecting p-channel. However, if the source has a 0, the 0 will not pass through to the drain even if the transistor is turned on, because the PMOS does not create an n-channel. Instead, only a weak 0 will pass through to the drain. On the other hand, when the gate is a 1 (or any value other than a 0), the transistor is turned off, and the connection between the source and the drain is disconnected. In this case, the drain will always have a high-impedance Z value independent of the source value.

● ● ● ● ● ● ● ● ● ● ● ● ● ● ● ●

5.4 CMOS Circuits

CMOS circuits are built using only the NMOS and PMOS transistors. Because of the inherent properties of the NMOS and PMOS transistors, CMOS circuits are always built with two halves. One half will use one transistor type while the other half will use the other type, and when combined together to form the complete circuit, they will work in complement of each other. The NMOS transistor is used to output the 0 half of the truth table, while the PMOS transistor is used to output the 1 half of the truth table.

Furthermore, notice that the truth tables for these two transistors, shown in Figures 5.4(b) and 5.5(b), suggest that CMOS circuits essentially must deal with five logic values instead of two. These five logic values are 0, 1, Z (high-impedance), weak 0, and weak 1. Therefore, when the two halves of a CMOS circuit are combined together, there is a possibility of mixing any combinations of these five logic values.

Figure 5.6 summarizes the result of combining these logic values. Here, a 1 combined with another 1 does not give you a 2, but rather, just a 1! A short circuit results from connecting a 0 directly to a 1 (that is, connecting ground directly to V_{CC}). This is like sticking two ends of a wire into the two holes of an electrical outlet in the wall. You know the result, and you don't want to do it! Connecting a 0 with a weak 1, or a 1 with a weak 0 will also result in a short, but it may take a longer time

	0	1	Z	Weak 0	Weak 1
0	0	Short	0	0	Short
1	Short	1	1	Short	1
Z	0	1	Z	Weak 0	Weak 1
Weak 0	0	Short	Weak 0	Weak 0	Short
Weak 1	Short	1	Weak 1	Short	Weak 1

Figure 5.6
Result of combining the five possible logic values.

before you start to see smoke coming out. Any value combined with Z is just that value, since Z is nothing.

A properly designed CMOS circuit should always output either a 0 or a 1. Only the tri-state buffer also outputs the Z value. The other two values (weak 0 and weak 1) should not occur in any part of the circuit. The construction of several basic gates using the CMOS technology will now be shown.

5.4.1   CMOS Inverter

Half of the inverter truth table says that, given a 1, the circuit needs to output a 0. Therefore, the question to ask is which CMOS transistor (NMOS or PMOS) when given a 1 will output a 0? Looking at the two truth tables for the two transistors, we find that only the NMOS transistor outputs a 0. The PMOS transistor outputs either a 1 or a weak 0. A weak 0, as you recall from Section 5.1, is an undefined or an unwanted value. The next question to ask is how do we connect the NMOS transistor so that, when we input a 1, the transistor outputs a 0? The answer is shown in Figure 5.7(a) where the source of the NMOS transistor is connected to ground (to provide the 0 value), the gate is the input, and the drain is the output. When the gate is a 1, the 0 from the source will pass through to the drain output.

The complementary half of the inverter circuit is to output a 1 when given a 0. Again, from looking at the two transistor truth tables, we find that the PMOS transistor will do the job. This is expected, since we have used the NMOS for the first half, and the complementary second half of the circuit must use the other transistor. This time, the source is connected to V_{CC} to supply the 1 value, as shown in Figure 5.7(b). When the gate is a 0, the 1 from the source will pass through to the drain output.

To form the complete inverter circuit, we simply combine these two complementary halves together, as shown in Figure 5.7(c). When combining two halves of a CMOS circuit together, the one thing to be careful of is not to create any possible shorts in the circuit. We need to make sure that, for all possible combinations of 0's and 1's to all of the inputs, there are no places in the circuit where both a 0 and a 1 can occur at the same node at the same time.

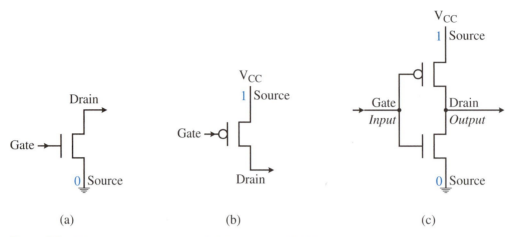

Figure 5.7 CMOS inverter circuit; (a) NMOS half; (b) PMOS half; (c) complete circuit.

For our CMOS inverter circuit, when the gate input is a 1, the bottom NMOS transistor is turned on while the top PMOS transistor is turned off. With this configuration, a 0 from ground will pass through the bottom NMOS transistor to the output, while the top PMOS transistor will output a high-impedance Z value. A Z combined with a 0 is still a 0, because a high impedance is of no value. Alternatively, when the gate input is a 0, the bottom NMOS transistor is turned off while the top PMOS transistor is turned on. In this case, a 1 from V_{CC} will pass through the top PMOS transistor to the output, while the bottom NMOS transistor will output a Z. The resulting output value is a 1. Since the gate input can never be both a 0 and a 1 at the same time, the output can only have either a 0 or a 1, and so, no short can result.

5.4.2 CMOS NAND Gate

Figure 5.8 shows the truth table for the NAND gate. Half of the truth table consists of the one 0 output while the other half of the truth table consists of the three 1 outputs. For the 0 half of the truth table, we want the output to be a 0 when both $A = 1$ and $B = 1$. Again, we ask the question: Which CMOS transistor when given a 1 will output a 0? Of course, the answer is again the NMOS transistor. This time, however, since there are two inputs, A and B, we need two NMOS transistors. We need to connect these two transistors so that a 0 is output only when both are turned on with a 1. Recall from Section 2.3 that the AND operation results from two binary switches connected in series. Figure 5.9(a) shows the two NMOS transistors connected in series, with the source of one connected to ground to provide the 0 value, and the drain of the other providing the output 0. The two transistor gates are connected to the two inputs, A and B, so that only when both inputs are a 1 will the circuit output a 0.

The complementary half of the NAND gate is to output a 1 when either $A = 0$ or $B = 0$. This time, two PMOS transistors are used. To realize the OR operation, the

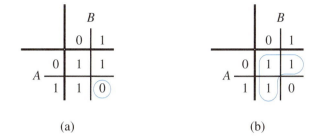

Figure 5.8
NAND-gate truth
table: (a) the 0
half; (b) the 1 half.

two transistors are connected in parallel with both sources connected to V_{CC} and both drains to the output, as shown in Figure 5.9(b). This way, only one transistor needs to be turned on for the circuit to output the 1 value.

The complete NAND-gate circuit is obtained by combining these two halves together, as shown in Figure 5.9(c). When both A and B are 1, the two bottom NMOS transistors are turned on while the two top PMOS transistors are turned off. In this configuration, a 0 from ground will pass through the two bottom NMOS transistors to the output, while the two top PMOS transistors will output a high-impedance Z value. Combining a 0 with a Z will result in a 0. Alternatively, when either $A = 0$ or $B = 0$ (or both equal 0) at least one of the bottom NMOS transistors will be turned off, thus outputting a Z. On the other hand, at least one of the top PMOS transistors

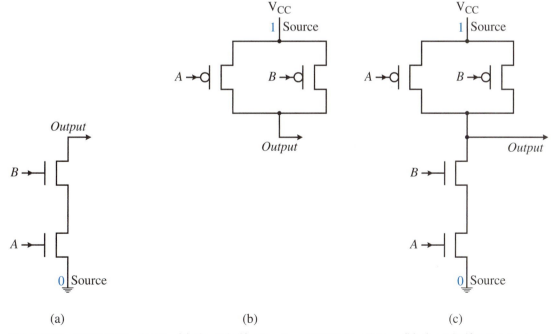

Figure 5.9 CMOS NAND circuit: (a) the 0 half using two NMOS transistors; (b) the 1 half using two PMOS transistors; (c) the complete NAND-gate circuit.

Figure 5.10
AND-gate truth
table: (a) the 0
half; (b) the 1 half.

will be turned on and a 1 from V_{CC} will pass through that PMOS transistor. The resulting output value will be a 1. Again, we see that no short circuit can occur.

5.4.3 CMOS AND Gate

Figure 5.10 shows the 0 and 1 halves of the truth table for the AND gate. We can proceed to derive this circuit in the same manner as we did for the NAND gate. For the 0 half of the truth table, we want the output to be a 0 when either $A = 0$ or $B = 0$. This means that we need a transistor that outputs a 0 when it is turned on also with a 0. This, being one of the main differences between the NAND gate and the AND gate, causes a slight problem. Looking again at the two transistor truth tables in Figures 5.4 and 5.5, we see that neither transistor fits this criterion. The NMOS transistor outputs a 0 when the gate is enabled with a 1, and the PMOS transistor outputs a 1 when the gate is enabled with a 0. If we pick the NMOS transistor, then we need to invert its input. On the other hand, if we pick the PMOS transistor, then we need to invert its output.

For this discussion, let us pick the PMOS transistor. To obtain the A or B operation, two PMOS transistors are connected in parallel. The output from these two transistors is inverted with a single NMOS transistor, as shown in Figure 5.11(a). When either A or B has a 0, that corresponding PMOS transistor is turned on, and a 1 from the V_{CC} source passes down to the gate of the NMOS transistor. With this NMOS transistor turned on, a 0 from ground is passed through to the drain output of the circuit.

For the 1 half of the circuit, we want the output to be a 1 when both $A = 1$ and $B = 1$. Again, we have the dilemma that neither transistor fits this criterion. To be complimentary with the 0 half, we will use two NMOS transistors connected in series. When both transistors are enabled with a 1, the output 0 needs to be inverted with a PMOS transistor, as shown in Figure 5.11(b).

Combining the two halves produces the complete AND-gate CMOS circuit shown in Figure 5.11(c). Instead of joining the two halves at the point of the output, the circuit connects together before inverting the signal to the output. The resulting AND-gate circuit is simply the circuit for the NAND gate followed by that of the INVERTER. From this discussion, we understand why (in practice) NAND gates are preferred over AND gates.

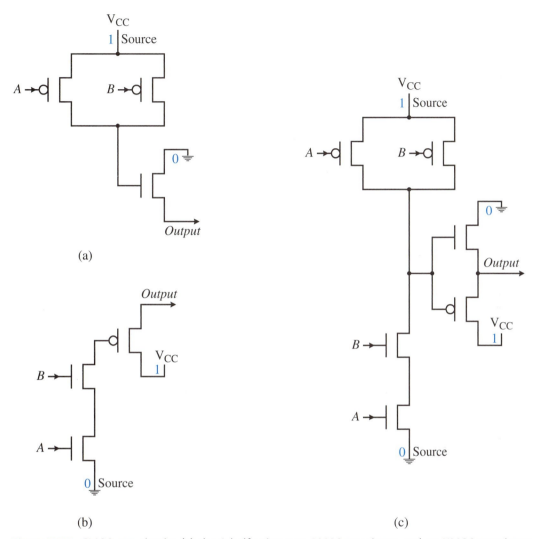

Figure 5.11   CMOS AND circuit: (a) the 0 half using two PMOS transistors and an NMOS transistor; (b) the 1 half using two NMOS transistors and a PMOS transistor; (c) the complete AND-gate circuit.

5.4.4   CMOS NOR and OR Gates

The CMOS NOR-gate and OR-gate circuits can be derived similarly to that of the NAND- and AND-gate circuits. Like the NAND gate, the NOR-gate circuit uses four transistors, whereas the OR-gate circuit uses six transistors.

5.4.5   Transmission Gate

The NMOS and PMOS transistors, when used alone as a control switch, can pass only a 0 or a 1, respectively. Very often, we like a circuit that is able to pass both a

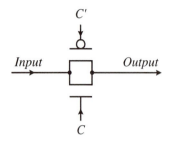

Figure 5.12
CMOS transmission-gate circuit.

0 and a 1 under a control signal. The transmission gate is such a circuit that allows both a 0 and a 1 to pass through when it is enabled. When it is disabled, it outputs the Z value.

The transmission gate uses the two complimentary transistors connected together, as shown in Figure 5.12. Both ends of the two transistors are connected in common. The top PMOS transistor gate is connected to the inverted control signal, C', while the bottom NMOS transistor gate is connected directly to the control signal, C. Hence, both transistors are enabled when the control signal $C = 1$, and the circuit is disabled when $C = 0$.

When the circuit is enabled, if the input is a 1, the 1 signal will pass through the top PMOS transistor, while the bottom NMOS transistor will pass through a weak 1. The final, combined output value will be a 1. On the other hand, if the input is a 0, the 0 signal will pass through the bottom NMOS transistor, while the top PMOS transistor will output a weak 0. The final, combined output value this time will be a 0. Therefore, in both cases, the output value is the same as the input value.

When the circuit is disabled with $C = 0$, both transistors will output the Z value. Thus, regardless of the input, there will be no output.

5.4.6 2-Input Multiplexer CMOS Circuit

CMOS circuits for larger components can be derived by replacing each gate in the circuit with the corresponding CMOS circuit for that gate. Since we know the CMOS circuit for the three basic gates (AND, OR, and NOT), this is a simple "copy-and-paste" operation.

For example, we can replace the gate-level 2-input multiplexer circuit shown in Figure 5.13(a) with the CMOS circuit shown in Figure 5.13(b). For this circuit, we simply replace the two AND gates with the two 6-transistor circuits for the AND gate, another 6-transistor circuit for the OR gate, and the 2-transistor circuit for the INVERTER; giving a total of 20 transistors for this version of the 2-input multiplexer.

However, since the NAND gate uses fewer transistors than the AND gate, we can convert the two-level OR-OF-ANDS circuit in Figure 5.13(a) to a two-level NAND-gate circuit shown in Figure 5.13(c). This conversion is based on the technology mapping technique discussed in Section 3.3. Performing the same "copy-and-paste" operation on this two-level NAND-gate circuit produces the CMOS circuit in Figure 5.13(d) that uses only 14 transistors.

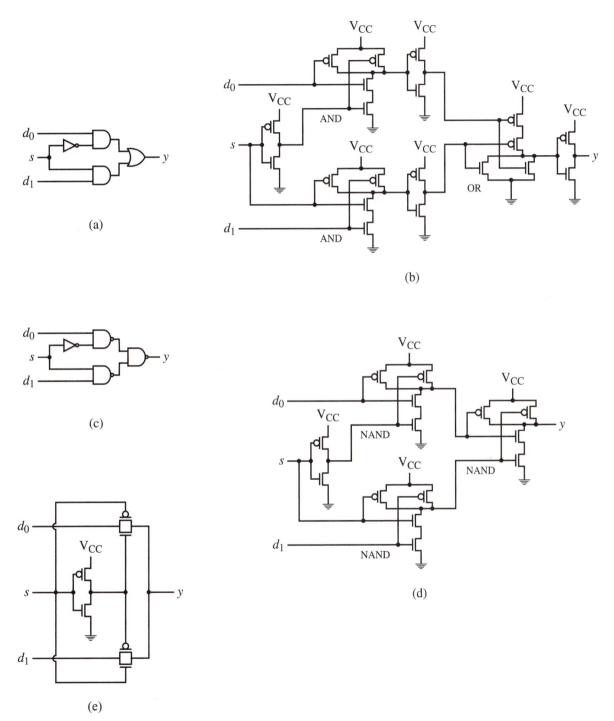

Figure 5.13   2-input multiplexer circuits: (a) gate-level circuit using AND and OR gates; (b) transistor-level circuit for part (a); (c) gate-level circuit using NAND gates; (d) transistor-level circuit for part (c); (e) transistor-level circuit using transmission gates.

We can do much better in terms of the number of transistors needed for the 2-input multiplexer circuit. From the original gate-level multiplexer circuit in Figure 5.13(a), we want to ask the question: What is the purpose of the two AND gates? The answer is that each AND gate acts like a control switch. When it is turned on by the select signal, s, the input passes through to the output. Well, the operation of the transmission gate is just like this, and it uses only two transistors. Hence, we can replace the two AND gates with two transmission gates. Furthermore, the AND gate outputs a 0 when it is disabled. In order for this 0 from the output of the disabled AND gate not to corrupt the data from the output of the other enabled AND gate, the OR gate is needed. If we connect the two outputs from the AND gates directly without the OR gate, a short circuit will occur when the enabled AND gate outputs a 1, because the disabled AND gate always outputs a 0. However, this problem disappears when we use two transmission gates instead of the two AND gates, because when a transmission gate is disabled, it outputs a Z value and not a 0. Thereby, we can connect the outputs of the two transmission gates directly without the need of the OR gate. The resulting circuit is shown in Figure 5.13(e), using only six transistors. The 2-transistor inverter is needed (just like in the gate-level circuit) for turning on only one switch while turning off the other switch at any one time.

5.4.7 CMOS XOR and XNOR Gates

The XOR circuit can be constructed using the same reasoning as for the 2-input multiplexer discussed in Section 5.4.6. First, we recall that the equation for the XOR gate is $AB' + A'B$. For the first AND term, we want to use a transmission gate to pass the A value. This transmission gate is enabled with the value B'. The resulting circuit for this first term is shown in Figure 5.14(a). For the second AND term, we want to use another transmission gate to pass the A' value and have the transmission gate enabled with the value B, resulting in the circuit shown in Figure 5.14(b). Combining the two partial circuits together gives us the complete XOR circuit shown in Figure 5.14(c). Again, as with the 2-input multiplexer circuit, it is not necessary to use an OR gate to connect the outputs of the two transmission gates together.

The CMOS XOR circuit shown in Figure 5.14(c) uses eight transistors: four transistors for the two transmission gates and another four transistors for the two inverters.

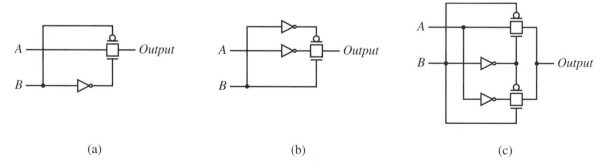

| (a) | (b) | (c) |

Figure 5.14 CMOS XOR gate circuit: (a) partial circuit for the term AB'; (b) partial circuit for the term $A'B$; (c) complete circuit.

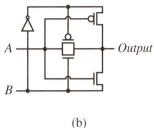

Figure 5.15
CMOS circuits using only six transistors for: (a) xor gate; (b) xnor gate.

(a)   (b)

However, with some ingenuity, we can construct the xor circuit with only six transistors, as shown in Figure 5.15(a). Similarly, the xnor circuit is shown in Figure 5.15(b). In the next section, we will perform an analysis of this xor circuit to see that it indeed has the same functionality as the xor gate.

5.5 Analysis of CMOS Circuits

The analysis of a CMOS circuit follows the same procedure as with the analysis of a combinational circuit, as discussed in Section 3.1. First, we must assume that the inputs to the circuit must have either a logic 0 or logic 1 value; that is, the input value cannot be a weak 0, a weak 1, or a Z. Then, for every combination of 0 and 1 to the inputs, trace through the circuit (based on the operations of the two CMOS transistors) to determine the value obtained at every node in the circuit. When two different values are merged together at the same point in the circuit, we will use the table in Figure 5.6 to determine the resulting value.

Example 5.1   Analyzing the xor CMOS circuit

Analyze the xor CMOS circuit shown in Figure 5.15. For this discussion, the words "top-right," "top-middle," "bottom-middle," and "bottom-right" are used to refer to the four transistors in the circuit.

Figure 5.16(a) shows the analysis of the circuit with the inputs $A = 0$ and $B = 0$. The top-right PMOS transistor is enabled with a 0 from input A; however, the source for this transistor is a 0 from input B, and this produces a weak 0 at the output of this transistor. In Figure 5.16, the arrow denotes that the transistor is enabled, and the label "w 0" at its output denotes that the output value is a weak 0. For the top-middle PMOS transistor, it is also enabled but with a 0 from input B. The source for this transistor is a 0 from A, and so, the output is again a weak 0. The bottom-middle NMOS transistor is enabled with a 1 from B'. Since the source is a 0 from A, this transistor outputs a 0. For the bottom-right NMOS transistor, the 0 from A disables it, and so, a Z value appears at its output. The outputs of these four transistors are joined together at the point of the circuit output. At this common point, two weak 0's, a 0, and a Z are combined together. Referring to Figure 5.6, combining these four values together results in an overall value of a 0. Hence, the circuit outputs a 0 for the input combination $A = 0$ and $B = 0$.

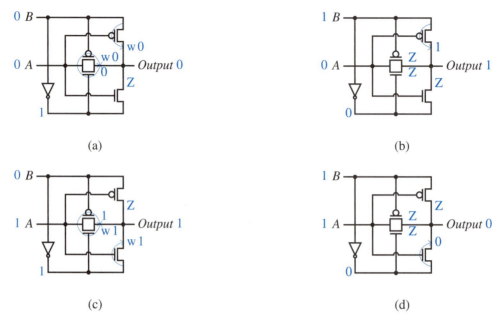

(a)   (b)

(c)   (d)

Figure 5.16 Analysis of the CMOS XOR-gate circuit: (a) shows the analysis for the inputs $AB = 00$. All the transistor outputs are annotated with the resulting output value. The letter "w" is used to signify that it is a weak value; (b) through (d) show the analysis for the remaining input combinations 01, 10, and 11, respectively.

Figure 5.16(b), (c), and (d) show the analysis of the circuit for the remaining three input combinations. The outputs for all four input combinations match exactly those of the 2-input XOR gate.

Example 5.2   **Analyzing a CMOS circuit with a short**

The CMOS circuit shown below is modified slightly from the XOR circuit from Example 5.1; the top-right PMOS transistor is replaced with a NMOS transistor. Let us perform an analysis of this circuit using the inputs $A = 1$ and $B = 0$.

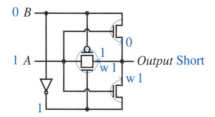

The top-right NMOS transistor is enabled with a 1 from input A. The source for this NMOS transistor is a 0 from input B, and so, it outputs a 0. The top-middle PMOS transistor is also enabled, but with a 0 from input B. The source for this

PMOS transistor is a 1 from input A, and so, it outputs a 1. It is not necessary to continue with the analysis of the remaining two transistors, because at the common output, we already have a 0 (from the top-right NMOS transistor) combining with a 1 (from the top-middle PMOS transistor) producing a short circuit.

• • • • • • • • • • • • • • •

*5.6   Using ROMs to Implement a Function

Memory is used for storing binary data. This stored data, however, can be interpreted as being the implementation of a combinational circuit. A combinational circuit expressed as a Boolean function in canonical form is implemented in the memory by storing data bits in appropriate memory locations. Any type of memory, such as ROM (read-only memory), RAM (random access memory), PROM (programmable ROM), EPROM (erasable PROM), EEPROM (electrically erasable PROM), and so on, can be used to implement combinational circuits. Of course, non-volatile memory is preferred, since you do want your circuit to stay intact after the power is removed.

In order to understand how combinational circuits are implemented in ROMs, we need to first understand the internal circuitry of the ROM. ROM circuit diagrams are drawn more concisely by the use of a new logic symbol to represent a logic gate. Figure 5.17 shows this new logic symbol for an AND gate and an OR gate with multiple inputs. Instead of having multiple input lines drawn to the gate, the input lines are replaced with just one line going to the gate. The multiple input lines are drawn perpendicular to this one line. To actually connect an input line to the gate, an explicit connection point (•) must be drawn where the two lines cross. For example, in Figure 5.17(a), the AND gate has only two inputs; whereas in (b), the OR gate has three inputs.

The circuit diagram for a 16×4 ROM having 16 locations, each being 4-bits wide, is shown in Figure 5.18(a). A 4-to-16 decoder is used to decode the four address lines, A_3, A_2, A_1, and A_0, to the 16 unique locations. Each output of the decoder is a location in the memory. Recall that the decoder operation is such that, when a certain address is presented, the output having the index of the binary address value will have a 1, while the rest of the outputs will have a 0.

Four OR gates provide the four bits of data output for each memory location. The area for making the connections between the outputs of the decoder with the inputs of the OR gates is referred to as the OR array. When no connections are made, the OR gates always will output a 0, regardless of the address input. With connections made as in Figure 5.18(b), the data output of the OR gates depends on the address selected. For the circuit in Figure 5.18(b), if the address input is 0000, then the decoder output

Figure 5.17
Array logic symbol
for: (a) AND gate;
(b) OR gate.

(a)   (b)

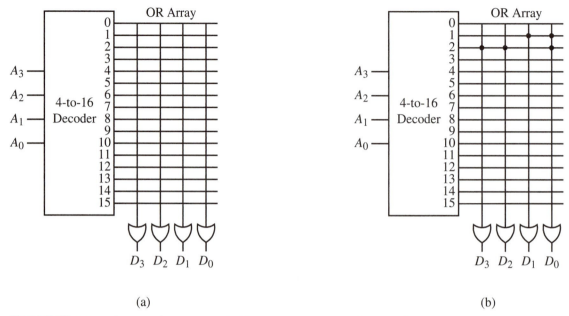

Figure 5.18 Internal circuit for a 16×4 ROM: (a) with no connections made; (b) with connections made.

line 0 will have a 1. Since there are no connections made between the decoder output line 0 and any of the four OR gate inputs, the four OR gates will output a 0. Therefore, the data stored in location 0 is 0000 in binary. If the address input is 0001, then the decoder output line 1 will have a 1. Since this line is connected to the inputs of the two OR gates for D_1 and D_0, therefore, D_1 and D_0 will both have a 1, while D_3 and D_2 will both have a 0. Hence, the data stored in location 1 is 0011. In the circuit of Figure 5.18(b), the value stored in location 2 is 1101.

A 16×4 ROM can be used to implement a 4-variable Boolean function as follows. The four variables are the inputs to the four address lines of the ROM. The 16 decoded locations become the 16 possible minterms for the 4-variable function. For each 1-minterm in the function, we make a connection between that corresponding decoder output line that matches that minterm number with the input of an OR gate. It does not matter which OR gate is used, as long as one OR gate is used to implement one function. Hence, up to four functions with a total of four variables can be implemented in a 16×4 ROM, such as the one shown in Figure 5.18(a). Larger sized ROMs, of course, can implement larger and more functions.

From Figure 5.18(b), we can conclude that the function associated with the OR gate output, D_0, is $F = \Sigma(1, 2)$. That is, minterms 1 and 2 are the 1-minterms for this function while the rest of the minterms are the 0-minterms. Similarly, the function for D_1 has only minterm 1 as its 1-minterm. The functions for D_2 and D_3 both have only minterm 2 as its 1-minterm.

ROMs are programmed during the manufacturing process and cannot be programmed afterwards. As a result, using ROMs to implement a function is only cost effective if a large enough quantity is needed. For small quantities, EPROMs or EEPROMs are preferred. Both EPROMs and EEPROMs can be programmed individually using an inexpensive programmer connected to the computer. The memory device is inserted into the programmer. The bits to be stored in each location of the memory device are generated by the development software. This data file is then transferred to the programmer, which then actually writes the bits into the memory device. Furthermore, both EPROMs and EEPROMs can be erased and reprogrammed with different data bits.

Example 5.3   **Using a 16 × 4 ROM to implement Boolean functions**

Implement the following two Boolean functions using the 16×4 ROM circuit shown in Figure 5.18.

$$F_1(w, x, y, z) = w'x'yz + w'xyz' + w'xyz + wx'y'z' + wx'yz' + wxyz'$$

$$F_2(w, x, y, z) = w'x'y'z' + w'x$$

For F_1, the 1-minterms are m_3, m_6, m_7, m_8, m_{10}, and m_{14}. For F_2, the 1-minterms are m_0, m_4, m_5, m_6, and m_7. Notice that in F_2, the term $w'x$ expands out to four minterms. The implementation is shown in the circuit connection below. We arbitrarily pick D_0 to implement F_1 and D_1 to implement F_2.

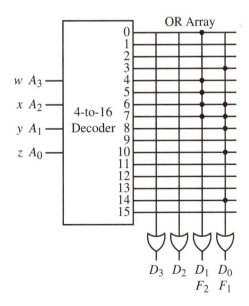

● ● ● ● ● ● ● ● ● ● ● ● ● ● ● ● ⋅ ⋅ ⋅

*5.7 Using PLAs to Implement a Function

Using ROMs or EPROMs to implement a combinational circuit is very wasteful because usually many locations in the ROM are not used. Each storage location in a ROM represents a minterm. In practice, only a small number of these minterms are the 1-minterms for the function being implemented. As a result, the ROM implementing the function usually is quite empty.

Programmable logic arrays (**PLAs**) are designed to reduce this waste by not having all of the minterms "built-in" as in ROMs but rather, by allowing the user to specify only the minterms that are needed. PLAs are designed specifically for implementing combinational circuits.

The internal circuit for a $4 \times 8 \times 4$ PLA is shown in Figure 5.19. The main difference between the PLA circuit and the ROM circuit is that, in the PLA circuit, an AND array is used instead of a decoder. The input signals are available both in the inverted and non-inverted forms. The AND array allows the user to specify only

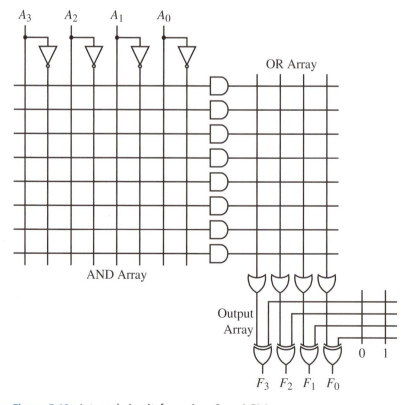

Figure 5.19 Internal circuit for a $4 \times 8 \times 4$ PLA.

the product terms needed by the function; namely, the 1-minterms. The OR array portion of the circuit is similar to that of the ROM, allowing the user to specify which product terms to sum together. Having four OR gates will allow up to four functions to be implemented in a single device.

In addition, the PLA has an output array which provides the capability to either invert or not invert the value at the output of the OR gate. This is accomplished by connecting one input of the XOR gate to either a 0 or a 1. By connecting one input of the XOR gate to a 1, the output of the XOR gate is the inverse of the other input. Alternatively, connecting one input of the XOR gate to a 0, the output of the XOR gate is the same as the other input. This last feature allows the implementation of the inverse of a function in the AND/OR arrays, and then finally getting the function by inverting it.

The actual implementation of a combinational circuit into a PLA device is similar to writing data bits into a ROM or other memory device. A PLA programmer connected to a computer is used. The development software allows the combinational circuit to be defined and then transferred and programmed into the PLA device.

Example 5.4   **Using a 4 × 8 × 4 PLA to implement a full adder circuit**

Implement the full adder circuit in a $4 \times 8 \times 4$ PLA. The truth table for the full adder from Section 4.2.1 is shown here.

x_i	y_i	c_i	c_{i+1}	s_i
0	0	0	0	0
0	0	1	0	1
0	1	0	0	1
0	1	1	1	0
1	0	0	0	1
1	0	1	1	0
1	1	0	1	0
1	1	1	1	1

In the PLA circuit shown next, the three inputs, x_i, y_i, and c_i, are connected to the PLA inputs, A_2, A_1, and A_0, respectively. The first four rows of the AND array implement the four 1-minterms of the function c_{i+1}, while the next three rows of the AND array implement the first three 1-minterms of the function s_i. The last minterm, m_7, is shared by both functions, and therefore, it does not need to be duplicated. The two functions, c_{i+1} and s_i, are mapped to the PLA outputs, F_1

and F_0, respectively. Since the two functions are implemented directly (i.e., not the inverse of the functions), the XOR gates for both functions are connected to 0.

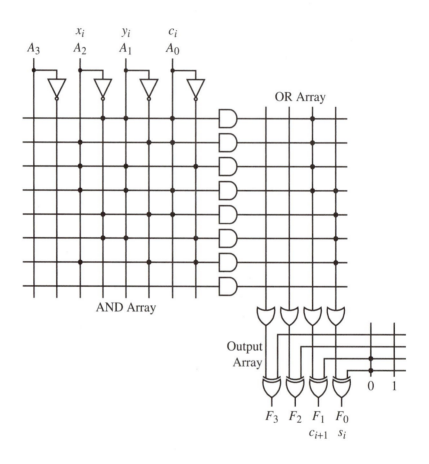

Example 5.5   **Using a 4 × 8 × 4 PLA to implement a function**

Implement the following function in a 4 × 8 × 4 PLA.

$$F(w, x, y, z) = \Sigma(0, 1, 3, 4, 5, 6, 9, 10, 11, 15)$$

This four-variable function has ten 1-minterms. Since the 4 × 8 × 4 PLA can accommodate only eight minterms, we need to implement the inverse of the function, which will have only six 1-minterms ($16 - 10 = 6$). The inverse of the function can then be inverted back to the original function at the output array by connecting one of the XOR inputs to a 1, as shown here.

$$F' = \Sigma(2, 7, 8, 12, 13, 14)$$

$$= w'x'yz' + w'xyz + wx'y'z' + wxy'z' + wxy'z + wxyz'$$

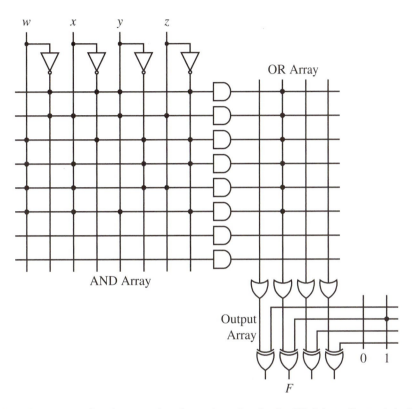

Another way to implement the above function in the PLA is to first minimize it. The following K-map shows that the function reduces to

$$F = w'y' + x'z + w'xz' + wyz + wx'y$$

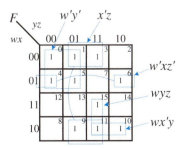

With only five product terms, the function can be implemented directly without having to be inverted, as shown in the following circuit.

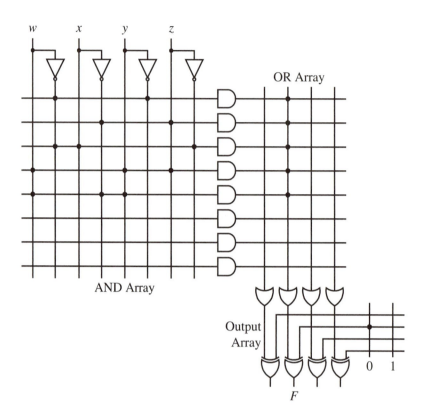

●●●●●●●●●●●●●

*5.8 Using PALs to Implement a Function

Programmable array logic (**PAL**®) devices are similar to PLAs, except that the OR array for the PAL is not programmable but rather, fixed by the internal circuitry. Hence, they are not as flexible in terms of implementing a combinational circuit.

The internal circuit for a 4×4 PAL is shown in Figure 5.20. The OR gate inputs are fixed; whereas, the AND gate inputs are programmable. Each output section is from the OR of the three product terms. This means that each function can have, at most, three product terms. To make the device a little bit more flexible, the output, F_3, is fed back to the programmable inputs of the AND gates. With this connection, up to five product terms are possible for one function.

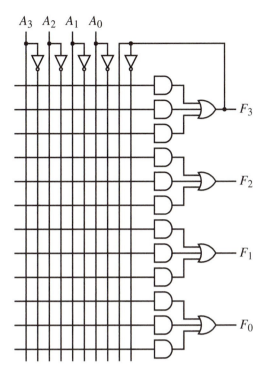

$$A_3 \ A_2 \ A_1 \ A_0$$

F_3

F_2

F_1

F_0

Figure 5.20
Internal circuit for a
4×4 PAL device.

Example 5.6

Using a 4 × 4 PAL to implement functions

Implement the following three functions given in sum-of-minterms format using the PAL circuit of Figure 5.20.

$$F_1(w, x, y, z) = w'x'yz + wx'yz'$$

$$F_2(w, x, y, z) = w'x'yz + wx'yz' + w'xy'z' + wxyz$$

$$F_3(w, x, y, z) = w'x'y'z' + w'x'y'z + w'x'yz' + w'x'yz$$

Function F_1 has two product terms, and it can be implemented directly in one PAL section. F_2 has four product terms, and so, it cannot be implemented directly. However, we note that the first two product terms are the same as F_1. Hence, by using F_1, it is possible to reduce F_2 from four product terms to three as shown here.

$$F_2(w, x, y, z) = w'x'yz + wx'yz' + w'xy'z' + wxyz$$

$$= F_1 + w'xy'z' + wxyz$$

F_3 also has four product terms, but these four product terms can be reduced to just one by minimizing the equation as shown here.

$$F_3(w, x, y, z) = w'x'y'z' + w'x'y'z + w'x'yz' + w'x'yz$$

$$= w'x'(y'z' + y'z + yz' + yz)$$

$$= w'x'$$

The connections for these three functions are shown in the following PAL circuit. Notice that, for functions F_1 and F_3, there are unused AND gates. Since there are no inputs connected to them, they output a 0, which does not affect the output of the OR gate.

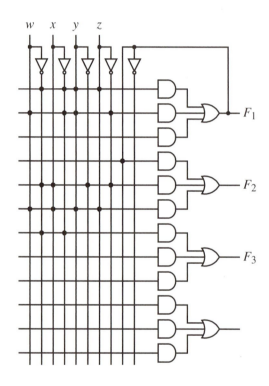

*5.9 Complex Programmable Logic Device (CPLD)

Using ROMs, PLAs, and PALs to implement a combinational circuit is fairly straightforward and easy to do. However, to implement a sequential circuit or a more complex combinational circuit may require more sophisticated and larger programming

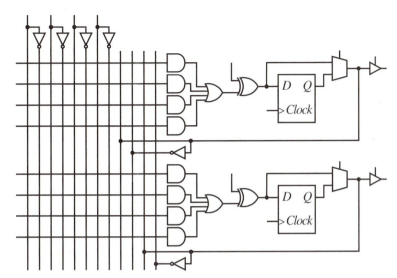

Figure 5.21
Circuit for the logic-array block with two macrocells.

devices. The **complex programmable logic device** (**CPLD**) is capable of implementing a circuit with upwards of 10,000 logic gates.

The CPLD contains many PAL-like blocks that are connected together using a programmable interconnect to form a matrix. The PAL-like blocks in the CPLD are called *macrocells*, as shown in Figure 5.21. Each macrocell has a programmable AND-fixed-OR array similar to a PAL device for implementing combinational logic operations. The XOR gate in the macrocell circuit, shown in Figure 5.21, will either invert or not invert the output from the combinational logic. Furthermore, a flip-flop (discussed in Chapter 6) is included to provide the capability of implementing sequential logic operations. The flip-flop can be bypassed using the multiplexer for combinational logic operations.

Groups of 16 macrocells are connected together to form the logic-array blocks. Multiple logic-array blocks are linked together using the programmable interconnect, as shown in Figure 5.22. Logic signals are routed between the logic-array blocks on the programmable interconnect. This global bus is a programmable path that can connect any signal source to any destination on the CPLD.

The input/output (I/O) blocks allow each I/O pin to be configured individually for input, output, or bi-directional operation. All I/O pins have a tri-state buffer that is controlled individually. The I/O pin is configured as an input port if the tri-state buffer is disabled; otherwise, it is an output port.

Figure 5.23 shows some of the main features of the Altera MAX7000 CPLD. Instead of needing a separate programmer to program the CPLD, all MAX devices support *in-system programmability* through the IEEE JTAG interface. This allows designers to program the CPLD after it is mounted on a printed circuit board. Furthermore, the device can be reprogrammed in the field. CPLDs are non-volatile; so (once they are programmed with a circuit) the circuit remains implemented in the device even when the power is removed.

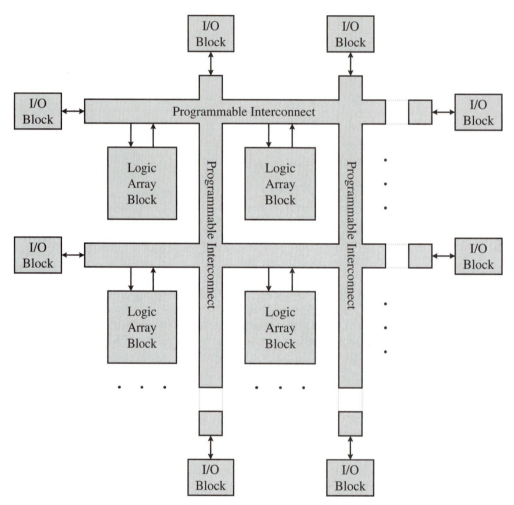

Figure 5.22 Internal circuit for a complex programmable logic device (CPLD).

Feature	MAX7000 CPLD	FLEX10K FPGA
Usable logic gates	10,000	250,000
Macrocells	512	N/A
Logic array blocks	32	1520
User I/O pins	212	470

Figure 5.23 Features of the Altera MAX7000 CPLD and the FLEX10K250 FPGA.

*5.10 **Field Programmable Gate Array (FPGA)**

Field programmable gate arrays (FPGAs) are complex programmable logic devices that are capable of implementing up to 250,000 logic gates and up to 40,960 RAM bits, as featured by the Altera FLEX10K250 FPGA chip (see Figure 5.23). The internal circuitry of the FLEX10K FPGA is shown in Figure 5.24. The device contains an embedded array and a logic array. The embedded array is used to implement memory functions and complex logic functions, such as microcontroller and digital-signal processing. The logic array is used to implement general logic, such as counters, arithmetic logic units, and state machines.

The embedded array consists of a series of embedded array blocks (EABs). When implementing memory functions, each EAB provides 2,048 bits, which can be used to create RAM, dual-port RAM, or ROM. EABs can be used independently, or multiple EABs can be combined to implement larger functions.

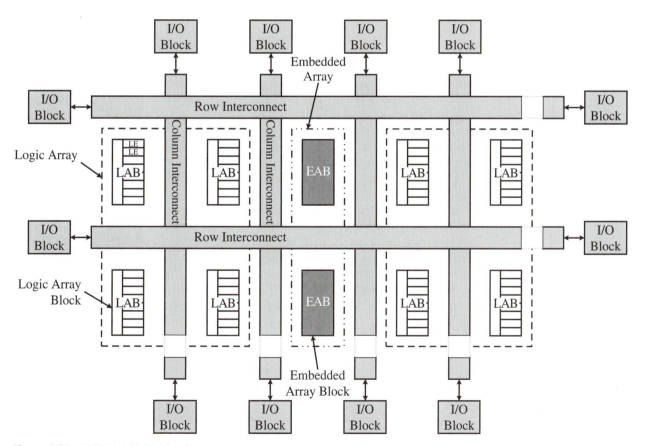

Figure 5.24   FLEX10K FPGA circuit.

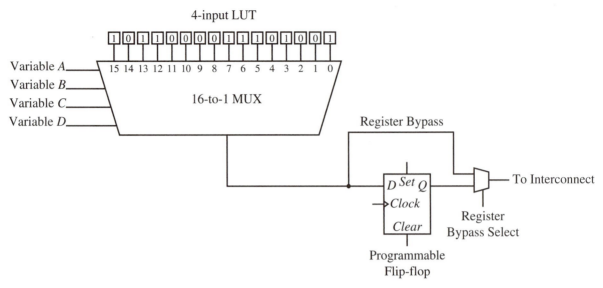

Figure 5.25 Logic element circuit with a 4-input LUT and a programmable register.

The logic array consists of multiple logic array blocks (LABs). Each LAB contains eight logic elements (LE) and a local interconnect. The LE shown in Figure 5.25 is the smallest logical unit in the FLEX10K architecture. Each LE consists of a 4-input look-up table (LUT) and a programmable flip-flop. The 4-input LUT is a function generator made from a 16-to-1 multiplexer that can quickly compute any function of four variables. (Refer to Section 4.9.1 on how multiplexers are used to implement Boolean functions.) The four input variables are connected to the four select lines of the multiplexer. Depending on the values of these four variables, the value from one of the 16 multiplexer inputs is passed to the output. There are 16 1-bit registers connected to the 16 multiplexer inputs to supply the multiplexer input values. Depending on the function to be implemented, the content of the 1-bit registers is set to a 0 or a 1. It is set to a 1 for all the 1-minterms of a 4-variable function, and to a 0 for all the 0-minterms. The LUT in the Figure 5.25 implements the 4-variable function $F(w, x, y, z) = \Sigma(0, 3, 5, 6, 7, 12, 13, 15)$. The programmable flip-flop can be configured for D, T, JK, or SR operations, and is used for sequential circuits. For combinational circuits, the flip-flop can be bypassed using the 2-to-1 multiplexer.

All of the EABs, LABs, and the I/O elements, are connected together via the FastTrack interconnect, which is a series of fast row and column buses that run the entire length and width of the device. The interconnect contains programmable switches so that the output of any block can be connected to the input of any other block.

Each I/O pin in an I/O element is connected to the end of each row and column of the interconnect and can be used as either an input, output, or bi-directional port.

● ● ● ● ● ● ● ● ● ● ● ● ● ● ● ● ●
5.11 Summary Checklist

- Voltage levels
- Weak 0, Weak 1
- NMOS
- NMOS truth table
- PMOS
- PMOS truth table
- High-impedance Z
- Transistor circuits for basic gates
- PLD
- ROM circuit implementation
- PLA circuit implementation
- PAL circuit implementation
- CPLD
- FPGA

● ● ● ● ● ● ● ● ● ● ● ● ● ● ● ● ●
5.12 Problems

P5.1. Draw the CMOS circuit for a 2-input NOR gate.

P5.2. Draw the CMOS circuit for a 2-input OR gate.

P5.3. Draw the CMOS circuit for a 3-input NAND gate.

P5.4. Draw the CMOS circuit for a 3-input NOR gate.

P5.5. Draw the CMOS circuit for an AND gate by using two NMOS transistors for the 0 half of the circuit and two PMOS transistors for the 1 half of the circuit.

P5.6. Draw the CMOS circuit for a 3-input AND gate.

P5.7. Derive the truth table for the following CMOS circuits. There are six possible values: 1, 0, Z, weak 1, weak 0, and short.

(a)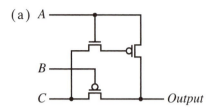

(b)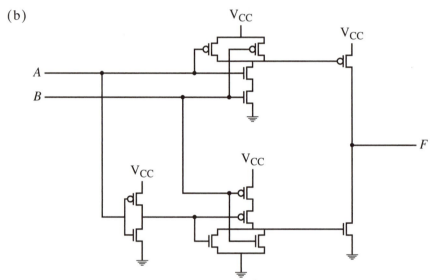

(c)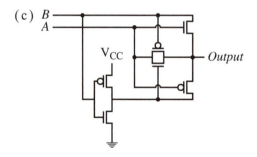

(d)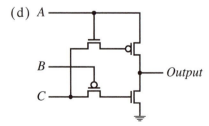

Figure P5.7

P5.8. Synthesize a CMOS circuit that realizes the following truth table.

A	B	Output
0	0	0
0	1	0
1	0	1
1	1	0

P5.9. Synthesize a CMOS circuit that realizes the following truth table having two inputs and one output. Use as few transistors as possible.

A	B	Output
0	0	0
0	1	0
1	0	1
1	1	Short

P5.10. Use one 16×4 ROM (4 address lines, 16 entries, 4 data lines) to implement the following functions. Label all of the lines clearly.

$$f_1 = w'xy'z + w'xz$$

$$f_2 = w$$

$$f_3 = xy' + xyz$$

P5.11. Use one 16×4 ROM (4 address lines, 16 entries, 4 data lines) to implement the following functions. Label all of the lines clearly.

$$f_1 = wx'y'z + wx'yz' + w'xy'z'$$

$$f_2 = xy + w'z + wx'y$$

P5.12. Use one $4 \times 8 \times 4$ PLA to implement the following two functions:

$$F_1(w, x, y, z) = wx'y'z + wx'yz' + wxy'$$

$$F_2(w, x, y, z) = wx'y + x'y'z$$

P5.13. Use one $4 \times 8 \times 4$ PLA to implement the following two functions:

$$F_1(w, x, y, z) = \Sigma(0, 2, 3, 4, 5, 6, 11, 12, 13, 14, 15)$$

$$F_2(w, x, y, z) = \Sigma(1, 2, 3, 5, 7, 9)$$

Latches and Flip-Flops

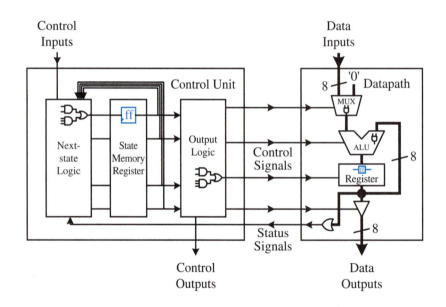

So far, we have been looking at the design of combinational circuits. We will now turn our attention to the design of **sequential circuits**. Recall that the outputs of sequential circuits are dependent on not only their current inputs (as in combinational circuits), but also on all their past inputs. Because of this necessity to remember the history of inputs, sequential circuits must contain memory elements.

The car security system from Section 2.9 is an example of a combinational circuit. In that example, the siren is turned on when the master switch is on and someone opens the door. If you close the door afterwards, then the siren will turn off immediately. For a more realistic car security system, we would like the siren to remain on even if you close the door after it was first triggered. In order for this modified system to work correctly, the siren must be dependent on not only the master switch and the door switch but also on whether the siren is currently on or off. In other words, this modified system is a sequential circuit that is dependent on both the current and on the past inputs to the system.

In order to remember this history of inputs, sequential circuits must have memory elements. Memory elements, however, are just like combinational circuits in the sense that they are made up of the same basic logic gates. What makes them different is in the way these logic gates are connected together. In order for a circuit to "remember" its current value, we have to connect the output of a logic gate directly or indirectly back to the input of that same gate. We call this a **feedback loop** circuit, and it forms the basis for all memory elements. Combinational circuits do not have any feedback loops.

Latches and **flip-flops** are the basic memory elements for storing information. Hence, they are the fundamental building blocks for all sequential circuits. A single latch or flip-flop can store only one bit of information. This bit of information that is stored in a latch or flip-flop is referred to as the **state** of the latch or flip-flop. Hence, a single latch or flip-flop can be in either one of two states: 0 or 1. We say that a latch or a flip-flop changes state when its content changes from a 0 to a 1 or vice versa. This state value is always available at the output. Consequently, the content of a latch or a flip-flop is the state value, and is always equal to its output value.

The main difference between a latch and a flip-flop is that for a latch, its state or output is constantly affected by its input as long as its enable signal is asserted. In other words, when a latch is enabled, its state changes immediately when its input changes. When a latch is disabled, its state remains constant, thereby, remembering its previous value. On the other hand, a flip-flop changes state only at the active edge of its enable signal, i.e., at precisely the moment when either its enable signal rises from a 0 to a 1 (referred to as the rising edge of the signal), or from a 1 to a 0 (the falling edge). However, after the rising or falling edge of the enable signal, and during the time when the enable signal is at a constant 1 or 0, the flip-flop's state remains constant even if the input changes.

In a microprocessor system, we usually want changes to occur at precisely the same moment. Hence, flip-flops are used more often than latches, since they all can be synchronized to change only at the active edge of the enable signal. This enable signal for the flip-flops is usually the global controlling clock signal.

Historically, there are basically four main types of flip-flops: SR, D, JK, and T. The major differences between them are the number of inputs they have and how their contents change. Any given sequential circuit can be built using any of these types of flip-flops (or combinations of them). However, selecting one type of flip-flop over another type to use in a particular sequential circuit can affect the overall size of the circuit. Today, sequential circuits are designed mainly with D flip-flops because of their ease of use. This is simply a tradeoff issue between ease of circuit design versus circuit size. Thus, we will focus mainly on the D flip-flop. Discussions about the other types of flip-flops can be found in Section 6.14.

In this chapter, we will look at how latches and flip-flops are designed and how they work. Since flip-flops are at the heart of all sequential circuits, a good understanding of their design and operation is very important in the design of microprocessors.

● ● ● ● ● ● ● ● ● ● ● ● ● ● ● ● ● ●

6.1   Bistable Element

Let us look at the inverter. If you provide the inverter input with a 1, the inverter will output a 0. If you do not provide the inverter with an input (that is neither a 0 nor a 1), the inverter will not have a value to output. If you want to construct a memory circuit using the inverter, you would want the inverter to continue to output the 0 even after you remove the 1 input. In order for the inverter to continue to output a 0, you need the inverter to self-provide its own input. In other words, you want the output to feed back the 0 to the input. However, you cannot connect the output of the inverter directly to its input, because you will have a 0 connected to a 1 and so creating a short circuit.

The solution is to connect two inverters in series, as shown in Figure 6.1. This circuit is called a **bistable element**, and it is the simplest memory circuit. The bistable element has two symmetrical nodes labeled Q and Q', both of which can be viewed as either an input or an output signal. Since Q and Q' are symmetrical, we can arbitrarily use Q as the state variable, so that the state of the circuit is the value at Q. Let us assume that Q originally has the value 0 when power is first applied to the circuit. Since Q is the input to the bottom inverter, therefore, Q' is a 1. A 1 going to the input of the top inverter will produce a 0 at the output Q, which is what we started off with. Hence, the value at Q will remain at a 0 indefinitely. Similarly, if we power-up the circuit with $Q = 1$, we will get $Q' = 0$, and again, we get a stable situation with Q remaining at a 1 indefinitely. Thus, the circuit has two stable states: $Q = 0$ and $Q = 1$; hence, the name "bistable."

We say that the bistable element has memory because it can remember its state (i.e., keep the value at Q constant) indefinitely. Unfortunately, we cannot change its state (i.e., cannot change the value at Q). We cannot just input a different value to

Figure 6.1
Bistable element circuit.

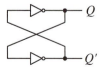

Q, because it will create a short circuit by connecting a 0 to a 1. For example, let us assume that Q is currently 0. If we want to change the state, we need to set Q to a 1, but in so doing we will be connecting a 1 to a 0, thus creating a short. Another way of looking at this problem is that we can think of both Q and Q' as being the primary outputs, which means that the circuit does not have any external inputs. Therefore, there is no way for us to input a different value.

6.2 SR Latch

In order to change the state for the bistable element, we need to add external inputs to the circuit. The simplest way to add extra inputs is to replace the two inverters with two NAND gates, as shown in Figure 6.2(a). This circuit is called an **SR latch**. In addition to the two outputs Q and Q', there are two inputs S' and R' for *set* and *reset*, respectively. Just like the bistable element, the SR latch can be in one of two

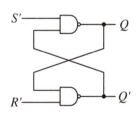

S'	R'	Q	Q_{next}	Q'_{next}
0	0	×	1	1
0	1	×	1	0
1	0	×	0	1
1	1	0	0	1
1	1	1	1	0

(a)   (b)

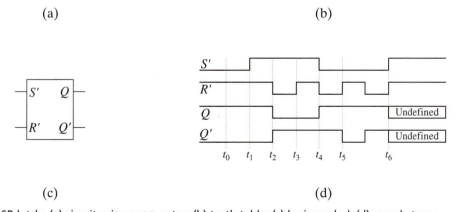

(c)   (d)

Figure 6.2   SR latch: (a) circuit using NAND gates; (b) truth table; (c) logic symbol; (d) sample trace.

states: a set state when $Q = 1$, or a reset state when $Q = 0$. Following the convention, the primes in S and R denote that these inputs are active-low (i.e., a 0 asserts them, and a 1 de-asserts them).

To make the SR latch go to the set state, we simply assert the S' input by setting it to 0 (and de-asserting R'). It doesn't matter what the other NAND gate input is, because 0 NAND anything gives a 1, hence $Q = 1$, and the latch is set. If S' remains at 0 so that Q (which is connected to one input of the bottom NAND gate) remains at 1, and if we now de-assert R' (i.e., set R' to a 1) then the output of the bottom NAND gate will be 0, and so, $Q' = 0$. This situation is shown in Figure 6.2(d) at time t_0. From this current situation, if we now de-assert S' so that $S' = R' = 1$, the latch will remain in the set state because Q' (the second input to the top NAND gate) is 0, which will keep $Q = 1$, as shown at time t_1. At time t_2, we reset the latch by making $R' = 0$ (and $S' = 1$). With R' being a 0, Q' will go to a 1. At the top NAND gate, 1 NAND 1 is 0, thus forcing Q to go to 0. If we de-assert R' next so that, again, we have $S' = R' = 1$, this time the latch will remain in the reset state, as shown at time t_3.

Notice the two times (at t_1 and t_3) when both S' and R' are de-asserted (i.e., $S' = R' = 1$). At t_1, Q is at a 1; whereas, at t_3, Q is at a 0. Why is this so? What is different between these two times? The difference is in the value of Q immediately before those times. The value of Q right before t_1 is 1; whereas, the value of Q right before t_3 is 0. When both inputs are de-asserted, the SR latch remembers its previous state. Previous to t_1, Q has the value 1, so at t_1, Q remains at a 1. Similarly, previous to t_3, Q has the value 0, so at t_3, Q remains at a 0.

If both S' and R' are asserted (i.e., $S' = R' = 0$), then both Q and Q' are equal to a 1, as shown at time t_4, since 0 NAND anything gives a 1. Note that there is nothing wrong with having Q equal to Q'. It is just because we named these two points Q and Q' that we don't like them to be equal. However, we could have used another name say, P instead of Q'.

If one of the input signals is de-asserted earlier than the other, the latch will end up in the state forced by the signal that is de-asserted later, as shown at time t_5. At t_5, R' is de-asserted first, so the latch goes into the set state with $Q = 1$, and $Q' = 0$.

A problem exists if both S' and R' are de-asserted at *exactly* the same time, as shown at time, t_6. Let us assume for a moment that both gates have exactly the same delay and that the two wire connections between the output of one gate to the input of the other gate also have exactly the same delay. Currently, both Q and Q' are at a 1. If we set S' and R' to a 1 at exactly the same time, then both NAND gates will perform a 1 NAND 1 and will both output a 0 at exactly the same time. The two 0's will be fed back to the two gate inputs at exactly the same time, because the two wire connections have the same delay. This time around, the two NAND gates will perform a 1 NAND 0 and will both produce a 1 again at exactly the same time. This time, two 1's will be fed back to the inputs, which again will produce a 0 at the outputs, and so on and on. This oscillating behavior, called the **critical race**, will continue indefinitely until one outpaces the other. If the two gates do not have exactly the same delay, then the situation is similar to de-asserting one input before the other, and so, the latch will go into one state or the other. However, since we do not know which is the faster gate, we do not know which state the latch will end up in. Thus, the latch's next state is undefined.

Of course, in practice, it is next to impossible to manufacture two gates and make the two connections with precisely the same delay. In addition, both S' and R' need to be de-asserted at exactly the same time. Nevertheless, if this circuit is used in controlling some mission-critical device, we don't want even this slim chance to happen.

In order to avoid this non-deterministic behavior, we must make sure that the two inputs are never de-asserted at the same time. Note that we do want the situation when both of them are de-asserted, as in times t_1 and t_3, so that the circuit can remember its current content. We want to de-assert one input after de-asserting the other, but just not de-asserting both of them at *exactly* the same time. In practice, it is very difficult to guarantee that these two signals are never de-asserted at the same time, so we relax the condition slightly by not having both of them asserted together. In other words, if one is asserted, then the other one cannot be asserted. Therefore, if both of them are never asserted simultaneously, then they cannot be de-asserted at the same time. A minor side benefit for not having both of them asserted together is that Q and Q' are never equal to each other. Recall that, from the names that we have given these two nodes, we do want them to be inverses of each other.

From the above analysis, we obtain the truth table in Figure 6.2(b) for the NAND implementation of the SR latch. In the truth table, Q and Q_{next} actually represent the same point in the circuit. The difference is that Q is the current value at that point, while Q_{next} is the new value to be updated in the next time period. Another way of looking at it is that Q is the input to a gate, and Q_{next} is the output from a gate. In other words, the signal Q goes into a gate, propagates through the two gates, and arrives back at Q as the new signal Q_{next}. Figure 6.2(c) shows the logic symbol for the SR latch.

The SR latch can also be implemented using NOR gates, as shown in Figure 6.3(a). The truth table for this implementation is shown in Figure 6.3(b). From the truth table, we see that the main difference between this implementation and the NAND implementation is that, for the NOR implementation, the S and R inputs are

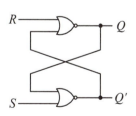

S	R	Q	Q_{next}	Q'_{next}
0	0	0	0	1
0	0	1	1	0
0	1	×	0	1
1	0	×	1	0
1	1	×	0	0

(a)   (b)   (c)

Figure 6.3 SR latch: (a) circuit using NOR gates; (b) truth table; (c) logic symbol.

active-high, so that setting S to 1 will set the latch, and setting R to 1 will reset the latch. However, just like the NAND implementation, the latch is set when $Q = 1$ and reset when $Q = 0$. The latch remembers its previous state when $S = R = 0$. When $S = R = 1$, both Q and Q' are 0. The logic symbol for the SR latch using NOR implementation is shown in Figure 6.3(c).

● ● ● ● ● ● ● ● ● ● ● ● ● ● ● ● ⋯

6.3 SR Latch with Enable

The SR latch is sensitive to its inputs all the time. In other words, Q will always change when either S or R is asserted. It is sometimes useful to be able to disable the inputs so that asserting them will not cause the latch to change state but to keep its current state. Of course, this is achieved by de-asserting both S and R. Hence, what we want is just one enable signal that will de-assert both S and R. The **SR latch with**

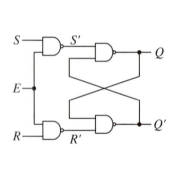

E	S	R	Q	Q_{next}	Q'_{next}
0	×	×	0	0	1
0	×	×	1	1	0
1	0	0	0	0	1
1	0	0	1	1	0
1	0	1	×	0	1
1	1	0	×	1	0
1	1	1	×	1	1

(a)   (b)

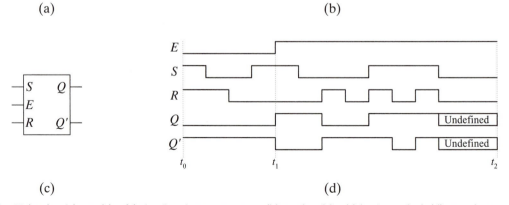

(c)   (d)

Figure 6.4 SR latch with enable: (a) circuit using NAND gates; (b) truth table; (c) logic symbol; (d) sample trace.

enable (also known as a **gated SR latch**) shown in Figure 6.4(a) accomplishes this by adding two extra NAND gates to the original NAND-gate implementation of the latch. These two new NAND gates are controlled by the enable input, E, which determines whether the latch is enabled or disabled. When $E = 1$, the circuit behaves like the normal NAND implementation of the SR latch, except that the new S and R inputs are active-high rather than active-low. When $E = 0$, then $S' = R' = 1$, and the latch will remain in its previous state, regardless of the S and R inputs. The truth table and the logic symbol for the SR latch with enable is shown in Figure 6.4(b) and (c), respectively.

A typical operation of the latch is shown in the sample trace in Figure 6.4(d). Between t_0 and t_1, $E = 0$, so changing the S and R inputs does not affect the output. Between t_1 and t_2, $E = 1$, and the trace is similar to the trace of Figure 6.4(d), except that the input signals are inverted.

• • • • • • • • • • • • • • • • • •

6.4 **D Latch**

Recall from Section 6.2 that the disadvantage with the SR latch is that we need to ensure that the two inputs, S and R, are never de-asserted at exactly the same time, and we said that we can guarantee this by not having both of them asserted. This situation is prevented in the **D latch** by adding an inverter between the original S' and R' inputs. This way, S' and R' will always be inverses of each other, and so, they will never be asserted together. The circuit using NAND gates and the inverter is shown in Figure 6.5(a). There is now only one input D (for *data*). When $D = 0$, then $S' = 1$ and $R' = 0$, so this is similar to resetting the SR latch by making $Q = 0$. Similarly, when $D = 1$, then $S' = 0$ and $R' = 1$, and Q will be set to 1. From this observation, we see that Q_{next} always gets the same value as the input D and is independent of the current value of Q. Hence, we obtain the truth table for the D latch, as shown in Figure 6.5(b).

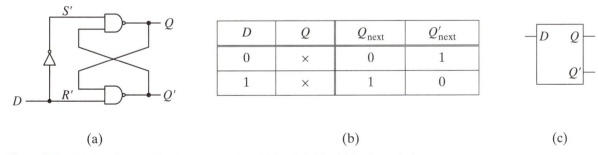

D	Q	Q_{next}	Q'_{next}
0	×	0	1
1	×	1	0

(a)   (b)   (c)

Figure 6.5   D latch: (a) circuit using NAND gates; (b) truth table; (c) logic symbol.

Comparing the truth table for the D latch shown in Figure 6.5(b) with the truth table for the SR latch shown in Figure 6.2(b), it is obvious that we have eliminated not just one, but three rows, where $S' = R'$. The reason for adding the inverter to the SR-latch circuit was to eliminate the row where $S' = R' = 0$. However, we still need to have the other two rows where $S' = R' = 1$ in order for the circuit to remember its current value. By not being able to set both S' and R' to 1, this D-latch circuit has now lost its ability to remember. Q_{next} cannot remember the current value of Q, instead, it will always follow D. The end result is like having a piece of wire where the output is the same as the input!

6.5 D Latch with Enable

In order to make the D latch remember the current value, we need to connect Q (the current state value) back to the input D, thus creating another feedback loop. Furthermore, we need to be able to select whether to loop Q back to D or input a new value for D. Otherwise, like the bistable element, we will not be able to change the state of the circuit. One way to achieve this is to use a 2-input multiplexer to select whether to feedback the current value of Q or pass an external input back to D. The circuit for the **D latch with enable** (also known as a **gated D latch**) is shown in Figure 6.6(a). The external input becomes the new D input, the output of the multiplexer is connected to the original D input, and the select line of the multiplexer is the enable signal E.

When the enable signal E is asserted ($E = 1$), the external D input passes through the multiplexer, and so Q_{next} (i.e., the output Q) follows the D input. On the other hand, when E is de-asserted ($E = 0$), the current value of Q loops back as the input to the circuit, and so Q_{next} retains its last value independent of the D input.

When the latch is enabled, the latch is said to be open, and the path from the input D to the output Q is transparent. In other words, Q follows D. Because of this characteristic, the D latch with enable circuit is often referred to as a **transparent latch**. When the latch is disabled, it is closed, and the latch remembers its current state. The truth table and the logic symbol for the D latch with enable are shown in Figure 6.6(b) and (c). A sample trace for the operation of the D latch with enable is shown in Figure 6.6(d). Between t_0 and t_1, the latch is enabled with $E = 1$, so the output Q follows the input D. Between t_1 and t_2, the latch is disabled, so Q remains stable even when D changes.

An alternative way to construct the D latch with enable circuit is shown in Figure 6.7. Instead of using the 2-input multiplexer, as shown in Figure 6.6(a), we start with the SR latch with enable circuit of Figure 6.4(a), and connect the S and R inputs together with an inverter. The functional operations of these two circuits are identical.

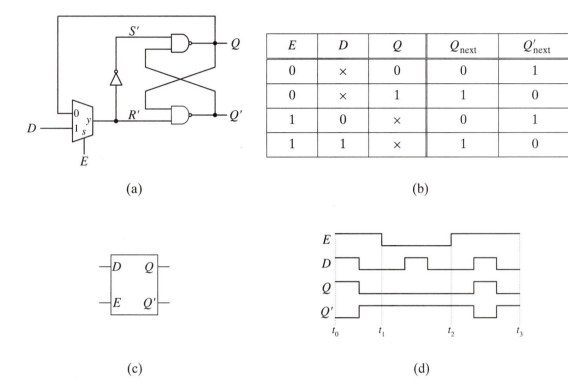

E	D	Q	Q_{next}	Q'_{next}
0	×	0	0	1
0	×	1	1	0
1	0	×	0	1
1	1	×	1	0

(a)   (b)

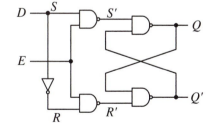

(c)   (d)

Figure 6.6   D latch with enable: (a) circuit; (b) truth table; (c) logic symbol; (d) sample trace.

Figure 6.7
D latch with enable
circuit using four
NAND gates.

6.6 **Clock**

Latches are known as **level-sensitive** because their outputs are affected by their inputs as long as they are enabled. Their memory state can change during this entire time when the enable signal is asserted. In a computer circuit, however, we do not want

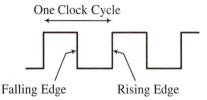

Figure 6.8
Clock signal.

the memory state to change at various times when the enable signal is asserted. Instead, we like to synchronize all of the state changes to happen at precisely the same moment and at regular intervals. In order to achieve this, two things are needed: (1) a synchronizing signal, and (2) a memory circuit that is not level-sensitive. The synchronizing signal, of course, is the **clock**, and the non-level-sensitive memory circuit is the flip-flop.

The clock is simply a very regular square wave signal, as shown in Figure 6.8. We call the edge of the clock signal when it changes from 0 to 1 the **rising edge**. Conversely, the **falling edge** of the clock is the edge when the signal changes from 1 to 0. We will use the symbol ↑ to denote the rising edge and ↓ for the falling edge. In a computer circuit, either the rising edge or the falling edge of the clock can be used as the synchronizing signal for writing data into a memory element. This edge signal is referred to as the **active edge** of the clock. In all of our examples, we will use the rising clock edge as the active edge. Therefore, at every rising edge, data will be clocked or stored into the memory element.

A **clock cycle** is the time from one rising edge to the next rising edge or from one falling edge to the next falling edge. The speed of the clock, measured in hertz (Hz), is the number of cycles per second. Typically, the clock speed for a microprocessor in an embedded system runs around 20 MHz, while the microprocessor in a personal computer runs upwards of 2 GHz and higher. A clock **period** is the time for one clock cycle (seconds per cycle), so it is just the inverse of the clock speed.

The speed of the clock is determined by how fast a circuit can produce valid results. For example, a two-level combinational circuit will have valid results at its output much sooner than, say, an ALU can. Of course, we want the clock speed to be as fast as possible, but it can only be as fast as the slowest circuit in the entire system. We want the clock period to be the time it takes for the slowest circuit to get its input from a memory element, operate on the data, and then write the data back into a memory element. More will be said on this in later sections.

Figure 6.9 shows a VHDL description of a clock-divider circuit that roughly cuts a 25 MHz clock down to 1 Hz.

● ● ● ● ● ● ● ● ● ● ● ● ● ● ● ● ●

6.7 D Flip-Flop

Unlike the latch, a flip-flop is not level-sensitive, but rather **edge-triggered**. In other words, data gets stored into a flip-flop only at the active edge of the clock. An **edge-triggered D flip-flop** achieves this by combining in series a pair of D latches.

```
LIBRARY IEEE;
USE IEEE.STD_LOGIC_1164.ALL;

ENTITY Clockdiv IS PORT (
    Clk25Mhz: IN STD_LOGIC;
    Clk: OUT STD_LOGIC);
END Clockdiv;

ARCHITECTURE Behavior OF Clockdiv IS
    CONSTANT max: INTEGER := 25000000;
    CONSTANT half: INTEGER := max/2;
    SIGNAL count: INTEGER RANGE 0 TO max;
BEGIN
    PROCESS
    BEGIN
      WAIT UNTIL Clk25Mhz'EVENT and Clk25Mhz = '1';
      IF count < max THEN
         count <= count + 1;
      ELSE
         count <= 0;
      END IF;
      IF count < half THEN
         Clk <= '0';
      ELSE
         Clk <= '1';
      END IF;
    END PROCESS;
END Behavior;
```

Figure 6.9 VHDL behavioral description of a clock-divider circuit.

Figure 6.10(a) shows a **positive edge triggered D flip-flop**, where two D latches are connected in series. A clock signal *Clk* is connected to the E input of the two latches: one directly and one through an inverter.

The first latch is called the *master* latch. The master latch is enabled when $Clk = 0$ because of the inverter, and so QM follows the primary input D. However, the signal at QM cannot pass over to the primary output Q, because the second latch (called the *slave* latch) is disabled when $Clk = 0$. When $Clk = 1$, the master latch is disabled, but the slave latch is enabled so that the output from the master latch, QM, is transferred to the primary output Q. The slave latch is enabled all the while that $Clk = 1$, but its content changes only at the rising edge of the clock, because once Clk is 1, the master latch is disabled, and the input to the slave latch, QM, will be constant. Therefore, when $Clk = 1$ and the slave latch is enabled, the primary output Q will not change because the input QM is not changing.

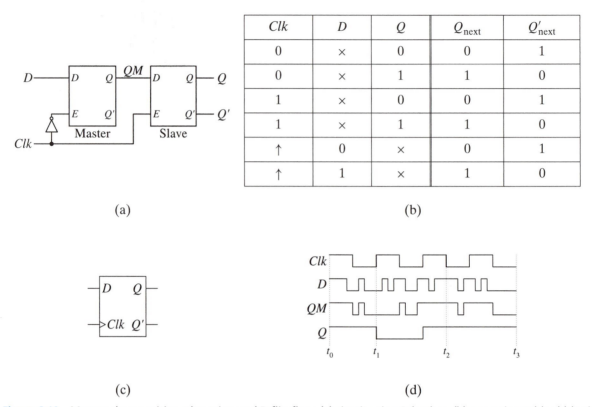

Clk	D	Q	Q_{next}	Q'_{next}
0	×	0	0	1
0	×	1	1	0
1	×	0	0	1
1	×	1	1	0
↑	0	×	0	1
↑	1	×	1	0

(a)   (b)

(c)   (d)

Figure 6.10   Master–slave positive edge-triggered D flip-flop: (a) circuit using D latches; (b) operation table; (c) logic symbol; (d) sample trace.

The circuit shown in Figure 6.10(a) is called a positive edge-triggered D flip-flop because the primary output Q on the slave latch changes only at the rising edge of the clock. If the slave latch is enabled when the clock is low (i.e., with the inverter output connected to the E of the slave latch), then it is referred to as a **negative edge-triggered** flip-flop. The circuit is also referred to as a **master–slave** D flip-flop because of the two D latches used in the circuit.

Figure 6.10(b) shows the operation table for the D flip-flop. The ↑ symbol signifies the rising edge of the clock. When *Clk* is either at 0 or 1, the flip-flop retains its current value (i.e., $Q_{next} = Q$). Q_{next} changes and follows the primary input D only at the rising edge of the clock. The logic symbol for the positive edge-triggered D flip-flop is shown in Figure 6.10(c). The small triangle at the clock input indicates that the circuit is triggered by the edge of the signal, and so it is a flip-flop. Without the small triangle, the symbol would be that for a latch. If there is a circle in front of the clock line, then the flip-flop is triggered by the falling edge of the clock, making it a negative edge-triggered flip-flop. Figure 6.10(d) shows a sample trace for the D flip-flop. Notice that when *Clk* = 0, *QM* follows *D*, and the output of the slave

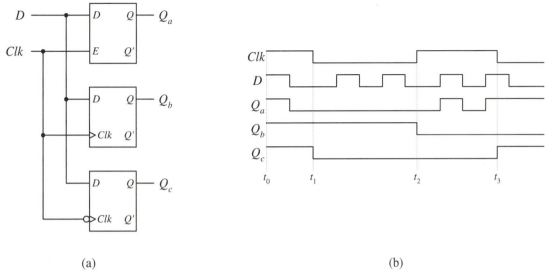

Figure 6.11 Comparison of a gated latch, a positive edge-triggered flip-flop, and a negative edge-triggered flip-flop: (a) circuit; (b) sample trace.

latch, Q, remains constant. On the other hand, when $Clk = 1$, Q follows QM, and the output of the master latch, QM, remains constant.

Figure 6.11 compares the different operations between a latch and a flip-flop. In Figure 6.11(a), we have a D latch with enable, a positive edge-triggered D flip-flop, and a negative edge-triggered D flip-flop, all having the same D input and controlled by the same clock signal. Figure 6.11(b) shows a sample trace of the circuit's operations. Notice that the gated D latch, Q_a, follows the D input as long as the clock is high (between times t_0 and t_1 and times t_2 and t_3). The positive edge-triggered flip-flop, Q_b, follows the D input only at the rising edge of the clock at time t_2, while the negative edge-triggered flip-flop, Q_c, follows the D input only at the falling edge of the clock at times t_1 and t_3.

*6.7.1 Alternative Smaller Circuit

Not all master–slave flip-flops are edge-triggered. For instance, using two SR latches to construct a master–slave flip-flop results in a flip-flop that is level-sensitive. Conversely, an edged-triggered D flip-flop can be constructed using SR latches instead of the master–slave D latches.

The circuit shown in Figure 6.12 shows how a positive edge-triggered D flip-flop can be constructed using three interconnected SR latches. The advantage of this circuit is that it uses only 6 NAND gates (26 transistors) as opposed to 11 gates (38 transistors) for the master–slave D flip-flop shown in Figure 6.10(a). The operation of the circuit is as follows. When $Clk = 0$, the outputs of gates 2 and 3 will be 1 (since 0 NAND $x = 1$). With $n_2 = n_3 = 1$, this will keep the output latch (comprising

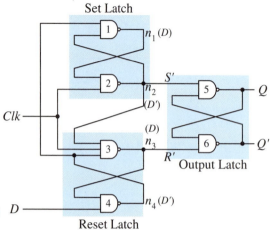

Figure 6.12
Positive edge-triggered D flip-flop.

of gates 5 and 6) in its current state. At the same time, $n_4 = D'$ since one input to gate 4 is n_3, which is a 1 (1 NAND $x = x'$). Similarly, $n_1 = D$ since $n_2 = 1$, and the other input to gate 1 is n_4, which is D' (again 1 NAND $x = x'$).

When Clk changes to 1, n_2 will be equal to D' because 1 NAND $n_1 = n_1'$, and $n_1 = D$. Similarly, n_3 will be equal to D when Clk changes to 1 because the other two inputs to gate 3 are both D'. Therefore, if $Clk = 1$ and $D = 0$, then n_2 (which is equal to D') will be 1 and n_3 (which is equal to D) will be 0. With $n_2 = 1$ and $n_3 = 0$, this will de-assert S' and assert R', thus resetting the output latch Q to 0. On the other hand, if $Clk = 1$ and $D = 1$, then n_2 (which is equal to D') will be 0 and n_3 (which is equal to D) will be 1. This will assert S' and de-assert R', thus setting the output latch Q to 1. So at the rising edge of the Clk signal, Q will follow D.

The setting and resetting of the output latch occurs only at the rising edge of the Clk signal, because once Clk is at a 1 and remains at a 1, changing D will not change n_2 or n_3. The reason, as noted in the previous paragraph, is that n_2 and n_3 are always inverses of each other. Furthermore, the following argument shows that both n_2 and n_3 will remain constant even if D changes. Let us first assume that n_2 is a 0. If $n_2 = 0$, then n_3 (the output of gate 3) will always be a 1 (since 0 NAND $x = 1$), regardless of what n_4 (the third input to gate 3) may be. Hence, if n_4 (the output of gate 4) cannot affect n_3, then D (the input to gate 4) also cannot affect either n_2 or n_3. On the other hand, if $n_2 = 1$, then $n_3 = 0$ ($n_3 = n_2'$). With a 0 from n_3 going to the input of gate 4, the output of gate 4 at n_4 will always be a 1 (0 NAND $x = 1$), regardless of what D is. With the three inputs to gate 3 being all 1's, n_3 will continue to be 0. Therefore, as long as $Clk = 1$, changing D will not change n_2 or n_3. And if n_2 and n_3 remain stable, then Q will also remain stable for the entire time that Clk is 1.

6.8 D Flip-Flop with Enable

So far, with the construction of the different memory elements, it seems like every time we add a new feature we have also lost a feature that we need. The careful reader will have noticed that, in building the D flip-flop, we have again lost the most important property of a memory element—it can no longer remember its current content! At every active edge of the clock, the D flip-flop will load in a new value. So how do we get it to remember its current value and not load in a new value?

The answer, of course, is exactly the same as what we did with the D latch, and that is by adding an enable input, E, through a 2-input multiplexer, as shown in Figure 6.13(a). When $E = 1$, the primary input D signal will pass to the D input of the flip-flop, thus updating the content of the flip-flop at the active edge. When $E = 0$, the current content of the flip-flop at Q is passed back to the D input of the flip-flop, thus keeping its current value. Notice that changes to the flip-flop value occur only at the active edge of the clock. Here, we are using the rising edge as the active edge. The operation table and the logic symbol for the D flip-flop with enable is shown in Figure 6.13(b) and (c), respectively.

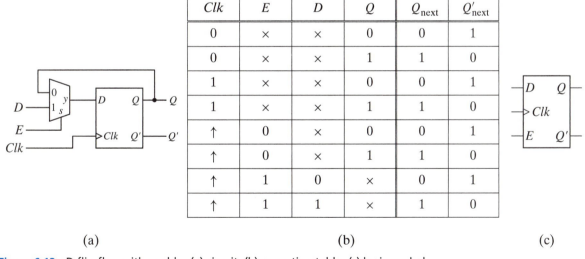

Clk	E	D	Q	Q_{next}	Q'_{next}
0	×	×	0	0	1
0	×	×	1	1	0
1	×	×	0	0	1
1	×	×	1	1	0
↑	0	×	0	0	1
↑	0	×	1	1	0
↑	1	0	×	0	1
↑	1	1	×	1	0

(a)　　　　　　　　　　　　　(b)　　　　　　　　　　　　　(c)

Figure 6.13　D flip-flop with enable: (a) circuit; (b) operation table; (c) logic symbol.

6.9 Asynchronous Inputs

Flip-flops (as we have seen so far) change states only at the rising or falling edge of a synchronizing clock signal. Many circuits require the initialization of flip-flops to a known state that is independent of the clock signal. Sequential circuits that change states whenever a change in input values occurs that is independent of the clock are referred to as **asynchronous** sequential circuits. **Synchronous** sequential circuits, on the other hand, change states only at the active edge of the clock signal. Asynchronous inputs usually are available for both flip-flops and latches, and they are used to either set or clear the storage element's content that is independent of the clock.

Figure 6.14(a) shows a gated D latch with asynchronous active-low *Set'* and *Clear'* inputs, and (b) is the logic symbol for it. Figure 6.14(c) is the circuit for the D

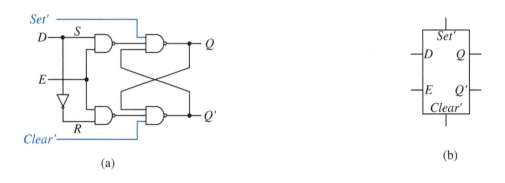

(a)　　　　　　　　　　　　　　　　(b)

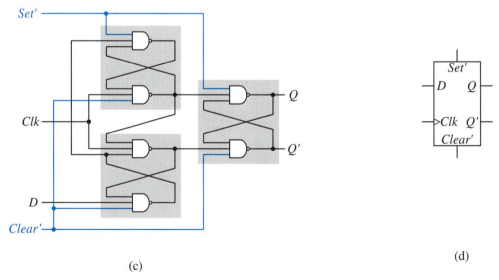

(c)　　　　　　　　　　　　　　　　(d)

Figure 6.14 Storage elements with asynchronous inputs: (a) D latch with active-low set and clear; (b) logic symbol for (a); (c) D edge-triggered flip-flop with active-low set and clear; (d) logic symbol for (c).

edge-triggered flip-flop with asynchronous *Set'* and *Clear'* inputs, and (d) is the logic symbol for it. When *Set'* is asserted (set to 0) the content of the storage element is set to 1 immediately (i.e., without having to wait for the next rising clock edge), and when *Clear'* is asserted (set to 0) the content of the storage element is set to 0 immediately.

6.10 Description of a Flip-Flop

Combinational circuits can be described with either a truth table or a Boolean equation. For describing the operation of a flip-flop or any sequential circuit in general, we use a characteristic table, a characteristic equation, a state diagram, or an excitation table, as discussed in the following subsections.

6.10.1   Characteristic Table

The **characteristic table** specifies the functional behavior of the flip-flop. It is a simplified version of the flip-flop's operational table by only listing how the state changes at the active clock edge. The table has the flip-flop's input signal(s) and current state (Q) listed in the input columns, and the next state (Q_{next}) listed in the output column. Q'_{next} is always assumed to be the inverse of Q_{next}, so it is not necessary to include this output column. The clock signal is also not included in the table, because it is a signal that we do not want to modify. Nevertheless, the clock signal is always assumed to exist. Furthermore, since all state changes for a flip-flop (i.e., changes to Q_{next}) occur at the active edge of the clock; therefore, it is not necessary to list the situations from the operation table for when the clock is at a constant value.

The characteristic table for the D flip-flop is shown in Figure 6.15(a). It has two input columns (the input signal D, and the current state Q) and one output column for Q_{next}. From the operation table for the D flip-flop shown in Figure 6.10(b), we see that there are only two rows where Q_{next} is affected during the rising clock edge. Hence, these are the only two rows inserted into the characteristic table.

The characteristic table is used in the analysis of sequential circuits to answer the question of what is the next state, Q_{next}, when given the current state, Q, and input signals (D in the case of the D flip-flop).

6.10.2   Characteristic Equation

The **characteristic equation** is simply the Boolean equation that is derived directly from the characteristic table. Like the characteristic table, the characteristic equation specifies the flip-flop's next state, Q_{next}, as a function of its current state, Q, and input signals. The D flip-flop characteristic table has only one 1-minterm, which results in the simple characteristic equation for the D flip-flop shown in Figure 6.15(b).

D	Q	Q_{next}
0	×	0
1	×	1

(a)

$$Q_{next} = D$$

(b)

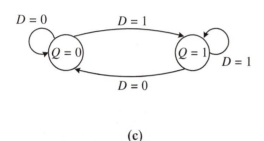

(c)

Q	Q_{next}	D
0	0	0
0	1	1
1	0	0
1	1	1

(d)

Figure 6.15   Description of a D flip-flop: (a) characteristic table; (b) characteristic equation; (c) state diagram; (d) excitation table.

6.10.3 State Diagram

A **state diagram** is a graph with nodes and directed edges connecting the nodes, as shown in Figure 6.15(c). The state diagram graphically portrays the operation of the flip-flop. The nodes are labeled with the states of the flip-flop, and the directed edges are labeled with the input signals that cause the transition to go from one state of the flip-flop to the next. Figure 6.15(c) shows the state diagram for the D flip-flop. It has two states, $Q = 0$ and $Q = 1$, which correspond to the two values that the flip-flop can contain. The operation of the D flip-flop is such that when it is in state 0, it will change to state 1 if the input D is a 1; otherwise, if the input D is a 0, then it will remain in state 0. Hence, there is an edge labeled $D = 1$ that goes from state $Q = 0$ to $Q = 1$, and a second edge labeled $D = 0$ that goes from state $Q = 0$ back to itself. Similarly, when the flip-flop is in state 1, it will change to state 0 if the input D is a 0; otherwise, it will remain in state 1. These two conditions correspond to the remaining two edges that go out from state $Q = 1$ in the state diagram.

6.10.4 Excitation Table

The **excitation table** is like the mirror image of the characteristic table by exchanging the input signal column(s) with the output (Q_{next}) column. The excitation table shows what the flip-flop's inputs should be in order to change from the flip-flop's current state to the next state desired. In other words, the excitation table answers the question of what the flip-flop's inputs should be when given the current state that the flip-flop is in and the next state that we want the flip-flop to go to. This table is used in the synthesis of sequential circuits.

Figure 6.15(d) shows the excitation table for the D flip-flop. As can be seen, this table can be obtained directly from the state diagram. For example, using the state diagram of the D flip-flop from Figure 6.15(c), if the current state is $Q = 0$ and we want the next state to be $Q_{next} = 0$, then the D input must be a 0, as shown by the label on the edge that goes from state 0 back to itself. On the other hand, if the current state is $Q = 0$ and we want the next state to be $Q_{next} = 1$, then the D input must be a 1.

● ● ● ● ● ● ● ● ● ● ● ● ● ● ● ●

*6.11 Timing Issues

So far in our discussion of latches and flip-flops, we have ignored timing issues and the effects of propagation delays. In practice, timing issues are very important in the correct design of sequential circuits. Consider again the D latch with enable circuit from Section 6.5 and redrawn in Figure 6.16(a). Signals from the inputs require some delay to propagate through the gates and finally to reach the outputs.

Assuming that the propagation delay for the inverter is 1 nanosecond (ns) and 2 ns for the NAND gates, the timing trace diagram would look like Figure 6.16(b) with the signal delays taken into consideration. The arrows denote which signal edge causes another signal edge. The number next to an arrow denotes the number of nanoseconds in delay for the resulting signal to change.

At time t_1, signal D drops to 0. This causes R to rise to 1 after a 1 ns delay through the inverter. The D edge also causes S' to rise to 1, but after a delay of 2 ns through the NAND gate. After that, R' drops to 0 at 2 ns after R rises to 1. This in turn causes Q' to rise to 1 after 2 ns, followed by Q dropping to 0.

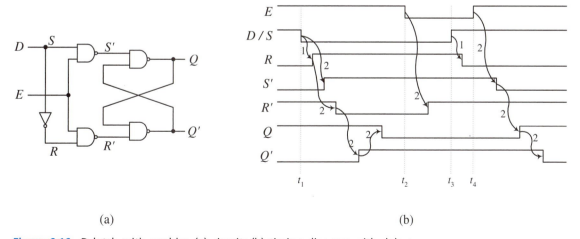

(a)   (b)

Figure 6.16 D latch with enable: (a) circuit; (b) timing diagram with delays.

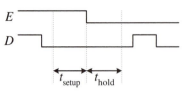

Figure 6.17
Setup and hold
times for the gated
D latch.

At time t_2, signal E drops to 0, disabling the circuit. As a result, when D rises to 1 at time t_3, both Q and Q' are not affected.

At time t_4, signal E rises to 1 and re-enables the circuit. This causes S' to drop to 0 after 2 ns. R' remains unchanged at 1, since the two inputs to the NAND gate, E and R, are 1 and 0, respectively. With S' asserted and R' de-asserted, the latch is set with Q rising to 1 at 2 ns after S' drops to a 0. This is followed by Q' dropping to 0 after another 2 ns.

Furthermore, for the D-latch circuit to latch in the data from input D correctly, there is a critical window of time right before and right after the falling edge of the enable signal, E, that must be observed. Within this time frame, the input signal, D, must not change. As shown in Figure 6.17, the time before the falling edge of E is referred to as the **setup time**, t_{setup}, and the time after the falling edge of E is referred to as the **hold time**, t_{hold}. The length of these two times is dependent on the implementation and manufacturing process and can be obtained from the component data sheet.

6.12 Designing a Car Security System—Version 2

In Section 2.9, we designed a combinational circuit for a car security system where the siren will come on when the master switch is on and either the door switch or the vibration switch is also on. However, as soon as both the door switch and the vibration switch are off, the siren will turn off immediately, even though the master switch is still on. In reality, what we really want is to have the siren remain on, even after both the door and vibration switches are off. In order to do so, we need to remember the state of the siren. In other words, for the siren to remain on, it should be dependent not only on whether the door or the vibration switch is on, but also on the fact that the siren is currently on.

We can use the state of a SR latch to remember the state of the siren (i.e., the output of the latch will drive the siren). The state of the latch is driven by the conditions of the input switches. The modified circuit, as shown in Figure 6.18, has an SR latch (in addition to its original combinational circuit) for remembering the current state of the siren. The latch is set from the output of the combinational circuit. The latch's reset is connected to the master switch so that the siren can be turned off immediately.

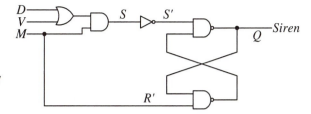

Figure 6.18
Modified car security system circuit with memory.

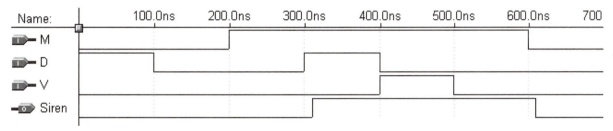

Figure 6.19   Sample timing trace of the modified car security system circuit with memory.

A sample timing trace of the operation of this circuit is shown in Figure 6.19. At time 0, the siren is off, even though the door switch is on, because the master switch is off. At time 300 ns, the siren is turned on by the door switch since the master switch is also on. At time 500 ns, both the door and the vibration switches are off, but the siren is still on because it was turned on previously. The siren is turned off by the master switch at time 600 ns.

• • • • • • • • • • • • • • •

6.13   VHDL for Latches and Flip-Flops

6.13.1   Implied Memory Element

VHDL does not have any explicit object for defining a memory element. Instead, the semantics of the language provide for signals to be interpreted as a memory element. In other words, the memory element is declared depending on how these signals are assigned.

Consider the VHDL code in Figure 6.20. If *Enable* is 1, then *Q* gets the value of *D*; otherwise, *Q* gets a 0. In this code, *Q* is assigned a value for all possible outcomes of the test in the IF statement. With this construct, a combinational circuit is produced.

If we remove the ELSE and the statement in the ELSE part, as shown in Figure 6.21, then we have a situation where no value is assigned to *Q* if *Enable* is not 1. The key point here is that the VHDL semantics stipulate that, in cases where the code does

```
LIBRARY IEEE;
USE IEEE.STD_LOGIC_1164.ALL;

ENTITY no_memory_element IS PORT (
   D, Enable: IN STD_LOGIC;
   Q: OUT STD_LOGIC);
END no_memory_element;

ARCHITECTURE Behavior OF no_memory_element IS
BEGIN
   PROCESS(D, Enable)
   BEGIN
      IF Enable = '1' THEN
         Q <= D;
      ELSE
         Q <= '0';
      END IF;
   END PROCESS;
END Behavior;
```

Figure 6.20
Sample VHDL
description of a
combinational
circuit.

not specify a value of a signal, the signal should retain its current value. In other words, the signal must remember its current value, and in order to do so, a memory element is implied.

```
LIBRARY IEEE;
USE IEEE.STD_LOGIC_1164.ALL;

ENTITY D_latch_with_enable IS PORT (
   D, Enable: IN STD_LOGIC;
   Q: OUT STD_LOGIC);
END D_latch_with_enable;

ARCHITECTURE Behavior OF D_latch_with_enable IS
BEGIN
   PROCESS(D, Enable)
   BEGIN
      IF Enable = '1' THEN
         Q <= D;
      END IF;
   END PROCESS;
END Behavior;
```

Figure 6.21
VHDL code for a D
latch with enable.

6.13.2   VHDL Code for a D Latch with Enable

Figure 6.21 shows the VHDL code for a D latch with enable. If *Enable* is 1, then *Q* gets the value of *D*. However, if *Enable* is not 1, the code does not specify what *Q* should be; therefore, *Q* retains its current value by using a memory element. This code produces a latch and not a flip-flop, because *Q* follows *D* as long as *Enable* is 1 and not only at the active edge of the *Enable* signal. The process sensitivity list includes both *D* and *Enable*, because either one of these signals can cause a change in the value of the *Q* output.

6.13.3   VHDL Code for a D Flip-Flop

Figure 6.22 shows the behavioral VHDL code for a positive edge-triggered D flip-flop. The only difference here is that *Q* follows *D* only at the rising edge of the clock, and it is specified here by the condition "*Clock'* EVENT AND *Clock* = '1'." The 'EVENT attribute refers to any changes in the qualifying *Clock* signal. Therefore, when this happens and the resulting *Clock* value is a 1, we have, in effect, a condition for a positive or rising clock edge. Again, the code does not specify what is assigned to *Q* when the condition in the IF statement is false, so it implies the use of a memory element. Note also that the process sensitivity list contains only the clock signal, because it is the only signal that can cause a change in the *Q* output.

Another way to describe a flip-flop is to use the WAIT statement instead of the IF statement, as shown in Figure 6.23. When execution reaches the WAIT statement, it stops until the condition in the statement is true before proceeding. The WAIT statement, when used in a process block for synthesis, must be the first statement in

```
LIBRARY IEEE;
USE IEEE.STD_LOGIC_1164.ALL;

ENTITY D_flipflop IS PORT (
    D, Clock: IN STD_LOGIC;
    Q: OUT STD_LOGIC);
END D_flipflop;

ARCHITECTURE Behavior OF D_flipflop IS
BEGIN
    PROCESS(Clock)                              -- sensitivity list is used
    BEGIN
        IF Clock'EVENT AND Clock = '1' THEN
            Q <= D;
        END IF;
    END PROCESS;
END Behavior;
```

Figure 6.22
Behavioral VHDL code for a positive edge-triggered D flip-flop using an IF statement.

```
                    LIBRARY IEEE;
                    USE IEEE.STD_LOGIC_1164.ALL;

                    ENTITY D_flipflop IS PORT (
                       D, Clock: IN STD_LOGIC;
                       Q: OUT STD_LOGIC);
                    END D_flipflop;

                    ARCHITECTURE Behavioral OF D_flipflop IS
                    BEGIN
                       PROCESS                -- sensitivity list is not used if WAIT is used
                       BEGIN
                          WAIT UNTIL Clock'EVENT AND Clock = '0'; -- negative edge triggered
                          Q = D;
                       END PROCESS;
                    END Behavioral;
```

Figure 6.23
Behavioral VHDL
code for a negative
edge-triggered D
flip-flop using a WAIT
statement.

the process. Note also that the process sensitivity list is omitted, because the WAIT statement implies that the sensitivity list contains only the clock signal.

Alternatively, we can write a structural VHDL description for the positive edge-triggered D flip-flop, as shown in Figure 6.24. This VHDL code is based on the circuit for a positive edge-triggered D flip-flop, as given in Figure 6.12.

```
                    -- define the operation of the 2-input NAND gate
                    LIBRARY IEEE;
                    USE IEEE.STD_LOGIC_1164.ALL;

                    ENTITY NAND_2 IS PORT (
                       I0, I1: IN STD_LOGIC;
                       O: OUT STD_LOGIC);
                    END NAND_2;

                    ARCHITECTURE Dataflow_NAND2 OF NAND_2 IS
                    BEGIN
                       O <= I0 NAND I1;
                    END Dataflow_NAND2;
```

Figure 6.24
Structural VHDL
code for a positive
edge-triggered D
flip-flop.
(continued on next page)

```
-- define the structural operation of the SR latch
LIBRARY IEEE;
USE IEEE.STD_LOGIC_1164.ALL;
ENTITY SRlatch IS PORT (
   SN, RN: IN STD_LOGIC;
   Q, QN: BUFFER STD_LOGIC);
END SRlatch;

ARCHITECTURE Structural_SRlatch OF SRlatch IS
   COMPONENT NAND_2 PORT (
      I0, I1 : IN STD_LOGIC;
      O : OUT STD_LOGIC);
   END COMPONENT;
BEGIN
   U1: NAND_2 PORT MAP (SN, QN, Q);
   U2: NAND_2 PORT MAP (Q, RN, QN);
END Structural_SRlatch;

-- define the operation of the 3-input NAND gate
LIBRARY IEEE;
USE IEEE.STD_LOGIC_1164.ALL;

ENTITY NAND_3 IS PORT (
   I0, I1, I2: IN STD_LOGIC;
   O: OUT STD_LOGIC);
END NAND_3;

ARCHITECTURE Dataflow_NAND3 OF NAND_3 IS
BEGIN
   O <= NOT (I0 AND I1 AND I2);
END Dataflow_NAND3;

-- define the structural operation of the D flip-flop
LIBRARY IEEE;
USE IEEE.STD_LOGIC_1164.ALL;

ENTITY positive_edge_triggered_D_flipflop IS PORT (
   D, Clock: IN STD_LOGIC;
   Q, QN: BUFFER STD_LOGIC);
END positive_edge_triggered_D_flipflop;
```

Figure 6.24
Structural VHDL code for a positive edge-triggered D flip-flop.
(continued on next page)

```
ARCHITECTURE StructuralDFF OF positive_edge_triggered_D_flipflop IS
   SIGNAL N1, N2, N3, N4: STD_LOGIC;

   COMPONENT SRlatch PORT (
      SN, RN: IN STD_LOGIC;
      Q, QN: BUFFER STD_LOGIC);
   END COMPONENT;

   COMPONENT NAND_2 PORT (
      I0, I1: IN STD_LOGIC;
      O: OUT STD_LOGIC);
   END COMPONENT;

   COMPONENT NAND_3 PORT (
      I0, I1, I2: IN STD_LOGIC;
      O: OUT STD_LOGIC);
   END COMPONENT;

BEGIN
   U1: SRlatch PORT MAP (N4, Clock, N1, N2);  -- set latch
   U2: SRlatch PORT MAP (N2, N3, Q, QN);      -- output latch
   U3: NAND_3 PORT MAP (N2, Clock, N4, N3);   -- reset latch
   U4: NAND_2 PORT MAP (N3, D, N4);           -- reset latch
END StructuralDFF;
```

Figure 6.24
Structural VHDL code for a positive edge-triggered D flip-flop.

The simulation trace for the positive edge-triggered D flip-flop is shown in Figure 6.25. In the trace, before the first rising edge of the clock at time 100 ns, both Q and Q' (QN) are undefined because nothing has been stored in the flip-flop yet. Immediately after this rising clock edge at 100 ns, Q gets the value of D, and QN gets the inverse. At 200 ns, D changes to 1, but Q does not follow D immediately but is delayed until the next rising clock edge at 300 ns. At the same time, QN drops to 0. At 400 ns, when D drops to 0, Q again follows it at the next rising clock edge at 500 ns.

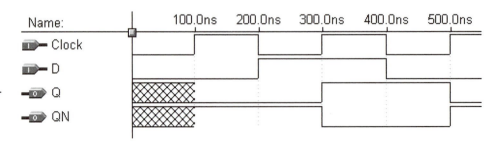

Figure 6.25
Simulation trace for the positive edge-triggered D flip-flop.

6.13.4   VHDL Code for a D Flip-Flop with Enable and Asynchronous Set and Clear

Figure 6.26 shows the VHDL code for a positive edge-triggered D flip-flop with enable and asynchronous active-high set and clear inputs. The two asynchronous inputs are checked independently of the clock event. When either the *Set* or the *Clear* input is asserted with a 1 (active-high), *Q* is set immediately to 1 or 0, respectively, independent of the clock. If *Enable* is asserted with a 1, then *Q* follows *D* at the rising edge of the clock; otherwise, *Q* keeps its previous content. Figure 6.27 shows the simulation trace for this flip-flop. Notice in the trace that when either *Set* or *Clear* is asserted (at 100 ns and 200 ns, respectively) *Q* changes immediately. However, when *Enable* is asserted at 400 ns, *Q* doesn't follow *D* until the next rising clock edge at 500 ns. Similarly, when *D* drops to 0 at 600 ns, *Q* doesn't change immediately but drops at the next rising edge at 700 ns. At 800 ns, when *D* changes to a 1, *Q* does not follow the change at the next rising edge at 900 ns, because *Enable* is now de-asserted.

Figure 6.26
Behavioral VHDL code for a positive edge-triggered D flip-flop with active-high enable and asynchronous set and clear inputs.

```
LIBRARY IEEE;
USE IEEE.STD_LOGIC_1164.ALL;

ENTITY d_ff IS PORT (
  Clock: IN STD_LOGIC;
  Enable: IN STD_LOGIC;
  Set: IN STD_LOGIC;
  Clear: IN STD_LOGIC;
  D: IN STD_LOGIC;
  Q: OUT STD_LOGIC);
END d_ff;

ARCHITECTURE Behavioral OF d_ff IS
BEGIN
   PROCESS(Clock,Set,Clear)
   BEGIN
      IF (Set = '1') THEN
         Q <= '1';
      ELSIF (Clear = '1') THEN
         Q <= '0';
      ELSIF (Clock'EVENT AND Clock = '1') THEN
         IF Enable = '1' THEN
            Q <= D;
         END IF;
      END IF;
   END PROCESS;
END Behavioral;
```

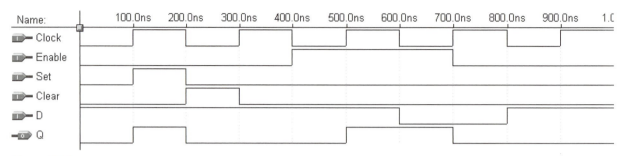

Figure 6.27 Simulation trace for the positive edge-triggered D flip-flop with active-high enable and asynchronous set and clear inputs.

*6.14 **Other Flip-Flop Types**

There are basically four main types of flip-flops: D, SR, JK, and T. The major differences in these flip-flop types are in the number of inputs they have and how they change states. Like the D flip-flop, each type can also have different variations, such as active-high or -low inputs, whether they change state at the rising or falling edge of the clock signal, and whether they have any asynchronous inputs. Any given sequential circuit can be built using any of these types of flip-flops or combinations of them. However, selecting one type of flip-flop over another type to use in a particular circuit can affect the overall size of the circuit. Today, sequential circuits are designed primarily with D flip-flops only because of their simple operation. Of the four flip-flop's characteristic equations, the characteristic equation for the D flip-flop is the simplest.

6.14.1 **SR Flip-Flop**

Like SR latches, SR flip-flops are useful in control applications where we want to be able to set or reset the data bit. However, unlike SR latches, SR flip-flops change their content only at the active edge of the clock signal. Similar to SR latches, SR flip-flops can enter an undefined state when both inputs are asserted simultaneously. When the two inputs are de-asserted, then the next state is the same as the current state. The characteristic table, characteristic equation, state diagram, circuit, logic symbol, and excitation table for the SR flip-flop are shown in Figure 6.28.

S	R	Q	Q_{next}	Q'_{next}
0	0	0	0	1
0	0	1	1	0
0	1	0	0	1
0	1	1	0	1
1	0	0	1	0
1	0	1	1	0
1	1	0	×	×
1	1	1	×	×

(a)

$$Q_{next} = S + R'Q$$

(b)

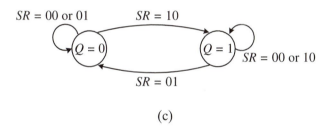

(c)

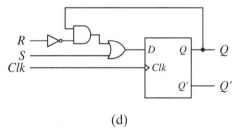

(d)

(e)

Q	Q_{next}	S	R
0	0	0	×
0	1	1	0
1	0	0	1
1	1	×	0

(f)

Figure 6.28   SR flip-flop: (a) characteristic table; (b) characteristic equation; (c) state diagram; (d) circuit; (e) logic symbol; (f) excitation table.

The SR flip-flop truth table shown in Figure 6.28(a) is for active-high set and reset signals. Hence, the flip-flop state, Q_{next}, is set to 1 when S is asserted with a 1, and Q_{next} is reset to 0 when R is asserted with a 1. When both S and R are de-asserted with a 0, the flip-flop remembers its current state. From the truth table, we get the following K-map for Q_{next}, which results in the characteristic equation shown in Figure 6.28(b).

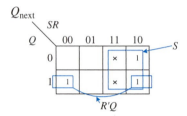

Notice that the SR flip-flop circuit shown in Figure 6.28(d) uses the D flip-flop. The signal for asserting the D input of the flip-flop is generated by the combinational circuit that is derived from the characteristic equation of the SR flip-flop, namely $D = Q_{\text{next}} = S + R'Q$.

6.14.2 JK Flip-Flop

The operation of the JK flip-flop is very similar to the SR flip-flop. The J input is just like the S input in the SR flip-flop in that, when asserted, it sets the flip-flop. Similarly, the K input is like the R input where it resets the flip-flop when asserted. The only difference is when both inputs, J and K, are asserted. For the SR flip-flop, the next state is undefined; whereas, for the JK flip-flop, the next state is the inverse of the current state. In other words, the JK flip-flop toggles its state when both inputs are asserted. The characteristic table, characteristic equation, state diagram, circuit, logic symbol, and excitation table for the JK flip-flop are shown in Figure 6.29.

6.14.3 T Flip-Flop

The T flip-flop has one input, T (which stands for toggle), in addition to the clock. When T is asserted ($T = 1$), the flip-flop state toggles back and forth at each active edge of the clock, and when T is de-asserted, the flip-flop keeps its current state. The characteristic table, characteristic equation, state diagram, circuit, logic symbol, and excitation table for the T flip-flop are shown in Figure 6.30.

J	K	Q	Q_{next}	Q'_{next}
0	0	0	0	1
0	0	1	1	0
0	1	0	0	1
0	1	1	0	1
1	0	0	1	0
1	0	1	1	0
1	1	0	1	0
1	1	1	0	1

(a)

$$Q_{next} = K'Q + JQ'$$

(b)

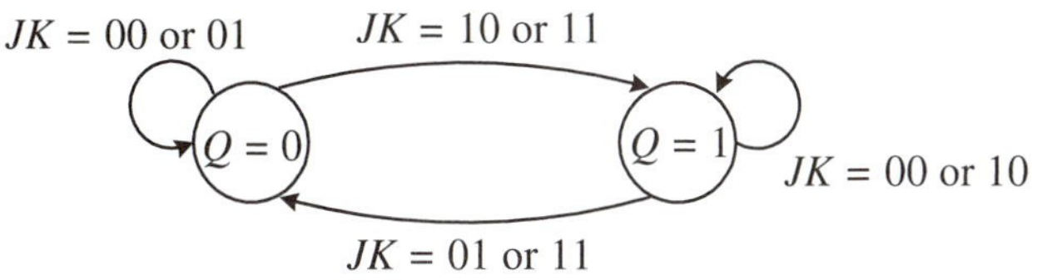

(c)

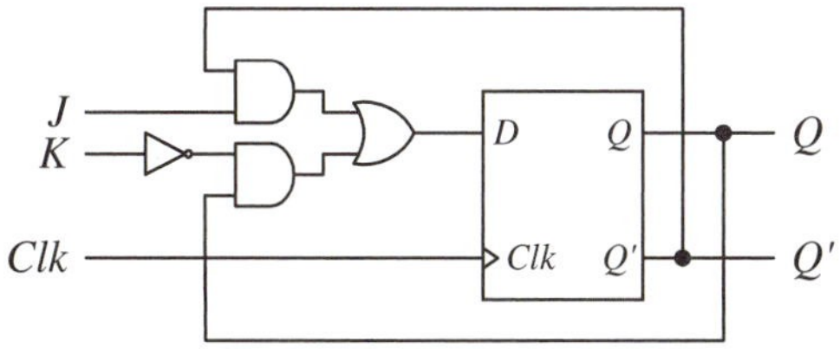

(d)

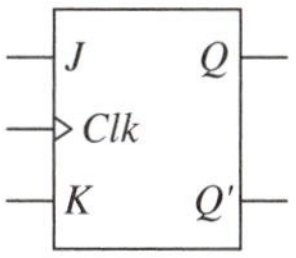

(e)

Q	Q_{next}	J	K
0	0	0	×
0	1	1	×
1	0	×	1
1	1	×	0

(f)

Figure 6.29 JK flip-flop: (a) characteristic table; (b) characteristic equation; (c) state diagram; (d) circuit; (e) logic symbol; (f) excitation table.

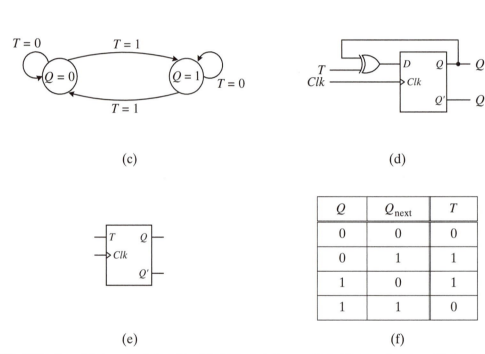

T	Q	Q_{next}	Q'_{next}
0	0	0	1
0	1	1	0
1	0	1	0
1	1	0	1

(a)

$$Q_{\text{next}} = T Q' + T' Q = T \oplus Q$$

(b)

(c)

(d)

(e)

Q	Q_{next}	T
0	0	0
0	1	1
1	0	1
1	1	0

(f)

Figure 6.30 T flip-flop: (a) characteristic table; (b) characteristic equation; (c) state diagram; (d) circuit; (e) logic symbol; (f) excitation table.

● ● ● ● ● ● ● ● ● ● ● ● ● · · · ·

6.15 Summary Checklist

- ■ Feedback loop
- ■ Bistable element
- ■ Latch

- Flip-flop
- Clock
  - Level-sensitive, active edge, rising/falling edge, clock cycle
- SR latch
- SR latch with enable
- D latch
- D latch with enable
- D flip-flop
  - Characteristic table, characteristic equation, state diagram, circuit, excitation table
- Asynchronous inputs
- VHDL implied memory element
- SR flip-flop
  - Characteristic table, characteristic equation, state diagram, circuit, excitation table
- JK flip-flop
  - Characteristic table, characteristic equation, state diagram, circuit, excitation table
- T flip-flop
  - Characteristic table, characteristic equation, state diagram, circuit, excitation table

● ● ● ● ● ● ● ● ● ● ● ● ● ● ● ●

6.16 Problems

P6.1. Draw an SR latch with enable similar to that shown in Figure 6.4, but use NOR gates to implement the SR latch. Derive the truth table for this circuit.

P6.2. Draw a D latch using NOR gates.

P6.3. Draw a D latch with enable similar to the circuit in Figure 6.6(a), but use NAND gates instead of the multiplexer.

P6.4. Draw a master–slave negative edge-triggered D flip-flop circuit.

P6.5. Derive the truth table for a negative edge-triggered D flip-flop.

P6.6. Draw the circuit for an SR flip-flop using SR latches.

P6.7. Derive the truth table for an SR flip-flop with enable.

P6.8. Write the behavioral VHDL code for an SR flip-flop with enable using the IF *clock*'EVENT statement.

P6.9. Do Problems P6.6, P6.7, and P6.8 for a JK flip-flop using SR latches.

P6.10. Do Problems P6.6, P6.7, and P6.8 for a T flip-flop using a JK flip-flop.

P6.11. Complete the timing diagram for the circuit in Figure P6.11. Assume that the signal delay through the NOR gates is 3 ns, and the delay through the NOT gate is 1 ns.

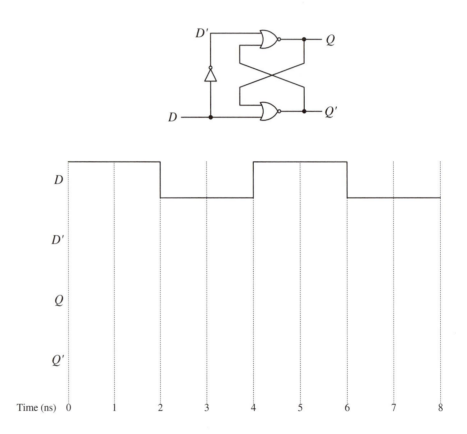

Figure P6.11

CHAPTER 7

Sequential Circuits

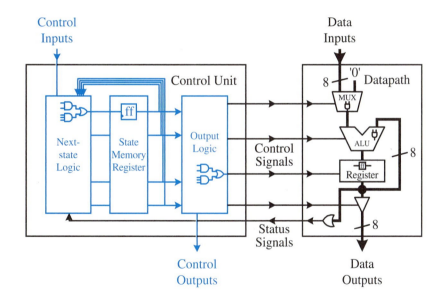

In the previous chapter, we looked at the design and operation of flip-flops—the most fundamental memory element used in microprocessor circuits. We saw that a single flip-flop is capable of remembering only one bit of information or one bit of history. In order for a sequential circuit to remember more inputs and a longer history, the circuit must contain more flip-flops. This collection of (D) flip-flops used to remember the complete history of past inputs is referred to as the **state memory**. The entire content of the state memory at a particular time forms a binary encoding that represents the complete history of inputs up to that time. We refer to this binary encoding in the state memory at one particular instance of time as the **state** of the system at that time.

The output signals from a sequential circuit are generated by the **output logic circuit**. Recall that the outputs of sequential circuits are dependent on their past and current inputs. Since all of the inputs are "remembered" in the form of states in the state memory, we can say that the outputs are dependent on the content of the state memory. Therefore, the output logic is simply a combinational circuit that is dependent on the content of the state memory and may or may not be dependent on the current inputs. The output signals that the output logic generates constitute the actions or operations that are performed by the sequential circuit. Hence, a sequential circuit can perform different operations in different states simply by generating different output signals.

If we want a sequential circuit to perform, say, four different operations, then we will need four states—one operation per state. Of course, if several operations can be performed in parallel, then we can assign them to one state. But for now, just to keep things simple, we simply will assign one operation per state. Furthermore, there might be an operation where we may want to repeat it for, say, a hundred times. Instead of assigning this same operation to one hundred different states, we will want to use just one state and have some form of looping capabilities to repeat that state a hundred times.

Thus, a sequential circuit operates by transitioning from one state to the next, generating different output signals. The part inside a sequential circuit that is responsible for determining what next state to go to is called the **next-state logic circuit**. Based on the current state that the system is in (i.e., the past inputs) and the current inputs, the next-state logic will determine what the next state should be. This statement, in fact, is equivalent to saying that the outputs are dependent on the past and current inputs, since a state is used to remember the past inputs, and it also determines the outputs to be generated. The next-state logic, however, is just a combinational circuit that takes the contents of the state memory flip-flops and the current inputs as its inputs. The outputs from the next-state logic are used to change the contents of the state memory flip-flops. The circuit changes state when the contents of the state memory change, and this happens at the active edge of every clock cycle, since values are written into a flip-flop at the active clock edge.

The speed at which a sequential circuit sequences through the states is determined by the speed of the clock signal. The state memory flip-flops are always enabled, so at every active edge of the clock, a new value is stored into the flip-flops. The limiting factor for the clock speed is in the time that it takes to perform all of the operations that are assigned to a particular state. All data operations assigned to a state must finish their operations within one clock period so that the results can be written into registers at the next active clock edge.

A sequential circuit is also known as a **finite state machine** (**FSM**) because the size of the state memory is finite, and therefore, the total number of different possible states is also finite. A sequential circuit is like a machine that operates by stepping through a sequence of states. Although there is only a finite number of different states, the FSM can go, however, to any of these states as many times as necessary. Hence, the sequence of states that the FSM can go through can be infinitely long.

The control unit inside the microprocessor is a finite state machine, therefore, in order to be able to construct a microprocessor, we need to understand the construction and operation of FSMs. In this chapter, we will first look at how to precisely describe the operation of a finite state machine using state diagrams. Next, we will look at the analysis and synthesis of finite state machines.

● ● ● ● ● ● ● ● ● ● ● ● ● ● ●

7.1 Finite State Machine (FSM) Models

In the introduction, we mentioned that the output logic circuit is dependent on the content of the state memory and may or may not be dependent on the current inputs. The fact that the output logic may or may not be dependent on the current inputs gives rise to two different FSM models.

Figure 7.1(a) shows the general schematic for the **Moore** FSM, where its outputs are dependent only on its current state (i.e., on the content of the state memory). Figure 7.1(b) shows the general schematic for the **Mealy** FSM, where its outputs are dependent on both the current state of the machine and the current inputs. The only difference between the two figures is that, for the Moore FSM, the output logic circuit only has the current state as its input; whereas, for the Mealy FSM, the output logic circuit has both the current state and the input signals as its inputs. In both models, we see that the inputs to the next-state logic are the primary input signals and the current state of the machine. The next-state logic circuit generates values to change the contents of the state memory. Since the state memory is made up of one or more D flip-flops and the content of the D flip-flop changes to whatever value is at its D-input at the next active clock edge, therefore, to change a state, the next-state logic circuit simply has to generate values for all of the D inputs for all of the flip-flops. These D-input values are referred to as the **excitation** values, since they "excite" or cause the D flip-flops to change states.

Figure 7.2(a) and (b) show a sample circuit of a Moore FSM and a Mealy FSM, respectively. The two circuits are identical except for their outputs. For the Moore FSM, the output circuit is a 2-input AND gate that gets its input values from the outputs of the two D flip-flops. Remember that the state of the FSM is represented by the content of the state memory, which is also the content of the flip-flops. The content

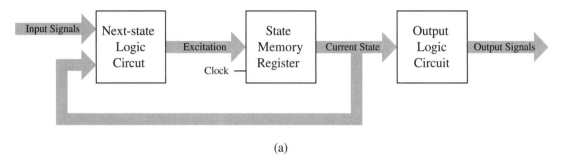

(a)

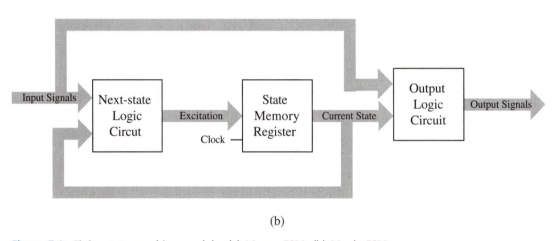

(b)

Figure 7.1 Finite state machine models: (a) Moore FSM; (b) Mealy FSM.

(or state) of a flip-flop is represented by the value at the Q (or Q') output. Hence, this circuit is dependent only on the current state of the machine.

For the Mealy FSM, the output circuit is a 3-input AND gate. In addition to getting its two inputs from the flip-flops, the third input to this AND gate is connected to the primary input, C. With this one extra connection, this output circuit is dependent on both the current state and the input, thus making it a Mealy FSM.

For both circuits, the state memory consists of two D flip-flops. Having two flip-flops, four different combinations of values can be represented. Hence, this finite state machine can be in any one of four different states. The state that this FSM will go to next depends on the value at the D inputs of the flip-flops.

Every flip-flop in the state memory requires a combinational circuit to generate a next-state value for its input(s). Since we have two D flip-flops (each having a D-input), the next-state logic circuit consists of two combinational circuits: one for input D_0 and one for D_1. The inputs to these two combinational circuits are the Q's,

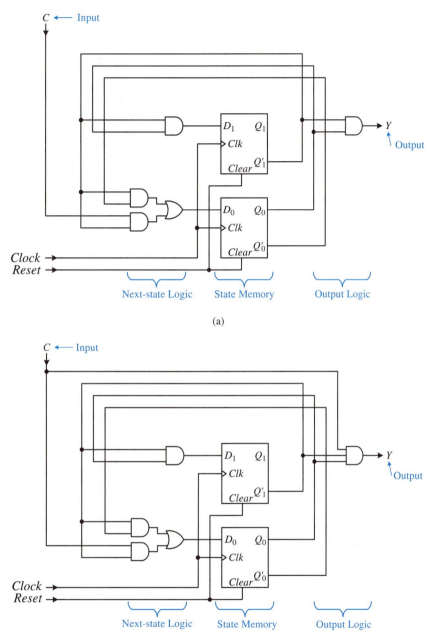

Figure 7.2
Sample finite state
machine circuits:
(a) Moore; (b) Mealy.

which represent the current state of the flip-flops and the primary input C. Notice that it is not necessary for the input C to be an input to all of the combinational circuits. In the sample circuit, only the bottom combinational circuit is dependent on the input C.

● ● ● ● ● ● ● ● ● ● ● ● ● ● ● ● ●● ●●

7.2 State Diagrams

State diagrams are used to precisely describe the operation of finite state machines. A state diagram is a deterministic graph with nodes and directed edges connecting the nodes. There is one node for every state of the FSM, and these nodes are labeled each with either its state name or encoding. For every state transition of the FSM, there is a directed edge connecting two nodes. The directed edge originates from the node for the current state that the FSM is transitioning from and goes to the node for the next state that the FSM is transitioning to. Edges may or may not have labels on them. Edges for unconditional transitions from one state to another will not have a label. In this case, only one edge can originate from that node. Conditional transitions from a state will have two outgoing edges for each input signal condition. The two edges from this state will have the corresponding input signal conditions labeled on them: one edge with the label for when the condition is true, and the other edge with the label for when the condition is false. If there is more than one input signal, then all of the possible input conditions must be labeled on the outgoing edges from the state. The state diagram is deterministic because from any node, it should show which is the next node to go to for any input combination. If an edge is not labeled, or if not all possible input conditions are labeled on the outgoing edges from the same state, then these missing conditions are don't-care conditions (see Section 3.4.2).

Figure 7.3(a) shows a sample state diagram having four states, one input signal, C, and one output signal, Y. The four states are labeled with the four encoded binary values 00, 01, 10, and 11. In this book, we will always use state 0 as the starting or reset state unless stated otherwise. There are three unconditional transitions (i.e., edges with no labels) from state 00 to 01, 10 to 00, and 11 to 00. There is one conditional transition from state 01 to either 10 or 11. For this conditional transition from state 01, if the condition $(C = 0)$ is true then the transition from 01 to 10 is made. Otherwise, if the condition $(C = 0)$ is false (i.e., $(C = 0)'$ is true or $(C = 1)$ is true) then the transition from 01 to 11 is made.

The output signal, Y, in Figure 7.3(a) is labeled inside or next to each node denoting that the output is dependent only on the current state. For example, when the FSM is in state 01, the output Y is set to 1; whereas, in state 11, Y is set to 0. Hence, this state diagram is for a Moore FSM.

The operation of the FSM based on the state diagram in Figure 7.3(a) goes as follows. After reset, the FSM starts from state 00. When it is in state 00, it outputs a 0 for Y. At the next rising clock edge, the FSM unconditionally transitions to state 01 and outputs a 1 for Y. Next, the FSM will either go to state 10 or 11 at the next rising clock edge, depending on the condition $(C = 0)$. If the condition $(C = 0)$ is true, then the FSM will go to state 10 and outputs a 0 for Y; otherwise, it will go to state 11 and also outputs a 0 for Y. From either state 10 or 11, the FSM will transition unconditionally back to state 00 at the next clock cycle. The FSM always will go to a new state at the beginning of the next active clock edge.

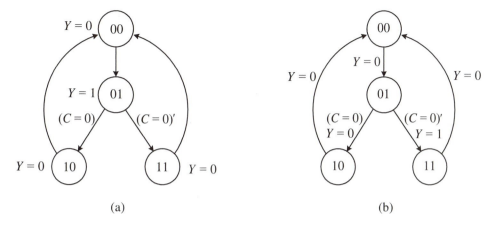

(a)   (b)

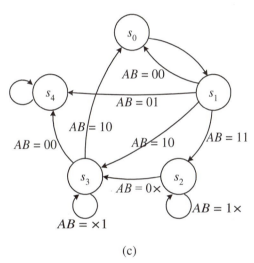

(c)

Figure 7.3 Sample state diagrams: (a) a Moore FSM with four states, one input signal C, and one output signal Y; (b) a Mealy FSM with four states, one input signal C, and one output signal Y; (c) an FSM with five states and two input signals, A and B.

Figure 7.3(b) shows a slightly different state diagram from the one in Figure 7.3(a). Instead of labeling the output signal Y inside or next to a node, it is labeled on the edges. What this means is that the output is dependent on both the current state (i.e., the state in which the edge originates from) and the input signal C. For example, when the FSM is in state 01, if the FSM takes the left edge for the condition $(C = 0)$ to state 10, then it will output a 0 for Y. However, if the FSM takes the right edge for the condition $(C = 0)'$ to state 11, then it will output a 1 for Y. Hence, this second state diagram is for a Mealy FSM.

Figure 7.3(c) shows a state diagram having five states, two input signals, and no output signals. In practice, all FSMs should have output signals; otherwise, they don't do anything useful. The five states in this state diagram are given the logical state names of s_0, s_1, s_2, s_3, and s_4. The two input signals are A and B. Again, we will use the state name with subscript 0, namely s_0, as the starting state. From state s_0, there is one unconditional edge going to state s_1. This unlabeled edge is equivalent to having the label $AB = \times\times$, meaning that this edge is taken for any combination of the two input signals. From state s_1, there are four outgoing edges labeled with the four different combinations of the two input signals. State s_2 has only two outgoing edges. However, the two labels on them cover the four possible input conditions, since B is don't-care in both cases. State s_3 has only three outgoing edges, but again, the labels on them cover all four input conditions.

As you can see, a state diagram is very similar to a computer program flowchart where the nodes are for the statements or data operations, and the edges are for the control of the program sequence. Because of this similarity, we should be able to convert any program to a state diagram. Example 7.1 shows how to convert a simple C-style pseudocode to a state diagram.

| Example 7.1 | **Converting pseudocode to a state diagram** |

Derive the state diagram based on the following pseudocode.

```
x = 5
WHILE (x ≠ 0) {
    OUTPUT x
    x = x - 1
    }
```

The pseudocode has three data-operation statements and one conditional test. Each data-operation statement is assigned to a node (state), as shown in Figure 7.4(a). Each node is given a name for the state and is annotated with the statement to be

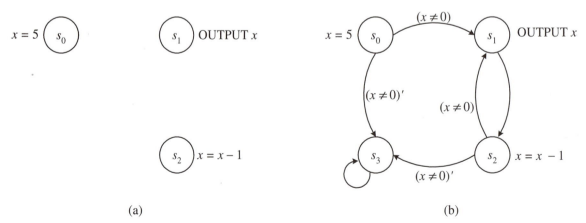

(a)   (b)

Figure 7.4   State diagram for Example 7.1: (a) data operations assigned to nodes; (b) complete state diagram with the transitional edges.

executed in that state. At this point, instead of labeling the nodes with the actual binary encoding for the state, it is better to just give it a name. The actual encoding of the state can be done later on during the synthesis process.

Next, we assign directional edges to the diagram based on the sequence of execution. Starting from state s_0 (where the statement $x = 5$ is executed), the program then tests for the condition $(x \neq 0)$. If the condition is true, then the output statement is executed; otherwise, the loop (and the program) is terminated. Referring to Figure 7.4(b), there are two outgoing edges from state s_0. The edge from s_0 to s_1 has the label $(x \neq 0)$, which means that if the condition $(x \neq 0)$ is true, then this edge is taken, and so, it will go to state s_1 to execute the output statement. On the other hand, if the condition is false, the loop will be terminated. Since there is no statement after the loop, we have to add an extra no-operation state, s_3, to the state diagram for it to go to. The edge from s_0 to s_3 is labeled $(x \neq 0)'$, meaning that the edge is taken when the condition $(x \neq 0)$ is false.

After executing the output statement, the decrement statement is executed. This sequence is reflected in the unconditional edge going from state s_1 to s_2. After executing the decrement statement in s_2, the condition $(x \neq 0)$ in the WHILE loop is again tested. If the condition is true, it will take the edge with the label $(x \neq 0)$ back to state s_1 to repeat the loop. If the condition is false, it will take the edge with the label $(x \neq 0)'$ to state s_3. From state s_3, it unconditionally loops back to itself; thus, going nowhere and doing nothing.

● ● ● ● ● ● ● ● ● ● ● ● ● ● ● ●

7.3 Analysis of Sequential Circuits

Very often, we are given a sequential circuit and need to know its operation. The **analysis of sequential circuits** is the process in which we are given a sequential circuit (such as the ones in Figure 7.2) and we want to obtain a precise description of the operation of the circuit by deriving the state diagram for it. The steps for the analysis of sequential circuits are as follows:

1. Derive the excitation equations from the next-state logic circuit.

2. Derive the next-state equations by substituting the excitation equations into the flip-flop's characteristic equations.

3. Derive the next-state table from the next-state equations.

4. Derive the output equations from the output logic circuit.

5. Derive the output table from the output equations.

6. Draw the state diagram from the next-state table and the output table.

The following sections will explain these steps in detail.

7.3.1 Excitation Equation

The **excitation equations** are the equations for the next-state logic circuit in the FSM. In other words, they are just the input equations to the state memory flip-flops in the FSM. Since the next-state logic is a combinational circuit, deriving the excitation equations is just an analysis of a combinational circuit, as discussed in Section 3.1.2. The next-state logic circuit that is derived by these equations "excites" the flip-flops by causing them to change states, hence the name "excitation equation." These equations provide the signals to the inputs of the flip-flops and are expressed as a function of the current state and the inputs to the FSM. The current state is determined by the current contents of the flip-flops (i.e., the flip-flops' output signals Q and Q'). There is one equation for each flip-flop's input.

The following are the two excitation equations for the two D flip-flops used in the circuit from Figure 7.2(a). Equation (7.1) is from the next-state logic circuit for the D_1 input of flip-flop 1, and Equation (7.2) is from the next-state circuit for the D_0 input of flip-flop 0.

$$D_1 = Q_1' Q_0 \tag{7.1}$$

$$D_0 = Q_1' Q_0' + C Q_1' \tag{7.2}$$

7.3.2 Next-State Equation

The **next-state equations** specify what the flip-flops' next state is going to be depending on two things: (1) the inputs to the flip-flops, and (2) the functional behavior of the flip-flops. The inputs to the flip-flops are provided by the excitation equations, as discussed in Section 7.3.1. The functional behavior of a flip-flop (as you recall from Section 6.10.2) is described formally by its characteristic equation. The characteristic equation tells us what Q_{next} ought to be (i.e., what the next state ought to be), depending on the current state and current inputs. Thus, to derive the next-state equations, we substitute the excitation equations into the corresponding flip-flops' characteristic equations.

For example, the characteristic equation for the D flip-flop (from Section 6.10.2 is

$$Q_{next} = D$$

Therefore, substituting the two excitation equations from Equations (7.1) and (7.2) into the characteristic equation for the D flip-flop will give us the following two next-state equations.

$$Q_{1next} = D_1 = Q_1' Q_0 \tag{7.3}$$

$$Q_{0next} = D_0 = Q_1' Q_0' + C Q_1' \tag{7.4}$$

7.3.3 Next-State Table

The **next-state table** is simply the truth table as derived from the next-state equations. It lists for every combination of the current state values (Q) and input values,

Current State	Next State	
$Q_1 Q_0$	$Q_{1next} Q_{0next}$	
	$C = 0$	$C = 1$
00	01	01
01	10	11
10	00	00
11	00	00

Figure 7.5
A next-state table with four states and one input signal, C.

what the next-state values (Q_{next}) should be. These next-state values are obtained by substituting the current state and input values into the appropriate next-state equations.

Figure 7.5 shows the next-state table as obtained from the two next-state equations, Equations (7.3) and (7.4). Having two flip-flops, Q_1 and Q_0, there are four encodings, $00, 01, 10$, and 11, for the current state. There is one input signal, C, with the two possible values, 0 and 1. The entries in the table are the next-state values Q_{1next} and Q_{0next}. For each entry, the leftmost bit is for the Q_1 flip-flop, and the rightmost bit is for the Q_0 flip-flop. These next-state values are obtained from substituting the current state values, $Q_1 Q_0$, and the input value, C, into the next-state equations, Equations (7.3) and (7.4).

For example, to get the Q_{1next} value for the top-left entry (the left bit in the blue entry), we substitute the current state values, $Q_1 = 0$ and $Q_0 = 0$, and the input value, $C = 0$, into Equation (7.3) giving

$$Q_{1next} = Q_1' Q_0$$
$$= 0' \cdot 0$$
$$= 1 \cdot 0$$
$$= 0$$

Substituting the same values into Equation (7.4) will give us the Q_{0next} value for that same top-left entry.

$$Q_{0next} = Q_1' Q_0' + C Q_1'$$
$$= 0' \cdot 0' + 0 \cdot 0'$$
$$= 1 + 0$$
$$= 1$$

The rest of the entries in the next-state table are obtained in the same manner by substituting the corresponding values for Q_1, Q_0, and C into the two next-state equations.

The top-left entry tells us that if the current state $Q_1 Q_0$ is 00 and the input signal C is 0, then the next state $Q_{1next} Q_{0next}$ that the FSM will go to is 01. From the current state 00, if the input signal C is 1, the next state is also 01. This means that the transition from state 00 to 01 does not depend on the input condition, C, so this is an unconditional transition. From state 01, there are two conditional transitions: the FSM will transition to state 10 if the condition $C = 0$ is true; otherwise, if $C = 1$, it will transition to state 11. From either state 10 or 11, the FSM will go to state 00 unconditionally.

7.3.4   Output Equation

The **output equations** are the equations derived from the combinational output logic circuit in the FSM. Depending on the type of FSM (Moore or Mealy), the output equations can be dependent on just the current state or on both the current state and the inputs.

For the Moore circuit of Figure 7.2(a), the output equation is

$$Y = Q_1' Q_0 \tag{7.5}$$

For the Mealy circuit of Figure 7.2(b), the output equation is

$$Y = C Q_1' Q_0 \tag{7.6}$$

A typical FSM will have many output signals, and so, there will be one equation for every output signal.

7.3.5   Output Table

The **output table** is the truth table that is derived from the output equations. The output tables for the Moore and Mealy FSMs are slightly different from each other. For the Moore FSM, the output table lists for every combination of the current state what the output values should be. Whereas, for the Mealy FSM, the output table lists for every combination of the current state *and* input values what the output values should be. These output values are obtained by substituting the current state and input values into the appropriate output equations.

Figure 7.6(a) and (b) shows the output tables for the Moore and Mealy FSMs as derived from the output equations in Equations (7.5) and (7.6), respectively, from Section 7.3.4. For the Moore FSM, the output signal Y is dependent only on the current state value $Q_1 Q_0$; whereas, for the Mealy FSM, the output signal Y is dependent on both the current state and input C.

7.3.6   State Diagram

The last step in the analysis of a sequential circuit is to derive the state diagram. The state diagram is obtained directly from the next-state table and the output table.

The next-state table from Figure 7.5 shows that there are four states in the state diagram. For each next-state entry in the table, there is a corresponding edge going

Current State $Q_1 Q_0$	Output Y
00	0
01	1
10	0
11	0

Current State $Q_1 Q_0$	Output Y	
	$C = 0$	$C = 1$
00	0	0
01	0	1
10	0	0
11	0	0

(a)   (b)

Figure 7.6   Output table: (a) for a Moore FSM; (b) for a Mealy FSM.

from that current state to that next state. The corresponding input condition is the label for that edge.

The state diagram shown in Figure 7.3(a) is derived from the next-state table from Figure 7.5 and the Moore output table from Figure 7.6(a). The state diagram shown in Figure 7.3(b) is derived from the same next-state table from Figure 7.5, but using the Mealy output table from Figure 7.6(b).

7.3.7   Analysis of a Moore FSM

We will now illustrate the complete process of analyzing FSMs with two examples.

Example 7.2   **Analyzing a Moore FSM**

Figure 7.7 shows a simple sequential circuit. Comparing this circuit with the general FSM schematic in Figure 7.1, we conclude that this is a Moore type FSM, since the output logic consists of a 2-input AND gate that is dependent only on the current state, $Q_1 Q_0$. We will follow the six steps described earlier to do a detailed analysis of this sequential circuit.

Step 1 of the analysis is to derive the excitation equations, which are the equations for the next-state logic circuit. These equations are dependent on the current state of the flip-flops, Q_1 and Q_0, and the input, C. One equation is needed for every data input of all of the flip-flops in the state memory. Our sample circuit has two flip-flops having the two inputs, D_1 and D_0, so we get the two excitation equations, as shown in Figure 7.8(a). These two equations are obtained from analyzing the two combinational circuits that provide the inputs, D_1 and D_0, to the two flip-flops. For this particular example, both of these combinational circuits are simple two-level sum-of-products circuits.

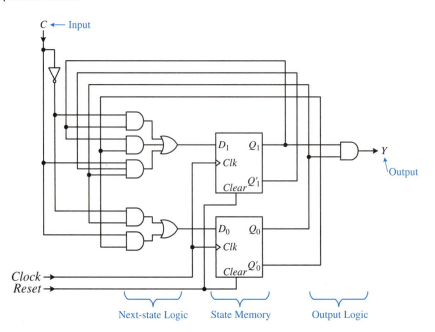

Figure 7.7
A simple Moore
finite state machine.

$$D_1 = C'Q_1 + Q_1Q_0' + CQ_1'Q_0$$
$$D_0 = C'Q_0 + CQ_0'$$

(a)

$$Q_{1next} = D_1 = C'Q_1 + Q_1Q_0' + CQ_1'Q_0$$
$$Q_{0next} = D_0 = C'Q_0 + CQ_0'$$

(b)

Current State	Next State	
Q_1Q_0	$Q_{1next}Q_{0next}$	
	$C = 0$	$C = 1$
00	00	01
01	01	10
10	10	11
11	11	00

(c)

$$Y = Q_1Q_0$$

(d)

Figure 7.8 Analysis of a Moore FSM: (a) excitation equations; (b) next-state equations; (c) next-state table; (d) output equation; (e) output table; (f) state diagram; (g) timing diagram.
(continued on next page)

Current State Q_1Q_0	Output Y
00	0
01	0
10	0
11	1

(e)

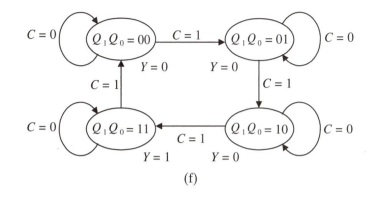

(f)

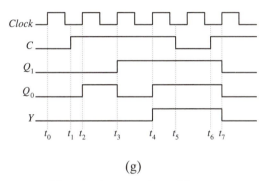

(g)

Figure 7.8   Analysis of a Moore FSM: (a) excitation equations; (b) next-state equations; (c) next-state table; (d) output equation; (e) output table; (f) state diagram; (g) timing diagram.

Step 2 is to derive the next-state equations. These equations tell us what the next state is going to be given the inputs to and the functional behavior of the flip-flops. One equation is needed for every flip-flop. The functional behavior of a flip-flop is described by its characteristic equation, which for the D flip-flop is $Q_{next} = D$. The inputs to the flip-flops are just the excitation equations derived from Step 1. Hence, we simply substitute the excitation equation into the characteristic equation for each flip-flop to obtain the next-state equation for that flip-flop. With two flip-flops in the example, we get two next-state equations: one for Q_{1next} and one for Q_{0next}. Figure 7.8(b) shows these two next-state equations.

Step 3 is to derive the next-state table. The next-state values in the table are obtained by substituting every combination of current state and input values into the next-state equations obtained in Step 2. In this example, there are two flip-flops, Q_1 and Q_0, and one input, C. Hence, the table will have eight next-state entries. There are two bits for every entry—the first bit is for Q_{1next}, and the second bit is for Q_{0next}.

For example, to find the Q_{1next} value for the current state $Q_1Q_0 = 00$ and $C = 1$ (the blue entry in Figure 7.8(c)), we substitute the values $Q_1 = 0$, $Q_0 = 0$, and $C = 1$ into the equation $Q_{1next} = C'Q_1 + Q_1Q_0' + CQ_1'Q_0 = (1' \cdot 0) + (0 \cdot 0') + (1 \cdot 0' \cdot 0)$ to get the value of 0. Similarly, we get Q_{0next} by substituting the same values for Q_1,

Q_0, and C into the equation $Q_{0next} = C'Q_0 + CQ_0' = (1' \cdot 0) + (1 \cdot 0')$ to get the value of 1. The resulting next-state table for this example is shown in Figure 7.8(c).

Step 4 is to derive the output equations from the output logic circuit. One output equation is needed for every output signal. For this example, there is only one output signal, Y, that is dependent only on the current state of the FSM. The output equation for Y, as derived from the circuit diagram, is shown in Figure 7.8(d).

Step 5 is to derive the output table. Just like the next-state table, the output table is obtained by substituting all possible combinations of the current state values into the output equation(s) for the Moore FSM. The output table for this Moore FSM example is shown in Figure 7.8(e).

Step 6 is to draw the state diagram, which is derived directly from the next-state and output tables. Every state in the next-state table will have a corresponding node labeled with the state encoding in the state diagram. For every next-state entry in the next-state table, there will be a corresponding directed edge. This edge originates from the node labeled with the current state and ends at the node labeled with the next-state entry. The edge is labeled with the corresponding input conditions.

For example, in the next-state table, when the current state Q_1Q_0 is 00, the next state $Q_{1next}Q_{0next}$ is 01 for the input $C = 1$. Hence, in the state diagram, there is a directed edge from node 00 to node 01 with the label $C = 1$. For a Moore FSM, the outputs are dependent only on the current state, thus the output values from the output table are included inside each node in the state diagram. The complete state diagram for this example is shown in Figure 7.8(f).

A sample timing diagram for the execution of the circuit is shown in Figure 7.8(g). The two D flip-flops used in the circuit are positive edge-triggered flip-flops, so they change their states at every rising clock edge. Initially, we assume that these two flip-flops, Q_1Q_0, are both in state 0. The first rising clock edge is at time t_0. Normally, the flip-flops will change state at this time, however, since C is a 0, the flip-flops' values remain constant. At time t_1, C changes to a 1, so that, at the next rising clock edge at time t_2, the flip-flop value Q_1Q_0 changes to 01. At the next two rising clock edges, t_3 and t_4, the value for Q_1Q_0 changes to 10, then to 11, respectively. At time t_4, when $Q_1Q_0 = 11$, the output Y also changes to a 1, since $Y = Q_1 \cdot Q_0$. At time t_5, input C drops back down to a 0, but the output Y remains at a 1. Q_1Q_0 remains the same (at 11) through the next rising clock edge, since C is 0. At time t_6, C changes back to a 1, and so at the next rising clock edge at time t_7, Q_1Q_0 increments again to 00, and the cycle repeats.

When $C = 1$, the FSM cycles through the four states in order repeatedly. When $C = 0$, the FSM stops at the current state until C is asserted again. If we interpret the four state encodings as a decimal number, then we can conclude that the circuit of Figure 7.7 is for a modulo-4 up counter that cycles through the four values 0, 1, 2, and 3. The input C enables or disables the counting.

7.3.8   Analysis of a Mealy FSM

Example 7.3 illustrates the process for performing an analysis of a Mealy FSM.

| **Example 7.3** | **Analyzing a Mealy FSM** |

Figure 7.9 shows a simple Mealy FSM. This circuit is exactly like the one in Figure 7.7, except that the output circuit, which in this example is just one 3-input AND gate, is dependent on not only the current state, $Q_1 Q_0$, but also on the input, C.

The analysis for this circuit goes exactly like the one for the Moore FSM in Example 7.2 up to creating the next-state table in Step 3. The only difference is in deriving the output equation and output table for Steps 4 and 5. For a Mealy FSM, the output equation is dependent on both the current state and the input value. Since the circuit has only one output signal, we obtain the output equation that is dependent on C, as shown in Figure 7.10(a). Figure 7.10(b) shows the resulting output table obtained by substituting all possible values for Q_1, Q_0, and C into the output equation.

For the state diagram, we cannot put the output value inside a node since the output value is dependent on the current state and the input value. Thus, the output value is placed on the edge that corresponds to the current state value and input value, as shown in Figure 7.10(c). The output signal, Y, is 0 for all edges, except for the one originating from state 11 having the input condition of $C = 1$. On this one edge, Y is a 1.

A sample timing diagram is shown in Figure 7.10(d). This diagram is exactly the same as the one for the Moore FSM shown in Figure 7.8(g) up to time t_5. At time t_5, input C drops to a 0, and so output Y also drops to a 0, since $Y = C \cdot Q_1 \cdot Q_0$. At time t_6, C rises back up to a 1, and so Y also rises to a 1 immediately. Since the output circuit is a combinational circuit, Y does not change at the active edge of the clock, but changes immediately when the inputs change. At time t_7, when $Q_1 Q_0$ changes to 00, Y again changes back to a 0.

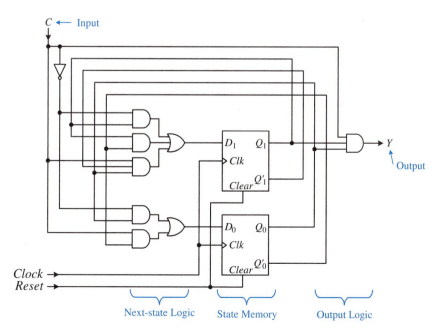

Figure 7.9
A simple Mealy
finite state machine.

$$Y = CQ_1Q_0$$

Current State	Output	
Q_1Q_0	Y	
	$C = 0$	$C = 1$
00	0	0
01	0	0
10	0	0
11	0	1

(a)   (b)

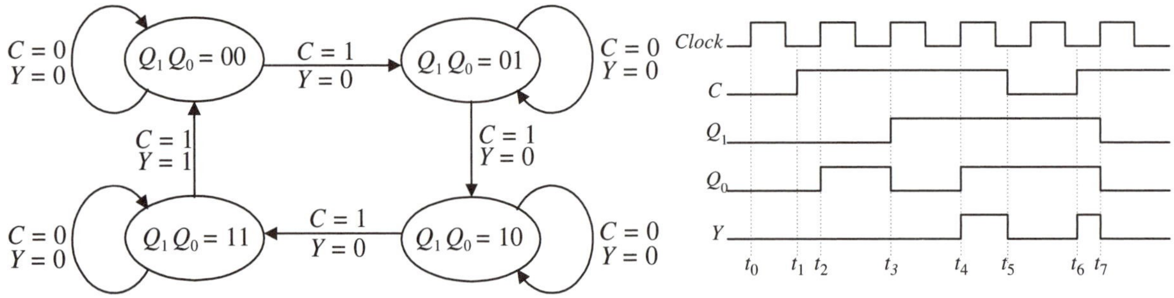

(c)   (d)

Figure 7.10   Analysis of a Mealy FSM: (a) output equation; (b) output table; (c) state diagram; (d) timing diagram.

Except for the difference in how this circuit generates the output signal Y, this Mealy FSM behaves exactly the same as the Moore FSM from Example 7.2 in the way that it changes from one state to the next. This, of course, is due to the fact that both next-state tables are identical. Thus, this Mealy FSM circuit is also a modulo-4 up counter.

● ● ● ● ● ● ● ● ● ● ● ● ● ● ● ●

7.4 Synthesis of Sequential Circuits

The **synthesis of sequential circuits** is just the reverse of the analysis of sequential circuits. In synthesis, we start with what is usually an ambiguous functional description of the circuit that we want. From this description, we need to come up with the precise operation of the circuit using a state diagram. The state diagram allows us to construct the next-state and output tables. From these two tables, we get the next-state and output equations, and finally, the complete FSM circuit.

During the synthesis process, there are many possible circuit optimizations in terms of the circuit size, speed, and power consumption that can be performed. Circuit optimization is discussed in Section 7.8. In this section, we will focus only on synthesizing a functionally correct sequential circuit.

The steps for the synthesis of sequential circuits are as follows:

1. Produce a state diagram from the functional description of the circuit.
2. Derive the next-state table from the state diagram.
3. Convert the next-state table to the implementation table.
4. Derive the excitation equations for each flip-flop input from the implementation table.
5. Derive the output table from the state diagram.
6. Derive the output equations from the output table.
7. Draw the FSM circuit diagram based on the excitation and output equations.

7.4.1 State Diagram

The first step in the sequential circuit synthesis process is to derive the state diagram for it. The circuit to be built usually is described using an ambiguous natural language. Not only does the language itself create uncertainties, in many cases the description of the circuit is also incomplete. This incomplete description arises when not all possible situations of an event or a behavior are specified. In order to translate an ambiguous description into a precise state diagram, the designer must have a full understanding of the functional behavior of the circuit in question. In addition, the designer may need some ingenuity and creativity to fill in the missing gaps. Meaningful assumptions need to be made and stated clearly, and ambiguous situations need to be clarified. This is the one step in the design process where there is no clear-cut answer for it. In this step, we rely on the knowledge and expertise of the designer to come up with a correct and meaningful state diagram.

Instead of using a natural language to describe the circuit, a more precise method can be used. Other ways to describe a circuit more precisely include the use of a hardware description language (such as VHDL), a state action table, or an ASM chart. The use of ASM charts and state action tables are described in Chapter 10.

In this section, we will construct an FSM circuit based on the C-style pseudocode shown in Figure 7.11. Do not try to interpret the logical execution of the code, because it does not perform anything meaningful. Furthermore, this section is not about optimizing the code by modifying it to make it shorter, although optimizing the code this way may produce a smaller FSM circuit. In this section, the focus is on learning how to convert any given pseudocode (as is) to an FSM circuit that realizes it. Section 7.8 discusses how to optimize sequential circuits.

The pseudocode shown in Figure 7.11 contains four signal assignment statements —two $Y = 0$, and two $Y = 1$. We assign one state to each of the four signal assignment statements. The first $Y = 0$ is assigned to state s_0, the second $Y = 0$ is assigned to state s_1, and so on, as shown in the pseudocode.

```
REPEAT {
    Y = 0              -- s₀
    IF (B = 0) THEN
        Y = 0          -- s₁
    ELSE
        Y = 1          -- s₂
    END IF
    Y = 1              -- s₃
}
```

Figure 7.11
C-style pseudocode
for synthesis.

After the first $Y = 0$ statement, the IF statement conditionally determines whether to execute the second $Y = 0$ statement or the $Y = 1$ statement. Hence, from state s_0, there is one edge going to state s_1, and one edge going to state s_2. The labels on these two edges are the conditions for the IF statement. The edge going to state s_1 has the label $(B = 0)$, and the edge going to state s_2 has the label $(B = 1)$. After executing either state s_1 or state s_2, state s_3 is executed, hence, there are two unconditional edges from these two states to s_3. Finally, because of the unconditional REPEAT loop, there is an unconditional edge from s_3 going back to state s_0. The resulting state diagram is shown in Figure 7.12(a).

7.4.2 Next-State Table

Given a state diagram, it is easy to derive both the next-state and output tables from it. Since the next-state and output tables and the state diagram portray the same

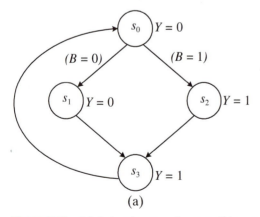

(a)

Current State	Next State	
$Q_1 Q_0$	$Q_{1next} Q_{0next}$	
	$B = 0$	$B = 1$
s_0 00	s_1 01	s_2 10
s_1 01	s_3 11	s_3 11
s_2 10	s_3 11	s_3 11
s_3 11	s_0 00	s_0 00

(b)

Figure 7.12 (a) A simple state diagram; (b) next-state table; (c) implementation table using D flip-flops; (d) excitation equations; (e) output table; (f) FSM circuit.
(continued on next page)

Current State	Implementation	
$Q_1 Q_0$	$D_1 D_0$	
	$B = 0$	$B = 1$
00	01	10
01	11	11
10	11	11
11	00	00

(c)

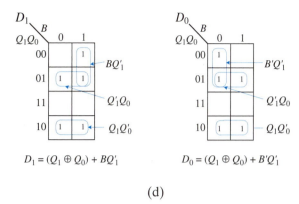

$$D_1 = (Q_1 \oplus Q_0) + BQ'_1$$

$$D_0 = (Q_1 \oplus Q_0) + B'Q'_1$$

(d)

Current State	Output
$Q_1 Q_0$	Y
s_0 00	0
s_1 01	0
s_2 10	1
s_3 11	1

(e)

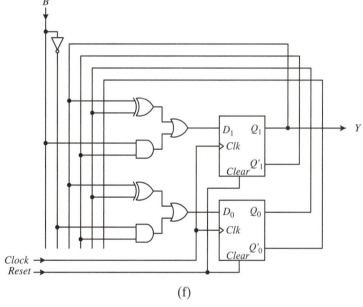

(f)

Figure 7.12 (a) A simple state diagram; (b) next-state table; (c) implementation table using D flip-flops; (d) excitation equations; (e) output table; (f) FSM circuit.

information but depicted in a different format, it requires only a straightforward translation from one to the other.

Figure 7.12(b) shows the next-state table for the state diagram shown in (a). The row labels are the current state and the column labels are the input conditions. The table entries are the next states. Translating directly from the state diagram from the current state, s_0, if the condition ($B = 0$) is true, then the next state is s_1. Correspondingly, in the next-state table, the entry for the intersection of the current state, s_0, and input $B = 0$ is s_1.

In the next-state table, the actual encoding for the states is also given. To encode the four states, two flip-flops, Q_1 and Q_0, are required. In the example, the encoding given to the four states, s_0, s_1, s_2, and s_3, is just the four different combinations of the two flip-flop values, 00, 01, 10, and 11, respectively. Using different encoding schemes can give different results in terms of circuit size, speed, and power consumption. This optimization technique is further discussed in Section 7.8.2.

7.4.3   Implementation Table

The implementation table is derived from the next-state table. Whereas, the next-state table is independent of the flip-flop type used, the implementation table is dependent on the choice of flip-flop used. An FSM can be implemented using any one of the four different types of flip-flops (as discussed in Section 6.11) or combinations of them. Using different flip-flops or combinations of flip-flops can produce different size circuits but with the same functionality. The current trend in microprocessor design is to use only D flip-flops because of their ease of use. We will, likewise, use only D flip-flops in our synthesis of sequential circuits. Section 7.8.3 discusses how sequential circuits are synthesized with other types of flip-flops.

The implementation table shows what the flip-flop inputs ought to be in order to realize the next-state table. In other words, it shows the necessary inputs for the flip-flops that will produce the next states as given in the next-state table. The next-state table answers the question: What is the next state of the flip-flop given the current state of the flip-flop and the input values? The implementation table, on the other hand, answers the question: What should the input(s) to the flip-flop be in order to realize the corresponding next state given in the next-state table?

The flip-flop inputs that we are concerned with are the synchronous inputs. For the D flip-flop, this is just the D input. For the other flip-flop types, they are the S and R inputs for the SR flip-flop; the J and K inputs for the JK flip-flop; and the T input for the T flip-flop. We do not consider the asynchronous inputs such as the *Set* and *Clear* inputs, nor do we consider the *Clock* input signal.

Hence, to derive the implementation table using D flip-flops, we need to determine the value that must be assigned to the D input so that it will result in the corresponding Q_{next} value as given in the next-state table. However, since the characteristic equation for the D flip-flop (i.e., the equation that describes the operation of the D flip-flop as given in Section 6.10.2) is

$$Q_{\text{next}} = D$$

therefore, the values for Q_{next} and D are the same.

Thus, the entries in the implementation table using D flip-flops are identical to the entries in the next-state table. The only difference between the two tables is in the meaning of the entries. In the next-state table shown in Figure 7.12(b), the label for the entries is Q_{next} for the next state to go to; whereas, in the implementation table shown in Figure 7.12(c), the label for the entries is D for the input to the D flip-flop. Since there are two flip-flops, Q_1 and Q_0, each having one input D, hence the implementation table has the two corresponding inputs, D_1 and D_0. The leftmost bit is for flip-flop 1, and the rightmost bit is for flip-flop 0. Note that if one of the other types of flip-flops is used, the two tables will not be the same as discussed in Section 7.8.3.

7.4.4 Excitation Equation and Next-State Circuit

Recall that the excitation equations are the equations for the flip-flops' synchronous inputs. There is one excitation equation for every input of every flip-flop. Remember that we do not include the asynchronous inputs and the clock input. The excitation equations are dependent on the current state encodings (i.e., the contents of the flip-flops) and the primary FSM input signals.

The excitation equations are what caused the flip-flops in the state memory to change state. The circuit that is derived from these equations is the next-state circuit in the FSM. The next-state circuit is a combinational circuit, and so deriving this circuit is the same as synthesizing any other combinational circuit, as discussed in Section 3.2.

The implementation table derived from the previous step is just the truth table for the excitation equations. For our example, we need two equations for the two flip-flop inputs, D_1 and D_0. In the example, extracting the leftmost bit in every entry in the implementation table will give us the truth table for D_1, and therefore, the excitation equation for D_1. Similarly, extracting the rightmost bit in every entry in the implementation table will give us the truth table and excitation equation for D_0. The truth table, in the form of a K-map, and the excitation equations for D_1 and D_0 are given in Figure 7.12(d).

7.4.5 Output Table and Equation

The output table and output equations are used to derive the output circuit in the FSM. The output table can be obtained directly from the state diagram. In the state diagram of Figure 7.12(a), the output signal, Y, is dependent only on the state. In states s_0 and s_1, Y is assigned the value 0. In states s_2 and s_3, Y is assigned a 1. The resulting output table is shown in Figure 7.12(e).

The output equation as derived from the output truth table is simply

$$Y = Q_1$$

7.4.6 FSM Circuit

Using Figure 7.2(a) as a template, our FSM circuit requires two D flip-flops for its state memory. The number of flip-flops to use was determined when the states were encoded. The type of flip-flops to use was determined when deriving the implementation table. The next-state circuit is drawn from the excitation equations, while the output circuit is drawn from the output equation. Connecting these three parts: state memory, next-state circuit, and output circuit, together produce the final FSM circuit shown in Figure 7.12(f).

7.4.7 Synthesis of Moore FSMs

We will now illustrate the synthesis of Moore FSMs with two examples. Example 7.4 illustrates the synthesis of a simple Moore FSM. Example 7.5 illustrates the synthesis of a Moore FSM that is more typical of what the control unit of a microprocessor is like.

Example 7.4 **Synthesizing a simple Moore FSM**

For our first synthesis example, we will design a modulo-6 up counter using D flip-flops having a count enable input, C, and an output signal, Y, that is asserted when the count is equal to five. The count is to be represented directly by the contents of the flip-flops.

Step 1 of the synthesis process is to construct the state diagram. From the above functional description, we need to construct a state diagram that will show the precise operation of the circuit. A modulo-6 counter counts from zero to five, and then back to zero. Since the count is represented by the flip-flop values and we have six different counts (from zero to five), we will need three flip-flops (Q_2, Q_1, Q_0) that will produce the sequence $000, 001, 010, 011, 100, 101, 000, \ldots$ when C is asserted; otherwise, when C is de-asserted, the counting stops. In other words, from state 000, which represents a count of zero, there will be an edge that goes to state 001 with the label $C = 1$. From state 001, there is an edge that goes to state 010 with the label $C = 1$, and so on. For the counting to stop at each count, there will be edges from each state that loop back to itself with the label $C = 0$. Furthermore, we want to assert Y in state 101, so in this state, we set Y to a 1. For the rest of the states, Y is set to a 0. Hence, we obtain the state diagram in Figure 7.13(a) for a modulo-6 up counter.

Step 2 is to derive the next-state table, which is a direct translation from the state diagram. We have three flip-flops, Q_2, Q_1, and Q_0, and one primary input, C. The current states for the flip-flops are listed down the rows, while the input is listed across the columns. The entries are the next states. For each entry in the next-state table, we need to determine what the next state is for each of the three flip-flops, so there are three bit values, Q_{2next}, Q_{1next}, and Q_{0next}, for each entry. For example, if the current state is $Q_2 Q_1 Q_0 = 010$ and the input is $C = 1$, then the next state, $Q_{2next} Q_{1next} Q_{0next}$, is 011. The next-state table is shown in Figure 7.13(b).

Step 3 is to convert the next-state table to its implementation table. Since, for the D flip-flop, the implementation table is the same as the next-state table, we simply can use the next-state table and just re-label the entry heading, as shown in Figure 7.13(c).

Step 4 is to derive the excitation equations for all of the flip-flop inputs in terms of the current state and the primary input. These equations are obtained directly from the implementation table. In the example, there are three flip-flops with the three inputs, D_2, D_1, and D_0, which correspond to the three bits in the entries in the implementation table. To derive the equation for D_2, we consider just the leftmost bit in each entry for the truth table for D_2. Looking at all the leftmost bits, there are four 1-minterms giving the canonical equation:

$$D_2 = C' Q_2 Q_1' Q_0' + C' Q_2 Q_1' Q_0 + C Q_2' Q_1 Q_0 + C Q_2 Q_1' Q_0'$$

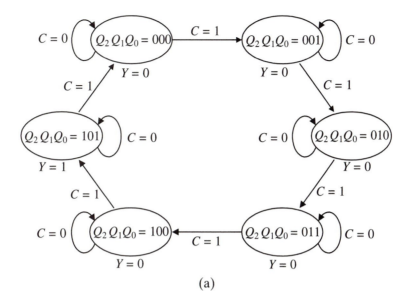

(a)

Current State	Next State	
$Q_2Q_1Q_0$	$Q_{2next} Q_{1next} Q_{0next}$	
	$C = 0$	$C = 1$
000	000	001
001	001	010
010	010	011
011	011	100
100	100	101
101	101	000

(b)

Current State	Implementation	
$Q_2Q_1Q_0$	$D_2D_1D_0$	
	$C = 0$	$C = 1$
000	000	001
001	001	010
010	010	011
011	011	100
100	100	101
101	101	000

(c)

Figure 7.13 Synthesis of a Moore FSM for Example 7.4: (a) state diagram; (b) next-state table; (c) implementation table; (d) K-maps and excitation equations; (e) output table and equation; (f) FSM circuit. *(continued on next page)*

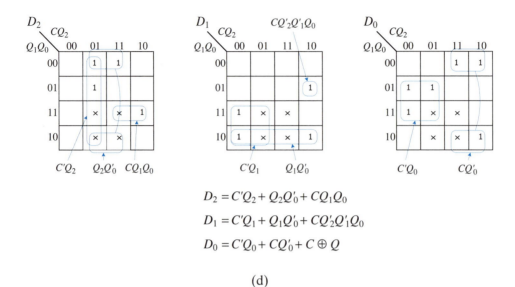

$$D_2 = C'Q_2 + Q_2Q'_0 + CQ_1Q_0$$

$$D_1 = C'Q_1 + Q_1Q'_0 + CQ'_2Q'_1Q_0$$

$$D_0 = C'Q_0 + CQ'_0 + C \oplus Q$$

(d)

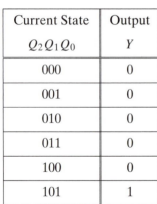

Current State	Output
$Q_2Q_1Q_0$	Y
000	0
001	0
010	0
011	0
100	0
101	1

$$Y = Q_2Q'_1Q_0$$

(e)

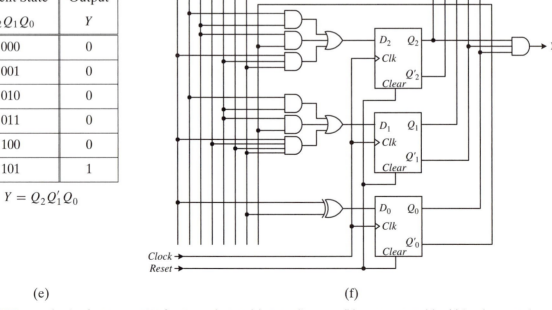

(f)

Figure 7.13 Synthesis of a Moore FSM for Example 7.4: (a) state diagram; (b) next-state table; (c) implementation table; (d) K-maps and excitation equations; (e) output table and equation; (f) FSM circuit.

Similarly, the equation for D_1 is derived from considering just the middle bit for all of the entries, and the equation for D_0 is derived from the rightmost bit. Since these equations will be used to construct the next-state circuit, they should be simplified. The three K-maps and simplified excitation equations for D_2, D_1, and D_0 are shown in Figure 7.13(d). In the K-maps, don't-care values are used for the unused state encodings 110 and 111.

Steps 5 and 6 are to derive the output table and equation. There is one equation for every output signal. Since the value of Y is labeled next to each node, it is therefore dependent only on the current state. From the state diagram, Y is asserted only in state 101, so Y has a 1 only in that current-state entry, while the rest of them are 0's. The output table and equation are shown in Figure 7.13(e).

Finally, we can draw the circuit for the FSM. We know that the circuit is a Moore FSM that uses three D flip-flops for its state memory having one primary input, C, and one output, Y. The next-state function circuit is derived from the three excitation equations for D_2, D_1, and D_0. The output function circuit is derived from the output equation for Y. The full circuit is shown in Figure 7.13(f).

Example 7.5   **Synthesizing a typical control unit Moore FSM**

In this example, we will synthesize a Moore FSM that is more typical of what the control unit of a microprocessor is like. We start with the state diagram, as shown in Figure 7.14(a). Each state is labeled with a state name, s_0, s_1, s_2, and s_3, and has two output signals, x and y. There are also two conditional status signals, *Start* and $(n = 9)$ labeled on four of the edges, while the rest of the edges do not have any conditions. From state s_0, the conditional edge labeled *Start* is taken when *Start* = 1; otherwise, the edge labeled *Start′* is taken. Similarly, from state s_2, the edge with the label $(n = 9)$ is taken when the condition is true (that is, when the value of variable n is equal to nine). If n is not equal to nine, then the edge with the label $(n = 9)′$ is taken.

Two flip-flops, Q_0 and Q_1, are needed in order to encode the four states. For simplicity, we will use the binary value of the index of the state name to be the encoding for that state. For example, the encoding for state s_0 is $Q_1 Q_0 = 00$ and the encoding for state s_1 is $Q_1 Q_0 = 01$, and so on.

From the above analysis, we are able to derive the next-state table, as shown in Figure 7.14(b). The four current states for $Q_1 Q_0$ are listed down the four rows. The four columns are for the four combinations of the two conditional signals, *Start* and $(n = 9)$. For example, the column with the value *Start*, $(n = 9) = 10$ means *Start* = 1 and $(n = 9) = 0$. The condition $(n = 9) = 0$ means that the condition $(n = 9)$ is false, which means that $(n = 9)′$ is true. The entries in the table are the next states, $Q_{1next} Q_{0next}$, for the two flip-flops.

For example, looking at the state diagram, from state s_2, we go back to state s_1 when the condition $(n = 9)′$ is true and independent of the *Start* condition. Hence, in the next-state table, for the current state row s_2 (10), the two next-state entries for when the condition $(n = 9)′$ is true is s_1 (01). The condition "$(n = 9)′$ is true" means $(n = 9) = 0$. This corresponds to the two columns with the labels 00 and 10, that is, *Start* can be either 0 or 1, while $(n = 9)$ is 0.

Using D flip-flops to implement the FSM, we get the implementation table shown in Figure 7.14(c). The implementation table and the next-state table are identical when D flip-flops are used since the characteristic equation for the D flip-flop is $Q_{next} = D$. The only difference between them is the meaning given to the entries. For the next-state table, the entries are the next state of the flip-flops, whereas for the implementation table, the entries are the inputs to the flip-flops. They are the input values necessary to get to that next state.

The excitation equations are derived from the implementation table. There is one excitation equation for every data input of every flip-flop used. Since we have two D flip-flops, we have two excitation equations: one for D_1 and the second for D_0.

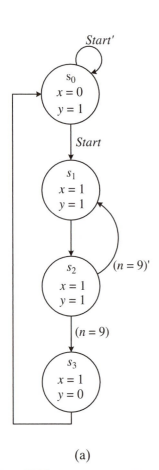

(a)

Current State	Next State			
$Q_1 Q_0$	$Q_{1next} Q_{0next}$			
	Start, $(n = 9)$			
	00	01	10	11
s_0 00	s_0 00	s_0 00	s_1 01	s_1 01
s_1 01	s_2 10	s_2 10	s_2 10	s_2 10
s_2 10	s_1 01	s_3 11	s_1 01	s_3 11
s_3 11	s_0 00	s_0 00	s_0 00	s_0 00

(b)

Current State	Implementation			
$Q_1 Q_0$	$D_1 D_0$			
	Start, $(n = 9)$			
	00	01	10	11
s_0 00	s_0 00	s_0 00	s_1 01	s_1 01
s_1 01	s_2 10	s_2 10	s_2 10	s_2 10
s_2 10	s_1 01	s_3 11	s_1 01	s_3 11
s_3 11	s_0 00	s_0 00	s_0 00	s_0 00

(c)

Figure 7.14 Synthesis of a Moore FSM for Example 7.5: (a) state diagram; (b) next-state table; (c) implementation table; (d) excitation equations and K-maps for D_1 and D_0; (e) output table; (f) output equations and K-maps; (g) FSM circuit.
(continued on next page)

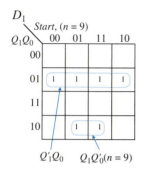

D_1
Start, $(n = 9)$

$D_1 = Q'_1 Q_0 + Q_1 Q'_0 (n = 9)$

D_0
Start, $(n = 9)$

$D_0 = Q_1 Q'_0 + Start Q'_0$

(d)

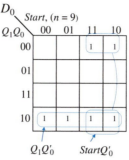

Current State	Output
$Q_1 Q_0$	$x\,y$
00	01
01	11
10	11
11	10

(e)

$x = Q_1 + Q_2$

$y = (Q_1 Q_0)'$

(f)

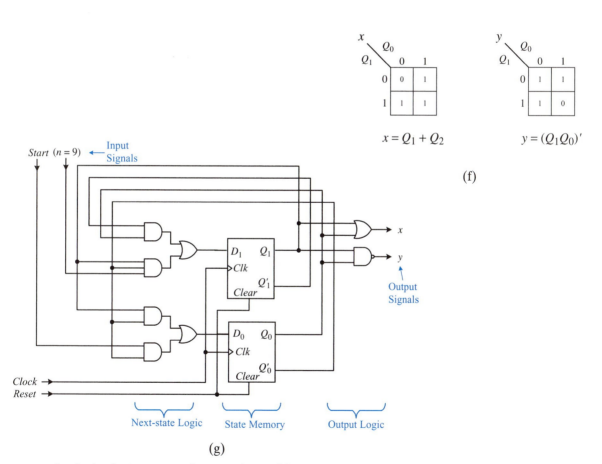

(g)

Figure 7.14   Synthesis of a Moore FSM for Example 7.5: (a) state diagram; (b) next-state table; (c) implementation table; (d) excitation equations and K-maps for D_1 and D_0; (e) output table; (f) output equations and K-maps; (g) FSM circuit.

The equations are dependent on the four variables, Q_1, Q_0, *Start*, and ($n = 9$). We look at the implementation table as one having two truth tables merged together: one truth table for D_1 and one for D_0. Since the two bits in the entries are ordered $D_1 D_0$, therefore, for the D_1 truth table, we look at only the leftmost D_1 bit in each entry, and, for the D_0 truth table, we look at only the rightmost D_0 bit. Extracting the two truth tables from the implementation table in this manner, we obtain the two K-maps and corresponding excitation equations for D_1 and D_0, as shown in Figure 7.14(d). The excitation equations allow us to derive the next-state combinational circuit.

The output table is obtained from the output signals given in the state diagram. The output table is just the truth table for the two output signals, x and y. The output signal equations derived from the output table are dependent on the current state, $Q_1 Q_0$. The output table, K-maps, and output equations are shown in Figure 7.14(e) and (f).

From the excitation and output equations, we easily can produce the next-state and output circuits, and the resulting FSM circuit shown in Figure 7.14(g).

7.4.8 Synthesis of a Mealy FSM

The next example illustrates the synthesis of a Mealy FSM. You will find that this process is almost identical to the synthesis of a Moore FSM with the one exception of deriving the output equations. The outputs for a Mealy FSM are dependent on both the current state and the input signals; whereas, for the Moore FSM, they are only dependent on the current state.

Example 7.6	**Synthesizing a Mealy FSM**

In this example, we will synthesize a Mealy FSM based on the state diagram shown in Figure 7.15(a) using D flip-flops. The four states already are encoded with the values of the two flip-flops. There are two conditional input signals, ($x = 0$) and ($x = y$). Since these are conditions, the equal sign means the test for equality. There is one output signal, A, which can be set to either a 0 or a 1 value. The equal sign here means assignment. Notice that what makes this a Mealy FSM state diagram is the fact that the outputs are associated with the edges and not the nodes.

Deriving the next-state and implementation tables for a Mealy FSM is exactly the same as for a Moore FSM. The next-state and implementation tables for this example are shown in Figure 7.15(b) and (c). The excitation equations and K-maps for D_1 and D_0 are shown in Figure 7.15(d).

The output table, as shown in Figure 7.15(e), is slightly different from the output tables for Moore FSMs. In addition to the output signal, A, being dependent on the current state, $Q_1 Q_0$, it is also dependent on the two input signals, ($x = 0$) and ($x = y$). Hence, the table has four columns for the four possible combinations of the two input signals. The entries in the table are the values for A.

Looking at the state diagram in Figure 7.15(a), we see that from state 00, output signal A is assigned the value 1 when the condition ($x = 0$) is true; otherwise, it is assigned a 0. Since the condition ($x = y$) is not labeled on these two edges going out from state 00, the output is independent to this condition from state 00.

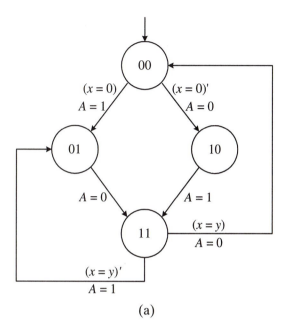

(a)

Current State	Next State			
$Q_1 Q_0$	$Q_{1next} Q_{0next}$			
	$(x = 0), (x = y)$			
	00	01	10	11
00	10	10	01	01
01	11	11	11	11
10	11	11	11	11
11	01	00	01	00

(b)

Figure 7.15   Synthesis of a Mealy FSM for Example 7.6: (a) state diagram; (b) next-state table; (c) implementation table; (d) excitation equations and K-maps for D_1 and D_0; (e) output table; (f) output equation and K-map; (g) FSM circuit.
(continued on next page)

Current State	Implementation			
$Q_1 Q_0$	$D_1 D_0$			
	$(x = 0), (x = y)$			
	00	01	10	11
00	10	10	01	01
01	11	11	11	11
10	11	11	11	11
11	01	00	01	00

(c)

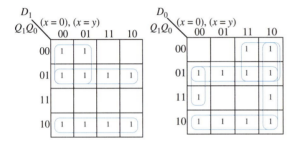

$$D_1 = Q'_1 Q_0 + Q_1 Q'_0 + Q'_1 (x = 0)'$$
$$= (Q_1 \oplus Q_0) + Q'_1 (x = 0)'$$

$$D_0 = Q'_1 Q_0 + Q_1 Q'_0 + Q'_1 (x = 0) + (x = 0)(x = y)' + Q_0(x = 0)'(x = y)'$$
$$= (Q_1 \oplus Q_0) + Q'_1 (x = 0) + (x = 0)(x = y)' + Q_0(x = 0)'(x = y)'$$

(d)

Figure 7.15 Synthesis of a Mealy FSM for Example 7.6: (a) state diagram; (b) next-state table; (c) implementation table; (d) excitation equations and K-maps for D_1 and D_0; (e) output table; (f) output equation and K-map; (g) FSM circuit.
(continued on next page)

Current State Q_1Q_0	Output A $(x = 0),\ (x = y)$			
	00	01	10	11
00	0	0	1	1
01	0	0	0	0
10	1	1	1	1
11	1	0	1	0

(e)

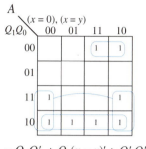

$A = Q_1Q_0' + Q_1(x = y)' + Q_1'Q_0'(x = 0)$

(f)

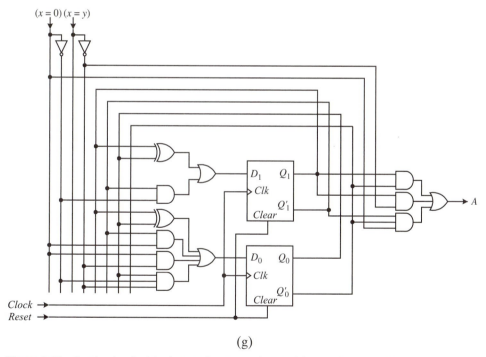

(g)

Figure 7.15   Synthesis of a Mealy FSM for Example 7.6: (a) state diagram; (b) next-state table; (c) implementation table; (d) excitation equations and K-maps for D_1 and D_0; (e) output table; (f) output equation and K-map; (g) FSM circuit.

Hence, in row 00, the two entries under the two columns with the labels 00 and 01 are both 0; whereas, the two entries under the two columns 10 and 11 are both 1.

Using the output table as the truth table, we are able to derive the K-map and output equation for A, as shown in Figure 7.15(f). Notice that the equation is also dependent on the two input signals.

Again, using the excitation and output equations, we are able to draw the final FSM circuit shown in Figure 7.15(g).

$\bullet\ \bullet\ \bullet\ \bullet\ \bullet\ \bullet\ \bullet\ \bullet\ \bullet\ \bullet\ \bullet\ \bullet\ \bullet\ \cdots$

7.5 Unused State Encodings and the Encoding of States

In a real world situation, the number of states used in the state diagram is most likely not a power of two. For example, the state diagram shown in Figure 7.13(a) for the modulo-6 counter uses six states. To encode six states, we need at least three flip-flops, since two flip-flops can encode only four different combinations. However, three flip-flops give eight different combinations. So two combinations are not used. The questions are: What do we do with these unused encodings? In the next-state table, what next-state values do we assign to these unused states? Do we just ignore them?

If the FSM can never be in any of the unused states, then it does not matter what their next states are. In this case, we can put don't-care values for their next states. The resulting next-state circuit may be smaller because of the don't-care values.

But what if, by chance, the FSM enters one of these unused states? The operation of the FSM will be unpredictable because we do not know what the next state is. Well, this is not exactly true because, even though we started with the don't-cares, we have mapped them to a fixed excitation equation. So, these unused states do have definite next states. It is just that these next states are not what we wanted. Hence, the resulting FSM operation will be incorrect if it ever enters one of the unused states. If this FSM is used in a mission-critical control unit, we do not want even this slight chance to occur.

One solution is to use the initialization or starting state as the next state for these unused state encodings. This way, the FSM will restart from the beginning if it ever enters one of these unused states.

So far, we have been using the sequential binary value to encode the states in order; for example, state s_0 is encoded as 00, state s_1 as 01, state s_2 as 10, and so on. However, there is no reason why we cannot use a different encoding for the states. In fact, we do want to use a different encoding if it will result in a smaller circuit.

Example 7.7 shows an FSM with an unused state encoding and the encoding of one state differently.

| Example 7.7 | **Synthesizing an FSM for a one-shot circuit** |

In this example, we will synthesize a FSM for the one-shot circuit first discussed in Section 3.5.1. Recall that the one-shot circuit outputs a single short pulse when given an input of arbitrary time length. In this FSM circuit, the length of the single short pulse will be one clock cycle. The state diagram for this circuit is shown in Figure 7.16(a).

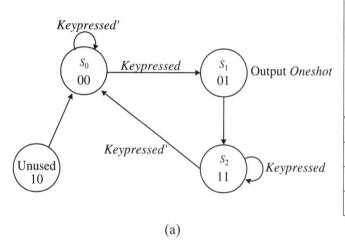

Current State	Next State	
Q_1Q_0	(Implementation)	
	$Q_{1next}Q_{0next}(D_1D_0)$	
	Keypressed	
	0	1
00	00	01
01	11	11
11	00	11
10 Unused	00	00

(a)   (b)

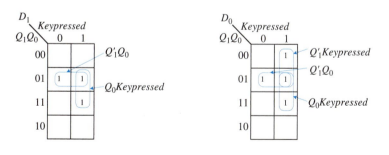

$$D_1 = Q'_1Q_0 + Q_0Keypressed \quad D_0 = Q'_1Keypressed + Q'_1Q_0 + Q_0Keypressed$$

Current State	Output
Q_1Q_0	Oneshot
00	0
01	1
11	0
10	0

(c)   (d)

Figure 7.16 FSM for one-shot circuit: (a) state diagram; (b) next-state (implementation) table; (c) excitation equations and K-maps for D_1 and D_0; (d) output table and output equation; (e) FSM circuit. *(continued on next page)*

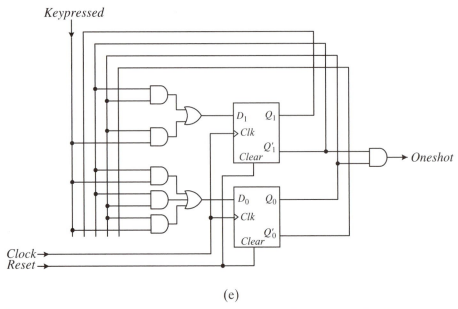

(e)

Figure 7.16 FSM for one-shot circuit: (a) state diagram; (b) next-state (implementation) table; (c) excitation equations and K-maps for D_1 and D_0; (d) output table and output equation; (e) FSM circuit.

State s_0, encoded as 00, is the reset state, and the FSM waits for a key press in this state. When a switch is pressed, the FSM goes to state s_1 (encoded as 01), and outputs a single short pulse. From s_1, the FSM unconditionally goes to state s_2 (encoded as 11), and turns off the one-shot pulse. Hence, the pulse only lasts for one clock cycle, regardless of how long the key is pressed. To break the loop and wait for another key press, the FSM has to wait for the release of the key in state s_2. When the key is released, the FSM goes back to state s_0 to wait for another key press.

This state diagram uses two bits to encode the three states; hence, state encoding 10 is not used. The state diagram shows that, if the FSM enters state 10, it unconditionally will go to the reset state 00 in the next clock cycle. Furthermore, we have encoded state s_2 as 11 instead of 10 for the index two.

The corresponding next-state table is shown in Figure 7.16(b). Using D flip-flops to implement this FSM, the implementation table, again, is like the next-state table. Therefore, we can use the next-state table directly to derive the two excitation equations for D_1 and D_0, as shown in Figure 7.16(c). The output table and output equation are shown in Figure 7.16(d) and finally, the complete FSM circuit in (e).

7.6 Designing a Car Security System—Version 3

We will revisit the car security system example from Chapters 2 and 6. Recall that in the first version (Chapter 2) the circuit is a combinational circuit. The problem with a combinational circuit is that once the alarm is triggered (by lets say, opening the door), the alarm can be turned off immediately by closing the door again. However, what we want is that once the alarm is triggered it should remain on even after closing the door, and the only way to turn it off is to turn off the master switch.

This requirement suggests that we need a sequential circuit instead where the output is dependent on not only the current input switch settings but also on the current state of the alarm. Version 2 of the car security system in Chapter 6 used an ad hoc approach to resolve this issue by adding a SR latch. In this section, we will use a more formal approach by designing a FSM for the car security system.

We start by deriving the state diagram for the system, as shown in Figure 7.17(a). In addition to the three input switches, M, D and V (for *Master*, *Door*, and *Vibration*), we need two states, 1 and 0, to depict whether the siren is on or off, respectively. If the siren is currently on (i.e., in the 1 state) then it will remain in that state as long as the master switch is still on, so it doesn't matter whether the door is now closed or open. This is represented by the edge that goes from state 1 and loops back to state 1 with the label $(MDV=1\times\times)$. From the on state, the only way to turn off the siren is to turn off the master switch. This is represented by the edge going from state 1 to state 0 with the label $(MDV=0\times\times)$. If the siren is currently off, it is turned on when the master switch is on, and either the door switch or the vibration switch is on. This is represented by the edge going from state 0 to state 1 with the labels $(MDV=101, 110,$ or $111)$. Finally, from the off state, the siren will remain off when either the master switch remains off or if the master switch is on but none of the other two switches are on. This is represented by the edge from state 0 looping back to state 0 with the labels $(MDV=0\times\times, 100)$.

The state diagram is translated to the corresponding next-state and implementation table using one D flip-flop, as shown in Figure 7.17(b). Again, the next-state and implementation tables are the same except that the entries for the next-state table are for the next states, and the entries for the implementation table are for the inputs to the flip-flop. Doing a 4-variable K-map on the implementation table gives us the excitation equation shown in Figure 7.17(c). The final circuit for this car security system is shown in Figure 7.17(d). The circuit uses one D flip-flop. The next-state circuit is derived from the excitation equation, which produces the signal for the D input of the flip-flop. The output of the flip-flop directly drives the siren.

7.7 VHDL for Sequential Circuits

Writing VHDL code for sequential circuits usually is done at the behavioral level. The advantage of writing behavioral VHDL code is that we do not need to manually

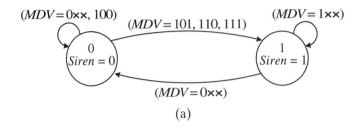

(a)

Current State	Next State (D Flip-flop Implementation)							
Q	$Q_{next}(D)$							
	M, D, V							
	000	001	010	011	100	101	110	111
0	0	0	0	0	0	1	1	1
1	0	0	0	0	1	1	1	1

(b)

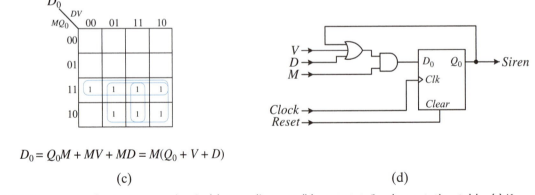

$$D_0 = Q_0 M + MV + MD = M(Q_0 + V + D)$$

(c) (d)

Figure 7.17 Car security system—version 3: (a) state diagram; (b) next-state/implementation table; (c) K-map and excitation equation; (d) circuit.

synthesize the circuit. The synthesizer automatically will produce the netlist for the circuit from the behavioral code.

In order to write the behavioral VHDL code for a sequential circuit, we need to use the information from the state diagram for the circuit. The main portion of the code contains two processes: a next-state-logic process, and an output-logic process. The edges (both conditional and unconditional) from the state diagram are used to derive the next-state-logic process, which will generate the next-state logic circuit.

The output signal information in the state diagram is used to derive the process for the output logic.

We will now illustrate the behavioral VHDL coding of sequential circuits with several examples.

| Example 7.8 | **Writing behavioral VHDL code for a Moore FSM** |

In this example, we will write the behavioral VHDL code for the Moore FSM of Example 7.2. The state diagram for the example from Figure 7.8 is repeated here in Figure 7.18. Since the synthesizer automatically will take care of the state encoding, the states only need to be labeled with their logical names. The behavioral VHDL code for this Moore FSM based on this state diagram and output table is shown in Figure 7.19.

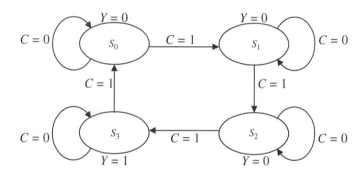

Figure 7.18
State diagram for
Example 7.8.

```
LIBRARY IEEE;
USE IEEE.STD_LOGIC_1164.ALL;

ENTITY MooreFSM IS PORT (
    clock: IN STD_LOGIC;
    reset: IN STD_LOGIC;
    C: IN STD_LOGIC;
    Y: OUT STD_LOGIC);
END MooreFSM;

ARCHITECTURE Behavioral OF MooreFSM IS
    TYPE state_type IS (s0, s1, s2, s3);
    SIGNAL state: state_type;
BEGIN
    next_state_logic: PROCESS (clock, reset)
    BEGIN
        IF (reset = '1') THEN
            state <= s0;
```

Figure 7.19
Behavioral VHDL
code of a Moore
FSM for Example 7.8.
(continued on next page)

```
                          ELSIF (clock'EVENT AND clock = '1') THEN
                             CASE state IS
                             WHEN s0 =>
                                IF C = '1' THEN
                                   state <= s1;
                                ELSE
                                   state <= s0;
                                END IF;
                             WHEN s1 =>
                                IF C = '1' THEN
                                   state <= s2;
                                ELSE
                                   state <= s1;
                                END IF;
                             WHEN s2 =>
                                IF C = '1' THEN
                                   state <= s3;
                                ELSE
                                   state <= s2;
                                END IF;
                             WHEN s3=>
                                IF C = '1' THEN
                                   state <= s0;
                                ELSE
                                   state <= s3;
                                END IF;
                             END CASE;
                          END IF;
                       END PROCESS;

                       output_logic: PROCESS (state)
                       BEGIN
                          CASE state IS
                          WHEN s0 =>
                             Y <= '0';
                          WHEN s1 =>
                             Y <= '0';
                          WHEN s2 =>
                             Y <= '0';
                          WHEN s3 =>
                             Y <= '1';
                          END CASE;
                       END PROCESS;

                       END Behavioral;
```

Figure 7.19
Behavioral VHDL
code of a Moore FSM
for Example 7.8.

The ENTITY section declares the primary I/O signals for the circuit. There are the global input *clock* and *reset* signals. The *clock* signal determines the speed in which the sequential circuit will transition from one state to the next. The *reset* signal initializes all of the state memory flip-flops to zero. In addition to the standard global *clock* and *reset* signals, the ENTITY section also declares all of the input and output signals. For this example, there is an input signal, C, and an output signal, Y; both of which are of type STD_LOGIC.

The ARCHITECTURE section starts out with using the TYPE statement to define the four states, s_0, s_1, s_2, and s_3, used in the state diagram. The SIGNAL statement declares the signal *state* to store the current state of the FSM. There are two processes in the ARCHITECTURE section that execute concurrently: the *next_state_logic* PROCESS, and the *output_logic* PROCESS. As the name suggests, the *next_state_logic* PROCESS defines the next-state-logic circuit that is inside the control unit, and the *output_logic* PROCESS defines the output logic circuit inside the control unit. The main statement within these two processes is the CASE statement that determines what the current state is.

In the *next_state_logic* PROCESS, the current state of the FSM is initialized to s_0 on reset. The CASE statement is executed only at the rising clock edge because of the test (*clock*'EVENT AND *clock* = '1') in the IF statement. Hence, the *state* signal is assigned a new state value at every rising clock edge. The new state value is, of course, dependent on the current state and input signals, if any. For example, if the current state is s_0, the case for s_0 is selected. From the state diagram, we see that when in state s_0, the next state is dependent on the input signal C. Hence, in the code, an IF statement is used. If C is 1 then the new state s_1 is assigned to the signal *state*, otherwise, s_0 is assigned to *state*. For the latter case, even though we are not changing the state value s_0, we still make that assignment to prevent the VHDL synthesizer from using a memory element for the *state* signal. Recall from Section 6.13.1 that VHDL synthesizes a signal using a memory element if the signal is not assigned a value for all possible cases. The rest of the cases in the CASE statement are written similarly based on the remaining edges in the state diagram.

In the *output_logic* PROCESS, all of the output signals must be assigned a value in every case. Again, the reason is that we do not want these output signals to come from memory elements. In the FSM model, the output circuit is a combinational circuit, and so, it should not contain any memory elements. For each state in the CASE statement in the *output_logic* PROCESS, the values assigned to each of the output signals are taken directly from the output table. For this example, there is only one output signal, Y.

A sample simulation trace of this sequential circuit is shown in Figure 7.20. In the simulation trace, between times 100 ns and 800 ns when R is de-asserted and C is asserted, the state changes at each rising clock edge (at times 300 ns, 500 ns, and 700 ns). At time 700 ns, when the current state is s_3, we see that the output signal, Y, is also asserted. At time 800 ns, input C is de-asserted; as a result, the FSM did not change state at the next rising clock edge at time 900 ns.

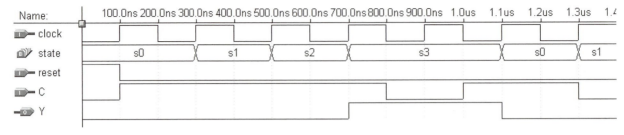

Figure 7.20 Simulation trace of a Moore FSM for Example 7.8.

Example 7.9

Writing behavioral VHDL code for a Mealy FSM

This example shows how a Mealy FSM is written using behavioral VHDL code. We will use the Mealy FSM from Example 7.3. The state diagram for this FSM is shown in Figure 7.10. This FSM is very similar to the one from the previous example, except that the generation of the output signal, Y, is also dependent on the input signal, C. The VHDL code is shown in Figure 7.21. In this code, we see that the *next_state_logic* PROCESS is identical to the previous FSM code. In the *output_logic* PROCESS, the only difference is in state s_3, where an IF statement is used to determine the value of the input signal, C. The output signal, Y, is assigned a value depending on the result of this test.

Figure 7.21
Behavioral VHDL
code for the Mealy
FSM of Example 7.9.
(continued on next page)

```
LIBRARY IEEE;
USE IEEE.STD_LOGIC_1164.ALL;

ENTITY MealyFSM IS PORT (
    clock: IN STD_LOGIC;
    reset: IN STD_LOGIC;
    C: IN STD_LOGIC;
    Y: OUT STD_LOGIC);
END MealyFSM;

ARCHITECTURE Behavioral OF MealyFSM IS
    TYPE state_type IS (s0, s1, s2, s3);
    SIGNAL state: state_type;
BEGIN
    next_state_logic: PROCESS (clock, reset)
    BEGIN
        IF (reset = '1') THEN
            state <= s0;
        ELSIF (clock'EVENT AND clock = '1') THEN
            CASE state is
```

```vhdl
            WHEN s0 =>.
               IF C = '1' THEN
                  state <= s1;
               ELSE
                  state <= s0;
               END IF;
            WHEN s1 =>
               IF C = '1' THEN
                  state <= s2;
               ELSE
                  state <= s1;
               END IF;
            WHEN s2 =>
               IF C = '1' THEN
                  state <= s3;
               ELSE
                  state <= s2;
               END IF;
            WHEN s3 =>
               IF C = '1' THEN
                  state <= s0;
               ELSE
                  state <= s3;
               END IF;
         END CASE;
      END IF;
   END PROCESS;


   output_logic: PROCESS (state, C)
   BEGIN
      CASE state IS
      WHEN s0 =>
         Y <= '0';
      WHEN s1 =>
         Y <= '0';
      WHEN s2 =>
         Y <= '0';
      WHEN s3 =>
         IF (C = '1') THEN
            Y <= '1';
         ELSE
            Y <= '0';
         END IF;
      END CASE;
   END PROCESS;
END Behavioral;
```

Figure 7.21
Behavioral VHDL
code for the Mealy
FSM of Example 7.9.

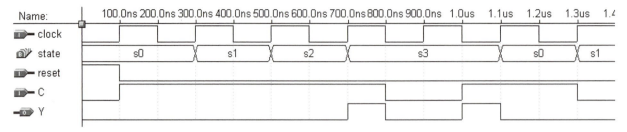

Figure 7.22 Simulation trace for the Mealy FSM of Example 7.9.

The simulation trace for this Mealy FSM is shown in Figure 7.22. Notice that the only difference between this trace and the one from the previous example is in the Y signal between times 800 ns and 1 μs. During this time period, the input signal C is de-asserted. In the previous trace, this has no effect on Y, however, for the Mealy FSM trace, Y is also de-asserted.

| Example 7.10 | **A Moore FSM using behavioral VHDL code** |

This is another example of a Moore FSM written using behavioral VHDL code. This FSM is from Example 7.5, and the state diagram for this example is shown in Figure 7.14. The behavioral VHDL code for this FSM is shown in Figure 7.23, and the simulation trace in Figure 7.24.

```
LIBRARY IEEE;
USE IEEE.STD_LOGIC_1164.ALL;

ENTITY MooreFSM IS PORT(
    clock: IN STD_LOGIC;
    reset: IN STD_LOGIC;
    start, neq9: IN STD_LOGIC;
    x,y: OUT STD_LOGIC);
END MooreFSM;

ARCHITECTURE Behavioral OF MooreFSM IS
    TYPE state_type IS (s0, s1, s2, s3);
    SIGNAL state: state_type;

BEGIN
    next_state_logic: PROCESS (clock, reset)
    BEGIN
        IF (reset = '1') THEN
            state <= s0;
```

Figure 7.23
Behavioral VHDL code for the Moore FSM of Example 7.10.
(continued on next page)

```
          ELSIF (clock'EVENT AND clock = '1') THEN
             CASE state IS
             WHEN s0 =>
                IF start = '1' THEN
                   state <= s1;
                ELSE
                   state <= s0;
                END IF;
             WHEN s1 =>
                state <= s2;
             WHEN s2 =>
                IF neq9 = '1' THEN
                   state <= s3;
                ELSE
                   state <= s1;
                END IF;
             WHEN s3 =>
                state <= s0;
             END CASE;
          END IF;
       END PROCESS;

       output_logic: PROCESS (state)
       BEGIN
          CASE state IS
          WHEN s0 =>
             x <= '0';
             y <= '1';
          WHEN s1 =>
             x <= '1';
             y <= '1';
          WHEN s2 =>
             x <= '1';
             y <= '1';
          WHEN s3 =>
             x <= '1';
             y <= '0';
          END CASE;
       END PROCESS;
    END Behavioral;
```

Figure 7.23
Behavioral VHDL
code for the Moore
FSM of
Example 7.10.

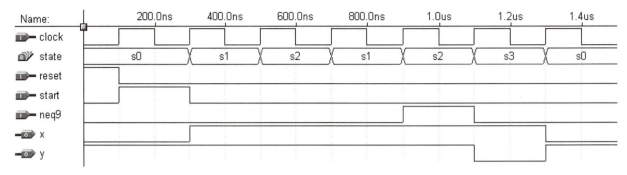

Figure 7.24 Simulation trace for the Moore FSM of Example 7.10.

*7.8 Optimization for Sequential Circuits

In designing any digital circuit, in addition to getting a functionally correct circuit, we like to optimize it for size, speed, and power consumption. In this section, we will discuss briefly some of the issues involved. A full treatment of optimization for sequential circuits is beyond the scope of this book.

Since sequential circuits also contain combinational circuit parts (the next-state logic and the output logic), these parts should also be optimized following the optimization procedures for combinational circuits, as discussed in Section 4.4. Some basic choices for sequential-circuit optimization include state reduction, state encoding, and choice of flip-flop types.

7.8.1 State Reduction

Sequential circuits with fewer states most likely will result in a smaller circuit, since the number of states directly translates to the number of flip-flops needed. Fewer flip-flops imply a smaller state memory for the FSM. Furthermore, fewer flip-flops also mean fewer flip-flop inputs, so the number of excitation equations needed is also reduced. This of course means that the next-state circuit will be smaller.

There are two levels in which we can reduce the number of states. At the pseudocode description level, we can try to optimize the code by shortening the code, if possible. We can also assign two or more data operations to the same state.

After obtaining a state diagram, we may still be able to reduce the number of states by removing equivalent states. If two states are equivalent, we can remove one of them and use instead the other equivalent state. The resulting FSM still will be functionally equivalent. Two states are said to be equivalent if the following two conditions are true:

1. Both states produce the same output for every input.

2. Both states have the same next state for every input.

7.8.2   State Encoding

When initially drawing the state diagram for a sequential circuit, it is preferred to keep the state names symbolic. However, these state names must be encoded eventually with a unique bit string. State encoding is the process of determining how many flip-flops are required to represent the states in the next-state table or state diagram and to assign a unique bit string combination to each named state. In all the examples presented so far, we have been using the straight binary encoding scheme, where n flip-flops are needed to encode 2^n states. For example, for four states, state s_0 gets the encoding 00, s_1 gets the encoding 01, s_2 gets 10, and s_3 gets the encoding 11. However, this scheme does not always lead to the smallest FSM circuit. Other encoding schemes are minimum bit change, prioritized adjacency, and one-hot encoding.

For the **minimum bit change** scheme, binary encodings are assigned to the states in such a way that the total number of bit changes for all state transitions is minimized. In other words, if every edge in the state diagram is assigned a weight that is equal to the number of bit changes between the source encoding and the destination encoding of that edge, this scheme would select the one that minimizes the sum of all of these edge weights.

For example, given the four-state state diagram shown in Figure 7.25(a), the minimum bit change scheme would use the encoding shown in (b) and not the encoding shown in (c). In both Figure 7.25(b) and (c), the number of bit changes between the encodings of two states joined by an edge is labeled on that edge. For example, in Figure 7.25(b), the number of bit changes between state $s_1 = 01$ and $s_2 = 11$ is 1. The encoding used in Figure 7.25(b) has a smaller sum of all of the edge weights than the encoding used in (c).

Notice that, even though the encoding of Figure 7.25(b) produces the smallest total edge weight, there are several other ways to encode these four states that will also produce the same total edge weight (for example, assigning 00 to s_1 instead of to s_0, 01 to s_2 instead of s_1, 11 to s_3, and 10 to s_0).

For the **prioritized adjacency** scheme, adjacent states to any state s are given certain priorities. Encodings are assigned to these adjacent states such that those

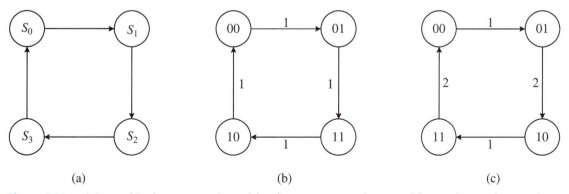

(a)   (b)   (c)

Figure 7.25   Minimum bit change encoding: (a) a four-state state diagram; (b) encoding with a total weight of four; (c) encoding with a total weight of six.

with a higher priority will have an encoding that has fewer bit changes from the encoding of state s than those adjacent states with a lower priority.

In the **one-hot encoding** scheme, each state is assigned one flip-flop. A state is encoded with its flip-flop having a 1 value, while all of the other flip-flops have a 0 value. For example, the one-hot encoding for four states would be 0001, 0010, 0100, and 1000.

7.8.3 Choice of Flip-Flops

An FSM can be implemented using any of the four types of flip-flops (SR, D, JK, and T, as discussed in Section 6.13) or any combinations of them. Using different flip-flops can produce a smaller circuit but with the same functionality. The decision as to what types of flip-flops to use is reflected in the implementation table. Whereas, the next-state table is independent of the flip-flop types used, the implementation table is dependent on these choices of flip-flops.

The implementation table answers the question of what the flip-flop inputs should be in order to realize the next-state table. In order to do this, we need to use the excitation table for the selected flip-flop(s). Recall that the excitation table is used to answer the question of what the inputs should be when given the current state that the flip-flop is in and the next state that we want the flip-flop to go to. So, to get the entries for the implementation table, we substitute the next-state values from the next-state table with the corresponding entry in the excitation table.

For example, if we have the following next-state table:

Current State	Next State	
$Q_1 Q_0$	$Q_{1next} Q_{0next}$	
	$C = 0$	$C = 1$
00	00	00
01	10	10
10	01	11
11	00	00

If we want to use the SR flip-flop to implement the circuit, we would convert the next-state table to the implementation table as follows. First, the next-state column headings from the next-state table, $Q_{1next} Q_{0next}$, are changed to the corresponding flip-flop input names, $S_1 R_1 S_0 R_0$. Since the SR flip-flop has two inputs, therefore, each next-state bit, Q_{next}, is replaced with two input bits, SR. This is done for all of the flip-flops used, as shown here:

Current State	Implementation	
$Q_1 Q_0$	$S_1 R_1 S_0 R_0$	
	$C = 0$	$C = 1$
00		
01	10__	
10		
11		

To derive the entries in the implementation table, we will need the excitation table for the SR flip-flop (from Section 6.13.1) shown here:

Q	Q_{next}	S	R
0	0	0	$\times$
0	1	1	0
1	0	0	1
1	1	$\times$	0

For example, if the current state for flip-flop one is $Q_1 = 0$ and the next state is $Q_{1\text{next}} = 1$, we would do a table look-up in the excitation table for $Q Q_{\text{next}} = 01$. The corresponding two input bits are $SR = 10$. Hence, we would replace the 1 bit for $Q_{1\text{next}}$ in the next-state table with the two input bits, $S_1 R_1 = 10$, in the same entry location in the implementation table. Proceeding in this same manner for all of the next-state bits in the next-state table entries, we obtain the complete implementation table.

Current State	Implementation	
$Q_1 Q_0$	$S_1 R_1 S_0 R_0$	
	$C = 0$	$C = 1$
00	$0 \times 0 \times$	$0 \times 0 \times$
01	1001	1001
10	0110	$\times 010$
11	0101	0101

Once we have the implementation table, deriving the excitation equations and drawing the next-state circuit are identical for all flip-flop types.

The output table and output equations are not affected by the change in flip-flop types, and so they remain exactly the same too.

Example 7.11 **Using T flip-flops to design a modulo-6 up counter**

In this example, we will design a modulo-6 up counter using T flip-flops. This is similar to Example 7.4 but using T flip-flops instead of D flip-flops. The next-state table for the modulo-6 up-counter, as obtained from Example 7.4, is shown in Figure 7.26(a).

Current State $Q_2 Q_1 Q_0$	Next State $Q_{2next} Q_{1next} Q_{0next}$	
	$C = 0$	$C = 1$
000	000	001
001	001	010
010	010	011
011	011	100
100	100	101
101	101	000

(a)

Q_{next}	Q'_{next}	T
0	0	0
0	1	1
1	0	1
1	1	0

(b)

Current State $Q_2 Q_1 Q_0$	Implementation $T_2 T_1 T_0$	
	$C = 0$	$C = 1$
000	000	001
001	000	011
010	000	001
011	000	111
100	000	001
101	000	101

(c)

Figure 7.26 Synthesis of an FSM for Example 7.11: (a) next-state table; (b) excitation table for the T flip-flop; (c) implementation table using T flip-flops; (d) K-maps and excitation equations; (e) FSM circuit. *(continued on next page)*

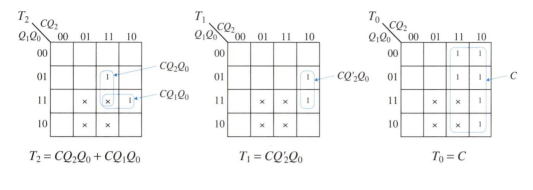

$$T_2 = CQ_2Q_0 + CQ_1Q_0 \qquad\qquad T_1 = CQ'_2Q_0 \qquad\qquad T_0 = C$$

(d)

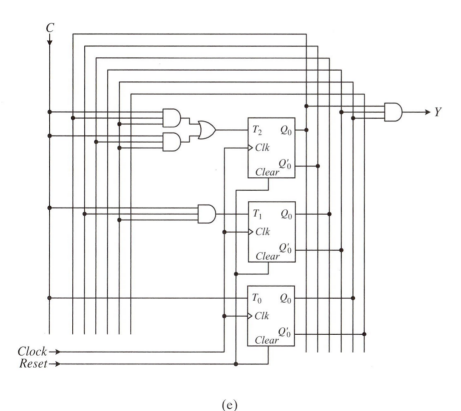

(e)

Figure 7.26   Synthesis of an FSM for Example 7.11: (a) next-state table; (b) excitation table for the T flip-flop;
(c) implementation table using T flip-flops; (d) K-maps and excitation equations; (e) FSM circuit.

The excitation table for the T flip-flop, as derived in Section 6.14.3, is shown in Figure 7.26(b).

The implementation table is obtained from the next-state table by substituting each next-state bit with the corresponding input bit of the T flip-flop. This is accomplished by doing a table look-up from the T flip-flop excitation table.

For example, in the next-state table for the current state $Q_2 Q_1 Q_0 = 010$ and the input $C = 1$, we want the next state $Q_{2next} Q_{1next} Q_{0next}$ to be 011. The corresponding entry in the implementation table shown in Figure 7.26(c) using T flip-flops would be $T_2 T_1 T_0 = 001$, because for flip-flop$_2$, we want its content to go from $Q_2 = 0$ to $Q_{2next} = 0$. The excitation table tells us that, to realize this change, the T_2 input needs to be a 0. Similarly, for flip-flop$_1$, we want its content to go from $Q_1 = 1$ to $Q_{1next} = 1$, and again, the T_1 input needs to be a 0 to realize this change. Finally, for flip-flop$_0$, we want its content to go from $Q_0 = 0$ to $Q_{0next} = 1$; this time, we need T_0 to be a 1. Continuing in this manner for all of the entries in the next-state table, we obtain the implementation table shown in Figure 7.26(c).

From the implementation table, we obtain the excitation equations just like before. For this example, we have the three input bits, T_2, T_1 and T_0, which results in the three equations. These equations are dependent on the four variables, Q_2, Q_1, Q_0, and C. The three K-maps and excitation equations for T_2, T_1, and T_0 are shown in Figure 7.26(d). The output equation is the same as before (see Figure 7.13(e)). Finally, the complete modulo-6 up-counter circuit is shown in Figure 7.26(e).

Comparing this circuit with the circuit from Example 7.3 shown in Figure 7.13(f) where D flip-flops are used, it is obvious that using T flip-flops for this problem results in a much smaller circuit than using D flip-flops.

● ● ● ● ● ● ● ● ● ● ● ● ● ● ● ●

7.9 Summary Checklist

- State diagram
- State encoding
- Output signal
- Conditional edge
- Next-state table
- Implementation table
- Excitation equation
- Output table
- Output equation
- Next-state logic
- State memory
- Output logic
- FSM circuit

- Unused state encoding

- Be able to derive the state diagram from an arbitrary pseudocode circuit description

- Be able to derive the next-state table from a state diagram

- Be able to derive the implementation table from a next-state table

- Be able to derive the excitation equations from an implementation table

- Be able to derive the output table from a state diagram

- Be able to derive the output equations from an output table

- Be able to derive the FSM circuit from the excitation and output equations

7.10 Problems

P7.1. Analyze the following FSMs and derive the state diagram for it:

(a) C is an input, and a and b are outputs.

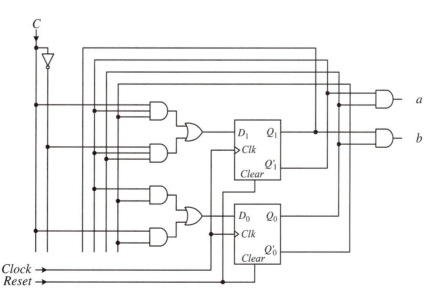

Figure P7.1(a)

(b) C is an input, and a and b are outputs.

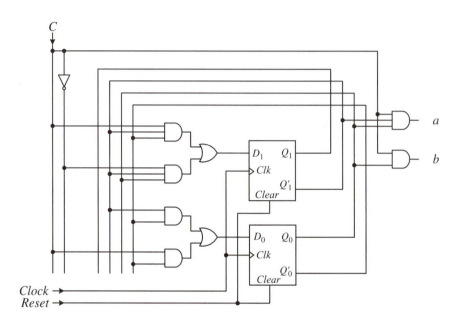

Figure P7.1(b)

(c) A and B are inputs, and X and Y are outputs.

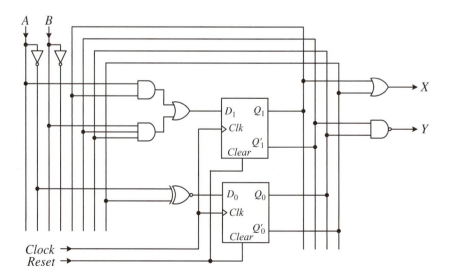

Figure P7.1(c)

(d) $(Z \neq 0)$ is an input, and *ClrX*, *LoadY*, *inZ*, *LoadX*, *stat1*, *LoadZ*, and *subtract* are outputs.

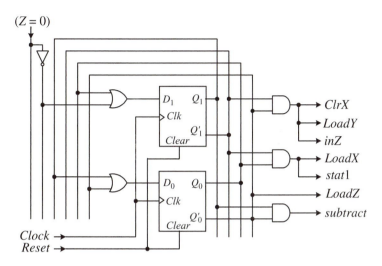

Figure P7.1(d)

(e) *Start* is an input, and *LoadN* and *LoadM* are outputs.

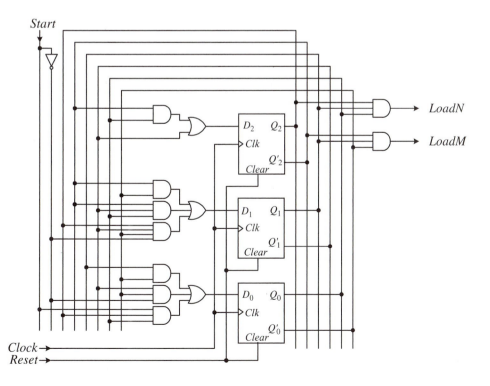

Figure P7.1(e)

P7.2. Analyze the following FSMs and derive the state diagram for it:

(a) C is an input, and a and b are outputs.

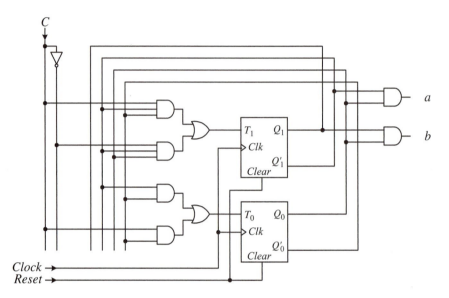

Figure P7.2(a)

(b) C is an input, and a and b are outputs.

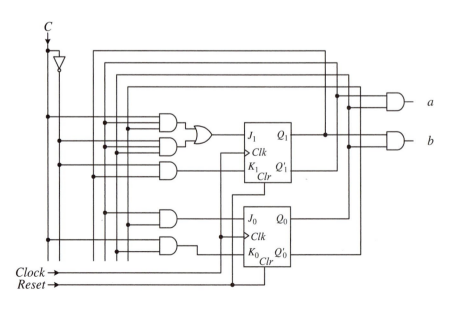

Figure P7.2(b)

(c) *C* is an input, and *a* and *b* are outputs.

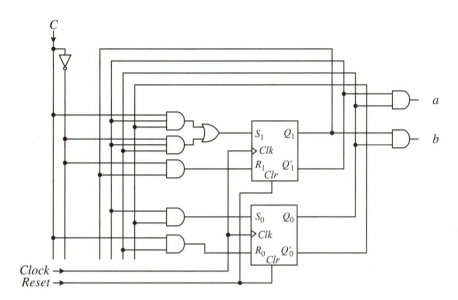

Figure P7.2(c)

(d) *C* is an input, and *a* and *b* are outputs.

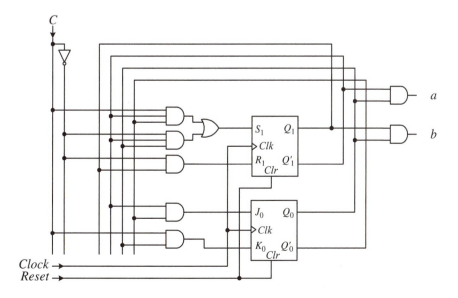

Figure P7.2(d)

P7.3. Synthesize a FSM circuit using D flip-flops for the following state diagrams:

(a) *A* is an input.

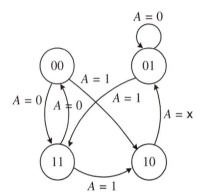

Figure P7.3(a)

(b) *J* and *K* are inputs, and *Q* is an output.

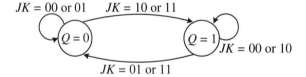

Figure P7.3(b)

(c) (*Z*≠0) is an input, and *YLoad*, *Xload*, *Zmux*, and *out* are outputs.

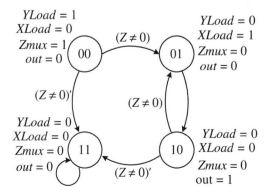

Figure P7.3(c)

(d) C is an input, and X is an output.

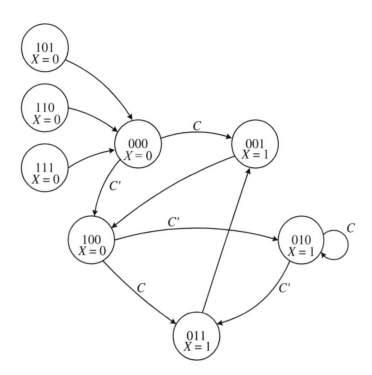

Figure P7.100(d)

(e) $(x = 0)$ and $(x = y)$ are inputs, and A is an output.

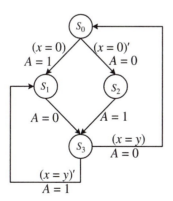

Figure P7.101(e)

P7.4. Use JK flip-flops to synthesize a FSM circuit for the state diagrams in Problem P7.3.

P7.5. Use SR flip-flops to synthesize a FSM circuit for the state diagrams in Problem P7.3.

P7.6. Use T flip-flops to synthesize a FSM circuit for the state diagrams in Problem P7.3.

P7.7. Use a JK flip-flop for flip-flop one, and a T flip-flop for flip-flop two to synthesize a FSM circuit for the state diagrams in Problem P7.3.

P7.8. Design a modulo-4 up-down counter using D flip-flops. The count is represented by the content of the flip-flops. The circuit has a *Count* signal and an *Up* signal. The counter counts when *Count* is asserted, and stops when *Count* is de-asserted. The *Up* signal determines the direction of the count. When *Up* is asserted, the count increments by one at each clock cycle. When *Up* is de-asserted, the count decrements by one at each clock cycle.

P7.9. Design a modulo-5 up-counter using D flip-flops similar to Problem P7.8, but without the *Up* signal.

P7.10. Design a modulo-5 up-down counter using D flip-flops similar to Problem P7.8.

P7.11. Design a modulo-4 up counter using T flip-flops.

P7.12. Design a FSM that counts the following decimal sequence.

$$3, 7, 2, 6, 3, 7, 2, 6, \ldots$$

The count is to be represented directly by the contents of the D flip-flops. The counting starts when the control input C is asserted and stops whenever C is de-asserted. Assume that the next state from all unused states is the state for the first count in the sequence (i.e., the state for 3).

P7.13. Design a counter that counts in the following sequence.

$$1, 4, 6, 7, 1, 4, 6, 7, \ldots$$

The count is to be represented directly by the contents of three D flip-flops. The counter is enabled by the input C. The count stops when $C = 0$. The next state from all unused states are undefined.

P7.14. Repeat Example 7.7, but with the following encodings. Which encoding results in the smallest FSM circuit?
(a) Encode state s_2 as 10 instead of 11, and encode the unused state as 11.
(b) Encode state s_0 as 0001, s_1 as 0010, s_2 as 0100, and the unused state as 1000. All remaining combinations are unused and their next state is undefined.

P7.15. Repeat Problem P7.13, but use a JK flip-flop, a D flip-flop, and a SR flip-flop in this order, starting from the most significant bit for the three flip-flops.

P7.16. Manually design and implement on the UP2 board the following FSM circuit. Make the LEDs in the 7-segment display move in a clockwise direction around in a circle (i.e., turn on and off the LED segments in this order: segment a, b, c, d, e, f, a, b, etc).

P7.17. Manually design and implement on the UP2 board the following FSM circuit. This is similar to Problem P7.16, but make one 7-segment LED display in a clockwise direction and the other in a counterclockwise direction.

P7.18. Manually design and implement on the UP2 board the following FSM circuit. This is similar to Problem P7.16, but make it so that each time a push-button switch is pressed, the display changes directions.

P7.19. Manually design and implement on the UP2 board the following FSM circuit. Input from the eight DIP switches. Output on the 7-segment the decimal number that represents the number of DIP switches that are in the on position.

P7.20. Manually design and implement on the UP2 board an FSM circuit for controlling three switches, T_1, T_2, and T_3, and three lights L_1, L_2, and L_3. Each light is turned on by the corresponding switch (for example, T_1 turns on L_1). Initially, all switches are off. The first switch that is pressed will turn on its corresponding light. When the first light is turned on, it will remain on, while the other two lights remain off, and they are unaffected by subsequent switch presses until reset.

P7.21. Design a FSM circuit for controlling a simple home security system. The operation of the system is as follows.

Inputs: Front gate switch (FS)
Motion detector switch (MS)
Asynchronous Reset switch (R)
Clear switch (C)

Outputs: Front gate melody (FM)
Motion detector melody (MM)

- When the reset switch (R) is asserted, the FSM goes to the initialization state (S_init) immediately. The encoding for the initialization state is zeros for all of the flip-flops.

- From state S_init, the FSM unconditionally goes to the wait state (S_wait).

- From state S_wait, the FSM waits for one of the four switches to be activated. All the switches are active-high, so when a switch is pressed or activated, it sends out a 1. The following actions are taken when a switch is pressed:

  - When FS is pressed, the FSM goes to state S_front. In state S_front, the front gate melody is turned on by setting $FM = 1$. The FSM remains in state S_front until the clear switch is pressed. Once the clear switch is pressed, the FSM goes back to S_wait.

  - When MS is activated, the FSM goes to state S_motion. In state S_motion, MM is turned on with a 1. $MM4$ will remain on for two more clock periods and then it will go back to S_wait.

  - From any state, as soon as the reset switch is pressed, the FSM immediately goes back to state S_init.

  - Pressing the clear switch only affects the FSM when it is in state S_front. The clear switch has no effect on the FSM when it is in any other states.

  - Any unused state encoding will have S_init as their next state.

Standard Sequential Components

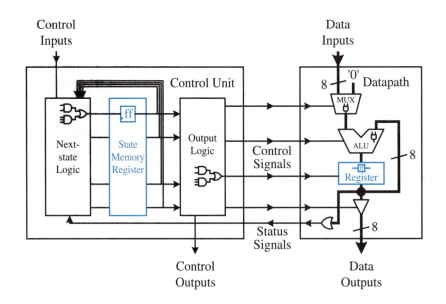

In a computer system, we usually want to store more than one bit of information. More over, we may want to group several bits together and consider them as one unit, such as an integer is made up of eight bits. In Chapter 6, we presented the circuits for latches and flip-flops for storing one bit of information. In this chapter, we will look at registers for storing multiple bits of information as a unit. Registers also are made more versatile by adding extra functionalities, such as counting and shifting, to it. We will also look at the design of counters and shift registers.

Very often, computer circuits may need to store several values at the same time. Instead of using several separate registers, we may want to combine these registers together. Register files and memories are like an array of registers for storing multiple values. In this chapter, we will also look at the construction of register files and memory circuits.

Similar to the standard combinational components, these sequential components are used in almost every digital circuit. Hence, rather than having to redesign them each time that they are needed, they usually are available in standard libraries.

• • • • • • • • • • • • • • • • •

8.1 Registers

When we want to store a byte of data, we need to combine eight flip-flops together and have them work together as a unit. A **register** is just a circuit with two or more D flip-flops connected together in such a way that they all work exactly the same way and are synchronized by the same clock and enable signals. The only difference is that each flip-flop in the group is used to store a different bit of the data.

Figure 8.1(a) shows a 4-bit register with parallel load and asynchronous clear. Four D flip-flops with active-high enable and asynchronous clear are used. Notice in the circuit that the control inputs, Clk, E, and $Clear$, for all of the flip-flops are connected, respectively, in common; so that when a particular input is asserted, all of the flip-flops will behave in exactly the same way. The 4-bit input data is connected to D_0 through D_3, while Q_0 through Q_3 serve as the 4-bit output data for the register. When the active-high load signal $Load$ is asserted (i.e., $Load = 1$), the data presented on the D lines is stored into the register (the four flip-flops) at the next rising edge of the clock signal. When $Load$ is de-asserted, the content of the register remains unchanged. The register can be asynchronously cleared (i.e., setting all of the Q_i's to 0 immediately, without having to wait for the next active clock edge) by asserting the $Clear$ line. The content of the register is always available on the Q output lines, so no control line is required for reading the data from the register. Figure 8.1(b) and (c) show the operation table and the logic symbol, respectively, for this 4-bit register.

Figure 8.2 shows the VHDL code for the 4-bit register with active-high $Load$ and $Clear$ signals. Notice that the coding is very similar to that for the single D flip-flop. The main difference is that the data inputs and outputs are 4-bits wide. A sample

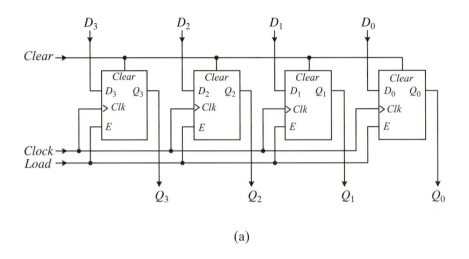

(a)

Clear	Load	Operation
1	×	Reset register to zero asynchronously
0	0	No change
0	1	Load in a value at rising clock edge

(b)

(c)

Figure 8.1   A 4-bit register with parallel load and asynchronous clear: (a) circuit; (b) operation table; (c) logic symbol.

simulation trace for the register is shown in Figure 8.3. At time 100 ns, even though *Load* is asserted, the register is not written with the *D* input value of 5, because *Clear* is asserted. Between times 200 ns and 400 ns, *Load* is de-asserted, so even though *Clear* is de-asserted, the register still is not loaded with the input value of 5. At time 400 ns, *Load* is asserted, but the input data is not loaded into the register immediately (as can be seen by *Q* being a 0). The loading occurs at the next rising edge of the clock at 500 ns, when *Q* changes to 5. At time 600 ns, *Clear* is asserted, and so, *Q* is reset to 0 immediately, without having to wait until the next rising clock edge at 700 ns.

```
LIBRARY IEEE;
USE IEEE.STD_LOGIC_1164.ALL;

ENTITY reg IS GENERIC (size: INTEGER := 3);-- size of the register
    PORT (
    Clock, Clear, Load: IN STD_LOGIC;
    D: IN STD_LOGIC_VECTOR(size DOWNTO 0);
    Q: OUT STD_LOGIC_VECTOR(size DOWNTO 0));
END reg;

ARCHITECTURE Behavior OF reg IS
BEGIN
    PROCESS(Clock, Clear)
    BEGIN
      IF Clear = '1' THEN
         Q <= (OTHERS => '0');
      ELSIF (Clock'EVENT AND Clock = '1') THEN
         IF Load = '1' THEN
            Q <= D;
         END IF;
      END IF;
    END PROCESS;
END Behavior;
```

Figure 8.2
VHDL code for a
4-bit register with
active-high *Load*
and *Clear* signals.

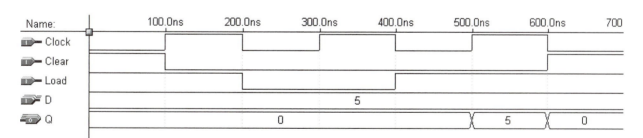

Figure 8.3 Sample simulation trace for the 4-bit register.

●　●　●　●　●　●　●　●　●　●　●　●　●　● ● ●

8.2 Shift Registers

Similar to the combinational shifter and rotator circuits, there are the equivalent
sequential shifter and rotator circuits. The circuits for the shift and rotate operations
are constructed exactly the same. The only difference in the sequential version is

that the operations are performed on the value that is stored in a register rather than directly on the input value. The main usage for a shift register is for converting from a serial-data input stream to a parallel-data output or vice versa. For a serial-to-parallel data conversion, the bits are shifted into the register at each clock cycle, and when all of the bits (usually eight bits) are shifted in, the 8-bit register can be read to produce the eight bit parallel output. For a parallel-to-serial conversion, the 8-bit register is first loaded with the input data. The bits are then individually shifted out, one bit per clock cycle, on the serial output line.

8.2.1 Serial-to-Parallel Shift Register

Figure 8.4(a) shows a 4-bit serial-to-parallel converter. The input data bits come in on the $Serial\_in$ line at a rate of one bit per clock cycle. When $Shift$ is asserted, the data bits are loaded in one bit at a time. In the first clock cycle, the first bit from the serial input stream, $Serial\_in$, gets loaded into Q_3, while the original bit in Q_3 is loaded into Q_2, Q_2 is loaded into Q_1, and so on. In the second clock cycle, the bit that is in Q_3 (i.e., the first bit from the $Serial\_in$ line) gets loaded into Q_2, while Q_3 is loaded with the second bit from the $Serial\_in$ line. This continues for four clock

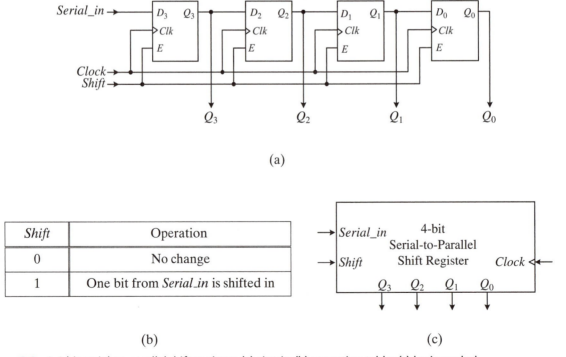

(a)

Shift	Operation
0	No change
1	One bit from $Serial\_in$ is shifted in

(b)

(c)

Figure 8.4 A 4-bit serial-to-parallel shift register: (a) circuit; (b) operation table; (c) logic symbol.

cycles until four bits are shifted into the four flip-flops, with the first bit in Q_0, second bit in Q_1, and so on. These four bits are then available for parallel reading through the output Q. Figure 8.4(b) and (c) show the operation table and the logic symbol, respectively, for this shift register.

The structural VHDL code for a 4-bit serial-to-parallel shift register is shown in Figure 8.5. The code is written at the structural level. The operation of a D flip-flop with enable is first defined. The ARCHITECTURE section for the *ShiftReg* entity uses four PORT MAP statements to instantiate four D flip-flops. These four flip-flops then are connected together using the internal signals, N_0, N_1, N_2, and N_3, such that the output of one flip-flop is connected to the input of the next flip-flop. These four internal signals also connect to the four output signals, Q_0 to Q_3, for the register output. Note that we cannot use the output signals, Q_0 to Q_3, to directly connect the four flip-flops together, since output signals cannot be read.

```
-- D flip-flop with enable
LIBRARY IEEE;
USE IEEE.STD_LOGIC_1164.ALL;

ENTITY D_flipflop IS
    PORT(D, Clock, E : IN STD_LOGIC;
    Q : OUT STD_LOGIC);
END D_flipflop;

ARCHITECTURE Behavior OF D_flipflop IS
BEGIN
    PROCESS(Clock)
    BEGIN
        IF (Clock'EVENT AND Clock = '1') THEN
            IF (E = '1') THEN
                Q <= D;
            END IF;
        END IF;
    END PROCESS;
END Behavior;

-- 4-bit shift register
LIBRARY IEEE;
USE IEEE.STD_LOGIC_1164.ALL;

ENTITY ShiftReg IS
    PORT(Serial_in, Clock, Shift : IN STD_LOGIC;
    Q : OUT STD_LOGIC_VECTOR(3 DOWNTO 0));
END ShiftReg;
```

Figure 8.5
Structural VHDL code for a 4-bit serial-to-parallel shift register.
(continued on next page)

```
ARCHITECTURE Structural OF ShiftReg IS
   SIGNAL N0, N1, N2, N3 : STD_LOGIC;
   COMPONENT D_flipflop PORT (D, Clock, E : IN STD_LOGIC;
      Q : OUT STD_LOGIC);
   END COMPONENT;


BEGIN
   U1: D_flipflop PORT MAP (Serial_in, Clock, Shift, N3);
   U2: D_flipflop PORT MAP (N3, Clock, Shift, N2);
   U3: D_flipflop PORT MAP (N2, Clock, Shift, N1);
   U4: D_flipflop PORT MAP (N1, Clock, Shift, N0);
   Q(3) <= N3;
   Q(2) <= N2;
   Q(1) <= N1;
   Q(0) <= N0;
END Structural;
```

Figure 8.5
Structural VHDL
code for a 4-bit
serial-to-parallel
shift register.

A sample simulation trace of the serial-to-parallel shift register is shown in Figure 8.6. At the first rising clock edge at time 100 ns, the *Serial_in* bit is a 0, so there is no change in the 4 bits of Q, since they are initialized to 0's. At the next rising clock edge at time 300 ns, the *Serial_in* bit is a 1, and it is shifted into the leftmost bit of Q. Hence, Q has the value of 1000. At time 500 ns, another 1 bit is shifted in, giving Q the value of 1100. At time 700 ns, a 0 bit is shifted in, giving Q the value of 0110. Notice that as bits are shifted in, the rightmost bits are lost. At time 900 ns, *Shift* is de-asserted, so the 1 bit in the *Serial_in* line is not shifted in. Finally, at time 1.1 μs, another 1 bit is shifted in.

8.2.2 Serial-to-Parallel and Parallel-to-Serial Shift Register

For both the serial-to-parallel and parallel-to-serial operations, we perform the same left-to-right shifting of bits through the register. The only difference between the two operations is whether we want to perform a parallel read after the shifting or a parallel write before the shifting. For the serial-to-parallel operation, we want to perform a parallel read after the bits have been shifted in. On the other hand, for the

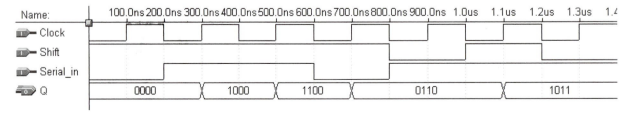

Figure 8.6 Sample simulation trace for the 4-bit serial-in-parallel-out shift register of Figure 8.5.

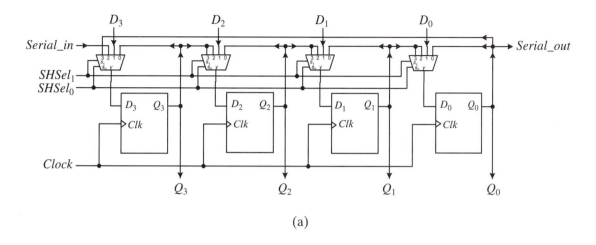

(a)

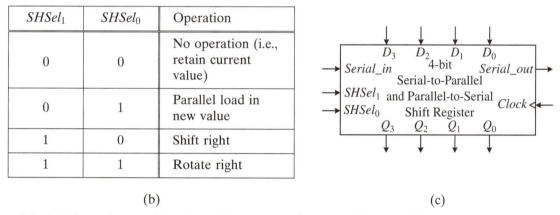

$SHSel_1$	$SHSel_0$	Operation
0	0	No operation (i.e., retain current value)
0	1	Parallel load in new value
1	0	Shift right
1	1	Rotate right

(b)   (c)

Figure 8.7   A 4-bit serial-to-parallel and parallel-to-serial shift register: (a) circuit; (b) operation table; (c) logic symbol.

parallel-to-serial operation, we want to perform a parallel write first and then shift the bits out as a serial stream.

We can implement both operations into the serial-to-parallel circuit from the previous section simply by adding a parallel load function to the circuit, as shown in Figure 8.7(a). The four multiplexers work together for selecting whether we want the flip-flops to retain the current value, load in a new value, or shift the bits to the right by one bit position. The operation of this circuit is dependent on the two select lines, $SHSel_1$ and $SHSel_0$, which control which input of the multiplexers is selected. The operation table and logic symbol are shown in Figure 8.7(b) and (c), respectively. The behavioral VHDL code and a sample simulation trace for this shift register are shown in Figures 8.8 and 8.9, respectively.

```
LIBRARY IEEE;
USE IEEE.STD_LOGIC_1164.ALL;

ENTITY shiftreg IS PORT (
    Clock: IN STD_LOGIC;
    SHSel: IN STD_LOGIC_VECTOR(1 DOWNTO 0);
    Serial_in: IN STD_LOGIC;
    D: IN STD_LOGIC_VECTOR(3 DOWNTO 0);
    Serial_out: OUT STD_LOGIC;
    Q: OUT STD_LOGIC_VECTOR(3 DOWNTO 0));
END shiftreg;

ARCHITECTURE Behavioral OF shiftreg IS
    SIGNAL content: STD_LOGIC_VECTOR(3 DOWNTO 0);
BEGIN
    PROCESS(Clock)
    BEGIN
        IF (Clock'EVENT AND Clock='1') THEN
            CASE SHSel IS
            WHEN "01" =>        -- load
                content <= D;
            WHEN "10" =>        -- shift right, pad with bit from Serial_in
                content <= Serial_in & content(3 DOWNTO 1);
            WHEN OTHERS =>
                NULL;
            END CASE;
        END IF;
    END PROCESS;

    Q <= content;
    Serial_out <= content(0);
END Behavioral;
```

Figure 8.8
Behavioral VHDL code for a 4-bit serial-to-parallel and parallel-to-serial shift register.

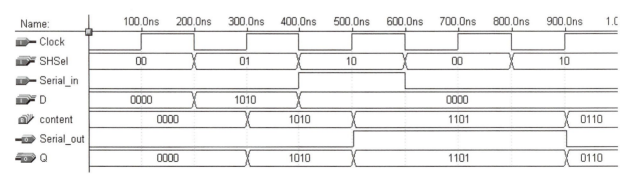

Figure 8.9 Sample trace for the 4-bit serial-to-parallel and parallel-to-serial shift register.

• • • • • • • • • • • • • • • • •

8.3 Counters

Counters, as the name suggests, are for counting a sequence of values. However, there are many different types of counters depending on the total number of count values, the sequence of values that it outputs, whether it counts up or down, and so on. The simplest is a modulo-n counter that counts the decimal sequence $0, 1, 2, \ldots$ up to $n-1$ and back to 0. Some typical counters are described next.

Modulo-n counter: Counts from decimal 0 to $n-1$ and back to 0. For example, a modulo-5 counter sequence in decimal is 0, 1, 2, 3, and 4.

Binary coded decimal (BCD) counter: Just like a modulo-n counter, except that n is fixed at 10. Thus, the sequence is always from 0 to 9.

n- bit binary counter: Similar to modulo-n counter, but the range is from 0 to 2^n-1 and back to 0, where n is the number of bits used in the counter. For example, a 3-bit binary counter sequence in decimal is 0, 1, 2, 3, 4, 5, 6, and 7.

Gray-code counter: The sequence is coded so that any two consecutive values must differ in only one bit. For example, one possible 3-bit gray-code counter sequence is 000, 001, 011, 010, 110, 111, 101, and 100.

Ring counter: The sequence starts with a string of 0 bits followed by one 1 bit, as in 0001. This counter simply rotates the bits to the left on each count. For example, a 4-bit ring counter sequence is 0001, 0010, 0100, 1000, and back to 0001.

We will now look at the design of several counters.

8.3.1 Binary Up Counter

An n-bit binary counter can be constructed using a modified n-bit register where the data inputs for the register come from an incrementer (adder) for an up counter, and a decrementer (subtractor) for a down counter. To get to the next up-count sequence from the value that is stored in a register, we simply have to add a 1 to it. We can use the full adder discussed in Section 4.2.1 as the input to the register, but we can do better. The full adder adds two operands plus the carry. But what we want is just to add a 1, so the second operand to the full adder is always a 1. Since the 1 can also be added in via the carry-in signal of the adder, we really do not need the second operand input. This modified adder that only adds one operand with the carry-in is called a **half adder** (HA). Its truth table is shown in Figure 8.10(a). We have a as the only input operand, c_{in} and c_{out} are the carry-in and carry-out signals, respectively, and s is the sum of the addition. In the truth table, we are simply adding a plus c_{in} to give the sum s and possibly a carry-out, c_{out}. From the truth table, we obtain the two equations for c_{out} and s shown in Figure 8.10(b). The HA circuit is shown in Figure 8.10(c) and its logic symbol in (d).

Several half adders can be daisy-chained together, just like with the full adders to form an n-bit adder. The single operand input a comes from the register. The initial carry-in signal, c_0, is used as the count enable signal, since a 1 on c_0 will result in incrementing a 1 to the register value, and a 0 will not. The resulting 4-bit binary up-counter circuit is shown in Figure 8.11(a), along with its operation table and logic

a	c_{in}	c_{out}	s
0	0	0	0
0	1	0	1
1	0	0	1
1	1	1	0

$$c_{out} = a\, c_{in}$$
$$s = a \oplus c_{in}$$

(a)　　　　　　　　(b)　　　　　　　　(c)　　　　　　　　(d)

Figure 8.10 Half adder: (a) truth table; (b) equations; (c) circuit; (d) logic symbol.

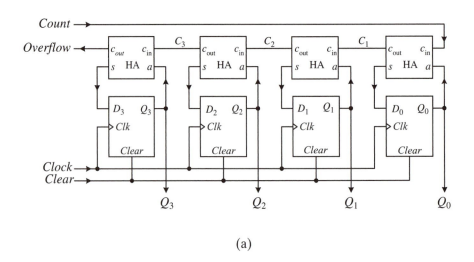

(a)

Clear	Count	Operation
1	×	Reset counter to zero
0	0	No change
0	1	Count up

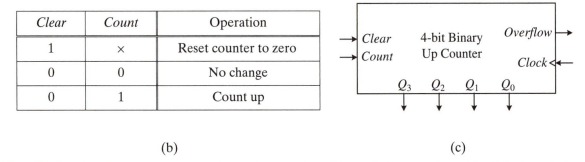

(b)　　　　　　　　　　　　　　　(c)

Figure 8.11 A 4-bit binary up counter with asynchronous clear: (a) circuit; (b) operation table; (c) logic symbol.

```
LIBRARY IEEE;
USE IEEE.STD_LOGIC_1164.ALL;
USE IEEE.STD_LOGIC_UNSIGNED.ALL;        -- need this to add STD_LOGIC_VECTORs

ENTITY counter IS PORT (
   Clock: IN STD_LOGIC;
   Clear: IN STD_LOGIC;
   Count: IN STD_LOGIC;
   Q : OUT STD_LOGIC_VECTOR(3 DOWNTO 0));
END counter;

ARCHITECTURE Behavioral OF counter IS
   SIGNAL value: STD_LOGIC_VECTOR(3 DOWNTO 0);
BEGIN
   PROCESS (Clock, Clear)
   BEGIN
     IF (Clear = '1') THEN
        value <= (OTHERS => '0');       -- 4-bit vector of 0, same as ''0000''
     ELSIF (Clock'EVENT AND Clock = '1') THEN
        IF (Count = '1') THEN
           value <= value + 1;
        END IF;
     END IF;
   END PROCESS;

  Q <= value;
END Behavioral;
```

Figure 8.12
Behavioral VHDL
code for a 4-bit
binary up counter.

symbol in (b) and (c). As long as *Count* is asserted, the counter will increment by 1 on each clock pulse until *Count* is de-asserted. When the count reaches $2^n - 1$ (which is equivalent to the binary number with all 1's), the next count will revert back to 0, because adding a 1 to a binary number with all 1's will result in an overflow on the *Overflow* bit, and all of the original bits will reset to 0. The *Clear* signal allows an asynchronous reset of the counter to 0.

The behavioral VHDL code for the 4-bit binary up counter is shown in Figure 8.12. The statement USE IEEE.STD_LOGIC_UNSIGNED.ALL is needed in order to perform additions on STD_LOGIC_VECTORS. The internal signal *value* is used to store the current count. When *Clear* is asserted, *value* is assigned the value "0000" using the expression OTHERS => '0'. Otherwise, if *Count* is asserted, then *value* will be incremented by 1 on the next rising clock edge. Furthermore, the count in *value* is assigned to the counter output, *Q*, using the concurrent statement, $Q <= value$, because it is outside the PROCESS block. A sample simulation trace is shown in Figure 8.13.

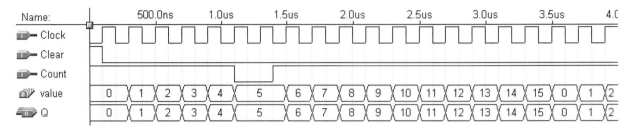

Figure 8.13 Simulation trace for the 4-bit binary up counter.

8.3.2 Binary Up-Down Counter

We can design an n-bit binary up-down counter just like the up counter, except that we need both an adder and a subtractor for the data input to the register. The **half adder-subtractor** (HAS) truth table is shown in Figure 8.14(a). The *Down* signal is to

Down	a	c_{in}	c_{out}	s
0	0	0	0	0
0	0	1	0	1
0	1	0	0	1
0	1	1	1	0
1	0	0	0	0
1	0	1	1	1
1	1	0	0	1
1	1	1	0	0

(a)

$$c_{out} = Down'\, a\, c_{in} + Down\, a'\, c_{in} = (Down \oplus a)c_{in}$$

$$s = Down'(a \oplus c_{in}) + Down(a \oplus c_{in}) = a \oplus c_{in}$$

(b)

(c)

(d)

Figure 8.14 Half adder-subtractor (HAS): (a) truth table; (b) equations; (c) circuit; (d) logic symbol.

select whether we want to count up or down. Asserting *Down* (setting to 1) will count down. The top half of the table is exactly the same as the HA truth table. For the bottom half, we are performing a subtraction of $a - c_{in}$, where s is the difference of the subtraction, and c_{out} is a 1 if we need to borrow. For example, for $0 - 1$, we need to borrow, so c_{out} is a 1. When we borrow, we get a 2; and $2 - 1 = 1$, so s is also a 1. The two resulting equations for c_{out} and s are shown in Figure 8.14(b). The circuit and logic symbol for the half adder-subtractor are shown in Figure 8.14(c) and (d).

We can simply replace the HAs with the HASes in the up-counter circuit to get the up-down counter circuit, as shown in Figure 8.15(a). Its operation table and logic symbol are shown in Figure 8.15(b) and (c). Again, the *Overflow* signal is asserted each time the counter rolls over from 1111 back to 0000.

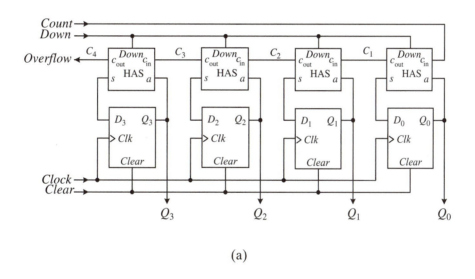

(a)

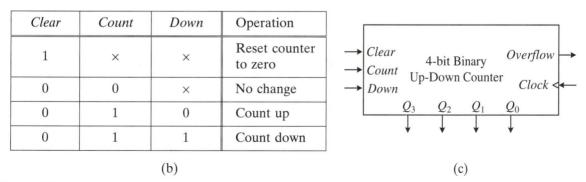

Clear	Count	Down	Operation
1	×	×	Reset counter to zero
0	0	×	No change
0	1	0	Count up
0	1	1	Count down

(b)

(c)

Figure 8.15 A 4-bit binary up-down counter with asynchronous clear: (a) circuit; (b) operation table; (c) logic symbol.

```vhdl
LIBRARY IEEE;
USE IEEE.STD_LOGIC_1164.ALL;

ENTITY udcounter IS PORT (
   Clock: IN STD_LOGIC;
   Clear: IN STD_LOGIC;
   Count: IN STD_LOGIC;
   Down: IN STD_LOGIC;
   Q: OUT INTEGER RANGE 0 TO 15);
END udcounter;

ARCHITECTURE Behavioral OF udcounter IS
BEGIN
   PROCESS (Clock, Clear)
      VARIABLE value: INTEGER RANGE 0 TO 15;
   BEGIN
     IF (Clear = '1') THEN
        value := 0;
     ELSIF (Clock'EVENT AND Clock='1') THEN
        IF (Count = '1') THEN
           IF (Down = '0') THEN
              value := value + 1;
           ELSE
              value := value - 1;
           END IF;
        END IF;
     END IF;
     Q <= value;
   END PROCESS;

END Behavioral;
```

Figure 8.16
Behavioral VHDL
code for a 4-bit
binary up-down
counter.

The VHDL code for the up-down counter, shown in Figure 8.16, is similar to the up-counter code but with the additional logic for the *Down* signal. If *Down* is asserted, then *value* is decremented by 1, otherwise it is incremented by 1. To make the code a little bit different, the counter output signal, Q, is declared as an integer that ranges from 0 to 15. This range, of course, is the range for a 4-bit binary value. Furthermore, the storage for the current count, *value*, is declared as a variable of type integer rather than as a signal. Notice also, that the signal assignment statement, $Q <= value$, is put inside the PROCESS block. Instead of being a concurrent statement

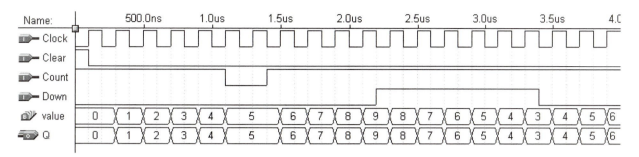

Figure 8.17   Simulation trace for the 4-bit binary up-down counter.

(when it was placed outside the PROCESS block in Figure 8.12), it is now a sequential statement. A sample simulation trace is shown in Figure 8.17.

8.3.3   Binary Up-Down Counter with Parallel Load

To make the binary counter more versatile, we need to be able to start the count sequence with any number other than zero. This is accomplished easily by modifying our counter circuit to allow it to load in an initial value. With the value loaded into the register, we can now count starting from this new value. The modified counter circuit is shown in Figure 8.18(a). The only difference between this circuit and the up-down counter circuit shown in Figure 8.15(a) is that a 2-input multiplexer is added between the s output of the HAS and the D_i input of the flip-flop. By doing this, the input of the flip-flop can be selected from either an external input value (if *Load* is asserted) or the next count value from the HAS output (if *Load* is de-asserted). If the HAS output is selected, then the circuit works exactly like before. If the external input is selected, then whatever value is presented on the input data lines will be loaded into the register. The operational table and logic symbol for this circuit are shown in Figure 8.18(b) and (c).

We have kept the *Clear* line, so that the counter can still be initialized to 0 at anytime. Notice that there is a timing difference between asserting the *Clear* line to reset the counter to 0, as opposed to loading in a 0 by asserting the *Load* line and setting the data input to 0. In the first case, the counter is reset to 0 immediately after the *Clear* is asserted, while the latter case will reset the counter to 0 at the next rising edge of the clock.

This counter can start with whatever value is loaded into the register, but it will always count up to $2^n - 1$, where n is the number of bits for the register. This is when the register contains all 1's. When the counter reaches the end of the count sequence, it will always cycle back to 0, and not to the initial value that was loaded in. However, we can add a simple comparator to this counter circuit so that the count sequence can start or end with any number in between and cycle back to the new starting value, as shown in the next section.

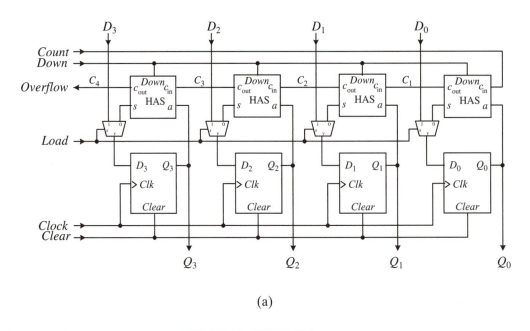

(a)

Clear	Load	Count	Down	Operation
1	×	×	×	Reset counter to zero
0	0	0	×	No change
0	0	1	0	Count up
0	0	1	1	Count down
0	1	×	×	Load value

(b)   (c)

Figure 8.18 A 4-bit binary up-down counter with parallel load and asynchronous clear: (a) circuit; (b) operation table; (c) logic symbol.

8.3.4 BCD Up Counter

A limitation with the binary up-down counter with parallel load is that it always counts up to $2^n - 1$ for an n-bit register and then cycles back to zero. If we want the count sequence to end at a number less than $2^n - 1$, we need to use an equality comparator to test for this new ending number. The comparator compares the current count value that is in the register with this new ending number. When the counter reaches this new ending number, the comparator asserts its output.

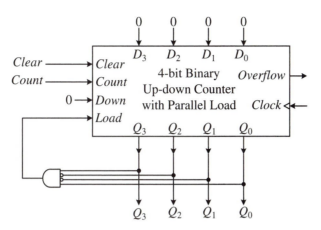

Figure 8.19
BCD up counter.

The counter can start from a number that initially is loaded in. However, if we want the count sequence to cycle back to this new starting number each time, we need to assert the *Load* signal at the end of each count sequence and reload this new starting number. The output of the comparator is connected to the *Load* line, so that when the counter reaches the ending number, it will assert the *Load* line and loads in the starting number. Hence, the counter can end at a new ending number and cycles back to a new starting number.

The binary coded decimal (BCD) up counter counts from 0 to 9 and then cycles back to 0. The circuit for it is shown in Figure 8.19. The heart of the circuit is just the 4-bit binary up-down counter with parallel load. A 4-input AND gate is used to compare the count value with the number 9. When the count value is 9, the AND-gate comparator outputs a 1 to assert the *Load* line. Once the *Load* line is asserted, the next counter value will be the value loaded in from the counter input D. Since D is connected to all 0's, the counter will cycle back to 0 at the next rising clock edge. The *Down* line is connected to a 0, since we only want to count up.

In order for the timing of each count to be the same, we must use the *Load* operation to load in the value 0, rather than using the *Clear* operation. If we connect the output of the AND gate to the *Clear* input instead of the *Load* input, we will still get the correct count sequence. However, when the count reaches 9, it will change to a 0 almost immediately, because when the output of the AND gate asserts the asynchronous *Clear* signal, the counter is reset to 0 right away and not at the next rising clock edge.

Example 8.1	**Constructing an up-counter circuit**

This example uses the 4-bit binary up-down counter with parallel load to construct an up-counter circuit that counts from 3 to 8 (in decimal), and back to 3.

The circuit for this counter, shown in Figure 8.20, is almost identical to the BCD up-counter circuit. The only difference is that we need to test for the number 8 instead of 9 as the last number in the sequence, and the first number to load in is a 3 instead of a 0. Hence, the inputs to the AND gate for comparing with the binary counter output is 1000, and the number for loading in is 0011.

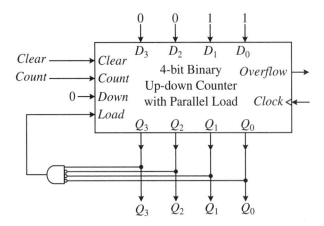

Figure 8.20
Counter for
Example 8.1.

8.3.5 BCD Up-Down Counter

We can get a BCD up-down counter by modifying the BCD up-counter circuit slightly. The counter counts from 0 to 9 for the up sequence and 9 down to 0 for the down sequence. For the up sequence, when the count reaches 9, the *Load* line is asserted to load in a 0 (0000 in binary). For the down sequence, when the count reaches 0, the *Load* line is asserted to load in a 9 (1001 in binary).

The BCD up-down counter circuit is shown in Figure 8.21. Two 5-input AND gates acting as comparators are used. The one labeled "Up" will output a 1 when *Down* is de-asserted (i.e., counting up), and the count is 9. The one label "Dn" will output a 1 when *Down* is asserted, and the count is 0. The *Load* signal is asserted

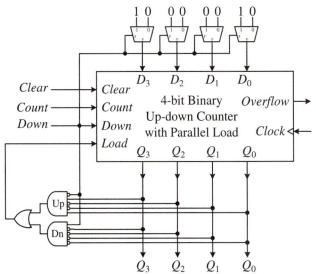

Figure 8.21
BCD up-down counter.

by either one of these two AND gates. Four 2-to-1 multiplexers are used to select which of the two starting values, 0000 or 1001, is to be loaded in when the *Load* line is asserted. The select lines for these four multiplexers are connected in common to the *Down* signal, so that when the counter is counting up, 0000 is loaded in when the counter wraps around, and 1001 is loaded in when the counter wraps around while counting down. It should be obvious that the two values, 0000 and 1001, can also be loaded in without the use of the four multiplexers.

Example 8.2   **Constructing an irregular sequence counter circuit**

This example uses the 4-bit binary up-down counter with parallel load to construct an up-down counter circuit that outputs the sequence: 2, 5, 9, 13, and 14, repeatedly.

   The 4-bit binary counter can only count numbers consecutively. In order to output numbers that are not consecutive, we need to design an output circuit that maps from one number to another number. The required sequence has five numbers, so we will first design a counter to count from 0 to 4. The output circuit will then map the numbers: 0, 1, 2, 3, and 4 to the required output numbers: 2, 5, 9, 13, and 14, respectively.

   The inputs to the output circuit are the four output bits of the counter: Q_3, Q_2, Q_1, and Q_0. The outputs from this circuit are the modified four bits: O_3, O_2, O_1, and O_0, for representing the five output numbers. The truth table and the resulting output equations for the output circuit are shown in Figure 8.22(a) and (b), respectively. The easiest way to see how the output equations are obtained is to use a K-map and put in all of the don't-cares. The complete counter circuit is shown in Figure 8.22(c).

Decimal Input	Q_3	Q_2	Q_1	Q_0	Decimal Output	O_3	O_2	O_1	O_0
0	0	0	0	0	2	0	0	1	0
1	0	0	0	1	5	0	1	0	1
2	0	0	1	0	9	1	0	0	1
3	0	0	1	1	13	1	1	0	1
4	0	1	0	0	14	1	1	1	0
Rest of the Combinations						×	×	×	×

(a)

Figure 8.22   Counter for Example 8.2.
(continued on next page)

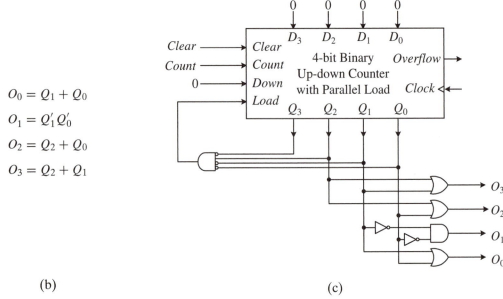

$$O_0 = Q_1 + Q_0$$

$$O_1 = Q_1' Q_0'$$

$$O_2 = Q_2 + Q_0$$

$$O_3 = Q_2 + Q_1$$

(b)   (c)

Figure 8.22 Counter for Example 8.2.

8.4 Register Files

When we want to store several numbers concurrently in a digital circuit, we can use several individual registers in the circuit. However, there are times when we want to treat these registers as a unit, similar to addressing the individual locations of an array or memory. So, instead of having several individual registers, we want to have an array of registers. This array of registers is known as a **register file**. In a register file, all of the respective control signals for the individual registers are connected in common. Furthermore, all of the respective data input and output lines for all of the registers are also connected in common. For example, the *Load* lines for all of the registers are connected together, and all of the D_3 data lines for all of the registers are connected together. So the register file has only one set of input lines and one set of output lines for all of the registers. In addition, address lines are used to specify which register in the register file is to be accessed.

In a microprocessor circuit requiring an ALU, the register file usually is used for the source operands of the ALU. Since the ALU usually takes two input operands, we like the register file to be able to output two values from possibly two different locations of the register file at the same time. So, a typical register file will have one write port and two read ports. All three ports will have their own enable and address lines. When the read enable line is de-asserted, the read port will output a 0. On the other hand, when the read enable line is asserted, the content of the register specified

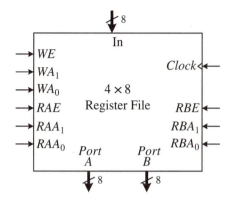

Figure 8.23
Logic symbol for a
4 × 8 register file.

by the read address lines is passed to the output port. The write enable line is used to load a value into the register specified by the write address lines.

The logic symbol for a 4×8 register file (four registers, each being 8-bits wide) is shown in Figure 8.23. The 8-bit write port is labeled *In*, and the two 8-bit read ports are labeled *Port A* and *Port B*. *WE* is the active-high write enable line. To write a value into the register file, this line must be asserted. The WA_1 and WA_0 are the two address lines for selecting the write location. Since there are four locations in this register file, two address lines are needed. The *RAE* line is the read enable line for *Port A*. The two read address select lines for *Port A* are RAA_1 and RAA_0. For *Port B*, we have the *Port B* enable line, *RBE*, and the two address lines, RBA_1 and RBA_0.

The register circuit from Figure 8.1 does not have any control for the reading of the data to the output port. In order to control the output of data, we can use a 2-input AND gate to enable or disable each of the data output lines, Q_i. We want to control all the data output lines together, therefore, one input from all of the 2-input AND gates are connected in common. When this common input is set to a 0, all the AND gates will output a 0. When this common input is set to a 1, the output for all of the AND gates will be the value from the other input. An alternative to using AND gates to control the read ports is to use tri-state buffers. Instead of outputting a 0 when disabled, the tri-state buffers will have a high impedance.

Our register file has two read ports, that is, two output controls for each register. So, instead of having just one 2-input AND gate per output line, Q_i, we need to connect two AND gates to each output line: one for *Port A*, and one for *Port B*. An 8-bit wide register file cell circuit will have eight AND gates for *Port A* and another eight AND gates for *Port B*, as shown in Figure 8.24. *AE* and *BE* are the read enable signals for *Port A* and *Port B*, respectively. For each read port, the read enable signal is connected in common to one input of all of the eight AND gates. The second input from each of the eight AND gates connects to the eight output lines, Q_0 to Q_7.

For a 4×8 register file, we need to use four 8-bit register file cells. In order to select which register file cell we want to access, three decoders are used to decode the addresses: WA_1, WA_0, RAA_1, RAA_0, RBA_1, and RBA_0. One decoder is used for the write addresses, WA_1 and WA_0; one for the *Port A* read addresses, RAA_1 and RAA_0; and one for the *Port B* read addresses, RBA_1 and RBA_0. The decoders' outputs are used to assert the individual register file cell's write line, *Load*, and read

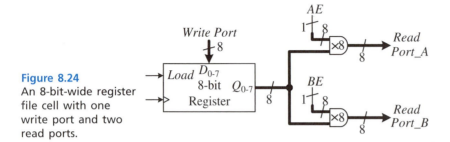

Figure 8.24
An 8-bit-wide register
file cell with one
write port and two
read ports.

enable lines, AE and BE. The complete circuit for the 4×8 register file is shown in
Figure 8.25. The respective read ports from each register file cell are connected to
the external read port through a 4-input $\times$ 8-bit OR gate.

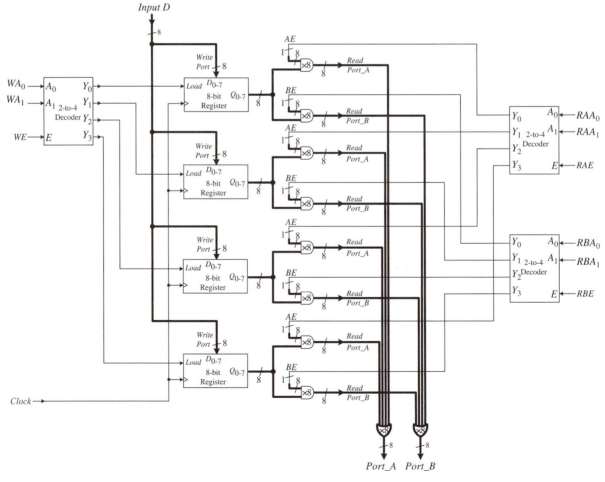

Figure 8.25 A 4×8 register file with one write port and two read ports.

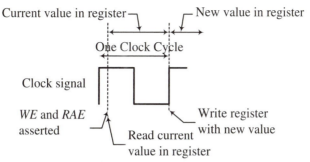

Figure 8.26
Read and write timings for a register file cell.

For example, to read from Register 3 through *Port B*, the *RBE* line has to be asserted, and the *Port B* address lines, RBA_0 and RBA_1, have to be set to 11 (for Register 3). The data from Register 3 will be available immediately on *Port B*. To write a value to Register 2, the write address lines, WA_0 and WA_1, are set to 10, and then the write enable line, *WE*, is asserted. The data at input *D* is then written into Register 2 at the next active (rising) clock edge. Since all three decoders can be enabled at the same time, the two read operations and the write operation all can be asserted together.

In terms of the timing issues, the data on the read ports are available immediately after the read enable line is asserted, whereas, the write occurs at the next active (rising) edge of the clock. Because of this, the same register can be accessed for both reading and writing at the same time; that is, the read and write enable lines can be asserted at the same time using the same read and write address. When this happens, then the value that is currently in the register is read through the read port, and a new value will be written into the register at the next rising clock edge. This timing is shown in Figure 8.26. The important point to remember is that, when the read and write operations are performed at the same time on the same register, the read operation always reads the current value stored in the register and never the new value that is to be written in by the write operation. The new value written in is available only after the next rising clock edge.

The VHDL code for the 4×8 register file is shown in Figure 8.27. The main code is composed of three processes: the write process and the two read port processes. These three processes are similar to three concurrent statements in that they are executed in parallel. The write process is sensitive to the clock, and because of the IF *clock* statement in the process, a write occurs only at the rising edge of the clock signal. The two read port processes are not sensitive to the clock but only to the read enable and read address signals. So the read data is available immediately when these lines are asserted. The function CONV_INTEGER(WA) converts the STD_LOGIC_VECTOR *WA* to an integer so that the address can be used as an index into the *RF* array.

```
LIBRARY IEEE;
USE IEEE.STD_LOGIC_1164.ALL;
USE IEEE.STD_LOGIC_UNSIGNED.ALL;                    -- needed for CONV_INTEGER()
```

Figure 8.27   VHDL code for a 4 × 8 register file with one write port and two read ports.
(continued on next page)

```
ENTITY regfile IS PORT(
   clock: IN STD_LOGIC;                                --clock
   WE: IN STD_LOGIC;                                   --write enable
   WA: IN STD_LOGIC_VECTOR(1 DOWNTO 0);                --write address
   D: IN STD_LOGIC_VECTOR(7 DOWNTO 0);                 --input
   RAE, RBE: IN STD_LOGIC;                             --read enable ports A & B
   RAA, RBA: IN STD_LOGIC_VECTOR(1 DOWNTO 0);          --read address port A & B
   PortA, PortB: OUT STD_LOGIC_VECTOR(7 DOWNTO 0));    --output port A & B
END regfile;
ARCHITECTURE Behavioral OF regfile IS
   SUBTYPE reg IS STD_LOGIC_VECTOR(7 DOWNTO 0);
   TYPE regArray IS ARRAY(0 TO 3) OF reg;
   SIGNAL RF: regArray;                                --register file contents
BEGIN
   WritePort: PROCESS (clock)
   BEGIN
      IF (clock'EVENT AND clock = '1') THEN
         IF (WE = '1') THEN
            RF(CONV_INTEGER(WA)) <= D;                 -- fn to convert from vector to integer
         END IF;
      END IF;
   END PROCESS;


   ReadPortA: PROCESS (RAA, RAE)
   BEGIN
      -- Read Port A
      IF (RAE = '1') THEN
         PortA <= RF(CONV_INTEGER(RAA));               -- fn to convert from vector to integer
      ELSE
         PortA <= (OTHERS => '0');
      END IF;
   END PROCESS;


   ReadPortB: PROCESS (RBE, RBA)
   BEGIN
      -- Read Port B
      IF (RBE = '1') THEN
         PortB <= RF(CONV_INTEGER(RBA));               -- fn to convert from vector to integer
      ELSE
         PortB <= (OTHERS => '0');
      END IF;
   END PROCESS;
END Behavioral;
```

Figure 8.27 VHDL code for a 4 × 8 register file with one write port and two read ports.

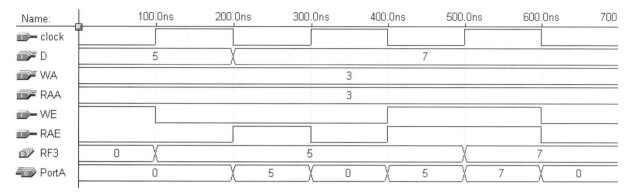

Figure 8.28   Sample simulation trace for the 4 × 8 register file.

A sample simulation trace is shown in Figure 8.28. In the simulation trace, both the write address, *WA*, and *Port A* read address, *RAA*, are set to Register 3. At 0 ns, the input data, *D*, is 5. With write enable, *WE*, asserted, the data 5 is stored into RF(3) at the next rising edge of the clock, which happens at 100 ns. When *RAE* is asserted at 200 ns, the data 5 from RF(3) is available on *Port A* immediately. At 400 ns, both *WE* and *RAE* are asserted at the same time. The current data 5 from RF(3) appears immediately on *Port A*. However, the new data 7 is written into RF(3) at 500 ns, the next rising clock edge. The new data 7 is available on *Port A* only after time 500 ns.

8.5 Static Random Access Memory

Another main component in a computer system is memory. This can refer to either random access memory (RAM) or read-only memory (ROM). We can make memory the same way we make the register file but with more storage locations. However, there are several reasons why we don't want to. One reason is that we usually want a lot of memory and we want it very cheap, so we need to make each memory cell as small as possible. Another reason is that we want to use a common data bus for both reading data from, and writing data to the memory. This implies that the memory circuit should have just one data port (and not two or three like the register file) for both reading and writing of data.

The logic symbol, showing all of the connections for a typical RAM chip is shown in Figure 8.29(a). There is a set of data lines, D_i, and a set of address lines, A_i. The data lines serve for both input and output of the data to the location that is specified by the address lines. The number of data lines is dependent on how many bits are used for storing data in each memory location. The number of address lines is dependent on how many locations are in the memory chip. For example, a 512-byte memory chip will have eight data lines (8 bits = 1 byte) and nine address lines ($2^9 = 512$).

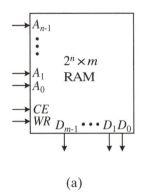

CE	WR	Operation
0	×	None
1	0	Read from memory location selected by address lines
1	1	Write to memory location selected by address lines

Figure 8.29
A $2^n \times m$ RAM chip:
(a) logic symbol;
(b) operation table.

(a)   (b)

In addition to the data and address lines, there are usually two control lines: chip enable (CE), and write enable (WR). In order for a microprocessor to access memory, either with the read operation or the write operation, the active-high CE line must first be asserted. Asserting the CE line enables the entire memory chip. The active-high WR line selects which of the two memory operations is to be performed. Setting WR to a 0 selects the read operation, and data from the memory is retrieved. Setting WR to a 1 selects the write operation, and data from the microprocessor is written into the memory. Instead of having just the WR line for selecting the two operations, read and write, some memory chips have both a read enable and a write enable line. In this case, only one line can be asserted at any one time. The memory location in which the read and write operations are to take place, of course, is selected by the value of the address lines. The operation of the memory chip is shown in Figure 8.29(b).

Notice in Figure 8.29(a) that the RAM chip does not require a clock signal. Both the read and write memory operations are not synchronized to the global system clock. Instead the data operations are synchronized to the two control lines, CE and WR. Figure 8.30(a) shows the timing diagram for a memory write operation. The write operation begins with a valid address on the address lines, followed immediately by the CE line being asserted. Shortly after, valid data must be present on the data lines, and then the WR line is asserted. As soon as the WR line is asserted, the data that is on the data lines is then written into the memory location that is addressed by the address lines.

A memory read operation also begins with setting a valid address on the address lines, followed by CE going high. The WR line is then pulled low, and shortly after, valid data from the addressed memory location is available on the data lines. The timing diagram for the read operation is shown in Figure 8.30(b).

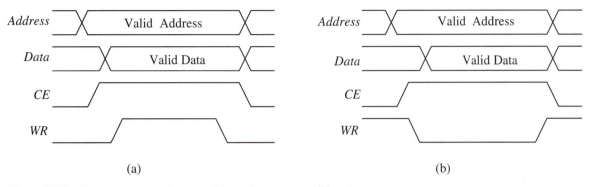

Figure 8.30   Memory timing diagram: (a) read operation; (b) write operation.

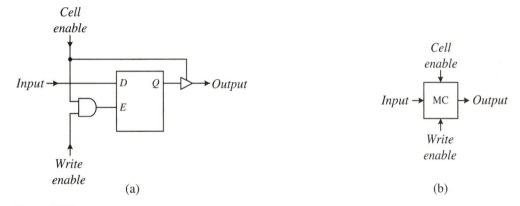

Figure 8.31   Memory cell: (a) circuit; (b) logic symbol.

Each bit in a static RAM chip is stored in a memory cell similar to the circuit shown in Figure 8.31(a). The main component in the cell is a D latch with enable. A tri-state buffer is connected to the output of the D latch so that it can be selectively read from. The *Cell enable* signal is used to enable the memory cell for both reading and writing. For reading, the *Cell enable* signal is used to enable the tri-state buffer. For writing, the *Cell enable* together with the *Write enable* signals are used to enable the D latch so that the data on the *Input* line is latched into the cell. The logic symbol for the memory cell is shown in Figure 8.31(b).

To create a 4×4 static RAM chip, we need sixteen memory cells forming a 4×4 grid, as shown in Figure 8.32. Each row forms a single storage location, and the number of memory cells in a row determines the bit width of each location. So all of the memory cells in a row are enabled with the same address. Again, a decoder is used to decode the address lines, A_0 and A_1. In this example, a 2-to-4 decoder is used to decode the four address locations. The *CE* signal is for enabling the chip,

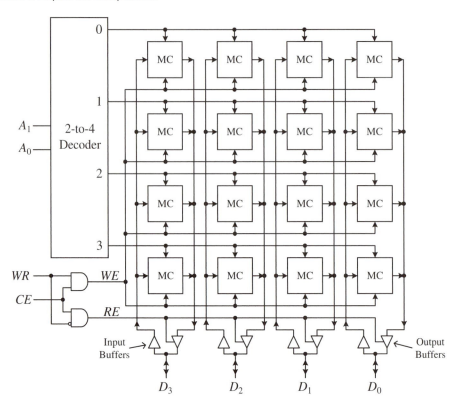

Figure 8.32
A 4 × 4 RAM chip circuit.

specifically to enable the read and write functions through the two AND gates. The internal *WE* signal, asserted when both the *CE* and *WR* signals are asserted, is used to assert the *Write enables* for all of the memory cells. The data comes in from the external data bus, D_i, through the input buffer and to the *Input* line of each memory cell. The purpose of using an input buffer for each data line is so that the external signal coming in only needs to drive just one device (the buffer) rather than having to drive several devices (i.e., all of the memory cells in the same column). Which row of memory cells actually gets written to will depend on the given address. The read operation requires *CE* to be asserted and *WR* to be de-asserted. This will assert the internal *RE* signal, which in turn will enable the four output tri-state buffers at the bottom of the circuit diagram. Again, the location that is read from is selected by the address lines.

The VHDL code for a 16×4 RAM chip is shown in Figure 8.33. The bi-directional data port, D, is declared as BUFFER so that it can be read from and written to. The actual memory content is stored in the variable *mem*, which is an array of size 16 of type STD_LOGIC_VECTOR.

```
LIBRARY IEEE;
USE IEEE.STD_LOGIC_1164.ALL;
USE IEEE.STD_LOGIC_ARITH.ALL;
USE IEEE.STD_LOGIC_UNSIGNED.ALL;              -- needed for CONV_INTEGER()


ENTITY memory IS PORT (
    CE, WR: IN STD_LOGIC;                      --chip enable, write enable
    A: IN STD_LOGIC_VECTOR(3 DOWNTO 0);        --address
    D: BUFFER STD_LOGIC_VECTOR(3 DOWNTO 0));   --data
END memory;


ARCHITECTURE Behavioral OF memory IS
BEGIN
    PROCESS (CE, WR)
        SUBTYPE cell IS STD_LOGIC_VECTOR(3 DOWNTO 0);
        TYPE memArray IS ARRAY(0 TO 15) OF cell;
        VARIABLE mem: memArray; --memory contents
        VARIABLE ctrl: STD_LOGIC_VECTOR(1 DOWNTO 0);
    BEGIN
        ctrl := CE & Wr;                       -- group signals for CASE decoding
        CASE ctrl IS
            WHEN "10" =>                       -- read
                D <= mem(CONV_INTEGER(A));     -- fn TO convert from bit vector TO integer
            WHEN "11" =>                       -- write
                mem(CONV_INTEGER(A)) := D;     -- fn TO convert from bit vector TO integer
            WHEN OTHERS = -- invalid or not enable
                D <= (OTHERS => 'Z');
        END CASE;
    END PROCESS;
END Behavioral;
```

Figure 8.33   VHDL code for a 16 × 4 RAM chip.

● ● ● ● ● ● ● ● ● ● ● ● ● ● ● ● ● ⋯

*8.6 Larger Memories

In general, there is always a need for larger memories. Because of product availability constraints, we need to construct these larger memories from multiple, smaller memory chips. Larger memory requirements can be for either more memory locations, wider bit widths for each location, or both.

8.6.1 More Memory Locations

For example, we may want to have a $1\,K \times 8$-bit memory built using multiple 256×8-bit memory chips. Using such small numbers is archaic, but you get the idea. In this case, we would need four of these 256×8-bit memory chips, since $1\,K = 4 \times 256$. A 256×8-bit memory chip has eight address lines, since $2^8 = 256$. To decode four chips, we need an additional two address lines to enable which of the four chips we want to address. Thus, we need a total of ten address lines with the first eight, A_0 to A_7, connected, respectively, in common directly to the eight address lines on the four chips, and the last two lines, A_8 and A_9, connected to the address inputs of a 2-to-4 decoder. The four outputs from the decoder are used to assert the chip enable, CE, line of the four memory chips, RAM_0 to RAM_3. The data lines and the write enable lines are all connected, respectively, in common. The circuit is shown in Figure 8.34(a).

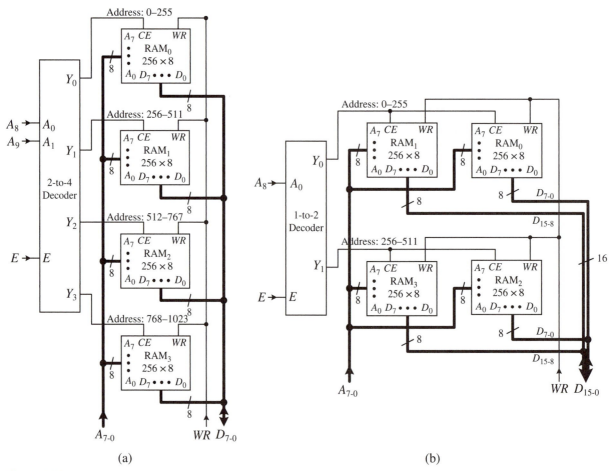

Figure 8.34 Larger memory made from smaller memory chips: (a) a $1\,K \times 8$-bit memory made from four 256×8-bit memory chips; (b) a 512×16-bit memory made from four 256×8-bit memory chips.

The 256-byte memory chip, RAM_0, is enabled when the address bits, A_8 and A_9, are 00. Hence, the address range for RAM_0 is from 0 to 255 (0000000000 to 0011111111 in binary). Similarly, RAM_1 is enabled when the address bits, A_8 and A_9, are 01. Hence, the address range for RAM_1 is from 256 to 511 (0100000000 to 0111111111 in binary). The address range for RAM_2 is from 512 to 767 (1000000000 to 1011111111 in binary), and the address range for RAM_3 is from 768 to 1023 (1100000000 to 1111111111 in binary).

A particular memory location is accessed as follows. If we want to write to memory location 717, which is binary 1011001101, the Y_2 line of the decoder would be asserted, since bits 8 and 9 are "10." This Y_2 line in turn asserts the CE line of the RAM_2 chip, while the remaining RAM chips are disabled. Finally, within the RAM_2 chip that is enabled, location 205, which is binary 11001101 from bits 0 to 7 of the original address, is selected. Location 205 in the third RAM chip is location 717 for the entire memory, since $256 + 256 + 205 = 717$.

8.6.2   Wider Bit Width

We may also want to have wider bit width for each memory location made from smaller ones. For example, we may want to have a memory that is 512 locations $\times$ 16-bits wide made from 256×8-bit memory chips. Again, we would need four 256-byte memory chips, but connected as shown in Figure 8.34(b). For 512 locations, only nine address lines are needed, with the first eight, A_0 to A_7, connected, respectively, in common directly to the eight address lines on the four chips, and the last line, A_8, connected to the address input of a 1-to-2 decoder. For a 16-bit wide data bus, we need to connect two 8-bit wide chips in parallel so that each two similar 8-bit wide location in the two chips can be combined together to form a 16-bit wide location. Since these two chips need to work together, their chip enable, CE, lines must be connected in common and asserted by the same output from the decoder.

Memory chips RAM_0 and RAM_2 are for storing the data bits D_0 to D_7, while memory chips RAM_1 and RAM_3 are for storing the data bits D_8 to D_{15}. The address range for RAM_0 and RAM_1 is from 0 to 255 (000000000 to 011111111 in binary), and the address range for RAM_2 and RAM_3 is from 256 to 511 (100000000 to 111111111 in binary).

| **Example 8.3** | **Building larger memory using smaller RAM chips** |

Build a 2 M-byte memory using 512 K-byte RAM chips.

A 512 K-byte RAM chip has 9 address lines, A_0 to A_8, because $2^9 = 512$ K. Since 4×512 K $= 2$ M, therefore, we need to use four 512 K-byte RAM chips. In order to select from these four RAM chips, we need two more address lines, A_9 and A_{10}. Hence, the system must have at least 11 address lines. The first 9 address lines, A_0 to A_8, are connected directly to the four RAM chips. The last two address lines, A_9 and A_{10}, are connected to a 2-to-4 decoder. The four outputs of the decoder are connected to the chip enables (CE) for the four RAM chips. The eight data lines and the write enable lines are all connected, respectively, in common. The circuit is shown in Figure 8.35.

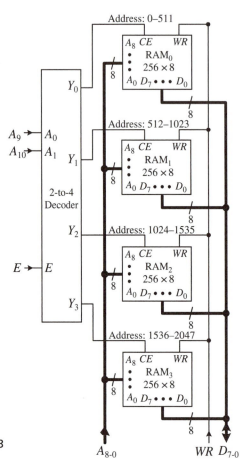

Figure 8.35

A 2 M-byte memory circuit for Example 8.3

Example 8.4

Determining an address range

What is the address range for the Y_5 line in Figure 8.36?

Y_5 is asserted when the address lines, A_{13}, A_{12}, A_{11}, and A_{10}, are 0101. The lowest address is when the ten low-order address bits, A_9 to A_0, are all 0's, and the highest address is when these ten bits are all 1's. Hence, the address range for Y_5 is from 01010000000000 to 01011111111111 in binary or 5120 to 6143 in decimal.

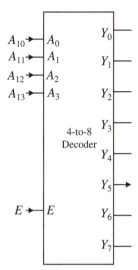

Figure 8.36
Address range for
Example 8.4.

8.7 Summary Checklist

- Registers
- Serial-to-Parallel Shift Registers
- Parallel-to-Serial Shift Registers
- Binary counters
- Binary up-down counters
- BCD counters
- BCD up-down counters
- Counters for random sequences
- Register files
- Random access memories (RAM)
- Building more memory locations using smaller RAM chips
- Building wider bit-width memories using smaller RAM chips

8.8 Problems

P8.1. The 4-bit binary up counter VHDL code shown in Figure 8.12 does not have the *Overflow* output signal. Modify the code to include the *Overflow* signal.

P8.2. For the BCD up counter circuit shown in Figure 8.19, what happens if the output of the AND gate comparator is connected to the *Clear* signal instead of to the *Load* signal? Will it produce the same waveform? Explain your observations.

P8.3. In the BCD up-down counter circuit shown in Figure 8.21, four 2-input multiplexers are used to select the correct value to be loaded in. Modify the circuit so that the multiplexers are not needed.

P8.4. Write the behavioral VHDL code for the BCD up-down counter.

P8.5. Use the 4-bit binary up-down counter with parallel load to construct an up-down counter circuit that counts from 0 to 7 decimal and back to 0.

P8.6. Use the 4-bit binary up-down counter with parallel load to construct an up-down counter circuit that counts from 5 to 13 decimal and back to 5.

P8.7. Use the 4-bit binary up-down counter with parallel load to construct an up-down counter circuit that outputs the sequence: 7, 12, 19, 36, 42, 58, and 57, repeatedly.

P8.8. Use the 4-bit binary up-down counter with parallel load to construct an up-down counter circuit that outputs the sequence: 4, 8, 5, 3, 16, and 7, repeatedly.

P8.9. Write the structural VHDL code for the BCD up-down counter based on the circuit diagram shown in Figure 8.21. Use the 4-bit binary up-down counter VHDL code as a component.

P8.10. What are the valid address ranges for the Y_5 and Y_7 lines in the following circuits of Figure P8.10?

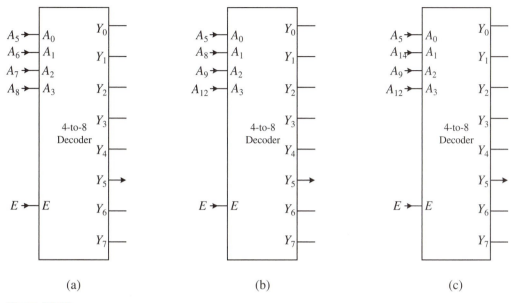

(a)   (b)   (c)

Figure P8.10

P8.11. Design a 32 M-byte memory using 4 M-byte RAM chips. Label all of the signals clearly.

P8.12. Design an 8 M-byte memory using 2 M × 4-bit RAM chips. Label all of the signals clearly.

P8.13. Manually design and implement on the UP2 board an FSM circuit for writing the value 13 into location 2 of a 4 × 8 register file, then read location 2 through *Port A*, and display the number as binary on the eight LEDs.

P8.14. Manually design and implement on the UP2 board the following FSM circuit for controlling a 4 × 8 register file. For input, use two DIP switches to specify the register file location and another eight DIP switches to specify the data input. Use a push-button for the write enable signal. For output, use the eight LEDs. The eight output LEDs continuously display the content of the current selected register file location. When the push-button is pressed, the data input is loaded into the selected location.

Datapaths

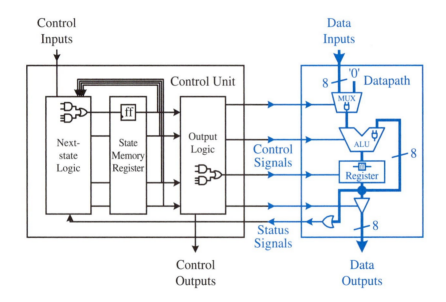

In Chapter 4, we learned how to design functional units for performing single, simple data operations, such as the adder for adding two numbers or the comparator for comparing two values. The next logical question to ask is how do we design a circuit for performing more complex data operations or operations that involve multiple steps? For example, how do we design a circuit for adding four numbers or a circuit for adding a million numbers? For adding four numbers, we can connect three adders together, as shown in Figure 9.1(a). However, for adding a million numbers, we really don't want to connect a million minus one adders together like that. Instead, we want a circuit with just one adder and to use it a million times. A **datapath** circuit allows us to do just that, that is, for performing operations involving multiple steps. Figure 9.1(b) shows a simple datapath using one adder to add as many numbers as we want. In order for this to be possible, a register is needed to store the temporary result after each addition. The temporary result from the register is fed back to the input of the adder so that the next number can be added to the current sum.

In this chapter, we will look at the design of datapaths. Recall that the datapath is the second main part in a microprocessor. The datapath is responsible for the manipulation of data. It includes (1) functional units such as adders, shifters, multipliers, ALUs, and comparators, (2) registers and other memory elements for the temporary storage of data, and (3) buses, multiplexers, and tri-state buffers for the transfer of data between the different components in the datapath, and the external world. From the microprocessor road map figure at the beginning of this chapter, we see that external data enters the datapath through the **data input** lines. Results from the datapath operations are provided through the **data output** lines. These signals serve as the primary input/output data ports for the microprocessor.

In order for the datapath to function correctly, appropriate **control signals** must be asserted at the right time. Control signals are needed for all of the select and control lines for all of the components used in the datapath. This includes all of the select

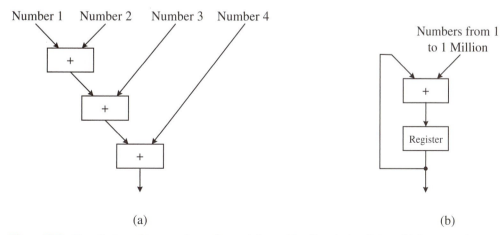

(a)   (b)

Figure 9.1 Circuits to add several numbers: (a) combinational circuit to add four numbers; (b) datapath to add one million numbers.

lines for multiplexers, ALUs, and other functional units having multiple operations; all of the read/write enable signals for registers and register files; address lines for register files; and enable signals for tri-state buffers. Thus, the operation of the datapath is determined by which control signals are asserted or de-asserted and at what time. In a microprocessor, these control signals are generated by the **control unit**.

Some of the control signals generated by the control unit are dependent on the data that is being manipulated within the datapath. (For example, the result of a conditional test with a number that is stored in a register.) Hence, in order for the control unit to generate these control signals correctly, the datapath needs to supply **status signals** to the control unit. These status signals are usually from the output of comparators. The comparator tests for a given logical condition between two data values in the datapath. These values are obtained either from memory elements or directly from the output of functional units, or are hardwired as constants. The status signals provide information for the control unit to determine what state to go to next. For example, in a conditional loop situation, the status signal provides the result of the condition being tested, and tells the control unit whether to repeat or exit the loop.

Since the datapath performs all of the functional operations of a microprocessor (and the microprocessor is for solving problems), the datapath must be able to perform all of the operations that are required to solve the given problem. For example, if the problem requires the addition of two numbers, the datapath, therefore, must contain an adder. If the problem requires the storage of three temporary variables, the datapath must have three registers. However, even with these requirements, there are still many options as to what actually is implemented in the datapath. For example, an adder can be implemented as a single adder circuit, or as part of the ALU. These functional units can be used many times. Registers can be separate register units or combined in a register file. Furthermore, two temporary variables can share the same register if they are not needed at the same time.

Datapath design is also referred to as **register-transfer level** (**RTL**) design. In a register-transfer level design, we look at how data is transferred from one register to another, or back to the same register. If the same data is written back to a register without any modifications, then nothing has been accomplished. Therefore, before writing the data to a register, the data usually passes through one or more functional units, and gets modified.

The sequence of RTL operations—read data from a register, modify data by functional units, and write result to a register—is referred to as a **register-transfer operation**. Every register-transfer operation must complete within one clock cycle (which is equivalent to one state of the FSM, since the FSM changes state at every clock cycle). Furthermore, in a single register-transfer operation, a functional unit cannot be used more than once. However, the same functional unit can be used more than once if it is used by different register-transfer operations. In other words, a functional unit can be used only once in the same clock cycle, but can be used again in a different clock cycle.

We will now look at how datapaths are designed, and how they are used to solve problems. First, we will look at the design of dedicated datapaths for solving single specific problems, and then we will look at general datapaths where they can be used for solving different problems.

9.1 Designing Dedicated Datapaths

The goal for designing a dedicated datapath is to build a circuit for solving a single specific problem. In this chapter, we will specify the problem in the form of an algorithm. We will use C-style pseudocodes to write the algorithms. The logical interpretation of the algorithm is irrelevant in what we are trying to do, so when given a certain segment of code, we will just take the code as is and will not optimize it in any manner.

In a register-transfer level design, we focus on how data move from register to register via some functional units where they are modified. In the design process, we need to decide on the following issues:

- What kind of registers to use, and how many are needed?
- What kind of functional units to use, and how many are needed?
- Can a certain functional unit be shared between two or more operations?
- How are the registers and functional units connected together so that all of the data movements specified by the algorithm can be realized?

Since the datapath is responsible for performing all of the data operations, it must be able to perform all of the data manipulation statements and conditional tests specified by the algorithm. For example, the assignment statement:

$$A = A + 3$$

takes the value that is stored in the variable A, adds the constant 3 to it, and stores the result back into A. Note that whatever the initial value of A is here is irrelevant since that is a logical issue. In order for the datapath to perform the data operation specified by this statement, the datapath must have a register for storing the value A. Furthermore, there must be an adder for performing the addition. The constant 3 can be hardwired into the circuit as a binary value.

The next question to ask is how do we connect the register, the adder, and the constant 3 together so that the execution of the assignment statement can be realized. Recall from Section 8.1 that a value stored in a register is available at the Q output of the register. Since we want to add $A + 3$, we connect the Q output of the register to the first operand input of the adder, and connect the constant 3 to the second operand input. We want to store the result of the addition back into A (i.e., back into the same register), therefore, we connect the output of the adder to the D input of the register, as shown in Figure 9.2(a).

The storing of the adder result into the register is accomplished by asserting the *Load* control signal of the register (i.e., asserting *Aload*). This *ALoad* signal is an example of what we have been referring to as the datapath control signal. This control signal controls the operation of this datapath. The control unit, which we will talk about in the next chapter, will control this signal by either asserting or de-asserting it.

The actual storing of the value into the register, however, does not occur immediately when *ALoad* is asserted. Since the register is synchronous to the clock signal, the actual storing of the value occurs at the next active clock edge. Because of this,

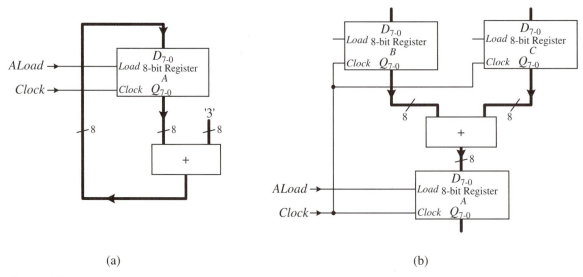

(a)   (b)

Figure 9.2 Sample datapaths: (a) for performing $A = A + 3$; (b) for performing $A = B + C$.

the new value of A is not available at the Q output of the register during the current clock cycle, but is available at the beginning of the next clock cycle.

As another example, the datapath shown in Figure 9.2(b) can perform the execution of the statement:

$$A = B + C$$

where B and C are two different variables stored in two separate registers, thus providing the two operand inputs to the adder. The output of the adder is connected to the D input of the A register for storing the result of the adder.

The execution of the statement is realized simply by asserting the $ALoad$ signal, and the actual storing of the value for A occurs at the next active edge of the clock. During the current clock cycle, the adder will perform the addition of B and C, and the result from the adder must be ready and available at its output before the end of the current clock cycle so that, at the next active clock edge, the correct value will be written into A. Since we are not writing any values to register B or C, we do not need to control the two $Load$ signals for them.

If we want a single datapath that can perform both of the statements:

$$A = B + C$$

and

$$A = A + 3$$

we will need to combine the two datapaths in Figure 9.2 together.

Since A is the same variable in the two statements, only one register for A is needed. However, register A now has two data sources: one from the first adder for $B + C$, and the second from the second adder for $A + 3$. The problem is that two or more data sources cannot be connected directly together to one destination, as

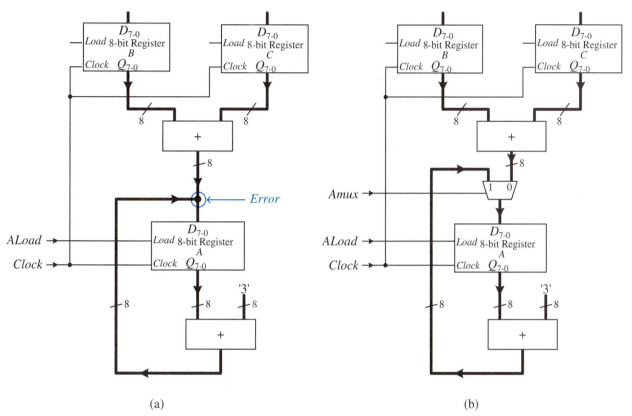

Figure 9.3 Datapath for performing $A = A + 3$ and $A = B + C$: (a) without multiplexer—wrong; (b) with multiplexer—correct.

shown in Figure 9.3(a) because their signals will collide, resulting in incorrect values. The solution is to use a multiplexer to select which of the two sources to pass to register A. The correct datapath using the multiplexer is shown in Figure 9.3(b).

Both statements assign a value to A, so $ALoad$ must be asserted for the execution of both statements. The actual value that is written into A, however, depends on the selection of the multiplexer. If $Amux$ is asserted, then the result from the bottom adder (i.e., the result from $A + 3$) is stored into A; otherwise, the result from the top adder is stored into A. Since the two adders are combinational circuits and the value from a register is always available at its output, the results from the two additions are always available at the two inputs of the multiplexer. But depending on the $Amux$ control signal, only one value will be passed through to the A register.

Notice that the datapath does not show which statement is going to be executed first. The sequence in which these two statements are executed depends on whether the signal $Amux$ is asserted first or de-asserted first. If this datapath is part of a microprocessor, then the control unit would determine when to assert or de-assert this $Amux$ control signal, since it is the control unit that performs the sequencing of datapath operations.

Furthermore, notice that these two statements cannot be executed within the same clock cycle. Since both statements write to the same register, and a register can only latch in one value at an active clock edge, only one result from one adder can be written into the register in one clock cycle. The other statement will have to be performed in another clock cycle, but not necessarily the next cycle.

Example 9.1   **Designing a dedicated datapath**

Design a datapath that can execute the two statements:

$$A = B + C$$

and

$$A = A + 3$$

using only one adder.

The only difference between this datapath and the one in Figure 9.3(b) is that it should use only one adder. So starting with this one adder, in order to execute the first statement, the first operand input to the adder is from register B, and the second operand input to the adder is from register C. However, to execute the second statement, the two input operands to the adder are register A and the constant 3. Since both input operands have two different sources, again we must use a multiplexer for each of them. The output of the two multiplexers will connect to the two adder input operands, as shown in Figure 9.4. For both statements, the result of the addition is stored in register A, therefore, the output of the adder connects to the input of the A register.

Notice that the two select lines for the two multiplexers can be connected together. This is possible because the two operands B and C for the first statement

Figure 9.4
Datapath for performing
$A = A + 3$ and
$A = B + C$ using only one adder.

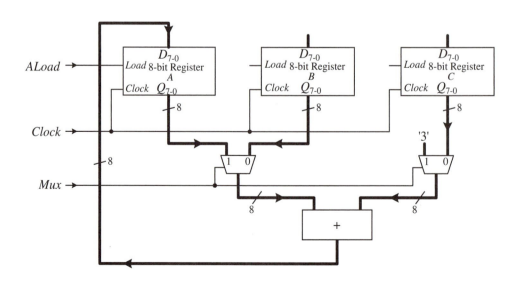

are connected to input 0 of the two multiplexers, respectively, and the two operands *A* and 3 for the second statement are connected to input 1 of the two multiplexers, respectively. Thus, de-asserting the *Mux* select signal will pass the two correct operands for the first statement, and likewise, asserting the *Mux* select signal will pass the two correct operands for the second statement. We want to reduce the number of control signals for the datapath as much as possible, because (as we will see in the next chapter) minimizing the number of control signals will minimize the size of the output circuit in the control unit.

9.1.1 Selecting Registers

In most situations, one register is needed for each variable used by the algorithm. However, if two variables are not used at the same time, then they can share the same register. If two or more variables share the same register, then the data transfer connections leading to the register and out from the register usually are made more complex, since the register now has more than one source and destination. Having multiple destinations is not too big of a problem, since we can connect all of the destinations to the same source.[1] However, having multiple sources will require a multiplexer to select one of the several sources to transfer to the destination. Figure 9.5 shows a circuit with a register having two sources—one from an external input and one from the output of an adder. A multiplexer is needed in order to select which one of these two sources is to be the input to the register.

After deciding how many registers are needed, we still need to determine whether to use a single register file containing enough register locations, separate individual registers, or a combination of both for storing the variables in. Furthermore, registers with built-in special functions, such as shift registers and counters, can also be used. For example, if the algorithm has a FOR loop statement, a single counter register can be used to not only store the count variable but also to increment the count. This way, not only do we reduce the component count, but the amount of datapath connections between components is also reduced. Decisions for selecting the type of registers to use will affect how the data transfer connections between the registers and functional units are connected.

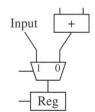

Figure 9.5
Circuit of a register with two sources.

---

[1] This is true only theoretically. In practice, there are fan-in (multiple sources with one destination) and fan-out (one source with multiple destinations) limits that must be observed.

9.1.2 Selecting Functional Units

It is fairly straightforward to decide what kind of functional units are required. For example, if the algorithm requires the addition of two numbers, then the datapath must include an adder. However, we still need to decide whether to use a dedicated adder, an adder–subtractor combination, or an ALU (which has the addition operation implemented). Of course, these questions can be answered by knowing what other data operations are needed by the algorithm. If the algorithm has only an addition and a subtraction, then you may want to use the adder–subtractor combination unit. On the other hand, if the algorithm requires several addition operations, do we use just one adder or several adders?

Using one adder may decrease the datapath size in terms of number of functional units, but it may also increase the datapath size because more complex data transfer paths are needed. For example, if the algorithm contains the following two addition operations:

$$a = b + c$$

$$d = e + f$$

Using two separate adders will result in the datapath shown in Figure 9.6(a); whereas, using one adder will require the use of two extra 2-to-1 multiplexers to select which register will supply the input to the adder operands, as shown in Figure 9.6(b). Furthermore, this second datapath requires two extra control signals for the two multiplexers. In terms of execution speed, the datapath on the left can execute both addition statements simultaneously within the same clock cycle, since they are independent of each other. However, the datapath on the right will have to execute these two additions sequentially in two different clock cycles, since there is only one adder available. The final decision as to which datapath to use is up to the designer.

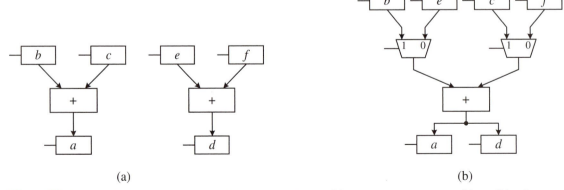

(a) (b)

Figure 9.6 Datapaths for realizing two addition operations: (a) using two separate adders; (b) using one adder.

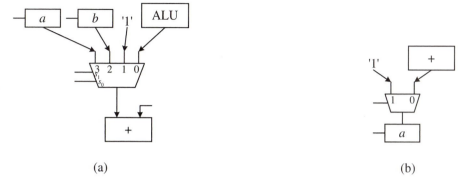

Figure 9.7
Examples of multiple sources using multiplexers: (a) an adder operand having four sources; (b) a register having two sources.

9.1.3 Data Transfer Methods

There are several methods in which the registers and functional units can be connected together so that the correct data transfers between the different units can be made.

Multiple Sources

If the input to a unit has more than one source, then a multiplexer can be used to select which one of the multiple sources to use. The sources can be from registers, constant values, or outputs from other functional units. Figure 9.7 shows two such examples. In Figure 9.7(a), the left operand of the adder has four sources: two from registers, one from the constant 1, and one from the output of an ALU. In Figure 9.7(b), register a has two sources: one from the constant 1 and one from the output of an adder.

Multiple Destinations

A source having multiple destinations does not require any extra circuitry. The one source can be connected directly to the different destinations, and all of the destinations where the data is not needed would simply ignore the data source. For example, in Figure 9.6(b), the output of the adder has two destinations: register a, and register d. If the output of the adder is for register a, then the *Load* line for register a is asserted, while the *Load* line for register d is not; and if the output of the adder is for register d, then the *Load* line for register d is asserted, while the *Load* line for register a is not. In either case, only the correct register will take the data while the other units simply will ignore the data.

This also works if one of the destinations is a combinational functional unit. In this case, the functional unit will take the source data and manipulates it. However, the output of the functional unit will not be used (that is, not stored in any register) so functionally, it doesn't matter that the functional unit worked on the source, because the result is not stored. However, it does require power for the functional unit to manipulate the data, so if we want to reduce the power consumption, we would want the functional unit to not manipulate the data at all. This, however, is a power optimization issue that is beyond the scope of this book.

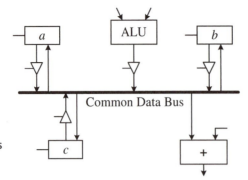

Figure 9.8
Multiple sources using tri-state buffers to share a common data bus.

Tri-State Bus

Another scheme where multiple sources and destinations can be connected to the same data bus is through the use of tri-state buffers. The point to note here is that, when multiple sources are connected to the same bus, only one source can output at any one time. If two or more sources output to the same bus at the same time, then there will be data conflicts. This occurs when one source outputs a 0 while another source outputs a 1. By using tri-state buffers to connect between the various sources and the common data bus, we want to make sure that only one tri-state buffer is enabled at any one time, while the rest of them are all disabled. Tri-state buffers that are disabled output high-impedance Z values, so no data conflicts can occur.

Figure 9.8 shows a tri-state bus with five units (three registers, an ALU, and an adder) connected to it. An advantage of using a tri-state bus is that the bus is bi-directional, so that data can travel in both directions on the bus. Connections for data going from a component to the bus need to be tri-stated, while connections for data going from the bus to a component need not be. Notice also that data input and output of a register both can be connected to the same tri-state bus; whereas, the input and output of a functional unit (such as the adder or ALU) cannot be connected to the same tri-state bus.

9.1.4 Generating Status Signals

Although it is the control unit that is responsible for the sequencing of statement execution, the datapath, however, must supply the results of the conditional tests for the control unit so that the control unit can determine what statement to execute next. Status signals are the results of the conditional tests that the datapath supplies to the control unit. Every conditional test that the algorithm has requires a corresponding status signal. These status signals usually are generated by comparators.

For example, if the algorithm has the following IF statement

$$\text{IF } (A = 0) \text{ THEN } \dots$$

the datapath must, therefore, have an equality comparator that compares the value from the A register with the constant 0, as shown in Figure 9.9(a). The output of the comparator is the status signal for the condition $(A = 0)$. This status signal is

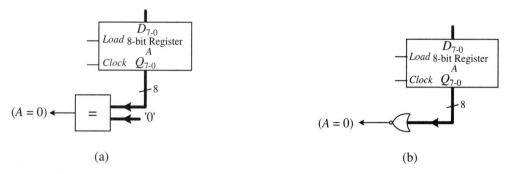

Figure 9.9 Comparators for generating the status signal ($A = 0$).

a 1 when the condition ($A = 0$) is true; otherwise, it is a 0. Recall from Section 4.10 that the circuit for the equality comparator with the constant 0 is simply a NOR gate, therefore, we can replace the black box for the comparator with just an 8-input NOR gate, as shown in Figure 9.9(b).

There are times when an actual comparator is not needed for generating a status signal. For example, if we want a status signal for testing whether a number is an odd number, as in the following IF statement

```
IF (A is an odd number) THEN ...
```

we can simply use the A_0 bit of the 8-bit number from register A as the status signal for this condition, since all odd numbers have a 1 in the zero bit position. The generation of this status signal is shown in Figure 9.10.

Figure 9.10
Comparators for
generating the
status signal (*A* is
an odd number).

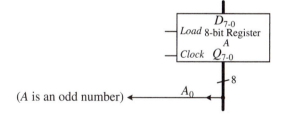

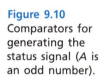

* * * * * * * * * * * * * * *

9.2 Using Dedicated Datapaths

Any given datapath will have a number of control signals. By asserting or de-asserting these control signals at different times, the datapath can perform different register-transfer operations. Since the execution of an operation requires the correct assertion and de-assertion of all of the control signals together, we would like to think of all of them as a unit rather than as individual signals. All of the control signals for a datapath, when grouped together, are referred to as a **control word**. Hence, a control word will have one bit for each control signal in the datapath. One register-transfer operation of a datapath, therefore, is determined by the values set in one control word, and so, we can specify the datapath operation simply by specifying the bit string for the control word. Each control word operation will take one clock cycle to perform. By combining multiple control words together in a certain sequence, the datapath will perform the specified operations in the order given.

Example 9.2   **Deriving control words for a datapath**

The datapath in Figure 9.4, having the two control signals $ALoad$ and Mux, was designed to execute the two statements: $A = A + 3$ and $A = B + C$. The control word for this datapath, therefore, has two bits—one for each control signal. The ordering of these two bits at this point is arbitrary; however, once decided, we need to be consistent with the order. The two control words for performing the two statements are shown in Figure 9.11.

Figure 9.11
Control words for the datapath in Figure 9.4 for performing the two statements:
$A = A + 3$ and
$A = B + C$.

Control Word	Instruction	$ALoad$	Mux
1	$A = A + 3$	1	1
2	$A = B + C$	1	0

Control word 1 specifies the control word bit string for executing the statement, $A = A + 3$. This is accomplished by asserting both the $ALoad$ and the Mux signals. Control word 2 is for executing the statement, $A = B + C$, by asserting $ALoad$ and de-asserting Mux.

In order for the datapath to operate automatically, the control unit will have to generate these control signals correctly at the appropriate time. In the following chapters, we will learn how to construct the control unit and then combine it with the datapath to form a microprocessor.

9.3 Examples of Dedicated Datapaths

We will now illustrate the design of dedicated datapaths with several examples. The datapaths produced in the examples are by no mean the only correct datapaths for solving each of the problems. Just like writing a computer program, there are many ways of doing it.

9.3.1   Simple IF-THEN-ELSE

Example 9.3

Simple IF-THEN-ELSE

In this example, we want to construct a 4-bit-wide dedicated datapath for solving the simple IF-THEN-ELSE algorithm shown in Figure 9.12. To create a datapath for the algorithm, we need to look at all of the data manipulation statements in the algorithm, since the datapath is responsible for manipulating data. These data manipulation instructions are the register-transfer operations. In most cases, one data manipulation instruction is equivalent to one register-transfer operation. However, some data manipulation instructions may require two or more register-transfer operations to realize.

Figure 9.12
Algorithm for solving the simple IF-THEN-ELSE problem of Example 9.3.

```
1  INPUT A
2  IF (A = 5) THEN
3      B = 8
4  ELSE
5      B = 13
6  END IF
7  OUTPUT B
```

The algorithm uses two variables, A and B; therefore, the datapath should have two 4-bit registers—one for each variable. Line 1 of the algorithm inputs a value into A. In order to realize this operation, we need to connect the data input signals to the input of register A, as shown in Figure 9.13. By asserting the $ALoad$ signal, the data input value will be loaded into register A at the next active clock edge.

Figure 9.13
Dedicated datapath for solving the simple IF-THEN-ELSE problem of Example 9.3.

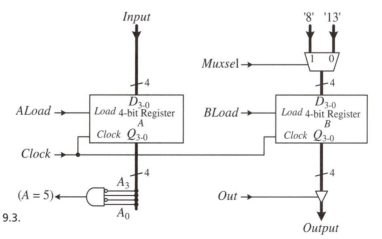

Line 2 of the algorithm tests the value of A with the constant 5. The datapath in Figure 9.13 uses a 4-input AND gate for the equality comparator with the four input bits connected as 0101 to the four output bits of register A. Since 5 in decimal is 0101 in binary, bits 0 and 2 are not inverted for the two 1's in the bit string, while bits 1 and 3 are inverted for the two 0's. With this connection, the AND gate will output a 1 when the input is a 5. The output of this comparator is the 1-bit status signal for the condition $(A = 5)$ that the datapath sends to the control unit.

Given the status signal for the comparison $(A = 5)$, the control unit will decide whether to execute line 3 or line 5 of the algorithm. This decision is done by the control unit and not by the datapath. The datapath is responsible only for all of the register-transfer operations. Lines 3 and 5 require loading either an 8 or a 13 into register B. In order to be able to select one data from several sources, a multiplexer is needed—in this case, a 2-to-1 multiplexer is used. One input of the multiplexer is connected to the constant 8 and the other to the constant 13. The output of the multiplexer is connected to the input of register B, so that one of the two constants can be loaded into the register. Again, which constant is to be loaded into the register is dependent on the condition in line 2. Knowing the result of the test from the status signal, the control unit will generate the correct signal for the multiplexer select line, *Muxsel*. The actual loading of the value into register B is accomplished by asserting the *BLoad* signal.

Finally, the algorithm outputs the value from register B in line 7. This is accomplished by connecting a tri-state buffer to the output of the B register. To output the value, the control unit asserts the enable line, *Out*, on the tri-state buffer, and the value from the B register will be passed to the data output lines.

Notice that the complete datapath shown in Figure 9.13 consists of two separate circuits. This is because the algorithm does not require the values of A and B to be used together. A question we might ask is whether we can connect the output of the comparator to the multiplexer select signal so that the status signal $(A = 5)$ directly controls *Muxsel*. Logically, this is alright, since if the condition $(A = 5)$ is true, then the status signal is a 1. Assigning a 1 to *Muxsel* will select the 1 input of the multiplexer, thus passing the constant 8 to register B. Otherwise, if the condition $(A = 5)$ is false, then *Muxsel* will get a 0 from the comparator, and the constant 13 will pass through the multiplexer. The advantage of doing this is that the datapath will generate one less status signal and requires one less control signal from the control unit. However, in some situations, we need to be careful with the timing when we use status signals to directly control the control signals.

Figure 9.14 shows the control words for performing the statements in Figure 9.12 using the datapath in Figure 9.13. Control word 1 executes the instruction INPUT A. To do this, the *ALoad* signal is asserted, and the data value at the input port will be loaded into the register at the next active clock edge. For this instruction, we do not need to load a value into the B register; hence, *BLoad* is de-asserted for this control word. Furthermore, because of this, it does not matter what the multiplexer outputs, so *Muxsel* can be a don't-care value. For control words 2 and 3, we want to load one of the two constants into B; therefore, *BLoad* is asserted for both of these control words, and the value for *Muxsel* determines which constant is loaded into B. When *Muxsel* is asserted, the constant 8 is passed to the input of the B register, and when

Control Word	Instruction	*ALoad*	*Muxsel*	*BLoad*	*Out*
1	INPUT *A*	1	×	0	0
2	*B* = 8	0	1	1	0
3	*B* = 13	0	0	1	0
4	OUTPUT *B*	0	×	0	1

Figure 9.14
Control words for solving the simple IF-THEN-ELSE problem of Example 9.3.

it is de-asserted, the constant 13 is passed to the register. Control word 4 asserts the *Out* signal to enable the tri-state buffer, thus outputting the value from the *B* register.

9.3.2 Counting 1 to 10

Example 9.4

Counting 1 to 10

Construct a 4-bit-wide dedicated datapath to generate and output the numbers from 1 to 10. The algorithm for this counting problem is shown in Figure 9.15.

Figure 9.15
Algorithm for solving the counting problem of Example 9.4.

```
1  i = 0
2  WHILE (i ≠ 10) {
3      i = i + 1
4      OUTPUT i
5      }
```

From the algorithm, we see that again we need a 4-bit register for storing the value for i. For line 3, an adder can be used for incrementing i. Both lines 1 and 3 write a value into i, thus providing two sources for the register. Our first inclination might be to use a 2-input multiplexer. However, notice that loading a 0 into a register is equivalent to clearing the register with the asynchronous *Clear* line, as long as the timing is correct. The resulting datapath is shown in Figure 9.16(a). For line 1, we assert the *Clear* signal to initialize i to 0, and for line 3, we assert the *iLoad* signal to load in the result from the adder, which adds a 1 to the current value of i. Asserting *Out* will output i. The status signal for the conditional test ($i \neq 10$) is realized by the 4-input NAND gate, where the four input bits of the NAND gate are connected to the four output lines from the register as 1010 binary for the constant decimal 10.

Alternatively, instead of using a separate register and adder, we can use a single 4-bit up counter to implement the entire algorithm, as shown in Figure 9.16(b). Again, initializing i to 0 is accomplished by asserting the *Clear* signal. To increment i for line 3, we simply assert the *Count* signal. Generating the status signal and outputting the count are the same as before.

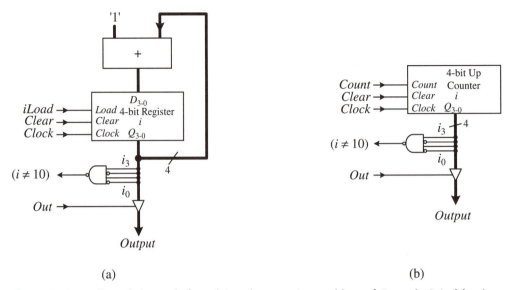

(a)   (b)

Figure 9.16 Dedicated datapath for solving the counting problem of Example 9.4: (a) using a separate adder and register; (b) using a single up counter.

The control words for the two different datapaths in Figure 9.16 are shown in Figure 9.17, respectively. For both cases, asserting the *Clear* signal will initialize i

Control Word	Instruction	*iLoad*	*Clear*	*Out*
1	$i = 0$	0	1	0
2	$i = i + 1$	1	0	0
3	OUTPUT i	0	0	1

(a)

Control Word	Instruction	*Count*	*Clear*	*Out*
1	$i = 0$	0	1	0
2	$i = i + 1$	1	0	0
3	OUTPUT i	0	0	1

(b)

Figure 9.17 Control words for solving the counting problem of Example 9.4: (a) using the datapath in Figure 9.16(a); (b) using the datapath in Figure 9.16(b).

to 0. To increment i for the datapath in Figure 9.16(a), we need to assert *iLoad*. This will load in the value from the output of the adder, which is the result for $i+1$, since one operand of the adder is i, and the other operand is the constant 1. For the datapath in Figure 9.16(b), we simply have to assert *Count* to increment i. The internal counter will increment the content in the register. Control word 3 asserts the *Out* signal, which asserts the enable signal on the tri-state buffer, thus passing the content of the register to the output port.

Note that control words 2 and 3 (corresponding to lines 3 and 4 in the algorithm, respectively) must be executed ten times in order to output the ten numbers. The looping in the algorithm is implemented in the control unit, and we will see in the next chapter how it is done.

9.3.3 Summation of n Down to 1

Example 9.5

Summation of n down to 1

Construct an 8-bit dedicated datapath to generate and add the numbers from n down to 1, where n is an 8-bit user-input number. The datapath should output the sum of the numbers when done and notify external devices that the calculation is completed by asserting a *Done* signal. The algorithm is shown in Figure 9.18.

Figure 9.18
Algorithm for
solving the
summation problem
of Example 9.5.

```
1   sum = 0
2   INPUT n
3   WHILE (n ≠ 0) {
4       sum = sum + n
5       n = n - 1
6       }
7   OUTPUT sum
```

We first note that we need to have two 8-bit registers with load function for storing the two variables, n and *sum*. The register for *sum* should also include a *Clear* function for initializing it to 0. The register for n should also be a down counter for decrementing n. A separate adder is used for the addition operation. The resulting dedicated datapath is shown in Figure 9.19.

For initializing *sum*, we simply can assert the *Clear* line. Asserting *nLoad* will input a value for n. Asserting *sumLoad* will load into register *sum* the value from the output of the adder, which is the summation of *sum* plus n. Decrementing n by 1 is accomplished by asserting *nCount*. Finally, asserting *Out* will enable the tri-state buffer, thus outputting the value from the *sum* register. The *Out* signal is also used as a *Done* signal to notify external devices that the calculation is completed. The comparator for the WHILE loop condition ($n \neq 0$) is an 8-input OR gate. The output of this OR gate is the status signal for the control unit.

The control words for this example are shown in Figure 9.20.

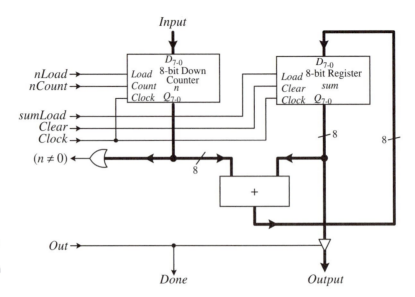

Figure 9.19
Dedicated datapath
for solving the
summation problem
of Example 9.5.

Control Word	Instruction	nLoad	nCount	sumLoad	Clear	Out
1	$sum = 0$	0	0	0	1	0
2	INPUT n	1	0	0	0	0
3	$sum = sum + n$	0	0	1	0	0
4	$n = n - 1$	0	1	0	0	0
5	OUTPUT sum	0	0	0	0	1

Figure 9.20   Control words for solving the summation problem of Example 9.5.

9.3.4 Factorial of n

Example 9.6

Factorial of n

Design an 8-bit dedicated datapath for evaluating the factorial of n. The factorial
of n is defined to be the product of $1 \times 2 \times 3 \times \ldots \times n$. Figure 9.21 shows the algorithm
for solving the factorial of n where n is an 8-bit user input number. The datapath
should notify external devices that the calculation is completed by asserting a *Done*
signal when the data at the output port is valid.

Figure 9.21
Algorithm for
solving the factorial
problem of
Example 9.6.

```
1  INPUT n
2  product = 1
3  WHILE (n > 1) {
4      product = product * n
5      n = n - 1
      }
6  OUTPUT product
```

After analyzing the algorithm, we conclude that the following registers and functional units are needed for the datapath:

- An 8-bit down counter with parallel load for storing the variable n and for decrementing n. The parallel load feature will allow for the input of n.
- An 8-bit register with load for storing the variable *product*.
- A multiply functional unit.
- A greater-than-one comparator for returning the status signal to the control unit.
- A tri-state buffer for output.

The complete dedicated datapath is shown in Figure 9.22. The input port is connected to the input of the n register. By asserting *nLoad*, the value at the input port is loaded into register n, thus realizing the instruction in line 1. Since this n register is also a down counter, by asserting *nCount*, we will have executed line 5.

Register *product* has two sources, one from line 2, and the second from line 4. A 2-to-1 multiplexer is used to select from these two sources using the *productMux* select line. For line 2, the constant 1 is connected to the 1 input of the multiplexer, so that, by asserting both *productMux* and *productLoad*, the constant 1 will be loaded

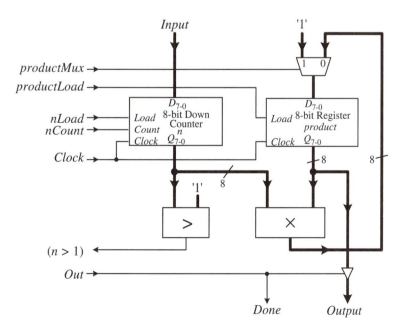

Figure 9.22
Dedicated datapath
for solving the
factorial problem
of Example 9.6.

into the register. For line 4, one operand of the multiply function is *product*, and the other operand is n. Hence, the output of the *product* register is connected to one input of the multiply unit. Since n is stored in the counter, therefore, the output of the counter is connected to the second input of the multiply unit. The result of the multiply is assigned back into the variable *product*. Thus, in the datapath, we needed a connection from the output of the multiply functional unit back to the input of the register where *product* is stored via the 2-to-1 multiplexer.

The tri-state buffer, connected to the output of the *product* register, is needed for the final output of the result. The *Out* signal for enabling the tri-state buffer is also used as the *Done* signal. Finally, the comparator generates the status signal for the condition $(n > 1)$ to the control unit. One input to the comparator comes from the counter where n is stored, and the other input is the constant 1.

The control words for evaluating the factorial using the dedicated datapath shown in Figure 9.22 has five control signals: *productMux*, *productLoad*, *nLoad*, *nCount*, and *Out*, as shown in Figure 9.23(a).

Control word 1 loads n into the counter register by asserting the *nLoad* signal. We are not loading into the *product* register, so the *productLoad* signal is de-asserted,

Control Word	Instruction	productMux	productLoad	nLoad	nCount	Out
1	INPUT n	×	0	1	0	0
2	$product = 1$	1	1	0	0	0
3	$product = product \times n$	0	1	0	0	0
4	$n = n - 1$	×	0	0	1	0
5	OUTPUT $product$	×	0	0	0	1

(a)

Control Word	Instruction	productMux	productLoad	nLoad	nCount	Out
1	INPUT n, $product = 1$	1	1	1	0	0
2	$product = product \times n$, $n = n - 1$	0	1	0	1	0
3	OUTPUT $product$	×	0	0	0	1

(b)

Figure 9.23 Control words for solving the factorial problem of Example 9.6: (a) using five control words; (b) using only three control words.

and therefore, it doesn't matter what the *productMux* signal is. Both *nCount* and *Out* are de-asserted, since we don't want to decrement *n*, and we are not done evaluating.

Control word 2 initializes *product* to 1. Since the constant 1 is connected to the 1 input of the multiplexer, we need to set *productMux* to 1. The *productLoad* signal has to be asserted to actually load the constant 1 into the register.

Control word 3 multiplies *product* with *n*. Since the value stored in a register is always available at its output, the combinational multiplier circuit constantly is performing the multiplication of these two numbers. The output from the multiplier is stored back into the *product* register simply by setting *productMux* to 0 and asserting the *productLoad* signal.

Control word 4 decrements *n* by asserting the *nCount* signal. Since we are using a down counter, asserting *nCount* will decrement the value stored in the register. Again, we are not loading a value into the *product* register, so the value of *product-Mux* does not matter.

Control word 5 outputs the value stored in the *product* register. This is accomplished simply by asserting the *Out* signal. The value stored in the *product* register, which is available at its output, is passed through the tri-state buffer to the output port. Since the *Out* signal is also connected to the *Done* signal, asserting *Out* will also assert *Done*.

Notice that with this dedicated datapath, several instructions can be performed in parallel, as shown by the control words in Figure 9.23(b). Specifically, inputting *n* and initializing *product* to 1 can be performed simultaneously by control word 1. In the datapath circuit, the path that goes from the data input port to the counter register for storing *n* is separated completely from the path that goes from the constant 1 to the register for storing *product*. Because of this, control word 1 can assert both the *productLoad* and the *nLoad* signals together. Asserting *productLoad* and *productMux* will load in the constant 1, while asserting *nLoad* will load in *n*.

Similarly, the multiplication and the decrement instructions can be performed together by control word 2. Here, *productLoad* is asserted and *productMux* is de-asserted to store the result from the multiplier, while *nCount* is asserted to decrement *n*. We will see in the next chapter that, by reducing the number of control words, the circuit for the control unit will be smaller.

9.3.5 Counting 0's and 1's

Example 9.7

Count 0's and 1's

In this example, we want to construct an 8-bit dedicated datapath for solving the following problem:

> Input an 8-bit number. Output a 1 if the number has the same number of 0's and 1's, otherwise, output a 0. (E.g., the number 10111011 will output a 0; whereas, the number 10100011 will output a 1.)

```
INPUT n
countbit = 0          // for counting the number of zero and one bits
counteight = 0        // for looping eight times
WHILE (counteight ≠ 8) {
   IF (n₀ = 1) THEN // test whether bit 0 of n is a 1
      countbit = countbit + 1
   ELSE
      countbit = countbit - 1
   END IF
   n = n >> 1      // shift n right one bit
   counteight = counteight + 1;
   }
IF (countbit = 0) THEN
   OUTPUT 1
ELSE
   OUTPUT 0
END IF
ASSERT Done
```

Figure 9.24
Algorithm for solving the count 0's and 1's problem of Example 9.7.

The algorithm for solving the problem is shown in Figure 9.24. The WHILE loop is executed eight times using the *counteight* variable for the 8 bits in the input number n. For each bit in n, if it is a 1, the variable *countbit* is incremented, otherwise, it is decremented. At the end of the WHILE loop, if *countbit* is equal to 0, then there are the same number of 0's and 1's in n.

After analyzing the algorithm, we conclude that the following registers and functional units are needed for the datapath:

- An 8-bit shifter with parallel load register for storing and shifting n.
- A 4-bit up counter for *counteight*.
- A 4-bit up-down counter for *countbit*.
- A "not equal to 8" comparator for looping eight times.
- An "equal to 0" comparator for testing with *countbit*.

The dedicated datapath for implementing the algorithm is shown in Figure 9.25. Notice that there are no connections between the shifter and the two counters; they are completely separate circuits. To extract bit 0 of n and test whether it is equal to a 1 or not, we only have to connect to the least significant bit, n_0, of the shifter, and no active component is necessary. To test for (*counteight* $\neq$ 8), we use a 4-input NAND gate with the three least significant input bits inverted. When *counteight* is equal to

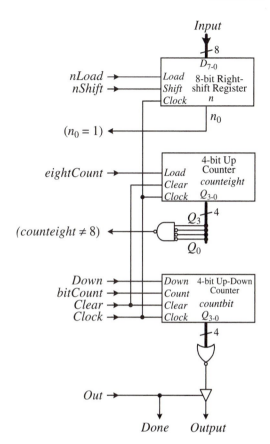

Figure 9.25
Dedicated datapath
for solving the
count 0's and 1's
problem of
Example 9.7.

eight, the NAND gate outputs a 0, which serves as the ending loop condition. Using the *Down* signal, the *countbit* register can be either incremented or decremented. Since register *countbit* is keeping track of the number of 0's and 1's, if it is a 0 at the end, it means that n has the same number of 0's and 1's. The NOR gate will output a 1 if *countbit* is a 0. Whether the output of the NOR gate actually is outputted will depend on whether the tri-state buffer is enabled or not. When the control unit asserts the *Out* signal to enable the tri-state buffer, this 1 signal also serves as the *Done* signal to inform the external user that the computation is completed. In other words, when the *Done* signal is asserted, the output value is the correct result for the computation.

The control words for this example are shown in Figure 9.26. Notice that there are several control words with don't-care values. For example, in control word 2 for initializing *countbit* to 0, it does not matter what the setting for the *Down* signal is since we are not incrementing or decrementing the count.

Control Word	Instruction	$nLoad$	$nShift$	$counteight$	$Down$	$countbit$	$Clear$	Out
1	INPUT n	1	0	×	×	×	×	0
2	$countbit = 0, counteight = 0$	0	0	0	×	0	1	0
3	$countbit = countbit + 1$	0	0	0	0	1	0	0
4	$countbit = countbit - 1$	0	0	0	1	1	0	0
5	$n = n >> 1$	0	1	0	×	0	0	0
6	$counteight = counteight + 1$	0	0	1	×	0	0	0
7	OUTPUT 1, OUTPUT 0, ASSERT $Done$	0	0	0	×	0	0	1

Figure 9.26 Control words for solving the count 0's and 1's problem of Example 9.7.

9.4 General Datapaths

As the name suggests, a general datapath is more general than dedicated datapath in the sense that it can be used to solve various problems instead of just one specific problem, as long as it has all of the required functional units and enough registers for storing all of the temporary data. The idea of using a general datapath is that we can use a "ready-made" datapath circuit to solve a given problem without having to modify it. The tradeoff is a time-versus-space issue. On one hand, we do not need the extra time to design a dedicated datapath. On the other hand, the general datapath may contain more features than what the problem requires, so it not only increases the size of the circuit but also consumes more power than necessary.

Figure 9.27 shows an example of a simple, general datapath. It contains one functional unit, the ALU, and one 8-bit register for storing data. The input to the A operand of the ALU can be either from an external data input or the constant 1, as selected by the multiplexer select line IE. The B operand of the ALU is always from the content of the register. The operation of the ALU is determined by the three control lines: ALU_2, ALU_1, and ALU_0, as defined in the ALU ALU operation table. The design of the ALU was discussed in Section 4.5. The register has a load capability for loading the output of the ALU into the register when the $Load$ signal line is asserted. The register can also be reset to 0 asynchronously by asserting the $Clear$ signal line. The content of the register can be passed to the external data output port by asserting the output enable line, OE, of the tri-state buffer. We assume here

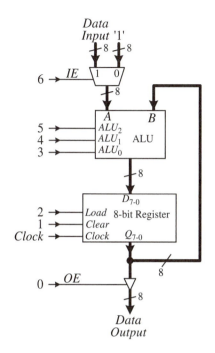

ALU_2	ALU_1	ALU_0	Operation
0	0	0	Pass through A
0	0	1	A AND B
0	1	0	A OR B
0	1	1	NOT A
1	0	0	$A + B$
1	0	1	$A - B$
1	1	0	$A + 1$
1	1	1	$A - 1$

Figure 9.27   A simple, general datapath circuit.

that the buses for transferring the data between the components are 8-bits wide. All the control lines, of course, are 1-bit wide.

There are seven control signals (numbers 0 to 6) for controlling the operations of this simple, general datapath. When grouped together, they form the control word for this datapath. Just like the dedicated datapath, various operations can be performed by this datapath either by asserting or de-asserting these control signals at different times.

For example, to load a value from the external data input to the register, we would set the control word as follows:

Control	IE	ALU_2	ALU_1	ALU_0	$Load$	$Clear$	OE
Word	6	5	4	3	2	1	0
Value	1	0	0	0	1	0	0

By setting $IE = 1$, we select the external input to pass through the multiplexer. From the ALU operation table, we see that setting the ALU control lines: ALU_2, ALU_1, and ALU_0, to 000 selects the pass-through operation, which passes the data from the A operand input of the ALU to the output of the ALU with no modifications.

Finally, setting $Load = 1$ loads the value from the output of the ALU into the register. Thus, we have stored the input value into the register. We do not want to clear the register nor output the value from the register, so both *Clear* and *OE* are set to 0's.

Note that the actual writing of the register occurs at the next active edge of the clock. Thus, the new value is not available from the register until the beginning of the next clock cycle. If we read from the register during the current clock cycle, we would get the original value that was in the register and not the new value that was just entered in. In other words, if we had set *OE* to 1 in the above control word, we would be reading the register in the current clock cycle and, thus, be outputting the original value found in the register rather than the new value from the data input. We will look at this important issue again in Section 9.6.

• • • • • • • • • • • • • • • •

9.5 Using General Datapaths

Using a general datapath is exactly the same as using a dedicated datapath. We simply have to assert (or de-assert) the appropriate control signals at the right time so that the data transfer between registers is correct to realize a data manipulation statement. The following example shows how we can use the general datapath shown in Figure 9.27 to solve a problem.

Example 9.8 **Using a general datapath to output the numbers from 1 to 10**

To see how a general datapath is used to perform a computation, let us write the control words to generate and output the numbers from 1 to 10 using the general datapath shown in Figure 9.27. This example, of course, is identical to Example 9.4, except that we are using a different datapath. Hence, only the control words will be different. The algorithm is repeated here in Figure 9.28(a), and the control words for using this general datapath to solve the problem are shown in Figure 9.28(b).

Control word 1 initializes i to 0 by asserting the *Clear* signal for the register. The ALU is not needed in this operation, so it doesn't matter what the inputs to the ALU are or the operation that is selected for the ALU to perform. Hence, the four control lines: *IE* (for selecting the input), ALU_2, ALU_1, and ALU_0 (for selecting the ALU operation) are all set to don't-cares. *Load* is de-asserted because we do not need to store the output of the ALU to the register. At this time, we also do not want to output the value from the register, so the output control line *OE* is also de-asserted.

Control word 2 increments i, so we need to add a 1 to the value that is stored in the register. Although, the ALU has an increment operation, we cannot use it because the ALU was designed such that the operation increments the A operand rather than the B operand, and this datapath is connected such that the output of the register goes to the B operand only. Now, we can modify the ALU to have an increment B operation, or we can modify the datapath so that the output of the register can be routed to the A operand of the ALU. However, both of these solutions require modifications to the datapath, and this defeats the purpose of using a general

```
1  i = 0
2  WHILE (i ≠ 10) {
3      i = i + 1
4      OUTPUT i
5      }
```

(a)

Control Word	Instruction	IE 6	ALU₂ALU₁ALU₀ 5–3	Load 2	Clear 1	OE 0
1	$i = 0$	×	× × ×	0	1	0
2	$i = i + 1$	0	100 (add)	1	0	0
3	OUTPUT i	×	× × ×	0	0	1

(b)

Figure 9.28   Generate and output the numbers from 1 to 10: (a) algorithm; (b) control words for the datapath in Figure 9.27 using three control words.

datapath. If we are going to modify this datapath, we may as well modify it even further to get a dedicated datapath. So instead, what we can do is to use the ALU addition operation (100) to increment the value stored in the register by one. We can get a 1 to the A operand by setting IE to 0, since the 0 input line of the multiplexer is tied to the constant 1. The B operand will have the register value. Finally, we need to load the result of the addition back into the register, so the $Load$ line is asserted.

Control word 3 outputs the value from the register. Again, we don't care about the inputs to the ALU and the operation of the ALU, so there is no new value to load into the register. We definitely do not want to clear the register. We simply want to output the value from the register, so we just assert OE by setting it to a 1.

Just like for the dedicated datapath, the general datapath must also provide the status signal for the condition ($i \neq 10$) to the control unit. Using this status signal, the control unit will determine whether to repeat control words 2 and 3 in the loop or to terminate. We must, therefore, add a comparator to this general datapath to test whether the condition ($i \neq 10$) is true or not. The output of this comparator is the status signal that the datapath sends back to the control unit. Since variable i is stored in the register, we should connect the output of the register to the input of this comparator, as shown in Figure 9.29.

The simulation trace for this counting example is shown in Figure 9.30. Notice that two cycles are needed for each count, as shown in the output signal in the trace— the first cycle for control word 2 where the output shows a Z value, and the second cycle for control word 3 where the actual count appears at the output. These two cycles are repeated ten times for the ten numbers. For example, at 500 ns (at the beginning of the first of the two clock cycles), $Load = 1$ and $OE = 0$. The current

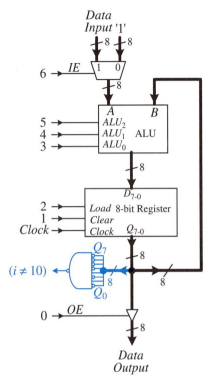

Figure 9.29
The comparator circuit for generating the status signal ($i \neq 10$) added to the general datapath.

content of the register is 1. Since $OE = 0$, so the output is Z. At 700 ns (the beginning of the second of the two clock cycles), the register is updated with the value 2. *Load* is de-asserted and OE is asserted, and the number 2 is output.

This simulation trace was obtained by manually asserting and de-asserting the seven datapath control signals at each clock cycle. This is only because we wanted

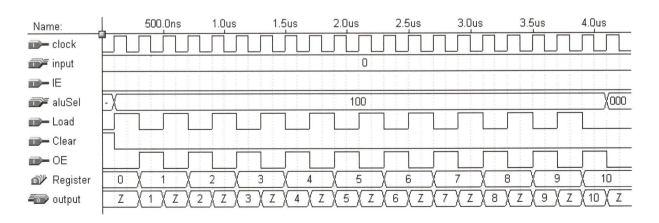

Figure 9.30 Simulation trace for Example 9.8 using the three control words shown in Figure 9.28(b).

to test out the datapath, and we have not yet constructed the control unit for generating these control signals automatically. What we really need to do eventually is to construct the control unit based on the control words from Figure 9.28(b). The control unit will generate the appropriate control signals for the datapath for each clock cycle.

• • • • • • • • • • • • • • • ··

9.6 A More Complex General Datapath

When a particular general datapath does not contain all of the functional units and/or registers needed to perform all of the required operations specified in the algorithm that we are trying to implement, then we need to select a more complex datapath. When working with general datapaths, the goal is to find the simplest and smallest datapath that matches the requirements of the problem as closely as possible. Example 9.9 shows the need for selecting a more complex datapath.

Example 9.9   **Determining the need for a complex datapath**

Let us use the simple datapath of Figure 9.27 to generate and add the numbers from n down to 1, where n is an 8-bit user input number, and output is the sum of these numbers. The algorithm for doing this is shown in Figure 9.31. The algorithm requires the use of two variables, n for the input that counts down to 0, and *sum* for adding up the total. This means that we need two registers in the datapath, unless we want the user to enter the numbers from n down to 1 manually and just use the one register to store the sum. Thus, we conclude that the datapath of Figure 9.27 cannot be used to implement this algorithm.

Figure 9.31
Algorithm to generate and sum the numbers from n down to 1.

```
1    sum = 0
2    INPUT n
3    WHILE (n ≠ 0) {
4        sum = sum + n
5        n = n - 1
6    }
7    OUTPUT sum
```

In order to implement the algorithm of Figure 9.31, we need a slightly more complex datapath that includes at least two registers. One possible datapath is shown in Figure 9.32(a). The main difference between this datapath and the previous one is that a register file (RF) with four locations is used instead of having just one register. The register file, as discussed in Section 8.4, has one write port and two read ports. To access a particular port, the enable line for that port must be asserted and the address for the location set up. The designated lines are WE for write enable, RAE for read *Port A* enable, and RBE for read *Port B* enable, $WA_{1,0}$ for the write address,

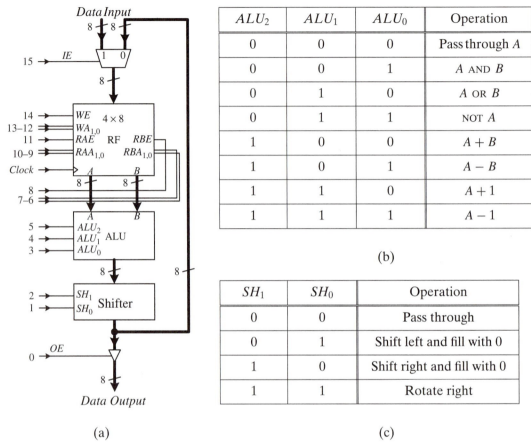

ALU_2	ALU_1	ALU_0	Operation
0	0	0	Pass through A
0	0	1	A AND B
0	1	0	A OR B
0	1	1	NOT A
1	0	0	$A + B$
1	0	1	$A - B$
1	1	0	$A + 1$
1	1	1	$A - 1$

(b)

SH_1	SH_0	Operation
0	0	Pass through
0	1	Shift left and fill with 0
1	0	Shift right and fill with 0
1	1	Rotate right

(a) (c)

Figure 9.32 Complex general datapath with a register file: (a) circuit; (b) ALU operations; (c) shifter operations.

$RAA_{1,0}$ for the read *Port A* address, and $RBA_{1,0}$ for the read *Port B* address. The read *Ports A* and *B* can be read simultaneously, and they are connected to the two input operands A and B of the ALU, respectively. The ALU operations are specified in Figure 9.32(b). The result of the ALU is passed through a shifter whose operations are specified in Figure 9.32(c). Although the shifter is not needed by the algorithm of Figure 9.31, it is available in this datapath. The output of the shifter is routed back to the register file via the multiplexer, or it can be output externally by enabling the output tri-state buffer. The datapath width is again 8-bits wide.

Example 9.10 **Using a complex general datapath to sum numbers**

The summation algorithm of Figure 9.31 can be implemented using the datapath shown in Figure 9.32. The control words for manipulating the datapath for this algorithm are shown in Figure 9.33.

Control Word	Instruction	IE 15	WE 14	$WA_{1,0}$ 13–12	RAE 11	$RAA_{1,0}$ 10–9	RBE 8	$RBA_{1,0}$ 7–6	$ALU_{2,1,0}$ 5–3	$SH_{1,0}$ 2–1	OE 0
1	$sum = 0$	0	1	00	1	00	1	00	101 (subtract)	00	0
2	INPUT n	1	1	01	0	××	0	××	× × ×	××	0
3	$sum = sum + n$	0	1	00	1	00	1	01	100 (add)	00	0
4	$n = n - 1$	0	1	01	1	01	0	××	111 (decrement)	00	0
5	OUTPUT sum	×	0	××	1	00	0	××	000 (pass)	00	1

Figure 9.33   Control words to generate and sum the numbers from n down to 1 using the datapath in Figure 9.32.

Control word 1 initializes sum to 0 by performing a subtraction where the two operands are the same. The location of the register file (RF) used for the two operands is arbitrary because it doesn't matter what the value is, as long as both operands get the same value. We use RF location 00 to store the value of the variable sum. Thus, we assert all three RF enable lines, and sets the RF write address and both read addresses to location 0. The shifter is not needed, so the pass-through operation is selected.

All of the operations specified by a control word are performed within one clock cycle. The timing for the operations of this control word is as follows. At the active edge of the clock, the FSM enters the state for this control word. The appropriate control signals for this control word to the datapath are asserted. Data from RF location 0 is read for both ports and passed to the ALU. Recall that the register file is constructed such that the data from reading a port is available immediately, and does not have to wait until the next active clock edge. Since both the ALU and the shifter are combinational circuits, they will operate within the same clock cycle. The result is written back to RF location 0 at the next active clock edge. Thus, the updated or new value in RF location 0 is not available until the beginning of the next clock cycle.

Control word 2 inputs the value n and stores it in RF location 1. To read from the input, we set $IE = 1$. To write n into RF location 1, we set $WE = 1$ and $WA_{1,0} = 01$. Both the ALU and the shifter are not used in this control word, so their select lines are set to don't-cares.

Control word 3 reads sum through $Port\ A$ by setting $RAE = 1$ and $RAA_{1,0} = 00$, and n through $Port\ B$ by setting $RBE = 1$ and $RBA_{1,0} = 01$. These two numbers are added together by setting the ALU select lines, $ALU_{2,1,0}$, to 100. The result of the addition passes through the shifter and the multiplexer, and then it is written back to RF location 0.

Control word 4 decrements n by 1 by using the decrement operation of the ALU (111). From RF location 01, n is read through $Port\ A$ and passes to the A operand of the ALU. The ALU decrements the A operand by 1, and the result is written back to RF location 01.

After decrementing n, the WHILE loop needs to test the value of n. In order for the control unit to loop correctly, the datapath must generate the correct status signal for the condition $(n \neq 0)$. The question is where do we connect the comparator inputs to in the datapath so that we can get the correct value for n? Since n is stored in location 01 of the register file, our first inclination might be to connect to *Port A* of the register file, which is also the input to the ALU. An alternative might be to connect to the output of the ALU. The difference between these two connection points is that, for the first connection, we would be using the value of n before the decrement, while for the second connection, we would be using the value of n after the decrement. Furthermore, if we look at these same two connection points at a different time, say, during the clock cycle for control word 3, then the values are not even for n, but rather for *sum*, since control word 3 reads *sum* through *Port A*, and the ALU calculates *sum+n*. Therefore, in order to know which is the correct connection point, we need to look at the state diagram for this algorithm to determine in which state is the status signal needed and what data is being passed through the datapath during that clock cycle. We will come back to this problem in the next chapter.

Control word 5 outputs the result that is stored in *sum* by reading from RF location 0 via *Port A* and passing it through the ALU and shifter. *OE* is asserted to enable the tri-state buffer for the output.

The simulation trace of the control words for $n = 10$ is shown in Figure 9.34. Again, the datapath control signals are set manually until n (stored in RF(1)) reaches 0, at which point *OE* is asserted, and the summation value 55 appears on the output.

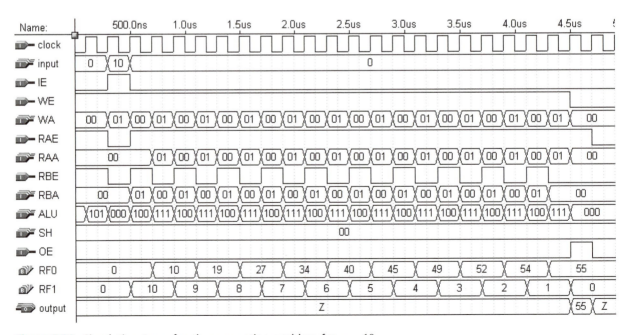

Figure 9.34 Simulation trace for the summation problem for $n = 10$.

Example 9.11 | **Using a general datapath to implement the multiplication of two numbers**

Write the control words for performing the multiplication of two unsigned numbers using the general datapath shown in Figure 9.32. The two numbers are entered in through the data input port. The multiplication algorithm is shown in Figure 9.35.

Figure 9.35
Algorithm to
multiply two
unsigned numbers.

```
1  prod = 0
2  INPUT A
3  INPUT B
4  WHILE (B ≠ 0) {
5      prod = prod + A
6      B = B - 1
7  }
8  OUTPUT prod
```

The control words for manipulating the datapath to perform the multiplication of two unsigned numbers are shown in Figure 9.36.

Control word 1 assigns a 0 (obtained from subtracting a number from itself) to $RF(0)$. $RF(0)$ is used for storing the variable *prod*.

Control words 2 and 3 input the two unsigned numbers A and B into $RF(1)$ and $RF(2)$, respectively. IE is set to 1 so that the value comes from *Data Input*, and because of this, it doesn't matter what the ALU and the shifter will do. WE is asserted for the write. $WA_{1,0}$ is set to the addresses 01 for variable A and 10 for variable B.

Control word 4 performs the addition of *prod* with A. Variable *prod*, stored in $RF(0)$, is read through *Port A*; and variable A, from $RF(1)$, is read through *Port B*. The ALU is selected for addition, and the result is passed through the shifter. With WE being asserted and $WA_{1,0}$ set to 00, the result is written back into $RF(0)$.

Control Word	Instruction	IE 15	WE 14	$WA_{1,0}$ 13–12	RAE 11	$RAA_{1,0}$ 10–9	RBE 8	$RBA_{1,0}$ 7–6	$ALU_{2,1,0}$ 5–3	$SH_{1,0}$ 2–1	OE 0
1	$prod = 0$	0	1	00	1	00	1	00	101 (subtract)	00	0
2	INPUT A	1	1	01	0	××	0	××	× × ×	××	0
3	INPUT B	1	1	10	0	××	0	××	× × ×	××	0
4	$prod = prod + A$	0	1	00	1	00	1	01	100 (add)	00	0
5	$B = B - 1$	0	1	10	1	10	0	××	111 (decrement)	00	0
6	OUTPUT $prod$	×	0	××	1	00	0	××	000 (pass)	00	1

Figure 9.36 Control words to multiply two unsigned numbers using the datapath in Figure 9.32.

Control word 5 decrements B. Variable B, from RF(2), is read through *Port A*. The ALU is selected for the decrement operation, and the result is again passed through the shifter. The result is written back into RF(2).

Control word 6 outputs the result stored in *prod* by reading RF(0) through *Port A*, and passing this value through the ALU and the shifter. The value is passed to the data output port by asserting *OE*.

9.7 Timing Issues

In Figure 9.28(b) of Example 9.8 to generate and output the numbers from 1 to 10, two control words are used for the addition and output operations. Control word 2 does the addition and writing of the result into the register, while control word 3 outputs the count stored in i. During the current clock cycle for control word 2, the operation starts with the constant '1' passing through the multiplexer, followed by the ALU performing the addition $i + 1$, as shown in Figure 9.37. The current value of i is available from the output of the register at the beginning of the current clock cycle. The result from the addition is available shortly before the beginning of the next clock cycle (next rising clock edge). Recall from Section 6.7 that a value is latched into a flip-flop at the active (rising) edge of the clock. So, although the *Load* signal is asserted at the beginning of the current clock cycle, a value is not written into the register until the next rising clock edge, which is at the beginning of the next clock cycle. Since the addition result is available before the next rising clock edge, it is this addition result that is written into the register at the beginning of the next clock cycle.

So for control word 2, we are doing both a read and a write of the same register, i, in one clock cycle for executing the instruction $i = i + 1$. The read is for the i that is on the right-hand side of the equal sign, and the write is for the i that is on the left-hand side of the equal sign. From the above analysis, we see that performing both a read and a write to the same register in the same clock cycle does not create

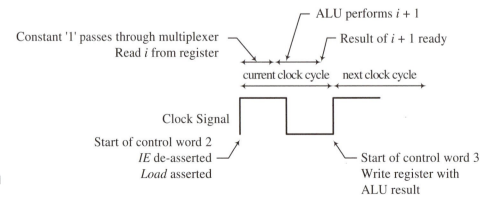

Figure 9.37
Read and write timings for a control word.

Control Word	Instruction	IE 6	$ALU_2ALU_1ALU_0$ 5–3	Load 2	Clear 1	OE 0
1	$i = 0$	×	× × ×	0	1	0
2	$i = i + 1$ & OUTPUT i	0	100 (add)	1	0	1

Figure 9.38   Counting algorithm using two control words for the datapath in Figure 9.27(a).

any data conflict because the reading occurs at the beginning of the current clock cycle and, therefore, is getting the current value that is in the register. The writing occurs at the beginning of the next clock cycle, which happens after the reading.

As a result, the value that is available at the output of the register in the current clock cycle is still the value before the write back, which is the value before the addition of $i + 1$. If we assert the OE signal in the current clock cycle to output the register value (that is, do the addition and output operations in the same clock cycle as shown in control word 2 of Figure 9.38), the output value would be the value *before* the addition and not the result from *after* the addition. According to the algorithm in Figure 9.28(a), the output should be for the value from after the addition. Because of this, Figure 9.28(b) uses control word 3 (starting at the next clock cycle) to output the value from after the addition.

The simulation trace for the two control words of Figure 9.38 (where the addition and output operations are both done in the same control word) is shown in Figure 9.39. There are two main differences between this simulation trace and the one in Figure 9.30. The first is that each count now requires only one clock cycle rather than two, because only control word 2 is repeated ten times. As a result, the time to count to ten is about half (2.3 μs versus 4.1 μs). The second thing is that the first output value is a 0 and not a 1 (as it should be). The first time that control word 2

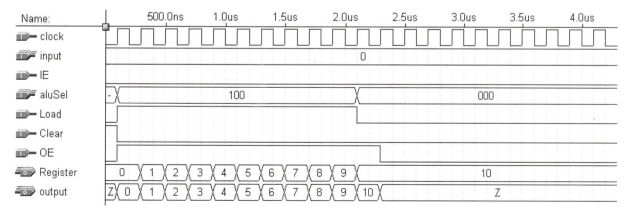

Figure 9.39   Simulation trace for using the two control words from Figure 9.38.

Control Word	Instruction	IE 6	$ALU_2 ALU_1 ALU_0$ 5–3	Load 2	Clear 1	OE 0
1	$i = 0$	×	× × ×	0	1	0
2	$i = i + 1$	0	100 (add)	1	0	0
3	$i = i + 1$ & OUTPUT i	0	100 (add)	1	0	1

Figure 9.40 Optimized control words for the counting algorithm using the datapath in Figure 9.27(a).

executes is the clock cycle between 100 ns and 300 ns. The incremented value (1) is not written into the register until at 300 ns, so when OE is asserted before 300 ns, the output value is a 0.

We certainly like the fact that it only requires half the time to execute the algorithm, but outputting an extra zero at the beginning is not what we wanted. There are several possible solutions, one of which is shown in Figure 9.40. In Figure 9.40, control signal OE is not asserted in control word 2, which is executed only once at the beginning before the loop. (Notice that control words 1 and 2 together are equivalent to assigning the constant 1 to i.) Subsequent executions of control word 3 will have OE asserted together with the addition, and this one we will repeat ten times. The corrected simulation trace for this set of control words is shown in Figure 9.41. Again, only 2.3 μs are needed for the completion of the algorithm, but this time, the count at the output starts at 1 rather than 0.

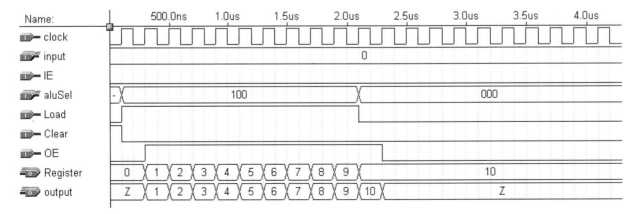

Figure 9.41 Corrected simulation trace for using the three control words from Figure 9.40.

● ● ● ● ● ● ● ● ● ● ● ● ● ●
9.8   VHDL for Datapaths

In modeling the datapath using VHDL, we need to work at the structural level. First, all of the components used in the datapath must be described individually. It doesn't matter whether these components are described at the behavioral, dataflow, or structural level. These components then are connected together in an enclosing entity using the structural level method.

9.8.1   Dedicated Datapath

Figure 9.42 shows the VHDL code for the dedicated datapath of Figure 9.25.

```
LIBRARY IEEE;
USE IEEE.STD_LOGIC_1164.ALL;
USE IEEE.STD_LOGIC_ARITH.ALL;

ENTITY Datapath IS PORT (
   Clock: IN STD_LOGIC;

   -- primary datapath input
   Input: IN STD_LOGIC_VECTOR(7 DOWNTO 0);

   -- control signals
   nLoad: IN STD_LOGIC;
   nShift: IN STD_LOGIC;
   eightCount: IN STD_LOGIC;
   Clear: IN STD_LOGIC;
   bitCount: IN STD_LOGIC;
   Down: IN STD_LOGIC;
   OutDone: IN STD_LOGIC;

   -- status signals
   eq8: OUT STD_LOGIC;
   N0eq1: OUT STD_LOGIC;

   -- primary datapath output
   Count: OUT STD_LOGIC_VECTOR(3 DOWNTO 0);   -- for debug
   Done: OUT STD_LOGIC;
   Output: OUT STD_LOGIC);
END Datapath;
```

Figure 9.42
VHDL code for the
datapath of
Figure 9.25.
(continued on next page)

```
ARCHITECTURE Structural OF Datapath IS
    COMPONENT shiftreg PORT (
        Clock: IN STD_LOGIC;
        SHSel: IN STD_LOGIC_VECTOR(1 DOWNTO 0);
        D: IN STD_LOGIC_VECTOR(7 DOWNTO 0);
        Q: OUT STD_LOGIC_VECTOR(7 DOWNTO 0));
    END COMPONENT;

    COMPONENT counter PORT (
        Clock: IN STD_LOGIC;
        Clear: IN STD_LOGIC;
        Count: IN STD_LOGIC;
        Down: IN STD_LOGIC;
        Q: OUT INTEGER RANGE 0 TO 15);
    END COMPONENT;

    SIGNAL SHSel: STD_LOGIC_VECTOR(1 DOWNTO 0);
    SIGNAL ShiftOut: STD_LOGIC_VECTOR(7 DOWNTO 0);
    SIGNAL CountbitOut: INTEGER RANGE 0 TO 15;
    SIGNAL CounteightOut: INTEGER RANGE 0 TO 15;
    SIGNAL Equal: STD_LOGIC;
    SIGNAL Up: STD_LOGIC;
BEGIN

    SHSel <= nShift & nLoad;
    U0: shiftreg PORT MAP(Clock,SHSel,Input,ShiftOut);
    N0eq1 <= ShiftOut(0);

    -- counteight
    Up <= '0';
    U1: counter PORT MAP(Clock,Clear,eightCount,Up,CounteightOut);
    eq8 <= '1' WHEN CounteightOut = 8 ELSE '0';
    Count <= CONV_STD_LOGIC_VECTOR(CounteightOut, 4);     -- for debug

    -- countbit
    U2: counter PORT MAP(Clock,Clear,bitCount,Down,CountbitOut);
    Equal <= '1' WHEN CountbitOut = 0 ELSE '0';
    Output <= Equal WHEN OutDone = '1' ELSE 'Z';
    Done <= OutDone;
END Structural;
```

Figure 9.42
VHDL code for the
datapath of
Figure 9.25.

9.8.2   General Datapath

Figure 9.43 and Figure 9.44 show the complete VHDL code for modeling the complex, general datapath circuit from Figure 9.32. Figure 9.43 lists the definitions of all of the components used in the datapath. The VHDL codes and the detail constructions of the components used in the datapath circuit have been discussed in previous chapters. Figure 9.44 shows the enclosing entity that connects these components together at the structural level to form the datapath.

```
-- 2-to-1 MUX
LIBRARY IEEE;
USE IEEE.STD_LOGIC_1164.ALL;

ENTITY mux2 IS PORT (
   S: IN STD_LOGIC;                               -- select line
   D1, D0: IN STD_LOGIC_VECTOR(7 DOWNTO 0);       -- data bus input
   Y: OUT STD_LOGIC_VECTOR(7 DOWNTO 0));          -- data bus output
END mux2;

ARCHITECTURE Behavioral OF mux2 IS
BEGIN
   PROCESS(S, D1, D0)
   BEGIN
      IF (S = '0') THEN
         Y <= D0;
      ELSE
         Y <= D1;
      END IF;
   END PROCESS;
END Behavioral;


-------------------------------------------------------------------------
-- Register File
LIBRARY IEEE;
USE IEEE.STD_LOGIC_1164.ALL;
USE IEEE.STD_LOGIC_UNSIGNED.ALL;
--USE IEEE.STD_LOGIC_ARITH.ALL;

ENTITY regfile IS PORT (
   clk: IN STD_LOGIC;                             --clock
   WE: IN STD_LOGIC;                              --write enable
   WA: IN STD_LOGIC_VECTOR(1 DOWNTO 0);           --write address
```

Figure 9.43

Components for the datapath of Figure 9.32.

(continued on next page)

```
            input: IN STD_LOGIC_VECTOR(7 DOWNTO 0);          --input
            RAE: IN STD_LOGIC;                               --read enable port A
            RAA: IN STD_LOGIC_VECTOR(1 DOWNTO 0);            --read address port A
            RBE: IN STD_LOGIC;                               --read enable port B
            RBA: IN STD_LOGIC_VECTOR(1 DOWNTO 0);            --read address port B
            Aout, Bout: OUT STD_LOGIC_VECTOR(7 DOWNTO 0));   --output port A & B
        END regfile;


        ARCHITECTURE Behavioral OF regfile IS
            SUBTYPE reg IS STD_LOGIC_VECTOR(7 DOWNTO 0);
            TYPE regArray IS ARRAY(0 TO 3) OF reg;
            SIGNAL RF: regArray;                             --register file contents
        BEGIN
            WritePort: PROCESS (clk)
            BEGIN
                IF (clk'EVENT AND clk = '1') THEN
                    IF (WE = '1') THEN
                        RF(CONV_INTEGER(WA)) <= input;
                    END IF;
                END IF;
            END PROCESS;


            ReadPortA: PROCESS (RAE, RAA)
            BEGIN
                IF (RAE = '1') THEN
                    Aout <= RF(CONV_INTEGER(RAA));           -- convert bit VECTOR to integer
                ELSE
                    Aout <= (OTHERS => '0');
                END IF;
            END PROCESS;


            ReadPortB: PROCESS (RBE, RBA)
            BEGIN
                IF (RBE = '1') THEN
                    Bout <= RF(CONV_INTEGER(RBA));           -- convert bit VECTOR to integer
                ELSE
                    Bout <= (OTHERS => '0');
                END IF;
            END PROCESS;
        END Behavioral;
```

Figure 9.43
Components for the datapath of Figure 9.32.
(continued on next page)

```
-------------------------------------------------------------------------
-- ALU
LIBRARY IEEE;
USE IEEE.STD_LOGIC_1164.ALL;
-- need the following to perform arithmetics on STD_LOGIC_VECTORs
USE IEEE.STD_LOGIC_UNSIGNED.ALL;

ENTITY alu IS PORT (
   ALUSel: IN STD_LOGIC_VECTOR(2 DOWNTO 0);          -- select for operations
   A, B: IN STD_LOGIC_VECTOR(7 DOWNTO 0);            -- input operands
   F: OUT STD_LOGIC_VECTOR(7 DOWNTO 0));             -- output
END alu;

ARCHITECTURE Behavior OF alu IS
BEGIN
   PROCESS(ALUSel, A, B)
   BEGIN
      CASE ALUSel IS
         WHEN "000" =>          -- pass A through
            F <= A;
         WHEN "001" =>          -- AND
            F <= A AND B;
         WHEN "010" =>          -- OR
            F <= A OR B;
         WHEN "011" =>          -- NOT
            F <= NOT A;
         WHEN "100" =>          -- add
            F <= A + B;
         WHEN "101" =>          -- subtract
            F <= A - B;
         WHEN "110" =>          -- increment
            F <= A + 1;
         WHEN OTHERS =>         -- decrement
            F <= A - 1;
      END CASE;
   END PROCESS;
END Behavior;


-------------------------------------------------------------------------
-- Shifter
LIBRARY IEEE;
USE IEEE.STD_LOGIC_1164.ALL;
```

Figure 9.43
Components for the datapath of Figure 9.32.
(continued on next page)

```
ENTITY shifter IS PORT (
    SHSel: IN STD_LOGIC_VECTOR(1 DOWNTO 0);          -- select for operations
    input: IN STD_LOGIC_VECTOR(7 DOWNTO 0);          -- input operands
    output: OUT STD_LOGIC_VECTOR(7 DOWNTO 0));        -- output
END shifter;

ARCHITECTURE Behavior OF shifter IS
BEGIN
    PROCESS(SHSel, input)
    BEGIN
        CASE SHSel IS
            WHEN "00" =>                             -- pass through
                output <= input;
            WHEN "01" =>                             -- shift right
                output <= input(6 DOWNTO 0) & '0';
            WHEN "10" =>                             -- shift left
                output <= '0' & input(7 DOWNTO 1);
            WHEN OTHERS =>                           -- rotate right
                output <= input(0) & input(7 DOWNTO 1);
        END CASE;
    END PROCESS;
END Behavior;


------------------------------------------------------------------------
-- Tri-state buffer
LIBRARY IEEE;
USE IEEE.STD_LOGIC_1164.ALL;

ENTITY TriStateBuffer IS PORT (
    E: IN STD_LOGIC;
    D: IN STD_LOGIC_VECTOR(7 DOWNTO 0);
    Y: OUT STD_LOGIC_VECTOR(7 DOWNTO 0));
END TriStateBuffer;

ARCHITECTURE Behavioral OF TriStateBuffer IS
BEGIN
    PROCESS (E, D)     -- get error message if no d
    BEGIN
        IF (E = '1') THEN
            Y <= D;
        ELSE
            Y <= (OTHERS => 'Z');    -- to get 8 Z values
        END IF;
    END PROCESS;
END Behavioral;
```

Figure 9.43
Components for the
datapath of
Figure 9.32.

```vhdl
LIBRARY IEEE;
USE IEEE.STD_LOGIC_1164.ALL;

ENTITY datapath IS PORT (
    clock: IN STD_LOGIC;
    input: IN STD_LOGIC_VECTOR (7 DOWNTO 0);
    IE, WE: IN STD_LOGIC;
    WA: IN STD_LOGIC_VECTOR (1 DOWNTO 0);
    RAE: IN STD_LOGIC;
    RAA: IN STD_LOGIC_VECTOR (1 DOWNTO 0);
    RBE: IN STD_LOGIC;
    RBA: IN STD_LOGIC_VECTOR (1 DOWNTO 0);
    ALUSel: IN STD_LOGIC_VECTOR (2 DOWNTO 0);
    SHSel: IN STD_LOGIC_VECTOR (1 DOWNTO 0);
    OE: IN STD_LOGIC;
    output: OUT STD_LOGIC_VECTOR(7 DOWNTO 0));
END datapath;

ARCHITECTURE Structural OF datapath IS

COMPONENT mux2 PORT (
    S: IN STD_LOGIC;                            -- select lines
    D1, D0: IN STD_LOGIC_VECTOR(7 DOWNTO 0);    -- data bus input
    Y: OUT STD_LOGIC_VECTOR(7 DOWNTO 0));       -- data bus output
END COMPONENT;

COMPONENT regfile PORT (
    clk: IN STD_LOGIC;                          -- clock
    WE: IN STD_LOGIC;                           -- write enable
    WA: IN STD_LOGIC_VECTOR(1 DOWNTO 0);        -- write address
    input: IN STD_LOGIC_VECTOR(7 DOWNTO 0);     -- input
    RAE: IN STD_LOGIC;                          -- read enable ports A & B
    RAA: IN STD_LOGIC_VECTOR(1 DOWNTO 0);       -- read address port A & B
    RBE: IN STD_LOGIC;                          -- read enable ports A & B
    RBA: IN STD_LOGIC_VECTOR(1 DOWNTO 0);       -- read address port A & B
    Aout, Bout: OUT STD_LOGIC_VECTOR(7 DOWNTO 0)); -- output port A & B
END COMPONENT;

COMPONENT alu PORT (
    ALUSel: IN STD_LOGIC_VECTOR(2 DOWNTO 0);    -- select for operations
    A, B: IN STD_LOGIC_VECTOR(7 DOWNTO 0);      -- input operands
    F: OUT STD_LOGIC_VECTOR(7 DOWNTO 0));       -- output
END COMPONENT;
```

Figure 9.44
Datapath of
Figure 9.32
constructed at the
structural level.
(continued on next page)

```
COMPONENT shifter PORT (
    SHSel: IN STD_LOGIC_VECTOR(1 DOWNTO 0);        -- select for operations
    input: IN STD_LOGIC_VECTOR(7 DOWNTO 0);        -- input operands
    output: OUT STD_LOGIC_VECTOR(7 DOWNTO 0));     -- output
END COMPONENT;

COMPONENT TriStateBuffer PORT (
    E: IN STD_LOGIC;
    D: IN STD_LOGIC_VECTOR(7 DOWNTO 0);
    Y: OUT STD_LOGIC_VECTOR(7 DOWNTO 0));
END COMPONENT;

SIGNAL muxout, rfAout, rfBout: STD_LOGIC_VECTOR(7 DOWNTO 0);
SIGNAL aluout, shiftout, tristateout: STD_LOGIC_VECTOR(7 DOWNTO 0);

BEGIN
    -- doing structural modeling here
    U0: mux2 PORT MAP(IE, input, shiftout, muxout);
    U1: regfile PORT MAP(clock,WE,WA,muxout,RAE,RAA,RBE,RBA,rfAout,rfBout);
    U2: alu PORT MAP(ALUsel, rfAout, rfBout, aluout);
    U3: shifter PORT MAP(SHSel,aluout,shiftout);
    U4: TriStateBuffer PORT MAP(OE, shiftout, tristateout);
    output <= tristateout;
END Structural;
```

Figure 9.44
Datapath of
Figure 9.32
constructed at the
structural level.

● ● ● ● ● ● ● ● ● ● ● ● ● ● ●
9.9 Summary Checklist

- Datapath
- Dedicated datapath
- Control signals
- Status signals
- Control word
- General datapath
- Register-transfer level design
- Timing issues (when a register is updated)
- Be able to design a dedicated datapath from an algorithm
- Be able to generate the status signals
- Be able to use a given datapath by deriving the control words for it to implement an algorithm

9.10 **Problems**

P9.1. Derive the truth table for the circuit in Figure P9.1. The truth table should only have columns for the control signal inputs and outputs. The data inputs, D_0 to D_3, are written in the table entries.

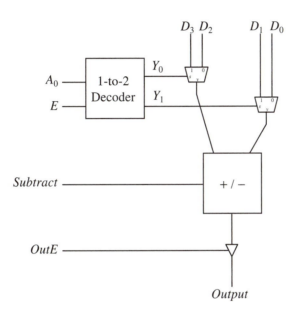

Figure P9.1

P9.2. Use schematic entry to implement the general datapath shown in Figure 9.27.

P9.3. Test the datapath from Problem P9.2 by performing a simulation on the algorithm shown in Figure 9.28.

P9.4. Implement the datapath from Problem P9.2 on the UP2 board. Perform sample tests to show that it operates correctly.

P9.5. Write the VHDL code for the general datapath shown in Figure 9.27.

P9.6. Use simulation to show that the synthesized circuits from Problem P9.2 and Problem P9.5 perform functionally identical to each other.

P9.7. Use schematic entry to implement the general datapath shown in Figure 9.32.

P9.8. Test the datapath from Problem P9.7 by performing a simulation on the algorithm shown in Figure 9.31.

P9.9. Implement the datapath from Problem P9.7 on the UP2 board. Perform sample tests to show that it operates correctly.

P9.10. Test the datapath from Problem P9.7 by performing a simulation on the algorithm shown in Figure 9.35.

P9.11. Use schematic entry to implement the datapath for Example 9.4.

P9.12. Simulate and test the datapath for Example 9.4.

P9.13. Write the VHDL code for the datapath for Example 9.4.

P9.14. Use schematic entry to implement the datapath for Example 9.5.

P9.15. Simulate and test the datapath for Example 9.5.

P9.16. Write the VHDL code for the datapath for Example 9.5.

P9.17. Use schematic entry to implement the datapath for Example 9.6.

P9.18. Simulate and test the datapath for Example 9.6.

P9.19. Write the VHDL code for the datapath for Example 9.6.

P9.20. Use schematic entry to implement the datapath for Example 9.7.

P9.21. Simulate and test the datapath for Example 9.7.

P9.22. Write the VHDL code for the datapath for Example 9.7.

P9.23. Derive and simulate the control words using the datapath shown in Figure 9.32 for inputting two 8-bit unsigned numbers, and then output the largest number.

P9.24. Derive and simulate the control words using the datapath shown in Figure 9.32 for inputting three 8-bit unsigned numbers, and then output the largest number, followed by the second largest.

P9.25. Derive and simulate the control words using the datapath shown in Figure 9.32 for inputting ten 8-bit unsigned numbers, and then output the sum of the numbers.

P9.26. Derive and simulate the control words using the datapath shown in Figure 9.32 for inputting multiple 8-bit unsigned numbers until a zero is entered, and then output the number of numbers entered.

P9.27. Derive and simulate the control words using the datapath shown in Figure 9.32 for inputting an 8-bit unsigned number, and then output the number of 1 bits in the number. For example, the number 11010110 has five 1 bits.

P9.28. Derive and simulate the control words using the datapath shown in Figure 9.32 for solving the Greatest Common Divisor (GCD) algorithm shown in Figure P9.28.

```
INPUT X
INPUT Y
WHILE (X ≠ Y) {
   IF (X > Y) THEN
      X = X - Y;
   ELSE
      Y = Y - X;
   END IF
   }
OUTPUT X
```

Figure P9.28

P9.29. Design an 8-bit dedicated datapath for the algorithm in Figure P9.29, and write the control words for it. Use only one adder-subtractor unit for all of the addition and subtraction operations. Label clearly all of the control and status signals.

```
w = 0
x = 0
y = 0
INPUT z
WHILE (z ≠ 0) {
   w = w - 2
   IF (z is an odd number) THEN
      x = x + 2
   ELSE
      y = y + 1
   END IF
   z = z - 1
   }
```

Figure P9.29

P9.30. Design a dedicated datapath for inputting two 8-bit unsigned numbers, and then output the largest number. The datapath should have only one input port and one output port. Label clearly all of the control and status signals.

P9.31. Design and simulate a dedicated datapath for inputting three 8-bit unsigned numbers, and then output the largest number, followed by the second largest.

P9.32. Design and simulate a dedicated datapath for inputting ten 8-bit unsigned numbers, and then output the sum of the numbers.

P9.33. Design and simulate a dedicated datapath for inputting multiple 8-bit unsigned numbers until a zero is entered, and then output the number of numbers entered.

P9.34. Design and simulate a dedicated datapath for inputting an 8-bit unsigned number, and then output the number of 1 bits in the number. For example, the number 11010110 has five 1 bits.

P9.35. Design and simulate a dedicated datapath for solving the algorithm in Figure P9.35. Use only one adder (i.e., no adder-subtractor and no ALU) for all the arithmetic operations. Include the circuits for generating all of the status signals. The datapath is 4-bits wide.

Figure P9.35

```
s1 = 0;
s2 = 0;
FOR(i=0; i ≠ 10; i++){
    INPUT j;
    IF (j is even) THEN
        s1++;
    ELSE
        s2++;
    END IF
}
OUTPUT s1;
OUTPUT s2;
```

P9.36. Design and simulate a dedicated datapath for solving the Greatest Common Divisor (GCD) algorithm shown in Figure P9.36. The two numbers X and Y are 8-bit unsigned numbers. Include the circuits for generating all of the status signals.

Figure P9.36

```
INPUT X
INPUT Y
WHILE (X ≠ Y) {
    IF (X > Y) THEN
        X = X - Y;
    ELSE
        Y = Y - X;
    END IF
}
OUTPUT X
```

P9.37. Design and simulate a dedicated datapath for implementing a stack of size 10. When the *Push* signal is asserted, the input value is pushed onto the stack. When the *Pop* signal is asserted, the value at the top of the stack is output.

CHAPTER 10

Control Units

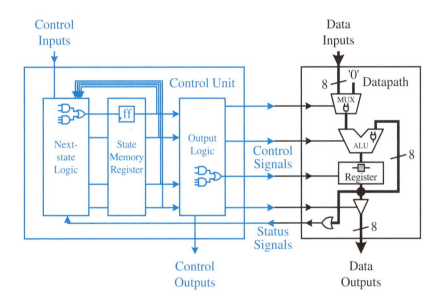

In the last chapter, we saw how a datapath is designed and how it is used to execute a particular algorithm by specifying the control words to manipulate the datapath at each clock cycle. In that chapter, we tested the datapath by setting the control word signals manually during simulation. However, to actually have the datapath automatically operate according to the control words, a control unit is needed to generate the control word signals automatically at each clock cycle.

The **control unit** inside the microprocessor is a finite state machine. By stepping through a sequence of states, the control unit controls the operations of the datapath. For each state that the control unit is in, the output logic that is inside the control unit will generate all of the appropriate control signals for the datapath to perform one data operation. These data operations are referred to as register-transfer operations. Each register-transfer operation consists of reading a value from a register, modifying the value by one or more functional units, and finally, writing the modified value back into the same or a different register.

Since the control unit is a finite state machine, the circuit for it is derived exactly as discussed in Section 7.3 having the next-state circuit, the state memory register, and the output logic circuit. However, the sequential circuits synthesized in Section 7.3 are stand-alone FSMs and were not used as part of larger circuits. What we want to do now is to synthesize these FSMs as control units for microprocessors. The main difference is in the way we define and use the input and output signals of the FSM. The FSM models shown in Figure 7.1 simply show that the FSM has input and output signals. For a control unit, we want to be more precise as to where the input signals come from, where the output signals go to, and what these signals are used for.

There are two types of input signals and two types of output signals. The **control inputs** are the primary external input signals to the control unit. These are the external signals for controlling the operation of the microprocessor. For example, a *Start* signal will tell the microprocessor to start executing, or a *Reset* signal will reset the state memory to the initialization state. The **status signals** are input signals from the datapath. Usually, comparators in the datapath generate these status signals for determining execution branches in an algorithm (for example, to test whether or not to repeat a loop). These two types of input signals together provide information for the next-state logic circuit in the FSM to determine what next state to go to in the next clock cycle.

The **control outputs** are the primary output signals from the microprocessor to the external world. For example, when a microprocessor is finished executing an algorithm, it outputs a *Done* signal to let the user know that it is done, and that the data being output by the datapath is valid. The **control signals** are probably the most important of all the input/output signals, since these are the signals that directly control the operations of the datapath. In every clock cycle, the control unit will generate a different set of control signals for the datapath to perform one register-transfer operation.

One of the first things that we do when constructing an FSM circuit is to derive the state diagram. When deriving the state diagram for the control unit, we have to be very careful with the timings of the register-transfer operations. The issue here is that, when we write a value into a register, that value is not available until the beginning of the next clock cycle. Hence, if we read from the register in the current clock cycle, we would be reading the old value in the register rather than the new value that is being written into the register.

When we were designing general FSMs in Chapter 7, this timing issue was never a problem, because we were not using the FSMs to control register-transfer operations in a datapath. They were stand-alone FSMs, and so their input and output signals are independent of each other.

However, the FSMs that we want to design in this chapter are for controlling register-transfer operations in a datapath. The output signals from these FSMs are control signals for the datapath, and some of them are used to load registers with new values. These new register values may be used by comparators for the testing of conditions. The results of these conditional tests are the status signals used by the control unit to determine what next state to go to. Finally, from the different states, different control signals are generated. Hence, for a control unit, the status (input) signals and the control (output) signals are dependent on each other. Thus, when status signals are generated, we need to make sure that they are from tests of the intended register value.

Timing problem issues are resolved during the state diagram derivation step. Once we have the correct timing state diagram, then deriving the actual control unit circuit is exactly the same as the process described in Chapter 7 for deriving an FSM. To help deal with timing issues, we sometimes prefer to use an **algorithmic state machine (ASM) chart** or a **state action table** to describe the behavior of a control unit rather than using a state diagram. The use of ASM charts and state action tables is discussed in Section 10.3.

- - - - - - - - - - - - - - - · · · · ·

10.1   Constructing the Control Unit

In Chapter 9, we learned how to construct and use a datapath to implement an algorithm. All data manipulation instructions in the algorithm are converted to control words, and each control word is executed in one clock cycle to perform one register-transfer operation. A control unit is used to generate the appropriate control signals in the control words so that the datapath can perform all of the required register-transfer operations automatically. These control signals are the output signals from the output logic circuit that is inside the FSM.

In addition to generating the control signals, the control unit is also needed to control the sequencing of the instructions in the algorithm. The datapath is responsible only for the manipulation of the data; it only performs the register-transfer operations. It is the control unit that determines when each register-transfer operation is to be executed and in what order. The sequencing done by the control unit is established during the derivation of the state diagram.

The state diagram shows what register-transfer operation is executed in what state and the sequencing of the execution of these operations. A state is created for each control word, and each state is executed in one clock cycle. The edges in the state diagram are determined by the sequence in which the instructions in the algorithm are executed. The sequential execution of instructions is represented by unconditional transitions between states (i.e., edges with no labels). Execution branches in the algorithm are represented by conditional transitions from a state with two outgoing edges: one with the label for when the condition is true and the other with the label for when the condition is false. If there is more than one condition, then all possible combinations of these conditions must be labeled on the outgoing edges from every state. These conditions are the status signals generated by the datapath, and passed to the next-state logic in the FSM.

Once the state diagram is derived, the actual construction of the control unit circuit follows the same procedure as for the FSM discussed in Section 7.3. We will now illustrate the construction of simple control units with several examples.

10.1.1   Counting 1 to 10

Example 10.1   **Deriving the control unit for the counting problem**

In this example, we will construct the state diagram and the control unit for controlling the dedicated datapath from Example 9.4 to generate and output the numbers from 1 to 10. The algorithm, dedicated datapath, and control words from Example 9.4 are repeated here in Figure 10.1 for convenience.

From the algorithm shown in Figure 10.1(a), we see that there are three data manipulation instructions: lines 1, 3, and 4. Line 2 is not a data manipulation statement, but rather, it is a control statement. From Example 9.4, we have derived the three corresponding control words for these three data manipulation instructions for controlling the dedicated datapath shown in Figure 10.1(b). These three control words are shown here in Figure 10.1(c).

We start by assigning these three control words to three separate states in the state diagram, as shown in Figure 10.2(a). The states are given the symbolic names s_0, s_1, s_2, and s_3 and are annotated with the control word and instruction that each is assigned to execute. In state s_0, we want the control unit to generate the control signals for control word 1 to execute the instruction $i = 0$. In state s_1, we want the control unit to generate the control signals for control word 2 to execute the instruction $i = i + 1$. State s_2 executes the instruction OUTPUT i. State s_3 is added simply for exiting the WHILE loop and halting the execution of the algorithm.

The sequence in which the states are connected follows the sequence of the instructions in the algorithm. The FSM starts from the reset state s_0, which initializes i to 0. After executing line 1 in the algorithm, the execution of line 3 depends on the condition in the WHILE loop. Since line 1 is executed in state s_0 and line 3 is executed in state s_1, transitioning from state s_0 to s_1 depends on the test condition $(i \neq 10)$. This condition is represented by the two outgoing edges from state s_0: one edge going to state s_1 with the label $(i \neq 10)$ for when the condition is true, and one edge going to state s_3 with the label $(i \neq 10)'$ for when the condition is false. The

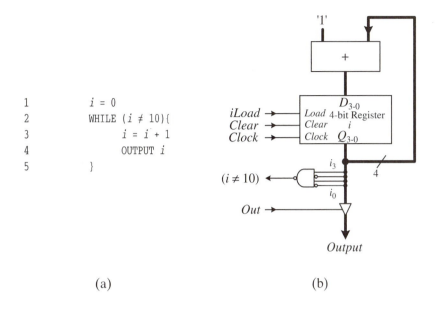

```
1          i = 0
2          WHILE (i ≠ 10){
3                i = i + 1
4                OUTPUT i
5          }
```

(a)

(b)

Control Word	Instruction	iLoad	Clear	Out
1	$i = 0$	0	1	0
2	$i = i + 1$	1	0	0
3	OUTPUT i	0	0	1

(c)

Figure 10.1
The counting problem for Example 10.1: (a) algorithm; (b) dedicated datapath; (c) control words.

execution of line 4 follows immediately after line 3; hence, there is an unconditional edge from state s_1 to s_2. From state s_2, there are the same two conditional edges as from state s_0 for testing whether to repeat the WHILE loop or not. If the condition $(i \neq 10)$ is true, then the FSM will go back to state s_1; otherwise, the FSM will exit the WHILE loop and go to state s_3. The FSM halts in state s_3 by having an unconditional edge going back to itself. No register-transfer operation is assigned to state s_3.

Having derived the state diagram, the actual construction of the control unit circuit is exactly the same as for constructing general FSM circuits. The next-state table is shown in Figure 10.2(b). Since there is a total of four states, two flip-flops are needed to encode them. For simplicity, the straight binary encoding scheme is used for encoding the states. Hence, state s_0 is encoded as $Q_1 Q_0 = 00$, state s_1 is encoded as $Q_1 Q_0 = 01$, and so on. In the next-state table, these four states are assigned to four rows, each labeled with the state name and their encoding.

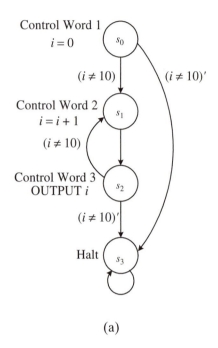

(a)

Current State	Next State	
$Q_1 Q_0$	$Q_{1next} Q_{0next}$	
	$(i \neq 10)'$	$(i \neq 10)$
s_0 00	s_3 11	s_1 01
s_1 01	s_2 10	s_2 10
s_2 10	s_3 11	s_1 01
s_3 11	s_3 11	s_3 11

(b)

Current State	Implementation	
$Q_1 Q_0$	$D_1 D_0$	
	$(i \neq 10)'$	$(i \neq 10)$
00	11	01
01	10	10
10	11	01
11	11	11

(c)

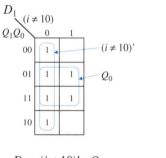

$$D_1 = (i \neq 10)' + Q_0$$

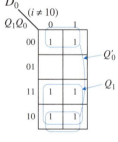

$$D_0 = Q_1 + Q'_0$$

(d)

Figure 10.2 Construction of the control unit for Example 10.1: (a) state diagram; (b) next-state table; (c) implementation table using D flip-flops; (d) K-maps and excitation equations; (e) output table; (f) output equations for the three control signals; (g) circuit.
(continued on next page)

Q_1Q_0	$iLoad$	$Clear$	Out
00	0	1	0
01	1	0	0
10	0	0	1
11	0	0	0

(e)

$$iLoad = Q_1'Q_0$$

$$Clear = Q_1'Q_0'$$

$$Out = Q_1Q_0'$$

(f)

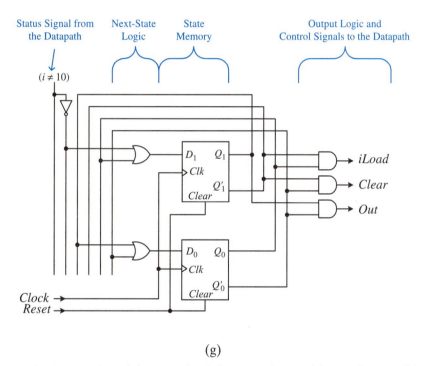

(g)

Figure 10.2   Construction of the control unit for Example 10.1: (a) state diagram; (b) next-state table; (c) implementation table using D flip-flops; (d) K-maps and excitation equations; (e) output table; (f) output equations for the three control signals; (g) circuit.

In addition to the four current states listed down the rows of the table, the next state of the FSM is also dependent on the status signal for the test condition $(i \neq 10)$. Thus, we have the last two columns in the table: one column with the label $(i \neq 10)'$ for when the condition is false and one column with the label $(i \neq 10)$ for when the condition is true.

The two flip-flops and one status signal give us a total of three variables (or 2^3 different combinations) to consider in the next-state table. Each next-state entry in the table is obtained from the state diagram by looking at the corresponding current state and the edges leading out from that state to see what the next state is. For example, looking at the state diagram shown in Figure 10.2(a), the edge with the label ($i \neq 10$) leading out from state s_0 goes to state s_1. Correspondingly, in the next-state table, the next-state entry at the intersection of row s_0 (00) and the column labeled ($i \neq 10$) has the value s_1 (01).

From the next-state table, we get the implementation table, as shown in Figure 10.2(c). Using D flip-flops to implement the FSM, the implementation table is the same as the next-state table because the characteristic equation for the D flip-flop is $Q_{next} = D$. The only difference between the two tables is that the bits in the entries mean something different. In the next-state table, the bits in the entries (labeled $Q_{1next} Q_{0next}$) are the next states for the FSM to go to. In the implementation table, the bits (labeled $D_1 D_0$) are the inputs necessary to realize those next states.

From the implementation table, we derive the excitation equations. The excitation equations are used to derive the next-state circuit for generating the inputs to the state memory flip-flops. Since we have used two D flip-flops, two excitation equations (one for D_1, and one for D_0) are needed, as shown in Figure 10.2(d). The two K-maps for these two excitation equations are obtained from extracting the corresponding bits from the implementation table. For example, the leftmost bit in each entry in the implementation table is for the D_1 K-map and equation, and the rightmost bit is for the D_0 K-map and equation. These two excitation equations are dependent on the three variables, Q_1, Q_0, and ($i \neq 10$), which represent the current state and status signal, respectively. Having derived the excitation equations, it is trivial to draw the next-state circuit based on these equations.

The output logic circuit for the FSM is derived from the control word signals and the states in which the control words are assigned to. Recall that the control signals control the operation of the datapath, and now we are constructing the control unit to control the datapath. So what the control unit needs to do is to generate and output the appropriate control signals in each state to execute the instruction that is assigned to that state. In other words, the control signals for controlling the operation of the datapath are simply the output signals from the output logic circuit in the FSM.

To derive the output table, we take the control word table and replace all of the control word numbers with the actual encoding of the state in which that control word is assigned to. For example, looking at the state diagram shown in Figure 10.2(a), control word 1 is assigned to state s_0. So in the output table, we put in the value 00 instead of the control word number 1. The value 00 is the encoding that we have given to state s_0, and it represents the current state value for the two flip-flops, Q_1, and Q_0. Since there is no control word or instruction assigned to the halting state, s_3 (11), all of the control signals for this state can be de-asserted. The output table and the resulting output equations are shown in Figure 10.2(e) and (f), respectively.

Once we have derived the excitation and output equations, we easily can draw the control unit circuit shown in Figure 10.2(g). The state memory simply consists of the two D flip-flops with asynchronous clear signals. All of the asynchronous clear signals are connected to the global *Reset* signal. Both the next-state logic circuit and the output logic circuit are combinational circuits and are constructed from the excitation equations and output equations, respectively.

10.1.2   Simple IF-THEN-ELSE

We will now construct the state diagram and the control unit for controlling the dedicated datapath from Example 9.3 for the simple IF-THEN-ELSE problem. The algorithm, dedicated datapath, and control words from Example 9.3 are repeated here in Figure 10.3 for convenience. Example 10.2 shows the naive way of creating a

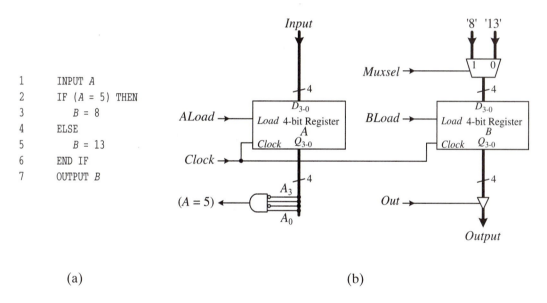

```
1       INPUT A
2       IF (A = 5) THEN
3           B = 8
4       ELSE
5           B = 13
6       END IF
7       OUTPUT B
```

(a)   (b)

Control Word	Instruction	*ALoad*	*Muxsel*	*BLoad*	*Out*
1	INPUT *A*	1	×	0	0
2	*B* = 8	0	1	1	0
3	*B* = 13	0	0	1	0
4	OUTPUT *B*	0	×	0	1

(c)

Figure 10.3   The IF-THEN-ELSE problem for Examples 10.2 and 10.3: (a) algorithm; (b) dedicated datapath; (c) control words.

state diagram by simply assigning one state for each control word, just like in Example 10.1. However, we will see from the example that the state diagram created this way for this problem is incorrect, because the status signal generated is wrong. Example 10.3 shows the derivation of the corrected state diagram and the construction of the control unit for the IF-THEN-ELSE problem.

| Example 10.2 | **Deriving an incorrect state diagram** |

In this example, we will derive the state diagram for the control unit for controlling the dedicated datapath from Example 9.3 for solving the IF-THEN-ELSE problem. From the algorithm shown in Figure 10.3(a), we see that there are four data manipulation instructions: lines 1, 3, 5, and 7. Line 2 is not a data manipulation instruction, but rather, it is a control statement. From Example 9.3, we have derived the four corresponding control words for these four data manipulation instructions for controlling the dedicated datapath shown in Figure 10.3(b). These four control words are repeated here in Figure 10.3(c).

Again, we start by assigning these four control words to four separate states in the state diagram, as shown in Figure 10.4(a). These four states are given the symbolic names s_input, s_equal, $s\_notequal$, and s_output, and are annotated with the control word and instruction that is assigned to them. For example, in state s_input, we want the control unit to generate the control word 1 signals for executing the instruction INPUT A, and in state s_equal, we want the control unit to generate the control word 2 signals for executing the instruction $B = 8$.

After the INPUT A instruction, the execution of the following two instructions, $B = 8$ and $B = 13$, depends on the condition of the IF statement. Since these three instructions are assigned to the three states: s_input, s_equal, and $s\_notequal$, respectively, transitioning from state s_input to the other two states depends on the condition ($A = 5$) in the IF statement. This conditional execution is represented by the two outgoing edges from state s_input: one edge going to state s_equal with the label ($A = 5$) for when the condition is true, and one edge going to state $s\_notequal$ with the label ($A = 5$)' for when the condition is false. The instruction, OUTPUT B, is executed unconditionally after executing either of the two instructions, $B = 8$ or $B = 13$; therefore, from either state s_equal or $s\_notequal$, there is an unconditional edge going to state s_output. The algorithm halts after executing OUTPUT B, so we make the FSM halt in state s_output by unconditionally going back to itself.

According to the algorithm, after inputting a value for A in state s_input, we need to test for the condition ($A = 5$). If the condition is true, we go to state s_equal to execute the instruction $B = 8$; otherwise, we go to state $s\_notequal$ to execute the instruction $B = 13$. From either state s_equal or $s\_notequal$, the next and final state is s_output. Let us assume that state s_input is executed in clock cycle 1. In clock cycle 2, either state s_equal or $s\_notequal$ is executed. State s_output is then executed in clock cycle 3.

There are two important points to understand and remember here:

1. At a rising clock edge, a register is loaded with a new value if its load signal is asserted

2. At every rising clock edge, the FSM enters a new state—the next state

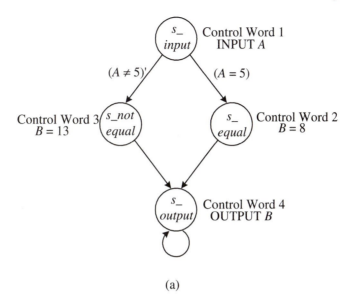

(a)

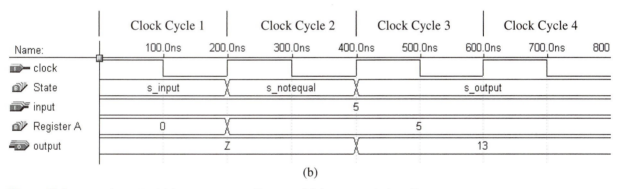

(b)

Figure 10.4 Example 10.2: (a) incorrect state diagram; (b) incorrect timing diagram.

As with all of our other examples, we use the rising clock edge as the active edge. The reason for point 1 is because we are using positive edge-triggered D flip-flops with enable in our registers. So if the flip-flop is enabled, then the input data will be stored into the flip-flop at the next rising clock edge. Point 2 is because we are also using positive edge-triggered D flip-flops in the state memory register inside the FSM. However, these flip-flops are always enabled without the need of an enable signal (see Section 6.7). Therefore, at every rising clock edge, a new value from the next-state logic circuit will be stored into the state memory register, and so, the FSM enters a new state at every rising clock edge.

If we construct the control unit based on the state diagram shown in Figure 10.4(a), then the following scenario can occur. At the first rising clock edge, the FSM enters state *s_input*. Shortly after entering state *s_input*, the FSM asserts the control signal *ALoad* to load in a value for variable *A*. Since a register is loaded at a rising clock edge and the first rising clock edge has passed, the value for A will be loaded into the register at the next rising clock edge (i.e., at the beginning of clock cycle 2). However, the FSM needs to go to the next state (either s_equal or $s\_notequal$) also at the beginning of clock cycle 2. In order for the FSM to know which of the two states to go to, the FSM must know the result of the test condition ($A = 5$) while it is still in state s_input. The dilemma here is that the status signal generated by the comparator for the test ($A = 5$) is needed in state s_input, and the test uses the input value of A. However, this value of A is not available until the beginning of the next clock cycle. Therefore, what the comparator is reading from the register in clock cycle 1 is the old (or current) value of A and not the new input value.

Figure 10.4(b) shows the timing diagram for this state diagram with the incorrect result. The user inputs a 5. The diagram shows that the value 5 is loaded into register A at time 200 ns (the beginning of clock cycle 2). Since the input is a 5, the test for ($A = 5$) should be true, and the output should be an 8. However, in the timing diagram, we see that the state changes to $s\_notequal$ at time 200 ns. This is because the conditional test is reading A with the old value of 0 rather than the new value of 5. Hence, the output of 13 is incorrect.

You may be wondering why the state diagram for Example 10.1 is correct, whereas the state diagram for Example 10.2 is incorrect. The first two lines of these two algorithms are very similar: The first line assigns a value to a variable (one with the constant zero, and the other with an input), and the second line is a conditional test with the value in the variable. So what causes the conditional test for Example 10.1 to be correct, but the conditional test for Example 10.2 to be incorrect? The reason is because the timing of these two variable assignments is different. In Example 10.1, the variable i is assigned a 0 by asserting the asynchronous *Clear* signal for the register. Because of the asynchronous clear, i will get a 0 immediately and will not have to wait until the next clock cycle; hence, the conditional test is correct. However, in Example 10.2, we use the $ALoad$ signal to load in a value for the register. Therefore, A will not get the new value until the beginning of the next clock cycle, but as the conditional test needs the value in the current clock cycle, hence, the test is incorrect.

Example 10.3 — Deriving a correct state diagram and the control unit

In Example 10.2, even though the comparator is testing for the condition ($A = 5$), it is not getting the correct value for A in the clock cycle that the test result needs. In order for the comparator to get the correct value for A, it needs to wait until the value is loaded into the register at the next clock cycle. One simple way to resolve this timing error is to add an extra state after inputting the value for A so that the value can be written into the register before it is read back out for the test. There is no control word assigned to this extra state. This modified state diagram is shown in Figure 10.5(a) and the corresponding timing diagram with the corrected result in (b). The same input value of 5 is loaded into the register at time 200 ns at the beginning of clock cycle 2. However, the reading of the register for the conditional test does

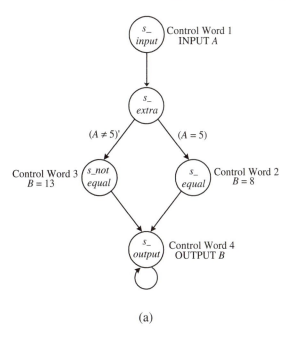

(a)

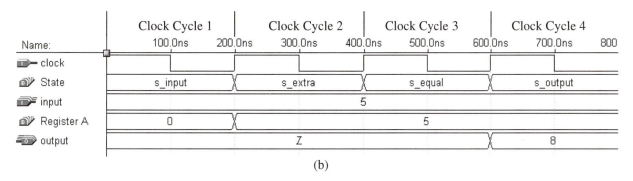

(b)

Figure 10.5   Example 10.3: (a) correct state diagram; (b) correct timing diagram.

not occur until shortly after time 200 ns. Hence, the test result is equal, and the FSM goes to state *s _equal* and outputs the correct value of 8.

Now that we have derived the correct state diagram, the actual construction of the control unit circuit is exactly the same as before. The next-state table is shown in Figure 10.6(b). Since there is a total of five states, three flip-flops are needed to encode them. For example, state *s_input* is encoded as $Q_2 Q_1 Q_0 = 000$, state *s_extra* is encoded as $Q_2 Q_1 Q_0 = 001$, and so on. The three remaining encodings (101, 110, and 111) are not used. In normal circumstances, the control unit should never get to one of the unused states. However, due to noises or glitches in the circuit, the FSM may end up in one of these unused states. Because of this, it is a good idea to set the

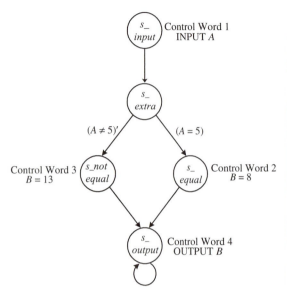

(a)

Current State	Next State	
$Q_2 Q_1 Q_0$	$Q_{2next} Q_{1next} Q_{0next}$	
	$(A = 5)'$	$(A = 5)$
s_input 000	s_extra 001	s_extra 001
s_extra 001	s_notequal 010	s_equal 011
s_notequal 010	s_output 100	s_output 100
s_equal 011	s_output 100	s_output 100
s_output 100	s_output 100	s_output 100
Unused 101	000	000
Unused 110	000	000
Unused 111	000	000

(b)

Current State	Next State	
$Q_2 Q_1 Q_0$	$D_2 D_1 D_0$	
	$(A = 5)'$	$(A = 5)$
000	001	001
001	010	011
010	100	100
011	100	100
100	100	100
101	000	000
110	000	000
111	000	000

(c)

$$D_2 = Q_2' Q_1 + Q_2 Q_1' Q_0'$$
$$D_1 = Q_2' Q_1' Q_0$$
$$D_0 = Q_2' Q_1' Q_0' + Q_2' Q_1' (A = 5)$$

(d)

Figure 10.6 Construction of the control unit for Example 10.3: (a) state diagram; (b) next-state table; (c) implementation table using D flip-flops; (d) excitation equations; (e) output table; (f) output equations for the four control signals; (g) circuit.
(continued on next page)

$Q_2Q_1Q_0$	Instruction	$ALoad$	$Muxsel$	$BLoad$	Out
000	INPUT A	1	×	0	0
001	No operation	0	×	0	0
010	$B = 8$	0	1	1	0
011	$B = 13$	0	0	1	0
100	OUTPUT B	0	×	0	1
101	No operation	0	×	0	0
110	No operation	0	×	0	0
111	No operation	0	×	0	0

(e)

$$ALoad = Q_2'Q_1'Q_0'$$
$$Muxsel = Q_2'Q_1Q_0'$$
$$BLoad = Q_2'Q_1$$
$$Out = Q_2Q_1'Q_0'$$

(f)

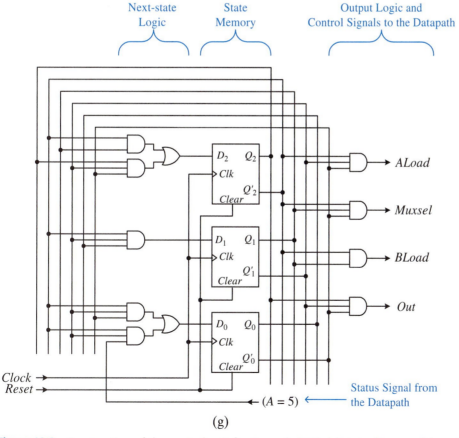

(g)

Figure 10.6   Construction of the control unit for Example 10.3: (a) state diagram; (b) next-state table; (c) implementation table using D flip-flops; (d) excitation equations; (e) output table; (f) output equations for the four control signals; (g) circuit.

next state for all of the unused states to the reset state. On the other hand, if you don't care what the next state is for these unused states, then don't-care values can be used, and thus may result in simpler excitation equations. In our next-state table, we have added three more rows for these three unused states. Their next state for all input conditions is the reset state, 000.

The corresponding implementation table (using D flip-flops) and the resulting excitation equations are shown in Figure 10.6(c) and (d), respectively. Note that the excitation equations listed in Figure 10.6(d) are not the only correct equations; there are other possible correct excitation equations. For instance, if don't-care values are used for the next states in the three unused states, then the excitation equations may be reduced further.

The output logic circuit for the FSM is based on the control word signals and the states in which the control words are assigned to. There is one output equation for each control word signal, and these equations are dependent only on the states of the FSM (i.e., the Q values of the flip-flops). To derive the truth tables for these output equations, simply take the control word table and replace all of the control word numbers with the actual encoding of the state (i.e., the Q values) in which that control word is assigned to. Having the output table, the output equations can be derived easily, as shown in Figure 10.6(e) and (f).

Once we have derived the excitation and output equations, we can draw the control unit circuit, as shown in Figure 10.6(g). The state memory simply consists of the three D flip-flops. Both the next-state logic circuit and the output logic circuit are combinational circuits, and are constructed from the excitation equations and output equations, respectively.

● ● ● ● ● ● ● ● ● ● ● ● ● ● ● ●
10.2 Generating Status Signals

In Chapter 9, we discussed how status signals are generated and provided several examples showing how it is done. In situations where a register value is to be tested, we have connected a comparator to the output of that register. However, there are situations when connecting a comparator to the output of a register will produce the wrong status signal because it is comparing with an incorrect value. In Examples 10.2 and 10.3, we saw that, if we read from a register that has just been written to in the same clock cycle, we will be reading the old value and not the new value. The solution we used in Example 10.3 was simply to insert an extra state to wait for the value to be written into the register and thus be made available at the output of the register.

What makes the problem even more difficult (but interesting) is that, during different clock cycles, different values may pass through the same point in the datapath. For example, the output of a 2-to-1 multiplexer can have two different values during two different times, because in one clock cycle, the select line for the multiplexer might be asserted, while in another clock cycle, the select line might be de-asserted—thus passing two different sources to the output. Another datapath component that can output different values at different times is the register file. By providing different read addresses to the register file, the register file will output values from different locations. Thus, we see that, in order to provide the correct value for the comparator to test, we have to look at not only the location in the datapath where the value is stored but also the clock cycle in which the value is needed by the comparator. This timing information is obtained from the state diagram. The state diagram tells us in what state a status signal is needed, when the FSM is in that state, what data manipulation instruction is being performed, and, therefore, what data is available at different points in the datapath. We will now illustrate this with two examples.

Example 10.4   **Generating a correct status signal by connecting the comparator to a different point in the datapath**

In Example 10.2, the state diagram shown in Figure 10.4(a) derived for the datapath shown in Figure 10.3(b) is wrong, because in state s_input (when the status signal for the comparison is needed), the comparator is not getting the correct input value. The disadvantage to the solution used in Example 10.3 is that an extra clock cycle is needed for the execution of the algorithm because of the extra state.

Another way to resolve this timing problem is to connect the comparator to a different point in the datapath instead of adding the extra state to the state diagram. The state diagram in Figure 10.4(a) shows that we need to use the input value right away in state s_input for the comparison ($A = 5$). The datapath shown in Figure 10.3(b) has the comparator connected to the output of register A. Because of this, we need an extra clock cycle to wait until the input value has been written into the register before it is available to be tested.

Instead of connecting the comparator to the output of the register, we can connect the comparator directly to the primary input (i.e., to the input of the register), as shown in Figure 10.7. This way, we do not have to wait for the input value to be written

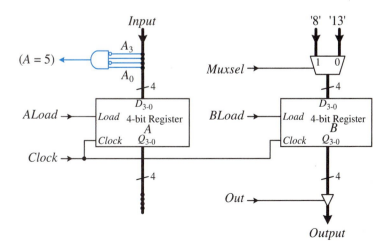

Figure 10.7
Generating the correct status signal for Example 10.4.

into the register first, but instead, the input value will be available immediately in state s_input. As a result, the comparator will have the correct value for performing the test in the right clock cycle, and we will not have to add an extra state to the state diagram.

Example 10.5   **Summation of n down to 1**

We will now derive the state diagram and generate the correct status signal for the general datapath from Example 9.10 for summing the numbers from n down to 1, where n is an input number. The algorithm, general datapath, and control words from Example 9.10 are repeated here in Figure 10.8. Figure 10.8(c) shows the five control words for controlling the general datapath shown in Figure 10.8(b) for performing the algorithm shown in Figure 10.8(a).

Figure 10.9(a) shows a first attempt to derive the state diagram for the algorithm in Figure 10.8(a). The five data manipulation instructions are assigned to five states. On reset, the FSM starts from state s_0, where it initializes sum to 0 and waits for the $Start$ signal. The $Start$ signal is for telling the control unit when the data input is ready and to begin the execution of the algorithm. (The $Start$ signal is similar in function to the $Enter$ key found on personal computer keyboards.) The remaining edges in the state diagram follow the execution sequence in the algorithm. State s_1 executes control word 2 to input the number for n. State s_2 executes control word 3 to sum n. State s_3 executes control word 4 to decrement n. Finally, state s_4 executes control word 5 to output the value for sum.

In state s_1, we input a value for n and immediately test for the condition $(n \neq 0)$. Similarly, in state s_3, we perform the instruction $n = n - 1$ and immediately test for the condition $(n \neq 0)$. However, we remember from Example 10.3 that we cannot test for a condition that involves a value that is being written into a register in the same clock cycle. If we try to solve this timing problem like we did in Example 10.3, we would get the state diagram shown in Figure 10.9(b), where an extra wait state, s_wait, is added. On the other hand, Example 10.4 shows that sometimes it is not necessary to add an extra wait state if we can find a point in the datapath to connect the comparator so that the correct status signal can be generated.

In terms of optimization, the state diagram in Figure 10.9(a) would be a better choice if the correct status signal can be generated because it requires one less state. So the problem is to find a location in the datapath to connect the comparator so that it will output the correct status signal for the test $(n \neq 0)$ during the clock cycle when it is needed. The state diagram in Figure 10.9(a) shows that the status signal $(n \neq 0)$ is needed in states s_1 and s_3. Therefore, we need to find a point in the datapath such that the correct value for n is available during the times when states s_1 and s_3 are being executed.

In state s_3, the instruction $n = n - 1$ is being executed. In order for the datapath to execute $n = n - 1$, the value for n must be present at the A operand input of the ALU (point C in Figure 10.8(b)), and the result for the decrement $n - 1$ will be available at the output of the ALU (point D in Figure 10.8(b)) during the same clock cycle. According to the state diagram in Figure 10.9(a), we want to test for n *after* the decrement. If we connect the comparator to the A operand input of the ALU

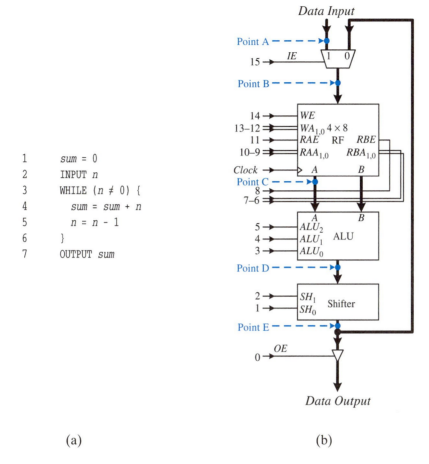

```
1       sum = 0
2       INPUT n
3       WHILE (n ≠ 0) {
4         sum = sum + n
5         n = n - 1
6       }
7       OUTPUT sum
```

(a)

(b)

Control Word	Instruction	IE 15	WE 14	WA$_{1,0}$ 13–12	RAE 11	RAA$_{1,0}$ 10–9	RBE 8	RBA$_{1,0}$ 7–6	ALU$_{2,1,0}$ 5–3	SH$_{1,0}$ 2–1	OE 0
1	sum = 0	0	1	00	1	00	1	00	101 (subtract)	00	0
2	INPUT n	1	1	01	0	××	0	××	×××	××	0
3	sum = sum + n	0	1	00	1	00	1	01	100 (add)	00	0
4	n = n − 1	0	1	01	1	01	0	××	111 (decrement)	00	0
5	OUTPUT sum	×	0	××	1	00	0	××	000 (pass)	00	1

(c)

Figure 10.8 The summation problem for Example 10.5: (a) algorithm; (b) general datapath; (c) control words.

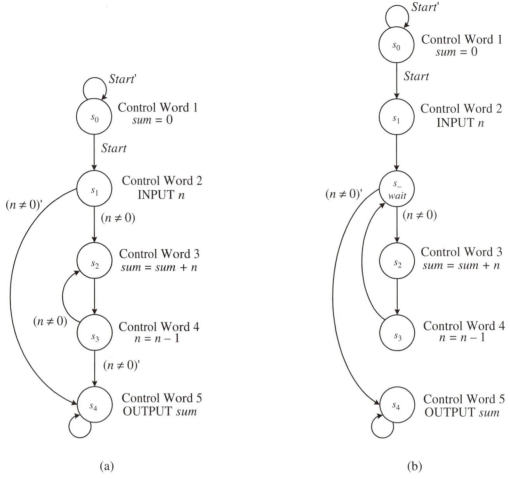

Figure 10.9 Two possible state diagrams for solving the summation problem for Example 10.5: (a) without an extra wait state; (b) with an extra wait state.

at point C, we would be comparing the value of n *before* the decrement during this clock cycle. Furthermore, the value for n is not even available in the next clock cycle at point C. The reason is that either state s_2 or s_4 will be executed in the next clock cycle, and for both of these states, the value for sum is needed. So in the next clock cycle, the value for sum, and not n, is available at point C. Since the result of the decrement is available at point D in the current clock cycle, the comparator should be connected to the output of the ALU at point D.

The problem now is that we also need n in state s_1. In state s_1, the instruction INPUT n is executed, and the value for n is being stored in the register file. Looking at Figure 10.8(b), we see that, in the clock cycle when the datapath executes the instruction INPUT n, n is available only at the primary data input of the datapath at point A and at the output of the multiplexer at point B. Since the writing of n to the register file occurs at the next rising clock edge, n is not going to be available at the output of the register file in that same clock cycle, let alone at the output of the ALU. So in order for the status signal ($n \neq 0$) to be correct in state s_1, we need to connect the comparator to either the primary data input at point A or to the output of the multiplexer at point B.

For state s_3, we need to connect the comparator to point D, but for state s_1, we need to connect the comparator to either point A or point B. However, in order to generate the correct status signal for both states, the comparator must be connected to a point that is correct for both cases. Notice that the execution of the instruction $n = n - 1$ in state s_3 requires the output of the ALU at point D be routed back to the output of the multiplexer at point B within the same clock cycle, so that the decremented value of n can be written into the register file at the next rising clock edge. This is possible because both the shifter and the multiplexer are combinational circuits. So, this new decremented value of n that we want for the comparison in state s_3 is not only available at point D, but also available at points E and B. We can therefore conclude that connecting the comparator to the output of the multiplexer at point B will generate the correct status signal for the comparison ($n \neq 0$) in both states, s_1 and s_3, as shown in Figure 10.10.

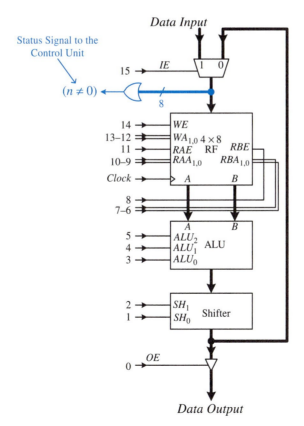

Figure 10.10
Generating the status signal ($n \neq 0$) in the datapath for Example 10.5.

One last point to note is that, since the comparator is a combinational circuit, it constantly is outputting a value. When the FSM is in a state other than s_1 or s_3, the comparator would be comparing another value with 0. For instance, in state s_2, the ALU is evaluating $sum + n$, so the comparator would be comparing the result of $sum + n$ with 0. However, this will not affect the correct operation of the control unit, because in state s_2, the next-state logic does not use this status signal to determine what the next state is. Therefore, even though the comparator is generating an incorrect status signal in state s_2, the control unit will not use it. When the status signal from the output of this comparator is needed in state s_1 or s_3, it will be the result of a correct comparison.

Having derived the correct state diagram shown in Figure 10.9(a) and the correct status signal from the datapath shown in Figure 10.10, we can now finish the construction of the control unit circuit, as shown in Figure 10.11. From the state diagram, we can derive the next-state table, as shown in Figure 10.11(b). Three flip-flops, Q_2, Q_1,

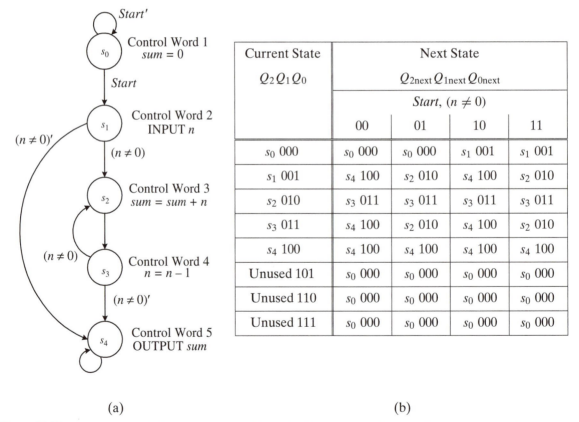

Current State	Next State			
$Q_2 Q_1 Q_0$	$Q_{2next} Q_{1next} Q_{0next}$			
	Start, $(n \neq 0)$			
	00	01	10	11
s_0 000	s_0 000	s_0 000	s_1 001	s_1 001
s_1 001	s_4 100	s_2 010	s_4 100	s_2 010
s_2 010	s_3 011	s_3 011	s_3 011	s_3 011
s_3 011	s_4 100	s_2 010	s_4 100	s_2 010
s_4 100	s_4 100	s_4 100	s_4 100	s_4 100
Unused 101	s_0 000	s_0 000	s_0 000	s_0 000
Unused 110	s_0 000	s_0 000	s_0 000	s_0 000
Unused 111	s_0 000	s_0 000	s_0 000	s_0 000

(a)   (b)

Figure 10.11 Construction of the control unit for Example 10.5: (a) state diagram; (b) next-state table; (c) implementation table; (d) K-maps and excitation equations; (e) output table; (f) output equations; (g) circuit. *(continued on next page)*

Current State	Implementation			
$Q_2Q_1Q_0$	$D_2D_1D_0$			
	Start, $(n \neq 0)$			
	00	01	10	11
000	000	000	001	001
001	100	010	100	010
010	011	011	011	011
011	100	010	100	010
100	100	100	100	100
101	000	000	000	000
110	000	000	000	000
111	000	000	000	000

(c)

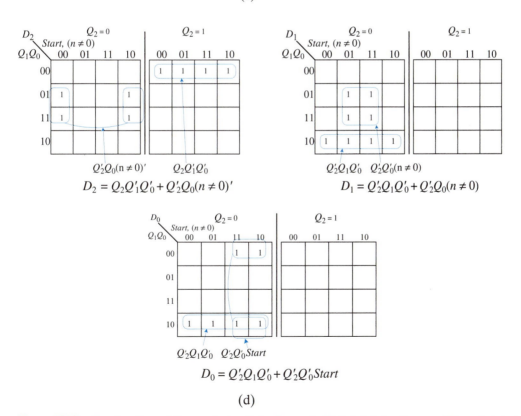

$$D_2 = Q_2Q_1'Q_0' + Q_2'Q_0(n \neq 0)'$$

$$D_1 = Q_2'Q_1Q_0' + Q_2'Q_0(n \neq 0)$$

$$D_0 = Q_2'Q_1Q_0' + Q_2'Q_0'Start$$

(d)

Figure 10.11   Construction of the control unit for Example 10.5: (a) state diagram; (b) next-state table; (c) implementation table; (d) K-maps and excitation equations; (e) output table; (f) output equations; (g) circuit. *(continued on next page)*

State $Q_2Q_1Q_0$	IE 15	WE 14	WA₁ 13	WA₀ 12	RAE 11	RAA₁ 10	RAA₀ 9	RBE 8	RBA₁ 7	RBA₀ 6	ALU₂ 5	ALU₁ 4	ALU₀ 3	SH₁ 2	SH₀ 1	OE 0
000	0	1	0	0	1	0	0	1	0	0	1	0	1	0	0	0
001	1	1	0	1	0	×	×	0	×	×	×	×	×	×	×	0
010	0	1	0	0	1	0	0	1	0	1	1	0	0	0	0	0
011	0	1	0	1	1	0	1	0	×	×	1	1	1	0	0	0
100	×	0	×	×	1	0	0	0	×	×	0	0	0	0	0	1
101	0	0	0	0	0	0	0	0	0	0	0	0	0	0	0	0
110	0	0	0	0	0	0	0	0	0	0	0	0	0	0	0	0
111	0	0	0	0	0	0	0	0	0	0	0	0	0	0	0	0

(e)

$$IE = Q_2'Q_1'Q_0$$
$$WE = Q_2'$$
$$WA_1 = 0$$
$$WA_0 = Q_2'Q_0$$
$$RAE = Q_2'Q_1 + Q_1'Q_0'$$
$$RAA_1 = 0$$
$$RAA_0 = Q_2'Q_1Q_0$$
$$RBE = Q_2'Q_0'$$

$$RBA_1 = 0$$
$$RBA_0 = Q_2'Q_1Q_0'$$
$$ALU_2 = Q_2'Q_0' + Q_2'Q_1$$
$$ALU_1 = Q_2'Q_1Q_0$$
$$ALU_0 = Q_2'Q_1'Q_0' + Q_2'Q_1Q_0$$
$$SH_1 = 0$$
$$SH_0 = 0$$
$$OE = Q_2Q_1'Q_0'$$

(f)

Figure 10.11   Construction of the control unit for Example 10.5: (a) state diagram; (b) next-state table; (c) implementation table; (d) K-maps and excitation equations; (e) output table; (f) output equations; (g) circuit. *(continued on next page)*

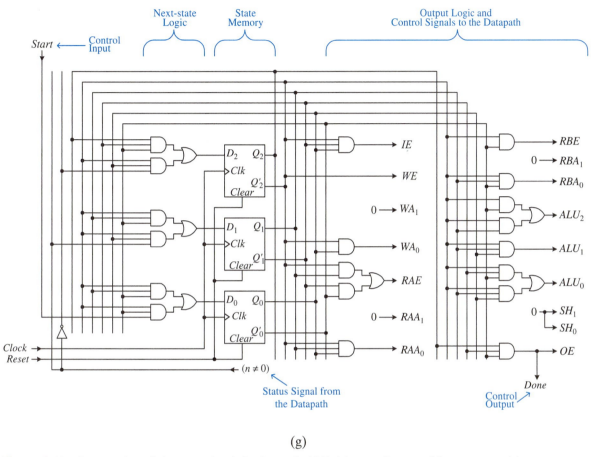

(g)

Figure 10.11   Construction of the control unit for Example 10.5: (a) state diagram; (b) next-state table; (c) implementation table; (d) K-maps and excitation equations; (e) output table; (f) output equations; (g) circuit.

and Q_0, are needed to encode the five states. In addition to the current states listed down the rows of the table, the next state of the FSM also is dependent on the control input signal *Start* and the status signal $(n \neq 0)$, from which we get the four columns. The labels for the four columns are the four combinations of the two input signals *Start* and $(n \neq 0)$. For example, the column labeled 10 means that the input signal *Start* is true (i.e., *Start* = 1) and the status signal $(n \neq 0)$ is false (i.e., $(n \neq 0) = 0$). Figure 10.11(c) shows the corresponding implementation table using *D* flip-flops. The derivation of the excitation equations (Figure 10.11(d)), output equations (Figure 10.11(f)), and finally, the complete control unit circuit (Figure 10.11(g)) follows the same procedure as for synthesizing a general FSM. The *Done* control output signal is to notify the external world that the execution of the algorithm has been completed and that the data at the *Data Output* is valid.

● ● ● ● ● ● ● ● ● ● ● ● ●● ●●●

10.3 Stand-Alone Controllers

So far in the discussion, the control unit is used as a controller for the datapath. However, the control unit can be independent of the datapath and used as a stand-alone controller for controlling an external device. There are many electronic devices that use a stand-alone control unit for controlling them (for example, the controllers for a microwave oven, a soda dispensing machine, a traffic light signal, and a pacemaker, to name just a few). All of these devices do not require any calculations or data manipulations, so they do not require the use of a datapath. The controllers in them are used just to turn on or off a switch or a light, or whatever external devices they are made to control.

For a traffic light controller, there are sets of three lights (green, yellow, and red) to turn on or off and sensors in the road to see if there are cars in one of the four directions. For a soda dispensing machine, there will be switches to sense the amount of money put in and the selection made. From these inputs, the controller will determine what switch to turn on to allow the selected soda to fall out.

We will now illustrate the construction of stand-alone controllers with several examples.

10.3.1 Rotating Lights

Example 10.6   **Stand-alone controller for rotating lights**

For a simple, interesting example, let us implement a controller that will make the six peripheral segments in a 7-segment LED light move around in a clockwise or counterclockwise direction, depending on the input of the switch W.

Since there are six light patterns, one for each LED being on and the rest of them off, we can use six states and assign one light pattern to one state. Each state will go to the next state in a sequential clockwise direction if the switch W is pressed and to the previous state in a sequential counterclockwise direction if W is not pressed. The resulting state diagram is similar to the one for a modulo-6 up-down counter and is shown in Figure 10.12(a). The six output signals (a, b, c, d, e, and f) correspond to the six peripheral segments in a 7-segment LED. A 1 to the segment will turn on that LED, while a 0 will turn it off.

The next-state table derived from the state diagram and the corresponding implementation table, using D flip-flops, are shown in Figure 10.12(b). The resulting excitation equations (as derived from the implementation table) are shown in Figure 10.12(c). The output table and the resulting output equations are shown in Figure 10.12(d) and (e), respectively. The final controller circuit is shown in Figure 10.12(f). To see the lights move, the six output signals are connected to the six segments of the 7-segment LED, and a push-button switch is connected to the W input, as shown in Figure 10.12(g).

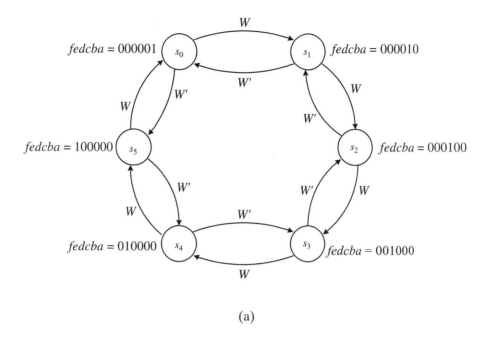

(a)

Current State	Next State (Implementation)	
$Q_2 Q_1 Q_0$	$Q_{2next} Q_{1next} Q_{0next}$ $(D_2 D_1 D_0)$	
	W'	W
000	101	001
001	000	010
010	001	011
011	010	100
100	011	101
101	100	000

(b)

Figure 10.12 Controller for Example 10.6: (a) state diagram; (b) next-state (implementation) table; (c) K-maps and excitation equations; (d) output table; (e) output equations; (f) controller circuit; (g) interface circuit. *(continued on next page)*

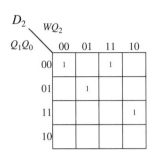

$D_2 = W'Q_2'Q_1'Q_0' + WQ_2Q_1'Q_0' + W'Q_2Q_1'Q_0 + WQ_2'Q_1Q_0$

Current State	Output
$Q_2Q_1Q_0$	*fedcba*
000	000001
001	000010
010	000100
011	001000
100	010000
101	100000

(d)

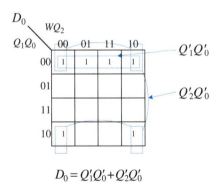

$D_1 = W'Q_2Q_1'Q_0' + WQ_2Q_1'Q_0 + W'Q_2'Q_1Q_0 + WQ_2'Q_1Q_0'$

$$f = Q_2Q_1'Q_0$$
$$e = Q_2Q_1'Q_0'$$
$$d = Q_2'Q_1Q_0$$
$$c = Q_2'Q_1Q_0'$$
$$b = Q_2'Q_1'Q_0$$
$$a = Q_2'Q_1'Q_0'$$

(e)

$D_0 = Q_1'Q_0' + Q_2'Q_0'$

(c)

Figure 10.12 Controller for Example 10.6: (a) state diagram; (b) next-state (implementation) table; (c) K-maps and excitation equations; (d) output table; (e) output equations; (f) controller circuit; (g) interface circuit. *(continued on next page)*

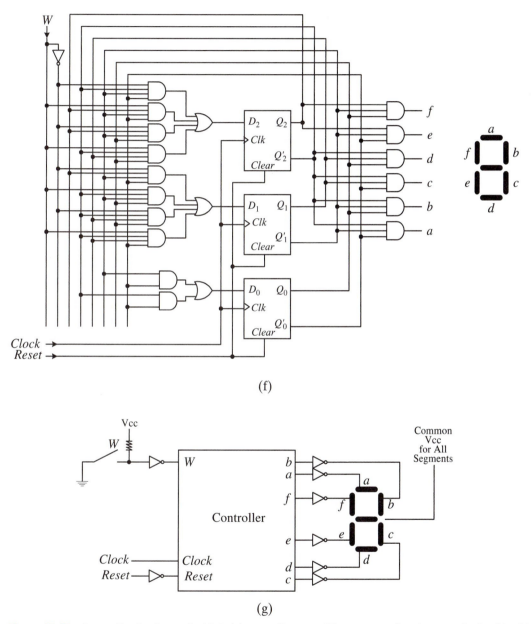

(f)

(g)

Figure 10.12   Controller for Example 10.6: (a) state diagram; (b) next-state (implementation) table; (c) K-maps and excitation equations; (d) output table; (e) output equations; (f) controller circuit; (g) interface circuit.

10.3.2 PS/2 Keyboard Controller

Example 10.7

Controller for a PS/2 keyboard

In this next example, we will design and implement a more realistic and practical controller, and that is for the PS/2 keyboard. The UP2 development board has a 6-pin mini-DIN connector for connecting either a PS/2 keyboard or a PS/2 mouse to it. We will design and implement the controller for the keyboard on the FLEX chip. The operation of the keyboard and the controller is as follows.

The communication between the keyboard and the controller (which will be on the FLEX chip) uses two signals, *KeyboardClock* and *KeyboardData*. When there is no activity, that is, when there is no key press on the keyboard, both *KeyboardClock* and *KeyboardData* are at a 1. When a key is pressed (or released), the keyboard sends a unique code for that key to the controller serially over the *KeyboardData* line. The serial data on the *KeyboardData* line is synchronized between the keyboard and the controller by clock pulses that the keyboard sends over the *KeyboardClock* line.

The data for each key that is sent over the *KeyboardData* line consists of eleven bits. These eleven bits are: a 0 for the start bit, eight data bits for the key code starting with the least significant bit to the most significant bit, an odd parity bit, and lastly, a 1 for the stop bit. Figure 10.13 lists some of the key codes generated by the keyboard when the corresponding key is pressed. When a key is released, a different code is generated. The odd parity bit is set such that the total number of 1 bits in the eight data bits plus the parity bit is an odd number.

Key	Key Code	Key	Key Code	Key	Key Code	Key	Key Code
1	16	A	1C	K	42	U	3C
2	1E	B	32	L	4B	V	2A
3	26	C	21	M	3A	W	1D
4	25	D	23	N	31	X	22
5	2E	E	24	O	44	Y	35
6	36	F	2B	P	4D	Z	1A
7	3D	G	34	Q	15	Esc	76
8	3E	H	33	R	2B	BS	66
9	46	I	43	S	1B	CR	5A
0	45	J	3B	T	2C	Ctrl	14

Figure 10.13 A partial list of key codes generated by the keyboard.

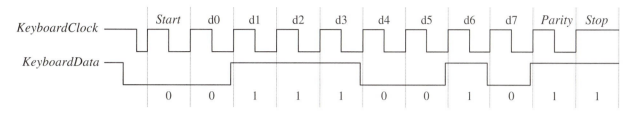

Figure 10.14   Sample timing diagram for the data transmission of the key code 4E.

Figure 10.14 shows a sample timing diagram for the data transmission of the key code 4E (01001110 in binary) for the hyphen key. Starting from the inactive state, where both the *KeyboardData* and *KeyboardClock* lines are at a 1, the transmission begins by setting the *KeyboardData* line to a 0 for the start bit. The keyboard then sends out the data and parity bit on the *KeyboardData* line at a rate of one bit per clock cycle on the *KeyboardClock* line. The clock pulses on the *KeyboardClock* line are generated by the keyboard. The parity bit for the key code 4E is a 1, since the eight data bits consist of an even number of 1 bits, therefore, to make the parity odd, the parity bit must be a 1.

The state diagram for our keyboard controller shown in Figure 10.15(a) is derived by following the timing diagram shown in Figure 10.14. In each of the eight data states, d0, d1, ..., d7, we will get one corresponding data bit from the *KeyboardData* input line. For example, suppose we use an 8-bit register named *Keycode* for storing the eight data bits. Then in state d0, we will assign *KeyboardData* to $Keycode_0$ (i.e., the 0th bit of *Keycode*), and in state d1, we will assign *KeyboardData* to $Keycode_1$, and so on for all eight data bits. This is possible because the transition of the FSM from one state to another is synchronized by the keyboard clock signal *KeyboardClock*. For simplicity, we will not check for the start bit, parity bit, nor the stop bit.

The next-state table and the implementation table using four D flip-flops to encode the eleven states are both shown in Figure 10.15(b). The excitation equations derived from the implementation table are shown in Figure 10.15(c).

This controller circuit actually does not control the keyboard because it does not generate control signals for the operation of the keyboard. Instead, it receives the serial data signals from the keyboard, and packages it into data bytes. The output of this controller is simply the data bytes, which represents the key code of the keys being pressed on the keyboard. The output table is shown in Figure 10.15(d) and the corresponding output equations in (e). In state d0, the bit on the *KeyboardData* line is loaded into bit 0 of the *Keycode* register; in state d1, the bit on the *KeyboardData* line is loaded into bit 1 of the *Keycode* register; and so on. Each bit of the *Keycode* register must, therefore, be able to load in the *KeyboardData* independently, and each load enable line is asserted by the corresponding state encoding. Hence, each of the output equations shown in Figure 10.15(e) is not implemented simply as a 5-input AND gate. Instead, a D flip-flop with enable is used for each output signal. The *D* input to the flip-flop is connected to the *KeyboardData* line. The load enable, *E*, on the flip-flop is asserted by ANDing the four *Q* values for that state.

Using the excitation equations for the next-state circuit, four D flip-flops for the state memory, and the output equations for the output circuit, we obtain the complete controller circuit, as shown in Figure 10.15(f). The implementation of the controller circuit using the FLEX chip is shown in Figure 10.15(g) and is available on the accompanying CD-ROM in the file UP2FLEX.GDF.

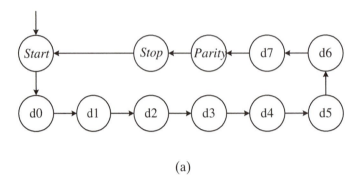

(a)

Current State		Next State (Implementation)
$Q_3 Q_2 Q_1 Q_0$		$Q_{3next} Q_{2next} Q_{1next} Q_{0next}$ $(D_3 D_2 D_1 D_0)$
0000	*Start*	0001
0001	d0	0010
0010	d1	0011
0011	d2	0100
0100	d3	0101
0101	d4	0110
0110	d5	0111
0111	d6	1000
1000	d7	1001
1001	*Parity*	1010
1010	*Stop*	0000

(b)

Figure 10.15 Controller for Example 10.7: (a) state diagram; (b) next-state (implementation) table: (c) K-maps and excitation equations; (d) output table; (e) output equations; (f) controller circuit; (g) interface circuit. *(continued on next page)*

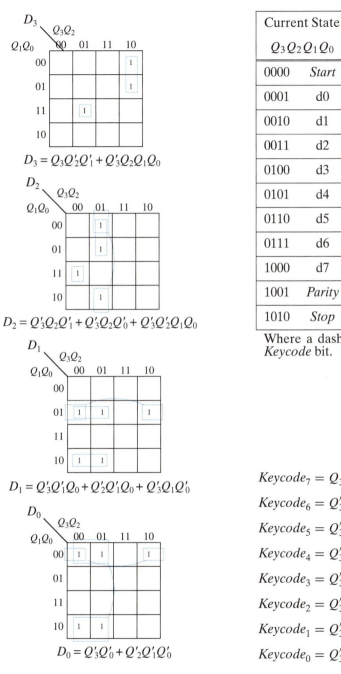

$D_3 = Q_3 Q_2' Q_1' + Q_3' Q_2 Q_1 Q_0$

$D_2 = Q_3' Q_2 Q_1' + Q_3' Q_2 Q_0' + Q_3' Q_2' Q_1 Q_0$

$D_1 = Q_3' Q_1' Q_0 + Q_2' Q_1' Q_0 + Q_3' Q_1 Q_0'$

$D_0 = Q_3' Q_0' + Q_2' Q_1' Q_0'$

(c)

Current State		Output
$Q_3 Q_2 Q_1 Q_0$		$Keycode_{7-0}$
0000	*Start*	
0001	d0	$------ KeyboardData$
0010	d1	$------ KeyboardData -$
0011	d2	$----- KeyboardData --$
0100	d3	$---- KeyboardData ---$
0101	d4	$--- KeyboardData ----$
0110	d5	$-- KeyboardData -----$
0111	d6	$- KeyboardData ------$
1000	d7	$KeyboardData -------$
1001	*Parity*	
1010	*Stop*	

Where a dash (–) means no change to that *Keycode* bit.

(d)

$Keycode_7 = Q_3 Q_2' Q_1' Q_0' KeyboardData$

$Keycode_6 = Q_3' Q_2 Q_1 Q_0 KeyboardData$

$Keycode_5 = Q_3' Q_2 Q_1 Q_0' KeyboardData$

$Keycode_4 = Q_3' Q_2 Q_1' Q_0 KeyboardData$

$Keycode_3 = Q_3' Q_2 Q_1' Q_0' KeyboardData$

$Keycode_2 = Q_3' Q_2' Q_1 Q_0 KeyboardData$

$Keycode_1 = Q_3' Q_2' Q_1 Q_0' KeyboardData$

$Keycode_0 = Q_3' Q_2' Q_1' Q_0 KeyboardData$

(e)

Figure 10.15 Controller for Example 10.7: (a) state diagram; (b) next-state (implementation) table: (c) K-maps and excitation equations; (d) output table; (e) output equations; (f) controller circuit; (g) interface circuit. *(continued on next page)*

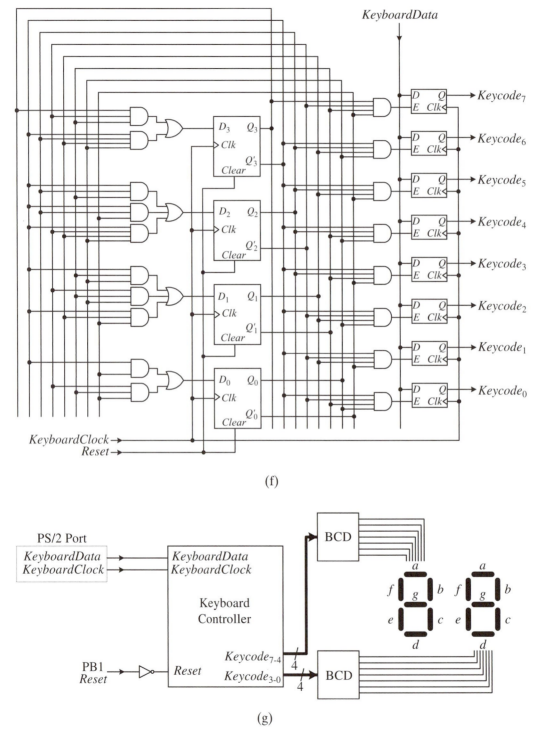

(f)

(g)

Figure 10.15 Controller for Example 10.7: (a) state diagram; (b) next-state (implementation) table: (c) K-maps and excitation equations; (d) output table; (e) output equations; (f) controller circuit; (g) interface circuit.

In practice, controllers usually are not designed manually like we have done. The purpose of this exercise is only to provide an understanding of how controllers are designed. Section 10.5 shows how controllers can be synthesized automatically using behavioral VHDL code.

10.3.3   VGA Monitor Controller

Most people will think that it is way beyond the beginner's level to design a VGA monitor controller. However, to the contrary, a simple VGA monitor controller can be constructed using a couple of standard sequential components. In Example 10.8, we will design and implement a VGA monitor controller using an FSM for the control unit. However, it turns out that the VGA monitor controller is so simple that it can be constructed without the need of a FSM. In Example 10.9, we will show how the VGA controller can be implemented without the FSM. The UP2 development board has a 15-pin Dsubconnector for connecting a standard VGA monitor to it. The signals from the Dsubconnector are connected directly to the FLEX FPGA chip on the board, so we will implement our controller circuits on the FLEX chip. In order to design the VGA controller, we need to first understand how the VGA monitor works.

The monitor screen for a standard VGA format contains 640 columns × 480 rows of picture elements called **pixels**, as shown in Figure 10.16. An image is displayed on the screen by turning on or off individual pixels. Although turning on just one pixel doesn't represent much, when many pixels are turned on, the combined pixels portray an image. The monitor continuously scans through the entire screen turning on or off one pixel at a time at a very fast speed. Although only one pixel is turned on at any one time, but since the monitor is scanning through the screen so fast, you get the impression that all of the pixels are on at the same time.

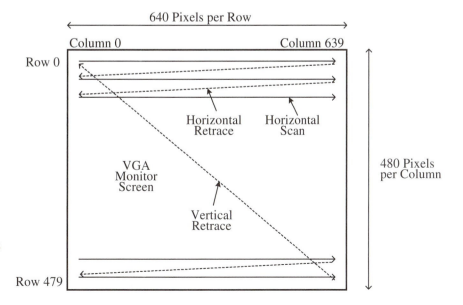

Figure 10.16
VGA monitor with 640 columns × 480 rows. The scan starts from row 0, column 0 and moves to the right and down until row 479, column 639.

Figure 10.16 shows that the scanning starts from row 0, column 0 at the top left corner and moves to the right until it reaches the last column in the row. When the scan reaches the end of a row, it retraces to the beginning of the next row. When the scan reaches the last pixel at the bottom-right corner of the screen, it retraces back to the top-left corner of the screen and repeats the scanning process again. In order to reduce flicker on the screen, the entire screen must be scanned 60 times per second or higher. During the horizontal and the vertical retraces, all of the pixels are turned off.

The VGA monitor is controlled by five signals: red, green, blue, horizontal synchronization, and vertical synchronization. The three color signals, referred to collectively as the RGB signal, are used to control the color of a pixel at a given location on the screen. These three color signals on the UP2 board are connected such that they can individually be turned on or off, hence each pixel can display only one of eight colors. In order to produce more colors, each analog color signal must be supplied with a voltage between 0 to 0.7 volts for varying the intensities of the colors. The horizontal and vertical synchronization signals are used to control the timing of the scan rate. The horizontal synchronization signal determines the time to scan a row, while the vertical synchronization signal determines the time to scan the entire screen. By manipulating these five signals, images are formed on the monitor screen.

The horizontal and vertical synchronization signals timing diagram is shown in Figure 10.17. When inactive, both synchronization signals are at a 1. The start of a row scan begins with the horizontal synchronization signal going low for 3.77 μsec, as shown in region B in Figure 10.17. This is followed by a 1.79 μsec high on the

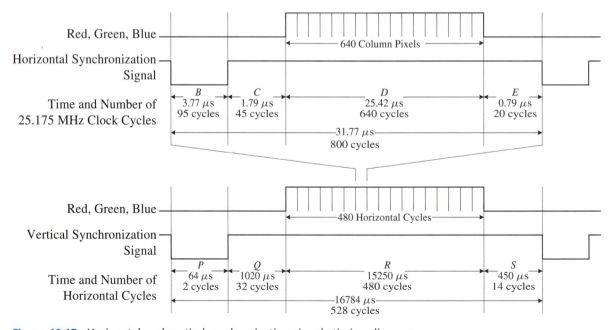

Figure 10.17 Horizontal and vertical synchronization signals timing diagram.

signal, as shown in region C. Next, the data for the three color signals are sent (one pixel at a time) for the 640 columns, as shown in region D for 25.42 μsec. Finally, after the last column pixel, there is another 0.79 μsec of inactivity on the RGB signal lines, as shown in region E, before the horizontal synchronization signal goes low again for the next row scan. The total time to complete one row scan is 31.77 μsec.

The timing for the vertical synchronization signal is analogous to the horizontal synchronization signal. The 64 μsec active-low vertical synchronization signal resets the scan to the top-left corner of the screen, as shown in region P, followed by a 1020 μsec high on the signal as shown in region Q. Next, there are the 480 row scans of 31.77 μsec each, giving a total of 15250 μsec, as shown in region R. Finally, after the last row scan, there is another 450 μsec, as shown in region S, before the vertical synchronization signal goes low again to start another complete screen scan starting at the top-left corner. The total time to complete one complete scan of the screen is 16784 μsec.

In order to get the monitor to operate properly, we simply have to get the horizontal and vertical synchronization signals timing correct, and then send out the RGB data for each pixel at the right column and row position. It turns out that it is fairly simple to get the correct timing for the two synchronization signals. The built-in clock crystal on the UP2 board has a frequency of 25.175 MHz. The clock period is then $1/25.175 \times 10^6$, which is about 0.0397 μsec per clock cycle. For region B in the horizontal synchronization signal, we need 3.77 μsec, which is approximately $3.77/0.0397 = 95$ clock cycles. For region C, we need 1.79 μsec, which is approximately 45 clock cycles. Similarly, we need 640 clock cycles for region D for the 640 columns of pixels and 20 clock cycles for region E. The total number of clock cycles needed for each row scan is, therefore, 800 clock cycles. Notice that with a 25.175 MHz clock, region D requires exactly 640 cycles, giving us the 640 columns per row. Hence, a different clock speed will produce a different screen resolution.

The vertical timings are multiples of the horizontal cycles. For example, region P is 64 μsec, which is approximately two horizontal cycles (2×31.77). The calculation for region R is 15250 μsec/31.77 μsec $= 480$. Of course, it has to be exactly 480 times, since we need to have 480 rows per screen.

In addition to generating the correct horizontal and vertical synchronization signals, the circuit needs to keep track of the current column within the D region and the current row within the R region of the scan in order to know when to turn on or off a specific pixel. To make a particular pixel green for example, you need to test the values of the column and row counts. If they are equal to the location of the pixel that you want to turn on, then you assert the green signal, and that pixel will be green.

Example 10.8	**A VGA controller—version 1**

To get the horizontal and vertical synchronization timing correct, we can design an FSM with 800 states running at a clock speed of 25.175 MHz. For the first 95 states, we will output a 0 for the horizontal synchronization signal H_Sync. For the next $45 + 640 + 20 = 705$ states, we will output a 1 for H_Sync. The problem with this, however, is that it is difficult to manually derive the circuit for an 800-state FSM. A simple solution around this difficulty is to use just two states: one for when H_Sync

is a 0 in region B and one for when it is a 1 in regions C, D, and E. We will then use a counter that runs at the same clock speed as the FSM to keep count of how many times we have been in a state. For the first state, we will stay there for 95 counts before going to the next state, and for the second state, we will stay there for 705 counts before going back to the first state. In the first state, we will output a 0 for H_Sync, and in the second state, we will output a 1 for H_Sync.

To help keep track of when the three color signals can be enabled, we will generate an additional H_Data_on signal the same way we generated the H_Sync signal. The H_Data_on signal is asserted in region D, and de-asserted in regions B, C, and E. Thus, for 640 counts of repeating in one state for region D, we will set H_Data_on to a 1 and to a 0 in a second state for the remaining 160 counts for regions B, C, and E.

Combining the two states for H_Sync and two states for H_Data_on together results in the final state diagram for the horizontal timing, as shown in Figure 10.18(a). This FSM has four states corresponding to the four regions, B, C, D, and E. The counter initially is set to 0 and increments by 1 at every clock cycle. In state H_B for region B, the FSM outputs a 0 for both H_Data_on and H_Sync. The FSM will stay in state H_B for 95 counts. The condition ($H\_cnt = B$) checks to see whether the counter is equal to B, where B is equal to 95. When the count is equal to 95, the FSM goes to state H_C, which corresponds to region C. In state H_C, the FSM outputs a 0 for H_Data_on and a 1 for H_Sync for 45 counts (i.e., until H_cnt is $B + C = 95 + 45 = 140$). When H_cnt reaches 140, the FSM goes to state H_D, and outputs a 1 for both H_Data_on and H_Sync. When H_cnt reaches $B + C + D = 95 + 45 + 640 = 780$, the FSM goes to state H_E, and outputs a 0 for H_Data_on and a 1 for H_Sync. The FSM stays in state H_E for 20 more counts until $H\_cnt = B + C + D + E = 95 + 45 + 640 + 20 = 800$, and then it goes back to state H_B. When the FSM goes back to state H_B, H_cnt is reset back to 0, and the process starts all over again for the next row scan.

The vertical synchronization timing is analogous to the horizontal synchronization timing, so we can do the same thing using a second counter and a second FSM. This second vertical FSM is identical to the horizontal FSM. The only difference is in the timing. Looking at the times for each region in the vertical synchronization signal in Figure 10.17, we see that the 64 μsec for region P is approximately two times the total horizontal scan time of 31.77 μsec each. The 1020 μsec for region Q is approximately 32 horizontal scan time ($1020/31.77 \approx 32$). For region R, it is 480 horizontal cycles, and for region S, it is approximately 14 horizontal cycles. Hence, the clock for both the vertical counter and the vertical FSM can be derived from the horizontal counter. The vertical clock ticks once for every 800 counts of the horizontal clock.

The next-state tables for the horizontal FSM and the vertical FSM are shown in Figure 10.18(c) and (d), respectively. Notice that besides the four count conditions, the two tables are the same. Hence, we can use the same circuit for both FSMs. The only difference is that their status signal inputs for the four count conditions come from different counter comparators. The output table is also the same for both FSMs and is shown in Figure 10.18(e). There are only two output signals to be generated: H_Data_on and H_Sync for the horizontal FSM, and V_Data_on and V_Sync for the vertical FSM. Finally, the FSM circuit is shown in Figure 10.18(f).

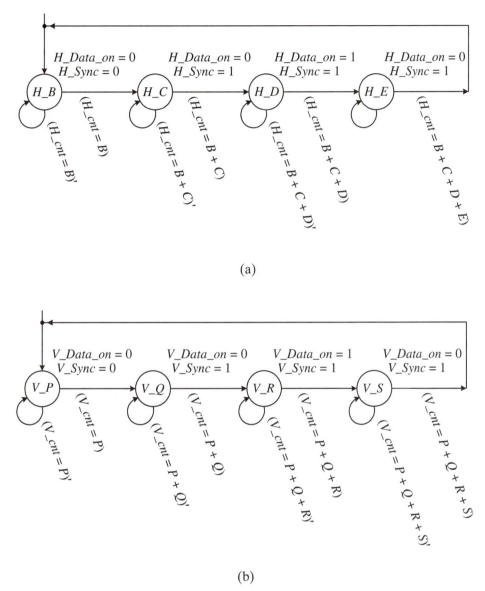

(a)

(b)

Figure 10.18 Controller for the VGA monitor: (a) state diagram for horizontal synchronization; (b) state diagram for vertical synchronization; (c) next-state table for horizontal synchronization; (d) next-state table for vertical synchronization; (e) output table; (f) FSM circuit for both the horizontal and vertical synchronization; (g) horizontal synchronization counter; (h) vertical synchronization counter; (i) complete circuit for the VGA controller. *(continued on next page)*

Current State Q_1Q_0		Next State (Implementation) $Q_{1next}Q_{0next}\ (D_1D_0)$							
		$(H\_cnt = B)$		$(H\_cnt = B + C)$		$(H\_cnt = B + C + D)$		$(H\_cnt = B + C + D + E)$	
		0	1	0	1	0	1	0	1
00	H_B	00	01	—	—	—	—	—	—
01	H_C	—	—	01	10	—	—	—	—
10	H_D	—	—	—	—	10	11	—	—
11	H_E	—	—	—	—	—	—	11	00

$$D_1 = Q_1'Q_0(H\_cnt = B + C) + Q_1Q_0' + Q_1Q_0(H\_cnt = B + C + D + E)'$$

$$D_0 = Q_1'Q_0'(H\_cnt = B) + Q_1'Q_0(H\_cnt = B + C)' + Q_1Q_0'(H\_cnt = B + C + D) + Q_1Q_0(H\_cnt = B + C + D + E)'$$

(c)

Current State Q_1Q_0		Next State (Implementation) $Q_{1next}Q_{0next}\ (D_1D_0)$							
		$(V\_cnt = P)$		$(V\_cnt = P + Q)$		$(V\_cnt = P + Q + R)$		$(V\_cnt = P + Q + R + S)$	
		0	1	0	1	0	1	0	1
00	V_P	00	01	—	—	—	—	—	—
01	V_Q	—	—	01	10	—	—	—	—
10	V_R	—	—	—	—	10	11	—	—
11	V_S	—	—	—	—	—	—	11	00

$$D_1 = Q_1'Q_0(V\_cnt = P + Q) + Q_1Q_0' + Q_1Q_0(V\_cnt = P + Q + R + S)'$$

$$D_0 = Q_1'Q_0'(V\_cnt = P) + Q_1'Q_0(V\_cnt = P + Q)' + Q_1Q_0'(V\_cnt = P + Q + R) + Q_1Q_0(V\_cnt = P + Q + R + S)'$$

(d)

Figure 10.18 Controller for the VGA monitor: (a) state diagram for horizontal synchronization; (b) state diagram for vertical synchronization; (c) next-state table for horizontal synchronization; (d) next-state table for vertical synchronization; (e) output table; (f) FSM circuit for both the horizontal and vertical synchronization; (g) horizontal synchronization counter; (h) vertical synchronization counter; (i) complete circuit for the VGA controller. *(continued on next page)*

Current State $Q_1 Q_0$	H_Data_on or V_Data_on	H_Sync or V_Sync
00	0	0
01	0	1
10	1	1
11	0	1

(e)

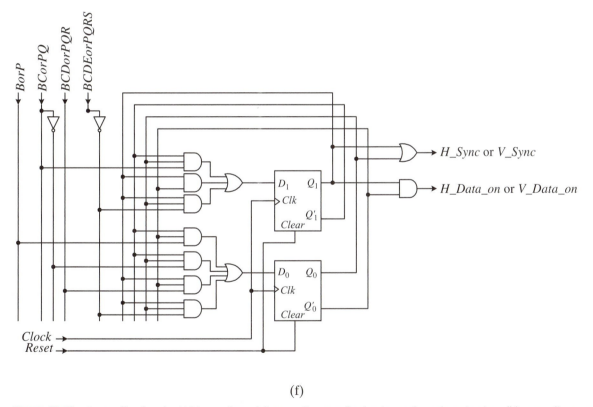

(f)

Figure 10.18 Controller for the VGA monitor: (a) state diagram for horizontal synchronization; (b) state diagram for vertical synchronization; (c) next-state table for horizontal synchronization; (d) next-state table for vertical synchronization; (e) output table; (f) FSM circuit for both the horizontal and vertical synchronization; (g) horizontal synchronization counter; (h) vertical synchronization counter; (i) complete circuit for the VGA controller. *(continued on next page)*

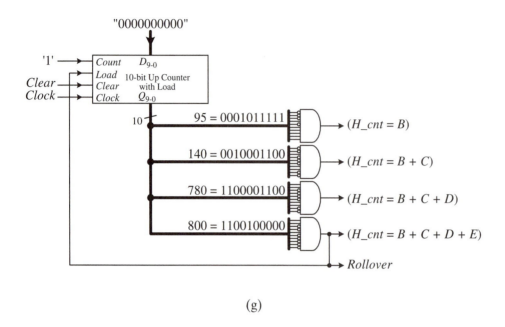

(g)

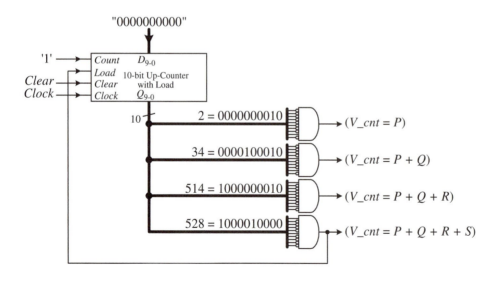

(h)

Figure 10.18 Controller for the VGA monitor: (a) state diagram for horizontal synchronization; (b) state diagram for vertical synchronization; (c) next-state table for horizontal synchronization; (d) next-state table for vertical synchronization; (e) output table; (f) FSM circuit for both the horizontal and vertical synchronization; (g) horizontal synchronization counter; (h) vertical synchronization counter; (i) complete circuit for the VGA controller. *(continued on next page)*

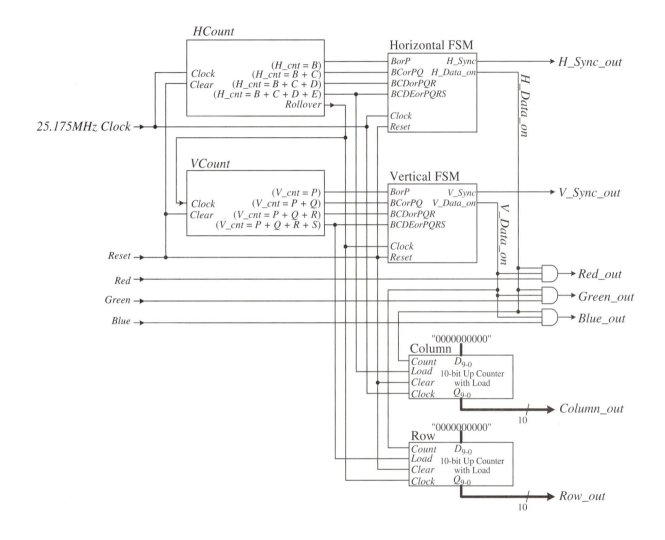

(i)

Figure 10.18   Controller for the VGA monitor: (a) state diagram for horizontal synchronization; (b) state diagram for vertical synchronization; (c) next-state table for horizontal synchronization; (d) next-state table for vertical synchronization; (e) output table; (f) FSM circuit for both the horizontal and vertical synchronization; (g) horizontal synchronization counter; (h) vertical synchronization counter; (i) complete circuit for the VGA controller.

We will need to use two instances of this FSM circuit: one for the horizontal FSM and one for the vertical FSM. The clock for the horizontal FSM is the 25.175 MHz clock, while the clock for the vertical FSM is derived from the rollover signal from the horizontal counter. The four status signals for the four counter conditions are generated from two counters: a horizontal counter and a vertical counter.

The horizontal counter, *HCount*, with the four comparators for ($H\_cnt = B$), ($H\_cnt = B + C$), ($H\_cnt = B + C + D$), and ($H\_cnt = B + C + D + E$) is shown in Figure 10.18(g). A 10-bit counter is needed for counting from 0 up to 800. A 10-input AND gate is used for each of the four comparators. The inputs to each AND gate is set to the equivalent binary value for $B = 95$, $B + C = 95 + 45 = 140$, $B + C + D = 95 + 45 + 640 = 780$, and $B + C + D + E = 95 + 45 + 640 + 20 = 800$, respectively. The counter cycles back to 0 by asserting the *Load* line when the count reaches 800 and loading in the value 0. The output from the comparator ($H\_cnt = 800$) is the counter rollover signal *Rollover* and is used as the vertical clock signal for the vertical counter, the vertical FSM, and the row counter.

The vertical counter, *VCount*, with the four comparators for ($V\_cnt = P$), ($V\_cnt = P + Q$), ($V\_cnt = P + Q + R$), and ($V\_cnt = P + Q + R + S$), is shown in Figure 10.18(h). The circuit for this counter is just like the horizontal counter, except that the comparator values are different, and we do not need to output a rollover clock signal. The clock for this counter is the vertical clock signal *Rollover* from the horizontal counter.

The complete VGA monitor controller circuit is shown in Figure 10.18(i). To make sure that the three RGB signals to the monitor are valid, they have to be turned on (if needed) only in regions *D* and *R*. Hence, the three color output signals, *Red_out*, *Green_out*, and *Blue_out*, are ANDed with *H_Data_on* and *V_Data_on*. For example, if the input *Red* signal is a 1, the output *Red_out* signal is a 1 only when the scan is within the regions *D* and *R*.

Finally, in order to turn on a specific pixel, the circuit needs to keep track of the current column within the *D* region and the current row within the *R* region of the scan. Two additional counters, *Column* and *Row*, are used for this purpose. Since they need to count only when the scan is in regions *D* and *R*, respectively, therefore, the *Count* input for the *Column* counter is asserted by the *H_Data_on* signal, while the *Count* input for the *Row* counter is asserted by the *V_Data_on* signal. Once the counts reach 640 and 480, respectively, they will have gone passed these two regions, and so the counters will not count. They will have to be reset to 0 anytime before the scan reaches the beginning of these two regions again. In the circuit, the two *Load* lines are asserted when the two respective counters roll back to 0. Finally, the *Column* counter clock is from the 25.175 MHz source, and the *Row* counter clock is from the vertical clock source *Rollover* derived from the horizontal counter.

To display something on the screen, you simply have to check the current column and row that the scan is at, and then assert the RGB signal if you want the pixel at that location to be turned on. For example, if you simply assert the *Red* signal continuously, all of the pixels will be red, and you will see the entire screen being

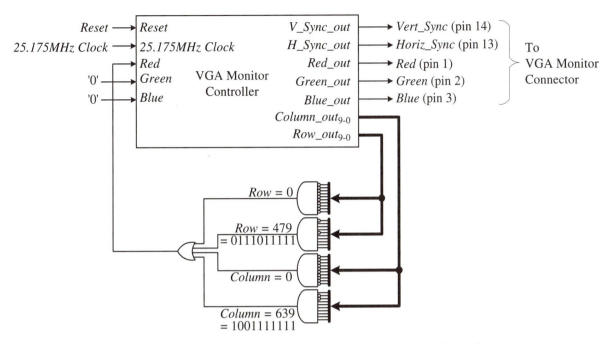

Figure 10.19 Circuit using the VGA controller to generate a red border around the entire screen.

red. On the other hand, if you just want the first row of pixels to be red, then you need to assert the *Red* signal only when the *Row* counter is 0. To get a red border around the screen, you would assert the *Red* signal when $Row = 0$, or $Row = 639$, or $Column = 0$, or $Column = 479$. Figure 10.19 shows the circuit to draw a red border around the entire screen using the VGA controller circuit from Figure 10.18(i).

Example 10.9 **A VGA controller—version 2**

If you understood the construction of the VGA controller in Example 10.8, you probably will have noticed that the controller circuit can be made much simpler. There are two reductions that we can make: (1) eliminate the column and row counters, and (2) eliminate the horizontal and vertical FSMs.

First, the column counter is always 140 (regions $B + C = 95 + 45 = 140$) plus the horizontal counter when the *H_Data_on* signal is asserted, and the row counter is always 34 (regions $P + Q = 2 + 32 = 34$) plus the vertical counter when the *V_Data_on* signal is asserted. Therefore, if we shift the horizontal count so that count 0 is at the beginning of region D, and the last count 800 is at the end of region C, then we do not need the extra column counter. Similarly, we can get rid of the row counter by shifting the vertical count so that count 0 is at the beginning of region R, and the last count 528 is at the end of region Q.

To eliminate the two FSMs, notice that the FSMs are used simply to assert the two pairs of signals: *H_Sync*, *H_Data_on* and *V_Sync*, *V_Data_on*. The assertion of these four signals is dependent only on the horizontal and vertical counters. Therefore, we can use just the horizontal and vertical counts to assert these signals directly. However, we need to keep these signals unchanged until after a certain count. An SR flip-flop can be used for each signal to set the signal to a 0 or a 1 at the right time. The SR flip-flop works just like the SR latch, except that the output changes only at the active (rising) edge of the clock. When S is asserted, the flip-flop is set, and the output $Q = 1$. When R is asserted, the flip-flop is reset, and the output $Q = 0$. For example, the *H_Sync_out* signal is a 1 when the count is at 95 and onward, and it is a 0 when the count is at 800, wraps back to 0, and onward. Hence, we connect the output of the comparator for $(H\_cnt = 95)$ to the S input of an SR flip-flop and the output of another comparator for $(H\_cnt = 800)$ to the R input. The output Q of the flip-flop is the *H_Sync_out* signal. As a result, the signal is set to a 1 when the count reaches 95, and it will remain at a 1 until it is reset to a 0 at count 800. The other three signals: *H_Data_on*, *V_Sync_out*, and *V_Data_on* are done similarly.

The complete version 2 of the VGA monitor controller circuit is shown in Figure 10.20. The modified horizontal counter, *HCount*, and the vertical counter, *VCount*, are shown in Figure 10.20(a) and (b) respectively. Notice that the four condition outputs for each of the two counters are different because we have shifted the counts so that the counting starts at the beginning of region D and R, for the horizontal and vertical counters respectively. The complete monitor circuit is shown in Figure 10.20(c).

We can implement and test out this second version of the monitor controller by using the same circuit shown in Figure 10.19, but with the controller replaced by this new version.

---

*10.4 ASM Charts and State Action Tables

The drawback to using a state diagram to describe the behavior of a sequential circuit is that it does not portray precisely the timing information, which is very important to the correct operation of the circuit. **Algorithmic state machine (ASM) charts** and **state action tables** are two alternate methods for describing sequential circuits more accurately and concisely. The state action table uses a tabular format, while the ASM chart uses a graphical flowcharting format.

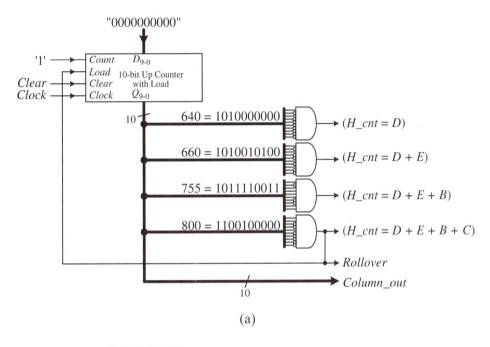

(a)

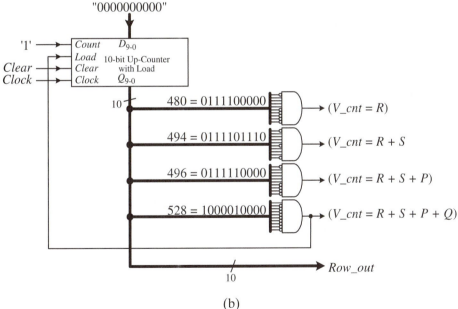

(b)

Figure 10.20   Second version of the VGA monitor controller for Example 10.9: (a) horizontal counter *HCount*; (b) vertical counter *VCount*; (c) complete controller circuit.
(continued on next page)

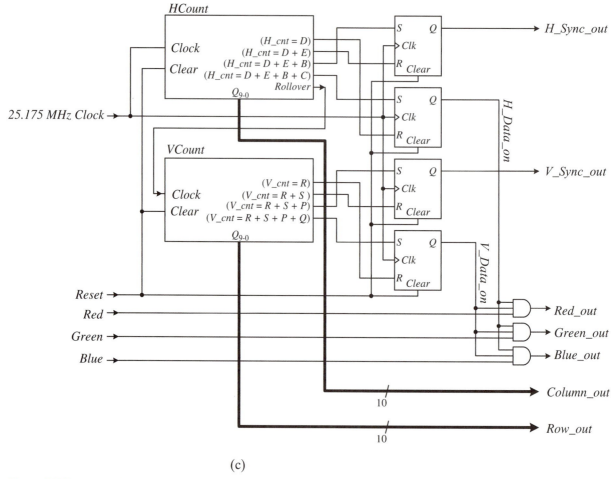

Figure 10.20 Second version of the VGA monitor controller for Example 10.9: (a) horizontal counter *HCount*; (b) vertical counter *VCount*; (c) complete controller circuit.

10.4.1 ASM Charts

Algorithmic state machine (ASM) charts are used to graphically portray the operations of a FSM more accurately. They are similar to flowcharts used in computer programming but use different shaped boxes, as shown in Figure 10.21, to describe a sequence of actions. The ASM chart portrays similar information as that in the state diagram. However, in addition to just describing a sequence of actions as in a state diagram, ASM charts also describe the timing relationships between the states.

The rectangle shown in Figure 10.21(a) is the **state box** for representing a state in an FSM. Each state box, therefore, is executed in one clock cycle and is similar to a node in a state diagram. A state in an FSM is for performing data manipulation and

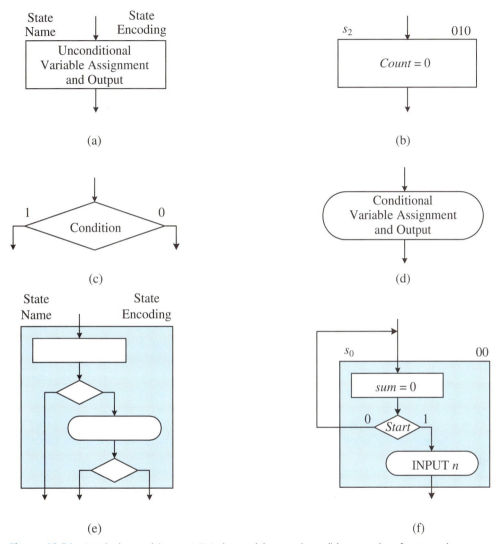

Figure 10.21 Symbols used in an ASM chart: (a) state box; (b) example of a state box;
(c) decision box; (d) condition box; (e) ASM block; (f) example of an ASM block.

input/output actions. Each state box contains unconditional (Moore type) register-transfer operations that are to be performed in that state. There is one path going into the state box and one path leading out of the state box. Outside the rectangle at the top-left corner is labeled with the symbolic state name, while the top-right corner is labeled with the binary state encoding (if known). For example, Figure 10.21(b) shows a state box with the register-transfer operation, $Count = 0$. The symbolic name given to the state is s_2, and the binary state encoding given to it is 010. When the FSM is in state s_2, the register for storing the variable $Count$ is reset to 0.

When a register-transfer operation is not specified for a specific register in a state box, it is assumed that that register content remains unchanged. A register changes its value in a state only when a register-transfer operation writes to that register in that state box.

The diamond-shaped box shown in Figure 10.21(c) is the **decision box**. The purpose of the decision box is to test whether the given condition written in the box is true or false. There is one path going into the decision box and two paths leading out of the box. The two paths coming out of the decision box are labeled 1 and 0. If the given condition is true, then the path labeled 1 is taken; otherwise, the path labeled 0 is taken. The decision box by itself does not represent a state, and no actions are written inside the box. Instead, it is used to determine the next state to go to. This is like the condition that is labeled on the edges in the state diagram.

The decision box is also used to describe a Mealy FSM where actions are performed depending on a condition of an input. In this case, the decision box is used in conjunction with the condition box (explained next), for performing conditional actions.

The oval-shaped **condition box** shown in Figure 10.21(d) is also used for performing register-transfer operations like the state box. However, unlike the state box, the condition box is used for conditional data manipulations in Mealy FSMs. The actual testing of the condition is not done within the condition box but rather in a decision box. The condition box by itself is not equivalent to one state. It must be used together with a decision box within an ASM block.

The **ASM block** shown in Figure 10.21(e), allows a state box and zero or more decision and condition boxes to be grouped together to form one state. All of the actions specified inside an ASM block are executed within one state or one clock cycle. Like the state box, the ASM block is labeled with the state name at the outside top-left corner and the state encoding at the top-right corner. The ASM block must start with one state box containing zero or more unconditional register-transfer operations. After the state box, there can be zero or more decision and condition boxes. The ASM block will have one entry point, and one or more exit paths leading to other states.

Figure 10.21(f) shows an example of an ASM block. The symbolic state name is s_0 using the binary encoding 00. When the FSM enters this state, the variable *sum* is initialized to 0 and the *Start* signal is tested. If there is no *Start* signal, that is, *Start* = 0, the FSM re-enters this state at the next clock cycle. If there is a *Start* signal, an input is performed to load the variable n. The actual writing of the registers *sum* and n with the new values occurs at the next active clock edge. In other words, the register *sum* is not zeroed until the next clock cycle, and the input value for n is not available until the next clock cycle.

| Example 10.10 | **Moore ASM chart** |

In this example, we will derive an ASM chart for a Moore type FSM based on the algorithm and datapath shown in Figure 10.3(a) and (b) for the IF-THEN-ELSE problem. Line 1 in the algorithm is put in a state box with the label *s_input* and encoding 000,

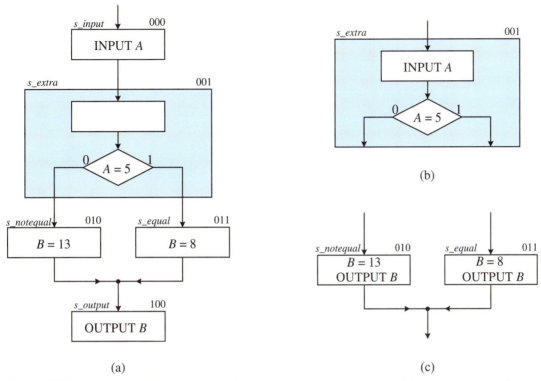

Figure 10.22 Moore ASM chart for Example 10.10: (a) correct ASM chart; (b) wrong ASM block; (c) wrong output.

as shown in Figure 10.22(a). A decision box is used for the conditional test in line 2 and must be put inside an ASM block. The test is dependent on the value from state *s_input*; therefore, it cannot be performed in the same state with line 1 but is assigned to state *s_extra*. In state *s_extra*, there is no action that can be performed unconditionally during this clock cycle; hence, the initial state box is empty. If we put both lines 1 and 2 in the same ASM block, as in Figure 10.22(b), then the timing will be wrong because we are again trying to read from the register for a value that has not been updated yet.

For a Moore FSM, actions are performed unconditionally; therefore, lines 3 and 5 in the algorithm are put into two separate state boxes. The two exit paths from the decision box go to these two state boxes: one for when the condition is true and the other for when the condition is false. Finally, both states, *s_notequal* and *s_equal*, go to state *s_output* to do the output. If we do the output in states *s_notequal* and *s_equal*, as shown in Figure 10.22(c), then again, the output will be wrong because we will be reading from register *B* before it is updated. If we want to do the output together with the assignment of *B* in the same state, then what we can do is output the constant value that is to be written into *B* rather than outputting the value from *B*.

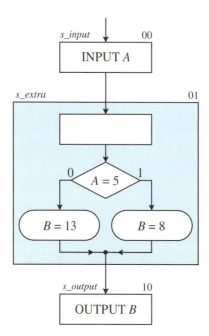

INPUT A

s_extra 01

0   A = 5   1

B = 13   B = 8

s_output ↓ 10

OUTPUT B

Figure 10.23
Mealy ASM chart
for Example 10.11.

Example 10.11

Mealy ASM chart

Let us derive the ASM chart for the same IF-THEN-ELSE algorithm and datapath used in Example 10.10, except for a Mealy FSM. The resulting ASM chart is shown in Figure 10.23. The main difference between this and the one in the previous example shown in Figure 10.22(a) is that the conditional assignments (lines 3 and 5 in the algorithm) are performed in the same state as the test in *s_extra*. The two conditional assignments are put in two separate oval condition boxes. The paths going to them are the same as before. Since the two condition boxes are put in the same ASM block, the ASM block now has only one exit path to state *s_output* to perform the output.

The advantage of this Mealy machine is not only the fact that it requires fewer states, thus resulting in a smaller control unit, but it also requires fewer clock cycles to execute the algorithm.

10.4.2 State Action Tables

The state action table is another way to describe the behavior of a sequential circuit. Whereas the ASM chart is like the state diagram, the state action table is similar to the next-state and output tables combined together. It has three columns: current state, next state, and datapath actions. The next state and datapath action entries can be with or without a condition. However, instead of enumerating all possible conditions and listing them out in columns, as with the next-state table, the condition is written in front of the next-state entries whenever it is needed. If no condition is written in front of a next-state entry, then it is an unconditional next state

Current State	Next State Condition, State	Datapath Actions Condition, Action
s_input	*s_extra*	INPUT *A*
s_extra	$(A = 5)$ *s_equal* $(A = 5)'$ *s_notequal*	
s_notequal	*s_output*	$B = 13$
s_equal	*s_output*	$B = 8$
s_output	*s_output*	OUTPUT *B*

Figure 10.24
A sample
Moore-type state
action table.

(i.e., the FSM will go to this next state unconditionally). If a next state is qualified with a condition, then the FSM will go to that state only if the condition is true. A branch in the state diagram corresponds to a next state with a condition, so the entry for that next state will have a condition.

The datapath actions are like the control word table information. Just like the next states, these datapath actions can be either conditional or unconditional. If an action is not qualified with a condition in front, then it is an unconditional action. Otherwise, it is a conditional action, and the action is performed only if the condition is true. All datapath actions for a Moore FSM will not have a condition, since Moore datapath actions are only dependent on the current state. A Mealy FSM will have conditional actions for some of its datapath action entries.

Figure 10.24 shows the Moore-type state action table for the corresponding ASM chart from Example 10.10. From current state *s_input*, the FSM unconditionally performs the action INPUT *A*, and then unconditionally goes to state *s_extra*. In state *s_extra*, the FSM will go to either state *s_notequal* or *s_equal*, depending on the condition $(A = 5)$. No action is performed in this state. In state *s_notequal*, the action $B = 13$ is performed, and then the FSM unconditionally transitions to state *s_output*. In state *s_equal*, the action $B = 8$ is performed, and then the FSM unconditionally transitions to state *s_output*. In state *s_output*, the FSM performs the action OUTPUT *B* and halts in that state by unconditionally going back to itself. All of the actions in this state action table are performed unconditionally; therefore, it is for a Moore type FSM. When the FSM enters a state, it always performs the unconditional action(s) assigned to that state first, and then determines what the next state should be. This is equivalent to the ASM block where the unconditional state box is always written before a decision box, if any.

Figure 10.25 shows the Mealy-type state action table for the corresponding ASM chart from Example 10.10. In current state *s_extra*, either one of the two conditional actions is performed, depending on the result of the conditional test. After performing the action, the FSM goes to state *s_output* unconditionally.

Current State	Next State Condition, State	Datapath Actions Condition, Action
s_input	*s_extra*	INPUT A
s_extra	*s_output*	$(A = 5)\ B = 8$ $(A = 5)'\ B = 13$
s_output	*s_output*	OUTPUT B

Figure 10.25
A sample Mealy-type state action table.

● ● ● ● ● ● ● ● ● ● ● ● ● ● ● ● ● ·

10.5 VHDL for Control Units

Figure 10.26 shows the behavioral VHDL code for the PS/2 keyboard controller from Example 10.7. The format follows exactly the VHDL code for Moore FSMs discussed in Section 7.6. Notice how closely the code for the FSM process follows the state diagram. Because of its simplicity, control units usually are synthesized this way rather than constructed manually, as in Example 10.7. For this example, the output process does not generate output signals to control the keyboard; rather, it reads in data signals from the keyboard and packages the eight data bits as a byte for output.

```
LIBRARY IEEE;
USE  IEEE.STD_LOGIC_1164.ALL;

ENTITY KeyboardCtrl IS PORT (
    Reset: IN STD_LOGIC;
    KeyboardClock: IN STD_LOGIC;
    KeyboardData: IN STD_LOGIC;
    keycode: OUT STD_LOGIC_VECTOR(7 DOWNTO 0));
END KeyboardCtrl ;

ARCHITECTURE Behavioral OF KeyboardCtrl IS
    TYPE state_type IS (s_start,s_d0,s_d1,s_d2,s_d3,s_d4,s_d5,s_d6,s_d7,s_parity,s_stop);
    SIGNAL state: state_type;
```

Figure 10.26
VHDL code for the
PS/2 keyboard
controller.
(continued on next page)

```
BEGIN

FSM: PROCESS(KeyboardClock, Reset)
BEGIN
   IF (Reset = '1') THEN
      state <= s_start;
   -- this FSM is driven by the keyboard clock signal
   ELSIF (KeyboardClock'EVENT AND KeyboardClock = '1') THEN
      CASE state is
      WHEN s_start =>
         state <= s_d0;
      WHEN s_d0 =>
         state <= s_d1;
      WHEN s_d1 =>
         state <= s_d2;
      WHEN s_d2 =>
         state <= s_d3;
      WHEN s_d3 =>
         state <= s_d4;
      WHEN s_d4 =>
         state <= s_d5;
      WHEN s_d5 =>
         state <= s_d6;
      WHEN s_d6 =>
         state <= s_d7;
      WHEN s_d7 =>
         state <= s_parity;
      WHEN s_parity =>
         state <= s_stop;
      WHEN s_stop =>
         state <= s_start;
      WHEN OTHERS =>
      END CASE;
   END IF;
END PROCESS;

output_logic: PROCESS (state)
BEGIN
   CASE state IS
   WHEN s_d0 =>
      keycode(0) <= KeyboardData;      -- read in data bit 0 from the keyboard
   WHEN s_d1 =>
      keycode(1) <= KeyboardData;      -- read in data bit 1 from the keyboard
   WHEN s_d2 =>
      keycode(2) <= KeyboardData;      -- read in data bit 2 from the keyboard
```

Figure 10.26
VHDL code for the
PS/2 keyboard
controller.
(continued on next page)

```
                              WHEN s_d3 =>
                                 keycode(3) <= KeyboardData;    -- read in data bit 3 from the keyboard
                              WHEN s_d4 =>
                                 keycode(4) <= KeyboardData;    -- read in data bit 4 from the keyboard
                              WHEN s_d5 =>
                                 keycode(5) <= KeyboardData;    -- read in data bit 5 from the keyboard
                              WHEN s_d6 =>
                                 keycode(6) <= KeyboardData;    -- read in data bit 6 from the keyboard
                              WHEN s_d7 =>
                                 keycode(7) <= KeyboardData;    -- read in data bit 7 from the keyboard
                              WHEN OTHERS =>
                              END CASE;
                           END PROCESS;
                        END Behavioral;
```

Figure 10.26
VHDL code for the
PS/2 keyboard
controller.

10.6 Summary Checklist

- Control unit
- Datapath
- Control signals
- Status signals
- Constructing a control unit
- Generating control signals
- Generating status signals
- Timing issues for a control unit
- ASM charts
- State action tables
- Be able to derive a state diagram from an algorithm
- Be able to derive output signals for a control unit from control words
- Be able to derive status signals from the conditions in an algorithm
- Be able to determine how a comparator for generating a status signal connects to the datapath
- Be able to derive a control unit (FSM) from a state diagram

● ● ● ● ● ● ● ● ● ● ● ● ● ● ● ●

10.7 **Problems**

P10.1. Derive the control unit for the counting problem from Example 10.1, except use the general datapath from Section 9.4.

P10.2. Use schematic entry to implement the control unit shown in Figure 10.2(g). Simulate and test the circuit.

P10.3. Implement and test the control unit for Problem P10.2 on the UP2 board.

P10.4. Use schematic entry to implement the control unit shown in Figure 10.6(g). Simulate and test the circuit.

P10.5. Implement and test the control unit for Problem P10.4 on the UP2 board.

P10.6. Use schematic entry to implement the control unit shown in Figure 10.11(g). Simulate and test the circuit.

P10.7. Implement and test the control unit for Problem P10.6 on the UP2 board.

P10.8. Derive the control unit for:
   (a) Problem P9.23
   (b) Problem P9.24
   (c) Problem P9.25
   (d) Problem P9.26
   (e) Problem P9.27
   (f) Problem P9.28

P10.9. Use schematic entry to implement the control units for Problem P10.8. Simulate and test the circuit.

P10.10. Derive the control unit for:
   (a) Problem P9.29
   (b) Problem P9.30
   (c) Problem P9.31
   (d) Problem P9.32
   (e) Problem P9.33
   (f) Problem P9.34
   (g) Problem P9.35.
   (h) Problem P9.36.

P10.11. Use schematic entry to implement the control units for Problem P10.10. Simulate and test the circuit.

P10.12. Instead of using an FSM circuit for the keyboard controller, as shown in Section 10.3.2, design and implement a keyboard controller circuit using just a serial-to-parallel shift register.

P10.13. Modify the keyboard controller circuit from Section 10.3.2 so that it will test for the correct parity bit.

P10.14. Modify the keyboard controller circuit from Problem P10.12 so that it will test for the correct parity bit.

P10.15. Construct and implement a circuit using the VGA controller from Section 10.3.3 to draw a blue triangle on the screen.

P10.16. Construct and implement a circuit using the VGA controller from Section 10.3.3 to draw a white square in the middle of the screen.

P10.17. Construct and implement a circuit using the VGA controller from Section 10.3.3 to draw a white square and a blue triangle on the screen.

P10.18. Using the keyboard controller from Section 10.3.2 and the VGA controller from Section 10.3.3, construct and implement a circuit that reads in a letter from the keyboard and displays it on the screen.

P10.19. Write VHDL code for the control units for Problem P10.8. Simulate and test the circuit.

P10.20. Write VHDL code for the control units for Problem P10.10. Simulate and test the circuit.

P10.21. Derive the ASM charts for the control units for Problem P10.8.

P10.22. Derive the ASM charts for the control units for Problem P10.10.

P10.23. Derive the state action tables for the control units for Problem P10.8.

P10.24. Derive the state action tables for the control units for Problem P10.10.

Dedicated Microprocessors

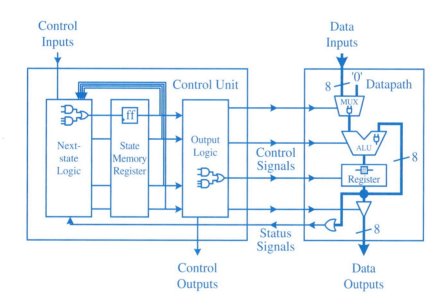

All microprocessors can be divided into two main categories: **general-purpose microprocessors** and **dedicated microprocessors**. General-purpose microprocessors are capable of performing a variety of computations. In order to achieve this goal, each computation is not hardwired into the processor, but rather, it is represented by a sequence of instructions in the form of a program that is stored in the memory and executed by the microprocessor. The program in the memory can be changed easily so that another computation can be performed. Because of the general nature of the processor, it is likely that in performing a specific computation not all of the resources available inside the general-purpose microprocessor are used.

Dedicated microprocessors, also known as **application-specific integrated circuits** (**ASIC**s), on the other hand, are dedicated to performing only one task. The instructions for performing that one task are, therefore, hardwired into the processor itself, and once manufactured, cannot be modified again. In other words, no memory is required to store the program because the program is built right into the microprocessor circuit itself. If the ASIC is customized completely, then only those resources that are required by the computation are included in the ASIC, so no resources are wasted. Another advantage of building the program instructions directly into the microprocessor circuit itself is that the execution speed of the program is many times faster than if the instructions are stored in memory.

The design of a microprocessor, whether it be a general-purpose microprocessor or a dedicated microprocessor, can be divided into two main parts: the datapath and the control unit, as shown in Figure 11.1. The **datapath** is responsible for all of the operations performed on the data. It includes (1) functional units such as adders, shifters, multipliers, ALUs, and comparators, (2) registers and other memory elements for the temporary storage of data, and (3) buses and multiplexers for the transfer of data between the different components in the datapath. External data can enter the datapath through the **data input** lines. Results from the computation can be returned through the **data output** lines.

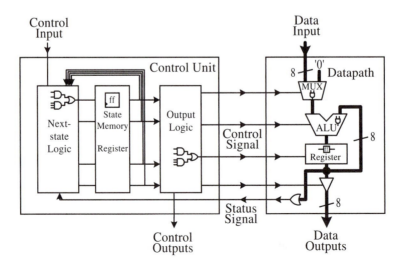

Figure 11.1
Schematic of a microprocessor.

The **control unit** (or **controller**) is responsible for controlling all of the operations of the datapath by providing appropriate **control signals** to the datapath at the appropriate times. At any one time, the control unit is said to be in a certain **state** as determined by the content of the **state memory**. The state memory is simply a register with one or more (D) flip-flops. The control unit operates by transitioning from one state to another—one state per clock cycle—and because of this behavior, the control unit is also referred to as a **finite state machine** (**FSM**). The **next-state logic** in the control unit will determine what state to go to next in the next clock cycle, depending on the current state that the FSM is in, the **control inputs**, and the **status signals**. In every state, the **output logic** that is in the control unit generates all of the appropriate control signals for controlling the datapath. The datapath, in return, provides status signals for the next-state logic. Status signals are usually from the output of comparators for testing branch conditions. Upon completion of the computation, the **control output** line is asserted to notify external devices that the value on the data output lines is valid.

In Chapters 9 and 10, you have learned how to design the datapath and the control unit separately. In this chapter, you will learn how to put them together to form a dedicated microprocessor. There are several abstraction levels at which a microprocessor can be designed.

At the first abstraction level, you manually construct the circuit for both the control unit and the datapath separately and then connect them together using the control and status signals. This manual process of constructing a microprocessor ties together everything that you have learned so far in this book. However, in a real situation, this method is not too practical, because the microprocessor usually will require many more states, input signals, and output signals. As a result, the manual construction process becomes much more complex. Furthermore, there are tools for automating this process. Nevertheless, being able to construct a microprocessor manually shows that you fully understand the theories and concepts of how a microprocessor is designed. Section 11.1 shows this complete dedicated microprocessor construction process. Section 11.2 provides several complete manual construction examples. All of the circuits are on the accompanying CD-ROM and can be downloaded onto the UP2 development board for testing.

After manually designing a microprocessor, we actually can implement the circuit either by manually drawing the circuit using a schematic editor such as the Schematic Editor in the MAX+plus II software on the accompanying CD-ROM (see Appendix A), or using VHDL to describe the circuit connections at the structural level (see Appendix B). Constructing a microprocessor manually this way uses the **FSM+D** (FSM *plus* datapath) model. In this model, both the FSM and the datapath circuits are constructed manually as separate units. The FSM and the datapath are connected together in an enclosing unit using the control and status signals.

The second abstraction level of microprocessor design also uses the FSM+D model. As before, you manually construct the datapath. However, instead of manually constructing the FSM, you use behavioral VHDL code to describe the operation of the FSM, as discussed in Sections 7.6 and 10.7. There will be a next-state process and an output process in the behavioral code. The next-state process will generate the next-state logic, and the output process will generate all of the control signals for driving the datapath. The final circuit for the FSM can then be synthesized automatically. The automatically synthesized FSM and the manually constructed datapath are then connected together (like before) in an enclosing unit using the control and status signals. In practice, this is probably the lowest abstraction level in which you would want to design a dedicated microprocessor. The advantage of using this FSM+D model is that you have full control as to how the datapath is built. Section 11.3 illustrates this process.

The third level of microprocessor design uses the **FSMD** (FSM *with* datapath) model. Using this model, you would design the FSM using behavioral VHDL code just like in the previous level. However, instead of constructing the datapath manually as a separate unit, all of the datapath operations are embedded within the FSM entity using the built-in VHDL operators. During the synthesis process, the synthesizer automatically will generate a separate FSM unit and a datapath unit, but these two units automatically will be connected together as one microprocessor by the synthesizer. The advantage of this model is that you do not have to manually design the datapath, but you still have full control as to what operation is executed in what state or in what clock cycle. In other words, you have control over the timing of the FSM circuit. Section 11.3.2 illustrates this process.

Finally, a microprocessor can be described algorithmically at the behavioral level using VHDL. This process synthesizes the full microprocessor with its control unit and datapath automatically. The advantage of designing microprocessors this way is that you do not need to know how to manually design a microprocessor. In other words, you do not need to know most of the materials presented in this book. Instead, you only need to know how to write VHDL codes. The disadvantage is that you do not have control over the timing of the circuit. You can no longer specify what register-transfer operation is executed in what clock cycle. Section 11.3.3 illustrates this process.

● ● ● ● ● ● ● ● ● ● ● ● ● ●

11.1   Manual Construction of a Dedicated Microprocessor

In Chapter 9, we described how a datapath is designed, and how it is used to execute a particular algorithm by specifying the control words to manipulate the datapath at each clock cycle. In that chapter, we tested the datapath by setting the control word signals manually. However, to actually have the datapath automatically operate according to the control words, a control unit is needed that will generate the control signals that correspond to the control words at each clock cycle.

The construction of the control unit was described in Chapter 10. Given the algorithm and the control words, we were able to derive the state diagram (from which we get the next-state table) and, finally, the next-state logic circuit for the control unit. The control words also serve as the output table; from this we get the output logic circuit. Combining the next-state logic circuit, the state memory, and the output logic circuit together produces the complete control unit circuit.

To form a complete dedicated microprocessor, we simply have to join the control unit and the datapath together using the control signals and the status signals. Recall that the control signals are generated by the control unit to control the operations of the datapath, while the status signals are generated by the datapath to inform the next-state logic in the control unit as to what the next state should be in the execution of the algorithm.

This method of manually constructing a dedicated microprocessor is referred to as the **FSM+D** (FSM *plus* datapath) model because the FSM and the datapath are constructed separately, and then they are joined together using the control and status signals. Example 11.1 shows this manual construction of a dedicated microprocessor.

Example 11.1

Constructing a dedicated microprocessor manually

In this example, we manually will construct the dedicated microprocessor for the summation algorithm of Example 9.10. Recall that the problem is to generate and add the numbers from n down to 1, where n is an input number. In Example 9.10, we derived the control words for solving this algorithm using the general datapath shown in Figure 9.32 and repeated here in Figure 11.2(a). Example 10.5 manually constructed the control unit circuit based on the state diagram and control words for the algorithm. The complete control unit circuit from Figure 10.11(g) is repeated here in Figure 11.2(b).

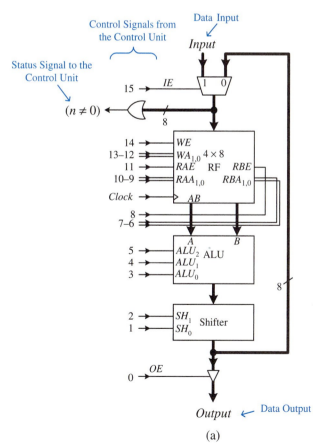

Figure 11.2

Microprocessor for the summation algorithm of Example 11.1:
(a) datapath;
(b) control unit;
(c) complete microprocessor circuit.

(continued on next page)

(a)

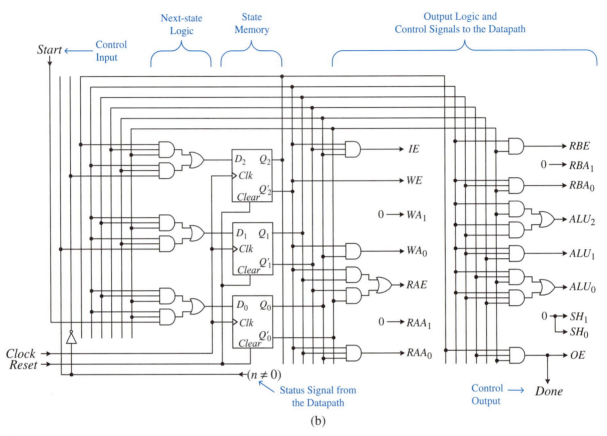

Figure 11.2 Microprocessor for the summation algorithm of Example 11.1: (a) datapath; (b) control unit; (c) complete microprocessor circuit.
(continued on next page)

The datapath circuit shown in Figure 11.2(a) uses 16 control signals. Correspondingly, the control unit circuit shown in Figure 11.2(b) generates the values for these 16 control signals. To combine the control unit and the datapath together, we simply connect the 16 control signals from the control unit to the corresponding 16 control signals on the datapath, as shown in Figure 11.2(c).

In addition to the 16 control signals, the control unit requires one status signal from the datapath. This status signal, $(n \neq 0)$, is generated by the OR gate comparator in the datapath and goes to the input of the next-state logic in the control unit.

The microprocessor has two types of inputs and two types of outputs. There is a control input signal, *Start*, that goes into the control unit and a control output signal, *Done*, that the control unit generates. Then there is the data input signal, *Input*, going into the datapath. For this example, this is where the value for n is entered in. Finally, there is the data output signal, *Output*, from which the datapath outputs the result of the computation. This complete dedicated microprocessor, as shown in Figure 11.2(c), of course does nothing more than sums the numbers from n down to 1 and outputs the result.

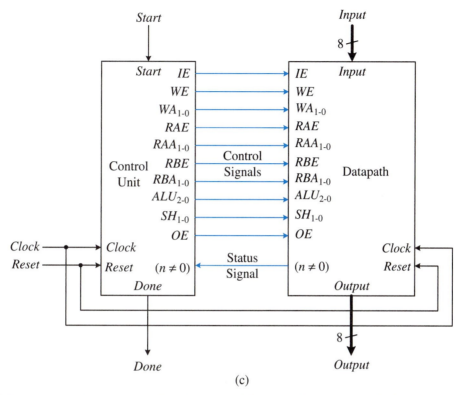

Figure 11.2   Microprocessor for the summation algorithm of Example 11.1: (a) datapath; (b) control unit; (c) complete microprocessor circuit.

11.2   Examples of Manual Designs of Dedicated Microprocessors

This section provides several complete examples of manually designing and constructing dedicated microprocessors. The complete schematic drawings of these microprocessors are available on the accompanying CD-ROM in their respective directories. All of the examples' microprocessor top-level schematic design files are named MP.GDF and can be used for simulation. The microprocessor circuits can also be downloaded onto the UP2 development board for testing. To download the design onto the UP2 development board, use either the top-level file named UP2MAX.GDF for the MAX chip, or UP2FLEX.GDF for the FLEX chip. In each respective directory on the CD-ROM, there is a README.TXT file that describes how to use the microprocessor. Furthermore, each example also has the VHDL codes for describing it using the FSMD model on the CD-ROM. The corresponding top-level VHDL files on the CD-ROM are MP.VHD, UP2MAX.VHD, and UP2FLEX.VHD.

Using the MAX+plus II software from the CD-ROM, you can input either the schematic drawings or the VHDL codes into the computer, synthesize and simulate the circuit, and finally download the circuit onto the PLD chip on the UP2 development board for testing. Appendix A provides a tutorial on how to use the Schematic Editor in the MAX+plus II software to input the schematic drawings. Appendix B provides a tutorial on entering VHDL code using the Text Editor. Finally, Appendix C shows how to synthesize a circuit, how to use the Floorplan Editor to map the I/O signals to the pins on the PLD, and how to implement the circuit on the PLD chip.

11.2.1 Greatest Common Divisor

Example 11.2

Designing a dedicated microprocessor to evaluate the GCD

In this example, we manually will design the complete dedicated microprocessor for evaluating the greatest common divisor (GCD) of two 8-bit unsigned numbers, X and Y. For example, $GCD(3, 5) = 1$, $GCD(10, 4) = 2$, and $GCD(12, 60) = 12$. The algorithm for solving the GCD problem is listed in Figure 11.3.

We first will design a dedicated datapath for the algorithm. Next, we will design the control unit for the datapath. We will use the Schematic Editor to implement the complete dedicated microprocessor. Finally, we will test it using simulation and on the UP2 board.

The algorithm shown in Figure 11.3 has five data manipulation statements in lines 1, 2, 5, 7, and 10. There are two conditional tests in lines 3 and 4. We can conclude that the datapath requires two 8-bit registers (one for variable X and one for variable Y) and a subtractor. The dedicated datapath is shown in Figure 11.4.

We need a 2-to-1 multiplexer for the input of each register, because we need to initially load each register with an input number and subsequently with the result from the subtractor. The two control signals, In_X and In_Y, select which of the two sources are to be loaded into the registers X and Y, respectively. The two control signals, $XLoad$ and $YLoad$, load a value into the respective register.

The bottom two multiplexers, selected by the same XY signal, determine the source to the two operands for the subtractor. When XY is asserted, then the value from register X will go to the left operand of the subtractor, and the value from register Y will go to the right operand. When XY is de-asserted, then Y goes to the

Figure 11.3
Algorithm for solving the GCD problem of Example 11.2.

```
1    INPUT X
2    INPUT Y
3    WHILE (X ≠ Y){
4        IF (X > Y) THEN
5            X = X - Y
6        ELSE
7            Y = Y - X
8        END IF
9        }
10   OUTPUT X
```

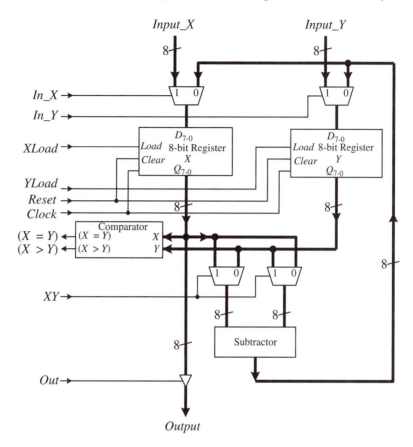

Figure 11.4
Dedicated datapath
for solving the
GCD problem of
Example 11.2.

left operand, and X goes to the right operand. Thus, this allows the selection of one
of the two subtraction operations, $X - Y$ or $Y - X$, to perform. Finally, a tri-state
buffer is used for outputting the result from register X. The *Out* control signal is
used to enable the tri-state buffer.

A comparator for testing the two conditions, equal-to and greater-than, is used
to generate the two needed conditional status signals. The comparator inputs are
directly from the two X and Y registers. There are two output signals, $(X = Y)$ and
$(X > Y)$, from the comparator. $(X = Y)$ is asserted if X is equal to Y, and $(X > Y)$
is asserted if X is greater than Y. The operation table and circuit for this comparator
was discussed in Section 4.10 and shown in Figure 4.29.

This dedicated datapath for solving the GCD problem requires six control signals,
$In\_X, In\_Y, XLoad, YLoad, XY$, and *Out*, and generates two status signals, $(X = Y)$,
and $(X > Y)$.

The state diagram for the GCD algorithm requires five states, as shown in Figure 11.5(a). Four states are used for the five data manipulation statements, since
only one state is used for performing both inputs. One "no-operation" state is used
for the conditional testing of the updated values of X and Y. This no-op state, 001,
is needed, since we need to test the conditions on the updated values of X and Y.

From state 001, we test for the two conditions, $(X = Y)$ and $(X > Y)$. If $(X = Y)$ is true, then the next state is 100. If $(X = Y)$ is false, then the next state is either 010 or 011, depending on whether the condition $(X > Y)$ is true or false, respectively.

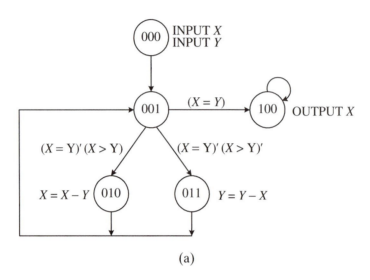

(a)

Current State $Q_2 Q_1 Q_0$	Next State (Implementation) $Q_{2next} Q_{1next} Q_{0next}$ $(D_2\ D_1\ D_0)$ $(X = Y), (X > Y)$			
	00	01	10	11
000	001	001	001	001
001	011	010	100	100
010	001	001	001	001
011	001	001	001	001
100	100	100	100	100
101 Unused	000	000	000	000
110 Unused	000	000	000	000
111 Unused	000	000	000	000

(b)

Figure 11.5 Control unit for solving the GCD problem of Example 11.2: (a) state diagram; (b) next-state (implementation) table; (c) K-maps and excitation equations; (d) control words and output table; (e) output equations; (f) circuit.
(continued on next page)

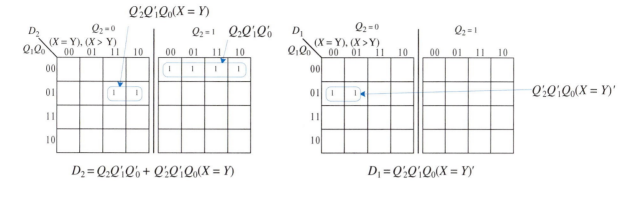

$$D_2 = Q_2Q_1'Q_0' + Q_2'Q_1'Q_0(X = Y)$$

$$D_1 = Q_2'Q_1'Q_0(X = Y)'$$

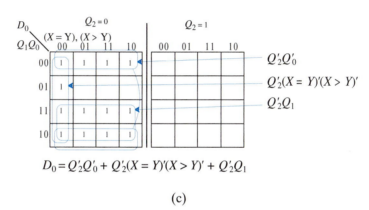

$$D_0 = Q_2'Q_0' + Q_2'(X = Y)'(X > Y)' + Q_2'Q_1$$

(c)

Control Word	State $Q_2\,Q_1\,Q_0$	Instruction	In_X	In_Y	$XLoad$	$YLoad$	XY	Out
0	000	INPUT X, INPUT Y	1	1	1	1	×	0
1	001	No operation	×	×	0	0	×	0
2	010	$X = X - Y$	0	×	1	0	1	0
3	011	$Y = Y - X$	×	0	0	1	0	0
4	100	OUTPUT X	×	×	0	0	×	1

(d)

Figure 11.5   Control unit for solving the GCD problem of Example 11.2: (a) state diagram; (b) next-state (implementation) table; (c) K-maps and excitation equations; (d) control words and output table; (e) output equations; (f) circuit.
(continued on next page)

$$In\_X = Q'_1$$

$$In\_Y = Q'_0$$

$$XLoad = Q'_2 Q'_0$$

$$YLoad = Q'_2 Q'_1 Q'_0 + Q'_2 Q_1 Q_0$$

$$XY = Q'_0$$

$$Out = Q_2$$

(e)

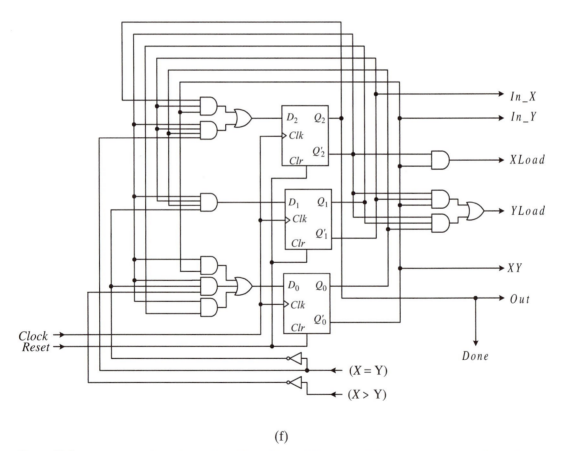

(f)

Figure 11.5 Control unit for solving the GCD problem of Example 11.2: (a) state diagram; (b) next-state (implementation) table; (c) K-maps and excitation equations; (d) control words and output table; (e) output equations; (f) circuit.

This state diagram does not have a *Start* signal, so in order for the resulting microprocessor to read the inputs correctly, we must first set up the input numbers and then assert the *Reset* signal to clear the state memory flip-flops to 0. This way, when the FSM starts executing from state 000, the two input numbers are ready to be read in.

The next-state table, as derived from the state diagram, is shown in Figure 11.5(b). The table requires five variables: three to encode the six states, Q_2, Q_1, and Q_0, and two for the status signals, $(X = Y)$ and $(X > Y)$. There are three unused state encodings: 101, 110 and 111. We have assumed that the next states from these three unused states are unconditionally back to state 000.

Using D flip-flops to implement the state memory, the implementation table is the same as the next-state table except that the values in the table entries are the inputs to the flip-flops (D_2, D_1, and D_0) instead of the flip-flops' outputs ($Q_{2\text{next}}$, $Q_{1\text{next}}$, and $Q_{0\text{next}}$). The K-maps and the excitation equations for D_2, D_1, and D_0 are shown in Figure 11.5(c).

The control words and output table, having the six control signals, are shown in Figure 11.5(d). State 000 performs both inputs of X and Y. The two multiplexer select lines, *In_X* and *In_Y*, must be asserted so that the data comes from the two primary inputs. The two numbers are loaded into the two corresponding registers by asserting the *XLoad* and *YLoad* lines. State 001 is for testing the two conditions, so no operations are performed. The no-op is accomplished by not loading the two registers and not outputting a value. For states 010 and 011, the *XY* multiplexer select line is used to select which of the two subtraction operations is to be performed. Asserting *XY* performs the operation $X - Y$; whereas, de-asserting *XY* performs the operation $Y - X$. The corresponding *In_X* or *In_Y* line is de-asserted to route the result from the subtractor back to the input of the register. The corresponding *XLoad* or *YLoad* line is asserted to store the result of the subtraction into the correct register. State 100 outputs the result from X by asserting the *Out* line.

The output equations, as derived from the output table, are shown in Figure 11.5(e). There is one equation for each of the six control signals. Each equation is dependent only on the current state (i.e., the current values in Q_2, Q_1, and Q_0). We have assumed that the control signals have don't-care values in all of the unused states.

The complete control unit circuit is shown in Figure 11.5(f). The state memory consists of three D flip-flops. The inputs to the flip-flops are the next-state circuits derived from the three excitation equations. The output circuits for the six control signals are derived from the six output equations. The two status signals, $(X = Y)$ and $(X > Y)$, come from the comparator in the datapath.

The final microprocessor can now be formed easily by connecting the control unit and the datapath together using the designated control and status signals, as shown in Figure 11.6. A sample simulation is shown in Figure 11.7. This complete microprocessor circuit is on the accompanying CD-ROM in the file MP.GDF and can be synthesized and simulated.

There is another top-level file called UP2MAX.GDF in the same directory for downloading and testing this microprocessor circuit on the UP2 development board using the MAX EPM7128S PLD.

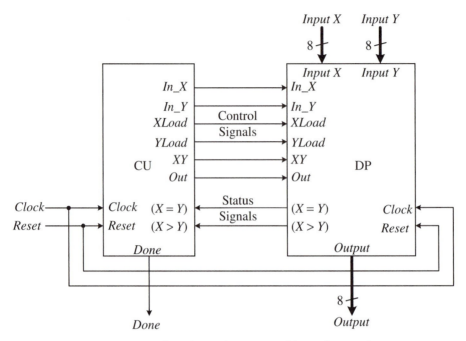

Figure 11.6 Microprocessor for solving the GCD problem of Example 11.2.

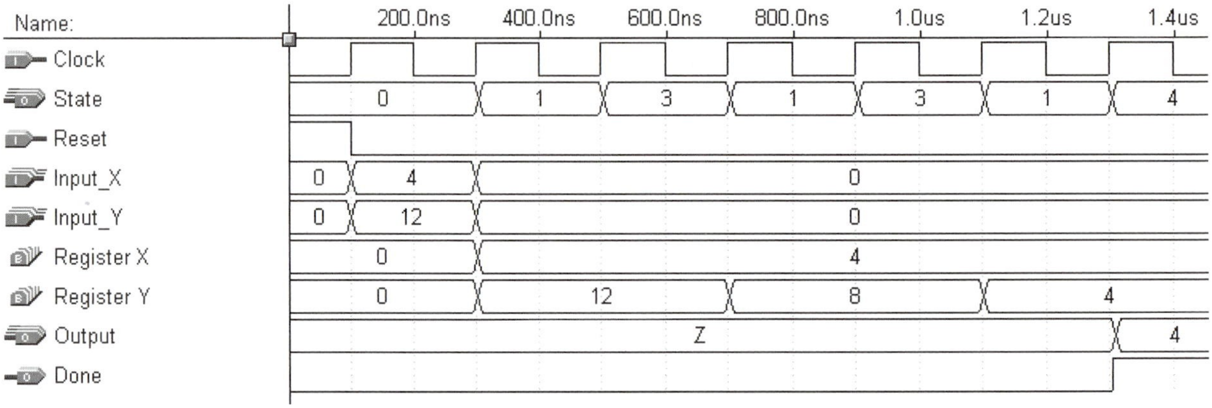

Figure 11.7 Sample simulation for the GCD problem for the two input numbers 4 and 12. The GCD of these two numbers is 4.

You need to use the Floorplan Editor in the MAX+plus II software to see how the I/Os are mapped to the EPM7128S pins. If you prefer, you can remap these pins. Note that the *Reset* signal must be mapped to pin 1 on the MAX chip; otherwise, you need to remove the Automatic Global Clear selection checkmark from the Compiler menu Assign / Global Project Logic Synthesis window, in order to compile this design file correctly.

Jumper wires between the PLD pins and the appropriate switches and the LEDs must be connected according to this floor plan. There are two sets of eight wires for the two input DIP switches. Another eight wires for the eight output LEDs, and one wire for the *Reset* button. The original *Done* signal is mapped to segment *a* on digit 2. You may want to refer to Section C.9 for making these connections.

To test this circuit, you need to first set up the two sets of eight DIP switches for the two binary unsigned input numbers. Then press the button connected to the *Reset* line. You should see the answer displayed as an 8-bit binary number on the eight LEDs. You should also see segment *a* on digit 2 light up for the *Done* signal.

11.2.2 Summing Input Numbers

| Example 11.3 | **Designing a dedicated microprocessor to sum the input of unsigned numbers** |

In this example, we manually will design the complete dedicated microprocessor for inputting many 8-bit unsigned numbers through one input port and then output the sum of these numbers. The algorithm continues to input numbers as long as the number entered is not a 0. Each number entered is also displayed on the output. When the number entered is a 0, the algorithm stops and outputs the sum of all of the numbers entered. The algorithm for solving this problem is listed in Figure 11.8.

We first will design a dedicated datapath for the algorithm. Next, we will design the control unit for the datapath. We will use the Schematic Editor to implement the complete dedicated microprocessor. Finally, we will test it using simulation and on the UP2 board.

The algorithm shown in Figure 11.8 has five data manipulation statements in lines 1, 3, 4, 8, and 10. There is one conditional test in line 5. The algorithm requires an adder and two 8-bit registers: one for variable X and one for variable *sum*. The dedicated datapath is shown in Figure 11.9.

Figure 11.8
Algorithm for solving the summing input numbers problem of Example 11.3.

```
1    sum = 0
2    BEGIN LOOP
3        INPUT X
4        sum = sum + X
5        IF (X = 0) THEN
6            EXIT LOOP
7        END IF
8        OUTPUT X
9    END LOOP
10   OUTPUT sum
```

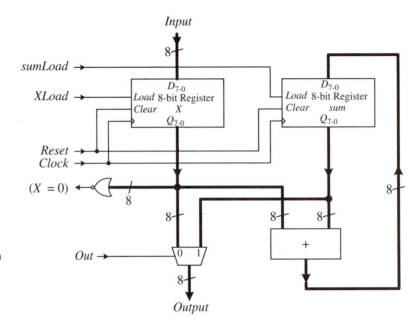

Figure 11.9
Dedicated datapath for solving the summing input numbers problem of Example 11.3.

Line 1 in the algorithm is performed by asserting the *Reset* signal, line 3 is performed by asserting the *XLoad* signal, and line 4 is performed by asserting the *sumLoad* signal. The algorithm continuously outputs either X or *sum*. The *Out* signal selects the 2-to-1 multiplexer for one of the two sources: register X or register *sum*. The output of the multiplexer is always available at the output. The conditional test $(X = 0)$ is generated by the 8-input NOR gate.

At first glance, this algorithm is very similar to the GCD algorithm in Example 11.2. However, because of the requirements of this problem, the actual hardware implementation of this microprocessor is slightly more difficult. Specifically, the requirement that many different numbers be input through one input port requires careful timing considerations and an understanding of how mechanical switches behave.

As a first try, we begin with the state diagram shown in Figure 11.10(a). Line 1 of the algorithm is performed by the asynchronous *Reset*, so it does not require a state to execute. Line 3 is performed in state 00, which is followed unconditionally by line 4 in state 01. The condition $(X = 0)$ is then tested. If the condition is true, the loop is exited, and the FSM goes to state 11 to output the value for *sum* and stays in that state until reset. If the condition is false, the FSM goes to state 10 to output X, and the loop repeats back to state 00.

One thing to notice is that in all of the states except for state 11, we need to output X. This is because in the datapath we do not have a tri-state buffer to disable the output. Instead, with the multiplexer, either one of two numbers is always being output. The behavior of this, nevertheless, is correct, according to the problem specifications.

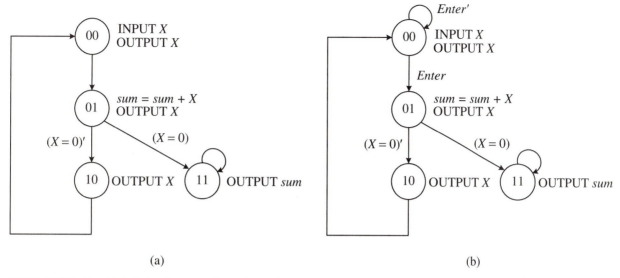

(a)   (b)

Figure 11.10   Incorrect state diagrams for solving the summing input numbers problem of Example 11.3.

As an exercise, you may want to derive the FSM circuit based on the state diagram shown in Figure 11.10(a), and perform a simulation to test its operation. In the simulation, the input numbers are assigned manually, so you should be able to obtain the correct simulation result.

However, if you implement this circuit in hardware, it will not work correctly. The reason is that the FSM cycles through the three loop states (00, 01, and 10) very fast because of the fast clock speed. As a result, the FSM will have gone through state 00 to input a number many times before you can even change the input to another number. Hence, the same number will be summed many times.

To correct this problem, we need to add another input signal that acts like the Enter switch. This way, the FSM will stay in state 00, waiting for the *Enter* signal to be asserted. This will give the user time to set up the input number before pressing the Enter switch. When the *Enter* signal is asserted, the FSM will exit state 00 with the new number to be processed. This modified state diagram is shown in Figure 11.10(b).

There is still a slight timing problem with this modified state diagram because of the fast clock speed. After pressing the Enter switch, and before you have time to release it, the FSM will have cycled through the complete loop and is back at state 00. But since you have not yet released the Enter switch, the FSM will continue on another loop with the same input number. What we need to do is to break the loop by waiting for the Enter switch to be released. This is shown in the state diagram in Figure 11.11(a). State 10 will wait for the Enter switch to be released before continuing on and looping back to state 00.

This last state diagram is correct. However, there might be a problem with the operation of the mechanical switch used for the *Enter* signal. When a mechanical switch is pressed, it usually goes on and off several times before settling down in

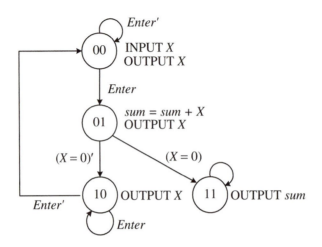

(a)

Current State Q_1Q_0	Next State (Implementation) $Q_{2next} Q_{1next} Q_{0next}$ (D_2 D_1 D_0)			
	Enter, $(X = 0)$			
	00	01	10	11
00	00	00	01	01
01	10	11	10	11
10	00	00	10	10
11	11	11	11	11

(b)

Figure 11.11 Control unit for solving the summing input numbers problem of Example 11.3: (a) state diagram; (b) next-state (implementation) table; (c) K-maps and excitation equations; (d) control words and output table; (e) output equations; (f) circuit.
(continued on next page)

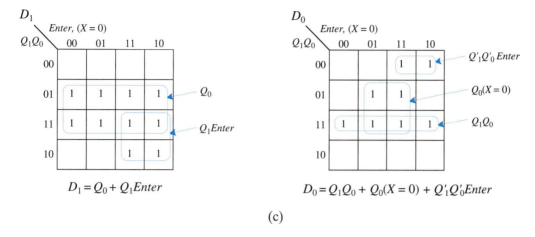

$$D_1 = Q_0 + Q_1 Enter$$

$$D_0 = Q_1 Q_0 + Q_0 (X = 0) + Q'_1 Q'_0 Enter$$

(c)

Control Word	State $Q_1 Q_0$	Instruction	XLoad	sumLoad	Out
0	00	INPUT X, OUTPUT X	1	0	0
1	01	$sum = sum - X$, OUTPUT X	0	1	0
2	10	OUTPUT X	0	0	0
3	11	OUTPUT sum	0	0	1

(d)

$$XLoad = Q'_1 Q'_0$$

$$sumLoad = Q'_1 Q_0$$

$$Out = Q_1 Q_0$$

(e)

Figure 11.11   Control unit for solving the summing input numbers problem of Example 11.3: (a) state diagram; (b) next-state (implementation) table; (c) K-maps and excitation equations; (d) control words and output table; (e) output equations; (f) circuit.
(continued on next page)

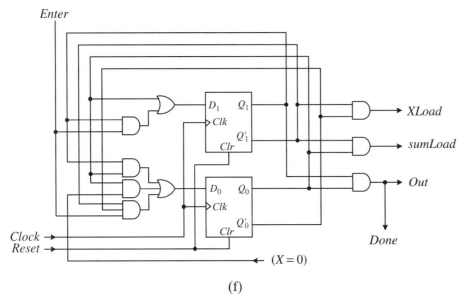

(f)

Figure 11.11 Control unit for solving the summing input numbers problem of Example 11.3: (a) state diagram; (b) next-state (implementation) table; (c) K-maps and excitation equations; (d) control words and output table; (e) output equations; (f) circuit.

the on position. This is referred to as the "debounce" problem. When the switch is fluctuating between the on and the off positions, the FSM can again be able to go through the loop many times. What we need to do is to debounce the switch. This, however, is not done in the FSM circuit itself but in the interface circuit between the FSM and the switch. We will come back to address this problem when building the interface circuit.

We will now construct the control unit circuit based on the state diagram shown in Figure 11.11(a). Four states are used for the five data manipulation statements. All of the states except for 11 will output X. State 00 inputs X and waits for the *Enter* signal. This allows the user to set up the input number and then press the Enter switch. When the Enter switch is pressed, the FSM goes to state 01, to sum X, and tests for the condition $(X = 0)$. If the condition is true, the FSM terminates in state 11 and outputs *sum*; otherwise, it goes to state 10 to wait for the *Enter* signal to be de-asserted by the user releasing the Enter switch. After exiting state 10, the FSM continues on to repeat the loop in state 00.

The next-state table, as derived from the state diagram, is shown in Figure 11.11(b). The table requires four variables: two to encode the four states, Q_1 and Q_0, and two for the status signals, *Enter* and $(X = 0)$.

Using D flip-flops to implement the state memory, the implementation table is the same as the next-state table, except that the values in the table entries are the inputs to the flip-flops, D_1 and D_0, instead of the flip-flops outputs, $Q_{1\text{next}}$ and $Q_{0\text{next}}$. The K-maps and the excitation equations for D_1 and D_0 are shown in Figure 11.11(c).

The control words and output table for the three control signals are shown in Figure 11.11(d). State 00 performs line 3 of the algorithm in Figure 11.8 by asserting *XLoad* and Line 8 by de-asserting *Out*. When *Out* is de-asserted, *X* is passed to the output. State 01 performs line 4 and line 8. Line 4 is executed by asserting *sumLoad*, and line 8 is executed by de-asserting *Out*. State 10 again performs line 8 by de-asserting *Out*. Finally, state 11 performs line 10 by asserting *Out*.

The output equations, as derived from the output table, are shown in Figure 11.11(e). There is one equation for each of the three control signals. Each equation is dependent only on the current state (i.e., the current values in Q_1 and Q_0).

The complete control unit circuit is shown in Figure 11.11(f). The state memory consists of two D flip-flops. The inputs to the flip-flops are the next-state circuits derived from the two excitation equations. The output circuits for the three control signals are derived from the three output equations. The status signal, $(X = 0)$, comes from the comparator in the datapath.

The final microprocessor can now be formed easily by connecting the control unit and the datapath together using the designated control and status signals, as shown in Figure 11.12. A sample simulation is shown in Figure 11.13. This complete microprocessor circuit is on the accompanying CD-ROM in the file MP.GDF and can be synthesized and simulated.

In order to implement the circuit on the UP2 development board, we need to connect the microprocessor's I/Os to the switches, LEDs, and clock source. The most important interface circuit for this problem is to debounce the Enter switch. A simple circuit to debounce a switch is to use a D flip-flop, as shown in Figure 11.14.

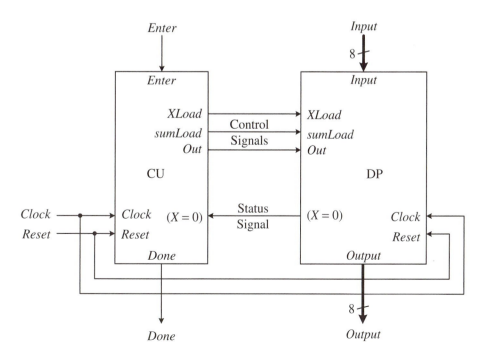

Figure 11.12
Microprocessor for solving the summing input numbers problem of Example 11.3.

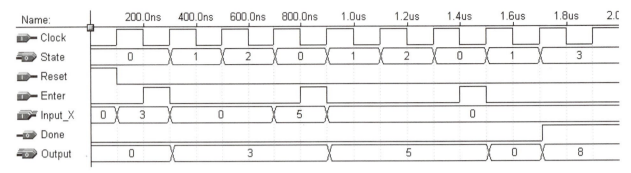

Figure 11.13 Sample simulation for the summing input numbers problem.

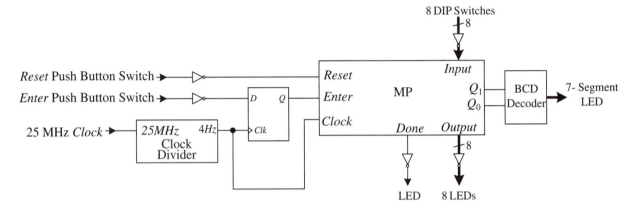

Figure 11.14 Interface to the summing input numbers microprocessor for the UP2 development board.

The clock frequency for the D flip-flop must be slow enough for the switch bounce to settle, so that the flip-flop will latch in a single value. The exact clock frequency is not too critical. The clock divider circuit used in the example slows down the on-board 25 MHz clock to approximately 4 Hz.

The top-level interface circuit is in the file UP2MAX.GDF for the MAX chip on the accompanying CD-ROM. This circuit also displays the current state value, Q_1 and Q_0, on the 7-segment LED. The FLEX version of this circuit is in the file UP2FLEX.GDF. It displays the sum output on the two 7-segment displays instead of using the eight discrete LEDs.

11.2.3 High-Low Guessing Game

Example 11.4 **Designing a dedicated microprocessor to play the high-low guessing game**

In this example, we manually will design the complete dedicated microprocessor for playing the high-low guessing game. The user picks a number between 0 and 99, and the computer will use the binary search algorithm to guess the number. After each guess, the user tells the computer whether the guess is high or low compared to the picked number. Two push-buttons, *hi_button* and *lo_button*, are used for the user to tell the computer whether the guess is too high, too low, or correct. The *hi_button* is pressed if the guess is too high, and the *lo_button* is pressed if the guess is too low. If the guess is correct, both buttons are pressed at the same time.

The algorithm for simulating this high-low guessing game is listed in Figure 11.15. The two boundary variables, *Low* and *High*, are initialized to 0 and 100, respectively. The loop between lines 3 to 11 will keep repeating until both buttons, *hi_button* and *lo_button*, are pressed. Inside the loop, line 4 calculates the next guess by finding the middle number between the lower and upper boundaries and assigns it to the variable *Guess*. Line 5 outputs this new *Guess*. Lines 6 to 10 checks which button is pressed. If the *lo_button* is pressed, that means the guess is too low, so line 7 changes the *Low* boundary to the current *Guess*. Otherwise, if the *hi_button* is pressed, that means the guess is too high, and line 9 changes the *High* boundary to the current *Guess*. The loop is then repeated with the calculation of the new *Guess* in line 4.

When both buttons are pressed, the condition in line 11 is true, and the loop is exited. Lines 12 to 15 simply cause the display to blink the correct guess by turning it on and off until either one of the buttons is pressed again.

We first will design a dedicated datapath for the algorithm. Next, we will design the control unit for the datapath. We will use the Schematic Editor to implement the complete dedicated microprocessor. Finally, we will test it using simulation and on the UP2 board.

The algorithm shown in Figure 11.15 has eight data manipulation operations in lines 1, 2, 4, 5, 7, 9, 13 and 14. The dedicated datapath for realizing this algorithm is shown in Figure 11.16. It requires three 8-bit registers (*Low*, *High*, and *Guess*)

```
1    Low = 0                                          // initialize Low
2    High = 100                                       // initialize High
3    REPEAT {
4       Guess = (Low + High) / 2                      // calculate guess using binary search
5       OUTPUT Guess
6       IF (lo_button = '1' AND hi_button = '0') THEN // low button pressed
7          Low = Guess
8       ELSE IF (lo_button = '0' AND hi_button = '1') THEN // high button pressed
9          High = Guess
10      END IF
11   } UNTIL (lo_button = '1' AND hi_button = '1')    // repeat until both buttons are pressed
12   WHILE (lo_button = '0' AND hi_button = '0')      // blink correct guess
13      OUTPUT Guess
14      turn off display
15   END WHILE
```

Figure 11.15   Algorithm for the high-low guessing game of Example 11.4.

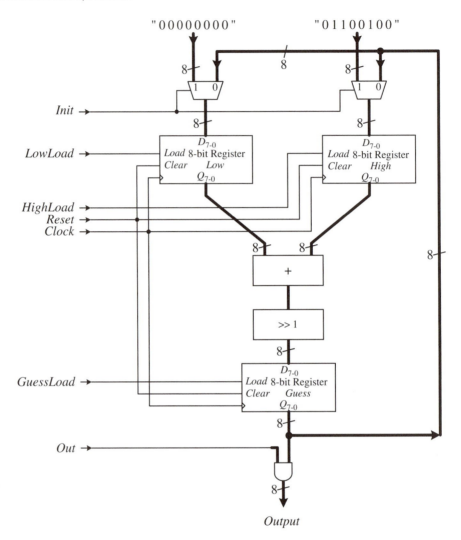

Figure 11.16
Dedicated datapath
for the high-low
guessing game of
Example 11.4.

for storing the low and high range boundary values and the guess, respectively. Two 2-to-1 multiplexers are used for the inputs to the *Low* and *High* registers to select between the initialization values for lines 1 and 2, and to the new *Guess* values for lines 7 and 9.

The only arithmetic operations needed are the addition and division-by-2 in line 4. Hence, the outputs from the two registers, *Low* and *High*, go to the inputs of an adder for the addition, and the output of the adder goes to a shifter. The division-by-2 is performed by doing a right shift of 1 bit. The result from the shifter is stored in the register *Guess*. Depending on the condition in line 6, the value in *Guess* is loaded into either the *Low* or the *High* register by asserting the corresponding load signal for that register.

A 2-input by 8-bit AND gate is used to control the output of the *Guess* number. One 8-bit set of inputs is connected to the output of the *Guess* register. The other 8-bit set of inputs is connected together in common to the output enable *Out* signal. By asserting *Out*, the data from *Guess* is passed to the output port. To blink the output display in lines 13 and 14, we just toggle the *Out* line.

The datapath shown in Figure 11.16 requires five control signals, *Init*, *LowLoad*, *HighLoad*, *GuessLoad*, and *Out*. The *Init* signal controls the two multiplexers to determine whether to load in the initialization values or the new guess. The three load signals, *LowLoad*, *HighLoad*, and *GuessLoad*, control the writing of the three respective registers. Finally, *Out* controls the output of the guess value.

The state diagram for this algorithm requires six states, as shown in Figure 11.17(a). State 000 is the starting initialization state. State 001 executes lines 4 and 5 by calculating the new guess and outputting it. State 001 also waits for the user keypress. If only the *lo_button* is pressed, then the FSM goes to state 010 to assign the guess as the new low value. If only the *hi_button* is pressed, then the FSM goes to state 011 to assign the guess as the new high value. If both buttons are pressed, then the FSM goes to state 100 to output the guess. From state 100, the FSM turns on and off the output by cycling between states 100 and 101 until a button is pressed. When a button is pressed from either state 100 or 101, the FSM goes back to the initialization state for a new game.

The output table showing the five output signals (*Init*, *LowLoad*, *HighLoad*, *GuessLoad*, and *Out*) to be generated in each state are shown in Figure 11.17(d). The corresponding output equations derived from the output table are shown in Figure 11.17(e).

Again, we will use D flip-flops to implement the state memory. Having six states, three flip-flops are needed with two unused states. Both the next-state table and the implementation table are shown in Figure 11.17(b). Recall that, when D flip-flops are used for the implementation, the next-state table and the implementation table are the same, because the characteristic equation for the D flip-flop is $Q_{next} = D$.

The implementation table is simply the truth table for the three variables, D_2, D_1, and D_0. Hence, the excitation equations for the three flip-flops, D_2, D_1, and D_0, are derived directly from the implementation table. The K-maps and equations for these three variables are shown in Figure 11.17(c).

Using the three excitation equations for deriving the next-state logic circuit, the three D flip-flops for the state memory, and the five output equations for deriving the output logic circuit, we get the complete control unit circuit for the high-low guessing game, as shown in Figure 11.17(f).

Connecting the control unit circuit shown in Figure 11.17(f) and the datapath circuit shown in Figure 11.16 together using the control and status signals produces the final microprocessor, as shown in Figure 11.18. This complete microprocessor circuit is on the accompanying CD-ROM in the file MP.GDF and can be synthesized and simulated.

Figure 11.19 shows the interface needed to test the high-low guessing game dedicated microprocessor on the UP2 development board. The top-level file UP2FLEX.GDF in the implementation directory can be used for downloading and testing of this circuit on the UP2 development board using the FLEX PLD.

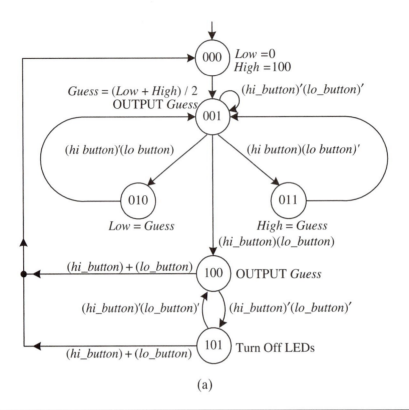

(a)

Current State $Q_2 Q_1 Q_0$	Next State (Implementation) $Q_{2next} Q_{1next} Q_{0next}$ (D_2 D_1 D_0)			
	hi_button, lo_button			
	00	01	10	11
000	001	001	001	001
001	001	010	011	100
010	001	001	001	001
011	001	001	001	001
100	101	000	000	000
101	100	000	000	000
110 Unused	000	000	000	000
111 Unused	000	000	000	000

(b)

Figure 11.17 Control unit for the high-low guessing game of Example 11.4: (a) state diagram; (b) next-state (implementation) table; (c) K-maps and excitation equations; (d) control words and output table; (e) output equations; (f) circuit.
(continued on next page)

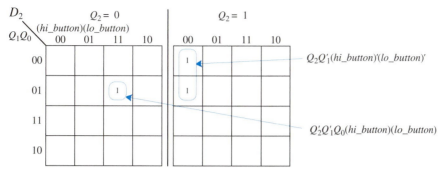

$$D_2 = Q_2Q_1'(hi\_button)'(lo\_button)' + Q_2'Q_1'Q_0(hi\_button)(lo\_button)$$

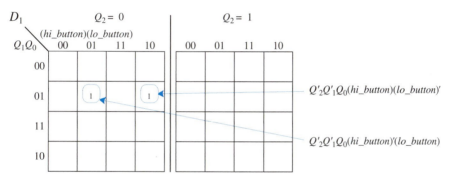

$$D_1 = Q_2'Q_1'Q_0(hi\_button)(lo\_button)' + Q_2'Q_1'Q_0(hi\_button)' (lo\_button)$$

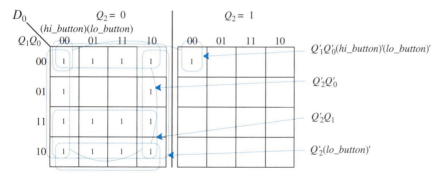

$$D_0 = Q_1'Q_0'(hi\_button)'(lo\_button)' + Q_2'Q_0' + Q_2'Q_1 + Q_2'(lo\_button)'$$

(c)

Figure 11.17   Control unit for the high-low guessing game of Example 11.4: (a) state diagram; (b) next-state (implementation) table; (c) K-maps and excitation equations; (d) control words and output table; (e) output equations; (f) circuit.
(continued on next page)

Control Word	State $Q_2Q_1Q_0$	Instruction	Init	High-Load	Low-Load	Guess-Load	Out
0	000	$Low = 0$, $High = 100$	1	1	1	0	1
1	001	$Guess = (Low + High) / 2$	0	0	0	1	1
2	010	$Low = Guess$	0	0	1	0	1
3	011	$High = Guess$	0	1	0	0	1
4	100	OUTPUT $Guess$	0	0	0	0	1
5	101	Turn off LEDs	0	0	0	0	0

(d)

$$Init = Q_2'Q_1'Q_0'$$
$$HighLoad = Q_2'Q_1'Q_0' + Q_2'Q_1Q_0$$
$$LowLoad = Q_2'Q_1'Q_0' + Q_2'Q_1Q_0'$$
$$GuessLoad = Q_2'Q_1'Q_0$$
$$Out = Q_2' + Q_0'$$

(e)

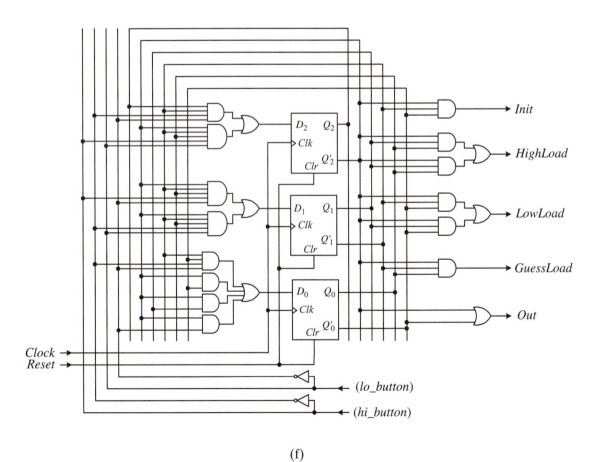

(f)

Figure 11.17 Control unit for the high-low guessing game of Example 11.4: (a) state diagram; (b) next-state (implementation) table; (c) K-maps and excitation equations; (d) control words and output table; (e) output equations; (f) circuit.

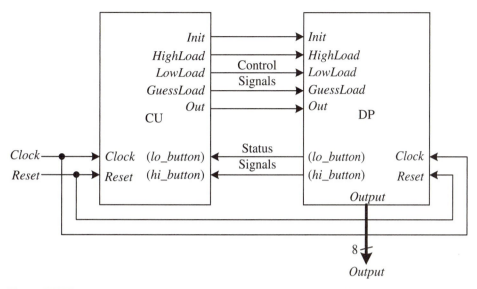

Figure 11.18   Microprocessor for the high-low guessing game of Example 11.4.

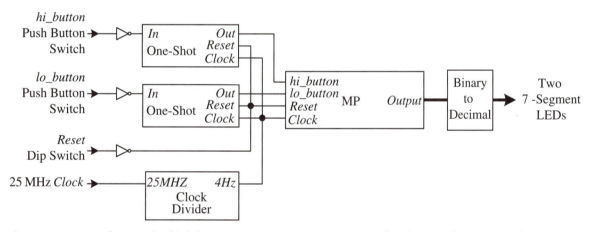

Figure 11.19   Interface to the high-low guessing game microprocessor for the UP2 development board.

11.2.4   Finding the Largest Number

Example 11.5   **Designing a dedicated microprocessor to find the largest number**

In this example, we will implement the classic computer program of finding the largest number from a set of input numbers. We will assume that the numbers, entered through one input port, are 8-bit unsigned numbers, and a 0 signals the end of the inputs. The current largest number is always displayed. The algorithm for solving this problem is listed in Figure 11.20.

```
1   Largest = 0                        // for storing the current largest number
2   INPUT X                            // enter first number
3   WHILE (X ≠ 0){
4       IF (X > Largest) THEN          // if new number greater?
5           Largest = X                // yes, remember new largest number
6       END IF
7       OUTPUT Largest
8       INPUT X                        // get next number
9   }
```

Figure 11.20
Algorithm for finding the largest number problem of Example 11.5.

We first will design a dedicated datapath for the algorithm. Next, we will design the control unit for the datapath. We will use the Schematic Editor to implement the complete dedicated microprocessor. Finally, we will test it using simulation and on the UP2 board.

The algorithm shown in Figure 11.20 has five data manipulation operations in lines 1, 2, 5, 7, and 8. It requires two registers: an 8-bit register for storing X and an 8-bit register for storing *Largest*. No functional unit for performing arithmetic is needed. The dedicated datapath is shown in Figure 11.21.

Line 1 in the algorithm is performed by asserting the *Reset* signal, lines 2 and 8 are performed by asserting the *XLoad* signal, and line 5 is performed by asserting the *LargestLoad* signal. The algorithm continuously outputs *Largest*. The conditional

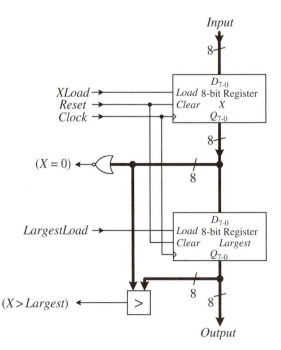

Figure 11.21
Dedicated datapath for the finding the largest number problem of Example 11.5.

test $(X = 0)$ is generated by the 8-input NOR gate, and the conditional test $(X > Largest)$ is generated by the greater-than comparator.

This algorithm is very similar to the summing input numbers problem in Example 11.3, especially the situation for handling the Enter switch for inputting each number through the same input port. The state diagram is shown in Figure 11.22(a).

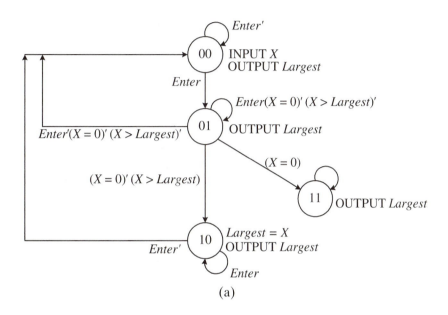

(a)

Current State	Next State (Implementation)							
$Q_1 Q_0$	$Q_{1\text{next}} Q_{0\text{next}}$ $(D_1\ D_0)$							
	Enter, $(X = 0)$, $(X > Largest)$							
	000	001	010	011	100	101	110	111
00	00	00	00	00	01	01	01	01
01	00	10	11	11	01	10	11	11
10	00	00	00	00	10	10	10	10
11	11	11	11	11	11	11	11	11

(b)

Figure 11.22   Control unit for finding the largest number problem of Example 11.5: (a) state diagram; (b) next-state (implementation) table; (c) K-maps and excitation equations; (d) control words and output table; (e) output equations; (f) circuit.
(continued on next page)

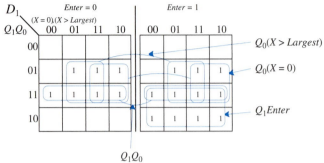

$$D_1 = Q_1Q_0 + Q_1 Enter + Q_0(X = 0) + Q_0(X > Largest)$$

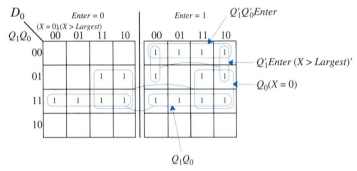

$$D_0 = Q_1Q_0 + Q_0(X = 0) + Q'_1Q'_0 Enter + Q'_1 Enter (X > Largest)'$$

(c)

Control Word	State Q_1Q_0	Instruction	XLoad	LargestLoad	Done
0	00	INPUT X, OUTPUT Largest	1	0	0
1	01	OUTPUT Largest	0	0	0
2	10	Largest $= X$, OUTPUT Largest	0	1	0
3	11	OUTPUT Largest	0	0	1

(d)

Figure 11.22 Control unit for finding the largest number problem of Example 11.5: (a) state diagram; (b) next-state (implementation) table; (c) K-maps and excitation equations; (d) control words and output table; (e) output equations; (f) circuit.
(continued on next page)

$$XLoad = Q_1' Q_0'$$

$$LargestLoad = Q_1 Q_0'$$

$$Done = Q_1 Q_0$$

(e)

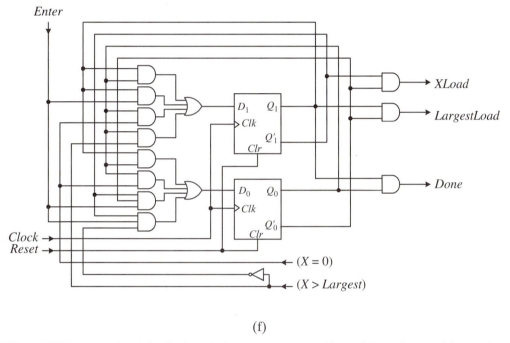

(f)

Figure 11.22   Control unit for finding the largest number problem of Example 11.5: (a) state diagram; (b) next-state (implementation) table; (c) K-maps and excitation equations; (d) control words and output table; (e) output equations; (f) circuit.

State 00 inputs X and waits for the *Enter* signal. This allows the user to set up the input number and then press the Enter switch. When the Enter switch is pressed, the FSM goes to state 01 to make two tests. If $(X = 0)$ is true, the FSM terminates in state 11. If $(X > Largest)$ is true, then the FSM goes to state 10 to assign X as the new largest number. If X is not 0 and not the largest, then the FSM will go back to state 00 to wait for another input number. Before going back to state 00 from both states 01 and 10, we need to wait for the release of the Enter switch as explained in Example 11.3.

The next-state table, as derived from the state diagram, is shown in Figure 11.22(b). The table requires five variables: two to encode the four states, Q_1 and Q_0, and three for the status signals, *Enter*, $(X = 0)$, and $(X > Largest)$.

Using D flip-flops to implement the state memory, the implementation table is the same as the next-state table, except that the values in the table entries are for the inputs to the flip-flops, D_1 and D_0. The K-maps and the excitation equations for D_1 and D_0 are shown in Figure 11.22(c).

The control words and output table for the three control signals are shown in Figure 11.22(d). State 00 performs lines 2 and 8 of the algorithm in Figure 11.20 by asserting the *XLoad* signal. All of the states output *Largest*, and this action does not require any control signals. State 10 performs line 5 by asserting *LargestLoad*. State 11 outputs a *Done* signal to inform the user that the FSM has stopped.

The output equations, as derived from the output table, are shown in Figure 11.22(e). There is one equation for each of the three control signals.

The complete control unit circuit is shown in Figure 11.22(f). The state memory consists of two D flip-flops. The inputs to the flip-flops are the next-state circuits derived from the two excitation equations. The output circuits for the three control signals are derived from the three output equations. The status signal $(X = 0)$ comes from the NOR-gate comparator in the datapath, and the status signal $(X > Largest)$ comes from the greater-than comparator in the datapath.

Connecting the control unit and the datapath together using the control and status signals produces the final microprocessor, as shown in Figure 11.23. A sample simulation is shown in Figure 11.24. This complete microprocessor circuit is on the accompanying CD-ROM in the file MP.GDF and can be synthesized and simulated.

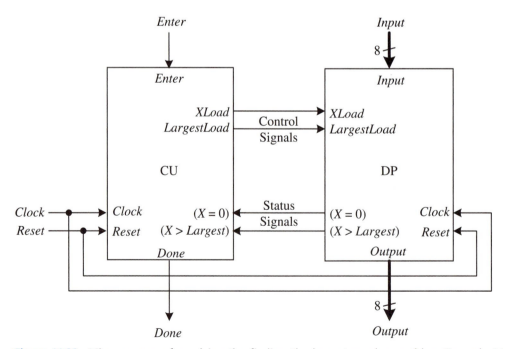

Figure 11.23 Microprocessor for solving the finding the largest number problem Example 11.5.

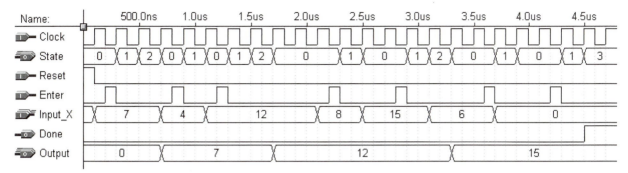

Figure 11.24   Sample simulation trace for the finding the largest number problem. The last output number (15) when *Done* is asserted is the largest of the six numbers entered (7, 4, 12, 8, 15, and 6).

The top-level interface circuit for testing this microprocessor on the UP2 development board using the MAX chip is shown in Figure 11.25 and is in the file UP2MAX.GDF on the accompanying CD-ROM. Like the circuit for Example 11.2, this circuit displays the largest number on the eight LEDs in binary and the current state value, Q_1Q_0, on the 7-segment LED. Again, a debounce circuit is needed for the Enter switch. The 8-bit number inputs are through the eight DIP switches. The FLEX version of this circuit is in the file UP2FLEX.GDF. For the FLEX version, instead of displaying the largest number on the eight discrete LEDs, it displays the number on the two 7-segment displays.

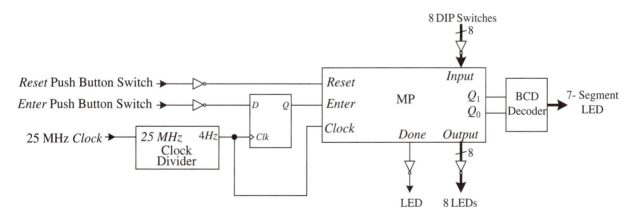

Figure 11.25   Interface to the finding the largest number microprocessor for the UP2 development board.

• • • • • • • • • • • • • • • • • •

11.3 VHDL for Dedicated Microprocessors

In the introduction to this chapter, we mentioned that we can use VHDL based on different models to automatically synthesize a microprocessor. The FSM+D model requires the manual construction of the datapath, but the FSM is described using VHDL and is synthesized automatically. The FSMD model also uses VHDL to describe the FSM, but in addition, the datapath operations are imbedded within the FSM VHDL code using built-in VHDL operators. Finally, the behavioral model uses VHDL to describe the behavior of the complete microprocessor algorithmically, and so, both the control unit and the datapath are synthesized automatically.

11.3.1 FSM+D Model

In Section 11.1, we constructed a dedicated microprocessor by manually constructing the circuits for a control unit and a datapath. These two components are then connected together to form the microprocessor. Instead of manually constructing the control unit circuit, an alternative way is to automatically synthesize it from behavioral VHDL code. A separate datapath circuit is still constructed manually like before. The two components then are joined together to form the complete microprocessor.

Writing the behavioral VHDL code for the control unit was discussed in Chapter 10. The method is again illustrated in Example 11.6. The final dedicated microprocessor is constructed using an enclosing VHDL entity written at the structural level that combines the VHDL code for both the control unit entity and the datapath entity.

| Example 11.6 | **Using the FSM+D model to synthesize a microprocessor for the summation algorithm** |

In this example, we will again construct the dedicated microprocessor for the summation algorithm of Example 11.1. However, unlike Example 11.1 where the circuit for the control unit is derived manually, we will use VHDL to automatically synthesize the control unit circuit. In order to write the behavioral VHDL code for the control unit, we need to use the information from the state diagram and the control words derived for the algorithm. The control words were derived manually in Example 9.10, and the state diagram was derived manually in Example 10.5. The state diagram and control words from these two examples are repeated in Figure 11.26(a) and (b), respectively for convenience.

The complete behavioral VHDL code for the control unit is shown in Figure 11.27. In the entity section, the output signals include the 16 control signals for controlling the datapath. There is one input signal, *neq0*, in the entity section, which is the status signal ($n \neq 0$) from the datapath. In addition to the control and status signals, there is the control input signal, *Start*, and the control output signal, *Done*. Finally, there are the two global input signals, *Clock* and *Reset*. The *Reset* signal is to reset all of the flip-flops in the state memory to zero.

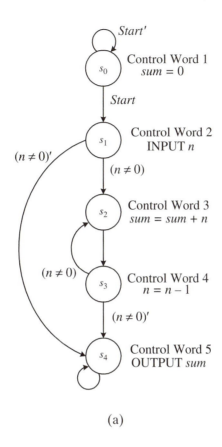

(a)

Control	Instruction	IE	WE	$WA_{1,0}$	RAE	$RAA_{1,0}$	RBE	$RBA_{1,0}$	$ALU_{2,1,0}$	$SH_{1,0}$	OE
Word		15	14	13–12	11	10–9	8	7–6	5–3	2–1	0
1	$sum = 0$	0	1	00	1	00	1	00	101 (subtract)	00	0
2	INPUT n	1	1	01	0	××	0	××	×××	××	0
3	$sum = sum + n$	0	1	00	1	00	1	01	100 (add)	00	0
4	$n = n - 1$	0	1	01	1	01	0	××	111 (decrement)	00	0
5	OUTPUT sum	×	0	××	1	00	0	××	000 (pass)	00	1

(b)

Figure 11.26   For Example 11.6: (a) state diagram; (b) control words; (c) datapath.
(continued on next page)

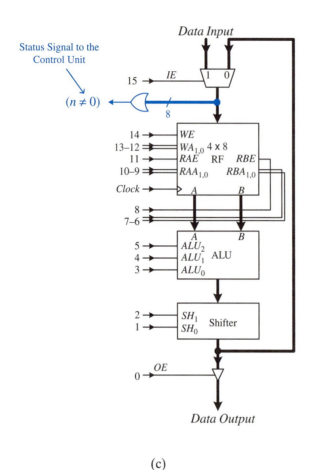

(c)

Figure 11.26 For Example 11.6: (a) state diagram; (b) control words; (c) datapath.

Figure 11.27
FSM+D model of the
control unit for the
summation
algorithm.
(continued on next page)

```
LIBRARY IEEE;
USE IEEE.STD_LOGIC_1164.ALL;

ENTITY fsm IS PORT (
    Clock, Reset, Start: IN STD_LOGIC;
    IE: OUT STD_LOGIC;
    WE: OUT STD_LOGIC;
    WA: OUT STD_LOGIC_VECTOR(1 DOWNTO 0);
```

```
              RAE: OUT STD_LOGIC;
              RAA: OUT STD_LOGIC_VECTOR(1 DOWNTO 0);
              RBE: OUT STD_LOGIC;
              RBA: OUT STD_LOGIC_VECTOR(1 DOWNTO 0);
              aluSel: OUT STD_LOGIC_VECTOR(2 DOWNTO 0);
              shSel: OUT STD_LOGIC_VECTOR(1 DOWNTO 0);
              OE: OUT STD_LOGIC;
              Done: OUT STD_LOGIC;
              neq0: IN STD_LOGIC);
        END fsm;

        ARCHITECTURE fsm_arc OF fsm IS
            TYPE state_type IS (s0, s1, s2, s3, s4);
            SIGNAL state: state_type;
        BEGIN
            next_state_logic: PROCESS(Reset, Clock)
            BEGIN
               IF (Reset = '1') THEN
                    state <= s0;
               ELSIF (Clock'EVENT AND Clock = '1') THEN
                  CASE state IS
                  WHEN s0 =>
                     IF (Start = '1') THEN state <= s1; ELSE state <= s0; END IF;
                  WHEN s1 =>
                     IF (neq0 = '1') THEN state <= s2; ELSE state <= s4; END IF;
                  WHEN s2 =>
                     state <= s3;
                  WHEN s3 =>
                     IF (neq0 = '1') THEN state <= s2; ELSE state <= s4; END IF;
                  WHEN s4 =>
                     state <= s4;
                  WHEN OTHERS =>
                     state <= s0;
                  END CASE;
               END IF;
            END PROCESS;

            output_logic: PROCESS(state)
            BEGIN
               CASE state IS
               WHEN s0 =>
                  IE<='0'; WE<='1'; WA<="00"; RAE<='1'; RAA<="00"; RBE<='1';
                  RBA<="00"; aluSel<="101"; shSel<="00"; OE<='0'; Done<='0';
```

Figure 11.27
FSM+D model of the control unit for the summation algorithm.
(continued on next page)

```
            WHEN s1 =>
                IE<='1'; WE<='1'; WA<="01"; RAE<='0'; RAA<="00"; RBE<='0';
                RBA<="00"; aluSel<="000"; shSel<="00"; OE<='0'; Done<='0';
            WHEN s2 =>
                IE<='0'; WE<='1'; WA<="00"; RAE<='1'; RAA<="00"; RBE<='1';
                RBA<="01"; aluSel<="100"; shSel<="00"; OE<='0'; Done<='0';
            WHEN s3 =>
                IE<='0'; WE<='1'; WA<="01"; RAE<='1'; RAA<="01"; RBE<='0';
                RBA<="00"; aluSel<="111"; shSel<="00"; OE<='0'; Done<='0';
            WHEN s4 =>
                IE<='0'; WE<='0'; WA<="00"; RAE<='1'; RAA<="00"; RBE<='0';
                RBA<="00"; aluSel<="000"; shSel<="00"; OE<='1'; Done<='1';
            WHEN others =>
                IE<='0'; WE<='0'; WA<="00"; RAE<='0'; RAA<="00"; RBE<='0';
                RBA<="00"; aluSel<="000"; shSel<="00"; OE<='0'; Done<='0';
            END CASE;
        END PROCESS;
    END fsm_arc;
```

Figure 11.27
FSM+D model of the
control unit for the
summation
algorithm.

The architecture section starts out with using the TYPE statement to define all of the states used in the state diagram. The SIGNAL statement declares the signal *state* to remember the current state of the FSM. There are two processes in the architecture section, the next-state logic process and the output logic process, that execute concurrently. As the name suggests, the next-state process defines the next-state logic circuit that is inside the control unit, and the output logic process defines the output logic circuit inside the control unit. The main statement within these two processes is the CASE statement that determines what the current state is. For the next-state process, the CASE statement is executed only at the rising clock edge because of the test (*Clock*'EVENT AND *Clock* = '1') in the IF statement. Hence, the *state* signal is assigned a new state value only at the rising clock edge. The new state value is, of course, dependent on the current state and input signals, if any. For example, in state s_0, the next state is dependent on the input signal *Start*, whereas, in state s_2, the next state is unconditionally s_3.

In the output process, all control signals are generated for every case (i.e., all of the control signals must be assigned a value in every case). Recall from Section 6.13.1 that VHDL synthesizes a signal using a memory element if the signal is not assigned a value in all possible cases. However, the output circuit is a combinational circuit, and it should not contain any memory elements. For each state in the CASE statement in the output process, the values assigned to each of the output signal are taken directly from the control words and output table.

For the datapath, we are using the same general datapath shown in Figure 9.32 as discussed in Section 9.6. For convenience, this datapath is repeated in Figure 11.26(c). The VHDL code for this datapath is listed in Figures 9.43 and 9.44. Figure 9.43 describes the individual components used in the datapath, and Figure 9.44 combines these components into the desired datapath. One extra concurrent signal assignment statement is added to the listing in Figure 9.44 to generate the status signal, $(n \neq 0)$. The name, *neq0*, is used for this status signal, and it is asserted if the output of the multiplexer is not equal to 0; otherwise, it is de-asserted.

Finally, Figure 11.28 combines the datapath and the control unit together using structural VHDL coding to produce the top-level entity *sum* for the microprocessor. The entity section specifies the primary I/O signals for the microprocessor. In addition to the global *Clock* and *Reset* signals, there are the four I/O signals shown in Figure 11.2. These are the control input *Start* signal, the control output *Done* signal, the datapath *Input*, and the datapath *Output*. The architecture section declares the two components, *fsm* and *datapath*, used in this module. These two components are connected together using the 16 control signals and one status signal, and are declared as internal signals using the SIGNAL declaration statement. Finally, two PORT MAP statements are used to actually connect the control unit and the datapath together.

```
LIBRARY IEEE;
USE IEEE.STD_LOGIC_1164.ALL;

ENTITY sum IS PORT (
    Clock, Reset, Start: IN STD_LOGIC;
    Input: IN STD_LOGIC_VECTOR(7 DOWNTO 0);
    Done: OUT STD_LOGIC;
    Output: OUT STD_LOGIC_VECTOR(7 DOWNTO 0));
END sum;

ARCHITECTURE Structural OF sum IS

COMPONENT fsm PORT (
    Clock, Reset, Start: IN STD_LOGIC;
    IE: OUT STD_LOGIC;
    WE: OUT STD_LOGIC;
    WA: OUT STD_LOGIC_VECTOR(1 DOWNTO 0);
    RAE: OUT STD_LOGIC;
    RAA: OUT STD_LOGIC_VECTOR(1 DOWNTO 0);
    RBE: OUT STD_LOGIC;
    RBA: OUT STD_LOGIC_VECTOR(1 DOWNTO 0);
    aluSel: OUT STD_LOGIC_VECTOR(2 DOWNTO 0);
    shSel: OUT STD_LOGIC_VECTOR(1 DOWNTO 0);
```

Figure 11.28   FSM+D model of the microprocessor for the summation algorithm. *(continued on next page)*

```
          OE: OUT STD_LOGIC;
          Done: OUT STD_LOGIC;
          neq0: IN STD_LOGIC);
      END COMPONENT;

      COMPONENT datapath PORT (
          Clock: IN STD_LOGIC;
          Input: IN STD_LOGIC_VECTOR(7 DOWNTO 0);
          IE, WE: IN STD_LOGIC;
          WA: IN STD_LOGIC_VECTOR(1 DOWNTO 0);
          RAE: IN STD_LOGIC;
          RAA: IN STD_LOGIC_VECTOR(1 DOWNTO 0);
          RBE: IN STD_LOGIC;
          RBA: IN STD_LOGIC_VECTOR(1 DOWNTO 0);
          aluSel : IN STD_LOGIC_VECTOR(2 DOWNTO 0);
          shSel: IN STD_LOGIC_VECTOR(1 DOWNTO 0);
          OE: IN STD_LOGIC;
          Output: OUT STD_LOGIC_VECTOR(7 DOWNTO 0);
          neq0: OUT STD_LOGIC);
      END COMPONENT;

      SIGNAL sIE, sWE: STD_LOGIC;
      SIGNAL sWA: STD_LOGIC_VECTOR(1 DOWNTO 0);
      SIGNAL sRAE: STD_LOGIC;
      SIGNAL sRAA: STD_LOGIC_VECTOR(1 DOWNTO 0);
      SIGNAL sRBE: STD_LOGIC;
      SIGNAL sRBA: STD_LOGIC_VECTOR(1 DOWNTO 0);
      SIGNAL sAluSel: STD_LOGIC_VECTOR(2 DOWNTO 0);
      SIGNAL sShSel: STD_LOGIC_VECTOR(1 DOWNTO 0);
      SIGNAL sOE: STD_LOGIC;
      SIGNAL sneq0: STD_LOGIC;

      BEGIN

          -- doing structural modeling here

          -- FSM control unit
          U0: fsm PORT MAP (Clock,Reset,Start,sIE,sWE,sWA,sRAE,sRAA,sRBE,sRBA,
                            sAluSel,sShSel,sOE,Done,sneq0);

          -- Datapath
          U1: datapath PORT MAP (Clock,Input,sIE,sWE,sWA,sRAE,sRAA,sRBE,sRBA,
                            sAluSel,sShSel,sOE,Output,sneq0);

      END Structural;
```

Figure 11.28   FSM+D model of the microprocessor for the summation algorithm.

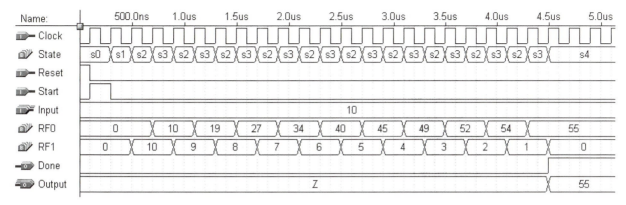

Figure 11.29   Simulation trace for the FSM+D summation algorithm with input $n = 10$.

A sample simulation trace is shown in Figure 11.29 for the input $n = 10$. After asserting the *Start* signal in state s_0, the input value 10 (decimal) is read in during state s_1. The value 10 is written to RF1 at the next rising clock edge, which also brings the FSM to state s_2. RF1 is for storing n that counts from 10 down to 0. RF0 is for storing *sum* and is updated after each count of n. When RF1 reaches zero, the sum 55 from RF0 is sent to the output, and the *Done* signal is asserted. At this point, the FSM stays in state s_4 until it is reset.

11.3.2   FSMD Model

When writing VHDL code using the FSM+D model, we need to manually construct the datapath module. The FSM module can be constructed either manually (as in Example 10.5) or synthesized automatically (as in Example 11.6). These two modules are then joined together at the structural level using the control signals and status signals. Although we can automatically synthesize the FSM, using the FSM+D method to build a microprocessor with a large datapath still requires a lot of work. We need to first manually construct the datapath, and then we need to connect the numerous control and status signals together between the two modules.

The **FSMD** (FSM *with* datapath) model is an abstraction used when we write VHDL code for a microprocessor. Instead of separating the FSM and the datapath into two different modules, we combine the FSM and the datapath together into the same entity. In other words, all of the data operations that need to be performed by the datapath actually are imbedded in the FSM coding itself as VHDL data manipulation statements. After all, there are built-in VHDL operators to perform data manipulations. As a result, when writing VHDL code for the FSMD model, we do not need to manually construct the datapath module, and therefore, there are no control and status signals to be connected together. The FSM module is written using behavioral VHDL code just like with the FSM+D model. Thus, the FSMD model is an abstraction for simplifying the VHDL code for a microprocessor.

Using the FSM+D model allows us to have full control as to what components are used in the datapath and how these components are used by the FSM. Whereas, using the FSMD model automates the datapath construction process. The synthesizer now decides what components are needed by the datapath. However, whether you use the FSM+D model or the FSMD model, the FSM is still constructed the same way, that is, using behavioral VHDL code. It is just that, for the FSMD model, the VHDL code for the FSM has extra data manipulation statements imbedded in it. Because the FSM is still written manually, you still have full control as to what instructions are executed in what state and the number of states needed to execute the algorithm. Example 11.7 shows the process to construct a microprocessor using the FSMD model.

Example 11.7

Using the FSMD model to synthesize a microprocessor for the summation algorithm

Figure 11.30 shows the complete microprocessor VHDL code using the FSMD model for the summation algorithm of Example 11.1. Notice the simplicity of this code as compared to the code for the FSM+D model shown in Example 11.6. Here, we have just one entity, which serves as both the top-level microprocessor entity and the FSMD entity. The FSMD entity includes both the control unit and the datapath. The architecture section is written similar to a regular FSM with the different cases for each of the states. Like before for the next-state process, the CASE statement selects the current state and determines the next state. But in addition to setting the next state, each case (state) also contains data operation statements such as $sum <= sum + n$. Because the control unit and the datapath are combined together into one module, therefore, the control signals and status signals are no longer needed for joining them. The primary output signals are still generated in the output process. A sample simulation trace is shown in Figure 11.31 for the input $n = 10$.

```
LIBRARY IEEE;
USE IEEE.STD_LOGIC_1164.ALL;
USE IEEE.STD_LOGIC_UNSIGNED.ALL;

ENTITY sum IS PORT (
    Clock, Reset, Start: IN STD_LOGIC;
    Input: IN STD_LOGIC_VECTOR(7 DOWNTO 0);
    Done: OUT STD_LOGIC;
    Output: OUT STD_LOGIC_VECTOR(7 DOWNTO 0));
END sum;

ARCHITECTURE FSMD OF sum IS
    TYPE state_type IS (s0, s1, s2, s3, s4);
    SIGNAL state: state_type;
```

Figure 11.30
FSMD model of the microprocessor for the summation algorithm.
(continued on next page)

```vhdl
                SIGNAL sum: STD_LOGIC_VECTOR(7 DOWNTO 0);
                SIGNAL n: STD_LOGIC_VECTOR(7 DOWNTO 0);

BEGIN
   next_state_logic: PROCESS(Reset, Clock)
   BEGIN
      IF(Reset = '1') THEN
         state <= s0;
      ELSIF(Clock'EVENT AND Clock = '1') THEN
         CASE state IS
         WHEN s0 =>
            IF (Start = '1') THEN state <= s1; ELSE state <= s0; END IF;
            sum <= (OTHERS => '0');
         WHEN s1 =>
            -- need to test with Input and not n because n has not been updated yet
            IF (Input /= 0) THEN state <= s2; ELSE state <= s4; END IF;
            n <= Input;
         WHEN s2 =>
            state <= s3;
            sum <= sum + n;
         WHEN s3 =>
            -- reading n in the following IF statement is BEFORE the decrement
            -- update, therefore, we need to compare with 1 and not 0
            n <= n - 1;
            IF (n /= 1) THEN state <= s2; ELSE state <= s4; END IF;
         WHEN s4 =>
            state <= s4;
         WHEN OTHERS =>
            state <= s0;
         END CASE;
      END IF;
   END PROCESS;

   output_logic: PROCESS(state)
   BEGIN
      CASE state IS
      WHEN s4 =>
         Done <= '1';
         Output <= sum;
      WHEN OTHERS =>
         Done <= '0';
         Output <= (OTHERS => 'Z');
      END CASE;
   END PROCESS;
END FSMD;
```

Figure 11.30
FSMD model of the
microprocessor for
the summation
algorithm.

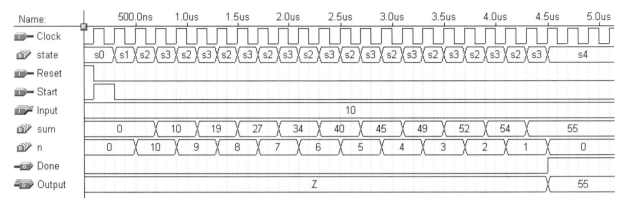

Figure 11.31 Simulation trace of the FSMD summation algorithm with input $n = 10$.

11.3.3 Behavioral Model

The complete microprocessor can also be designed by writing VHDL code in a truly behavioral style so that both the FSM and the datapath are synthesized automatically. Using the behavioral model to design a circuit is quite similar to writing computer programs using a high-level language. The synthesizer, like the compiler, will translate the VHDL behavioral description of the circuit automatically to a netlist. This netlist can then be programmed directly onto a FPGA chip.

Since the synthesizer automatically constructs both the FSM and the datapath, you have no control over what components are used in the datapath and what control words are executed in what state of the FSM. Not being able to decide what components are used in the datapath is not too big of a problem, because the synthesizer does do a good job in deciding that for you. The issue is with not being able to specify what control words are executed in what state of the FSM. This is purely a timing issue. In some timing-critical applications (such as communication protocols and real-time controls) we need to control exactly in what clock cycle a certain register-transfer operation is performed. In other words, we need to be able to assign a control word to a specific state of the FSM.

Behavioral VHDL code offers all of the basic language constructs that are available in most computer programming languages, such as variable assignments, FOR LOOPS, and IF-THEN-ELSES. These statements, written within a process block, are executed sequentially. Besides the fact that we have no control over which clock cycle each operation is performed in, using the behavioral model is very powerful and simple. If timing is not an issue for your circuit, then between the different models, this should be the method of choice.

Example 11.8 — **Using the behavioral model to synthesize a microprocessor for the summation algorithm**

Figure 11.32 shows the VHDL code using the behavioral model for the summation algorithm of Example 11.1. Note that some VHDL synthesizers (such as MAX+plus II) do not allow the use of loops that cannot be unrolled (i.e., loops with variables for

```vhdl
LIBRARY IEEE;
USE IEEE.STD_LOGIC_1164.ALL;

ENTITY sum IS PORT (
   Start: IN STD_LOGIC;
   Done: OUT STD_LOGIC;
   Output: OUT INTEGER);
END sum;

ARCHITECTURE Behavioral OF sum IS
BEGIN
   PROCESS
      VARIABLE n: INTEGER;
      VARIABLE sum: INTEGER;
   BEGIN
      IF (Start = '0') THEN
         Done <= '0';
         Output <= 0;
      ELSE
         sum := 0;
         n := 10;
         FOR i in n DOWNTO 1 LOOP
            sum := sum + i;
         END LOOP;
         Done <= '1';
         Output <= sum;
      END IF;
   END PROCESS;
END Behavioral;
```

Figure 11.32
Behavioral model of
the microprocessor
for the summation
algorithm.

their starting or ending value whose values are unknown at compile time). Hence, in the code, the value for the variable n cannot be from a user input. A sample simulation trace is shown in Figure 11.33 for $n = 10$.

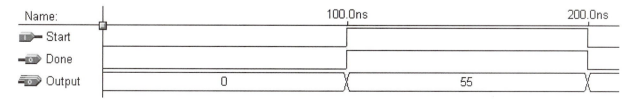

Figure 11.33   Simulation trace of the behvioral summation algorithm with $n = 10$.

11.4 Summary Checklist

- Be able to manually construct the complete circuit for a dedicated microprocessor from a given algorithm

- Be able to construct a dedicated microprocessor from a given algorithm using the FSM+D model with VHDL

- Be able to construct a dedicated microprocessor from a given algorithm using the FSMD model with VHDL

- Be able to construct a dedicated microprocessor from a given algorithm using the behavioral model with VHDL

11.5 Problems

P11.1. Manually construct a dedicated microprocessor to enter two 8-bit numbers, and output the larger of the two numbers. The two numbers are entered through two separate input ports.

P11.2. Manually construct a dedicated microprocessor to enter one 8-bit number. Output a 1 if the number has five 1 bits; otherwise, output a 0.

P11.3. Manually construct a dedicated microprocessor to enter two 8-bit numbers. Output a 1 if the two numbers together have five 1 bits; otherwise, output a 0.

P11.4. Manually construct a dedicated microprocessor to enter two numbers, and output the product of these two numbers.

P11.5. Manually construct a dedicated microprocessor to enter two numbers. Output a 1 if the first number is divisible by the second number; otherwise, output a 0.

P11.6. Manually construct a dedicated microprocessor to enter three numbers, and output the larger of the three numbers.

P11.7. Manually construct a dedicated microprocessor to enter three numbers. Output the three numbers in ascending order.

P11.8. Manually construct a dedicated microprocessor to evaluate the factorial of n. The algorithm is shown in Figure P11.8.

Figure P11.8

```
product = 1
INPUT n
WHILE (n > 1){
   product = product * n
   n = n - 1
   }
OUTPUT product
```

P11.9. Manually construct a dedicated microprocessor to enter several numbers until a 0 is entered. Output the largest and second largest of the numbers entered.

P11.10. Manually construct a dedicated microprocessor to input one 8-bit value, and then determine whether the input value has an equal number of 0 and 1 bits. The microprocessor outputs a 1 if the input value has the same number of 0's and 1's; otherwise, it outputs a 0. For example, the number 10111011 will produce a 0 output; whereas, the number 00110011 will produce a 1 output. The algorithm is shown in Figure P11.10.

Figure P11.10

```
1   Count = 0                  // for counting the number of 1 bits
2   INPUT N
3   WHILE (N ≠ 0){
4      IF (N(0) = 1) THEN      // least significant bit of N
5         Count = Count + 1
6      END IF
7      N = N >> 1              // shift N right one bit
8   }
9   OUTPUT (Count = 4)         // output 1 if the test (Count = 4) is true
```

P11.11. Assume that the control unit and datapath circuits in Figure P11.11 are used to construct a dedicated microprocessor. Determine the instructions being executed in each state of the FSM. Write out the complete algorithm that the resulting microprocessor will execute. In other words, write out the pseudocode for the algorithm. Briefly describe what the algorithm does.

P11.12. Write the VHDL code for Problems P11.1 to P11.10 using the FSM+D model.

P11.13. Write the VHDL code for Problems P11.1 to P11.10 using the FSMD model.

P11.14. Implement the microprocessor for Problems P11.1 to P11.10 on the UP2 development board.

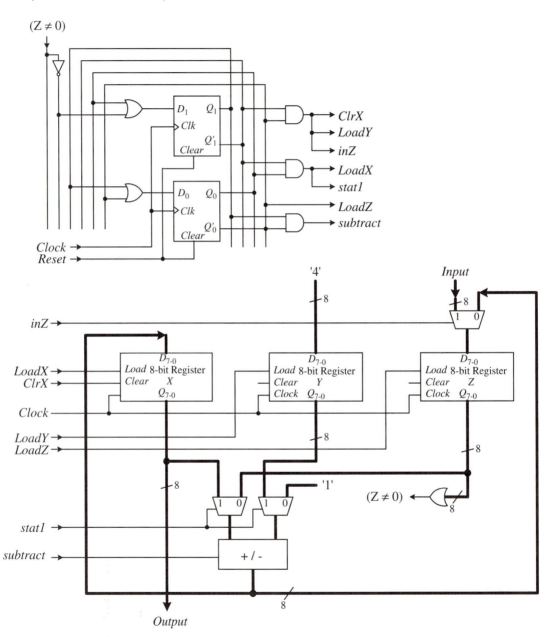

Figure P11.11

CHAPTER 12

General-Purpose Microprocessors

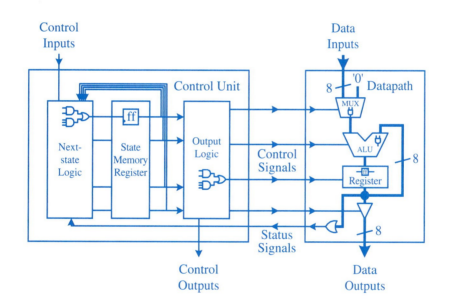

Unlike a dedicated or custom microprocessor that is capable of performing only one function, a general-purpose microprocessor, such as the Pentium® CPU, is capable of performing many different functions under the direction of program instructions. Given a different instruction set or program, the general-purpose microprocessor will perform a different function. However, a general-purpose microprocessor can also be viewed as a dedicated microprocessor, because it is made to perform only one function, and that is to execute the program instructions. In this sense, we can design and construct a general-purpose microprocessor in the same way that we constructed the dedicated microprocessors in the last chapter.

12.1 Overview of the CPU Design

A general-purpose microprocessor is often referred to as the **central processing unit (CPU)**. The CPU is simply a dedicated microprocessor that only executes software instructions. Figure 12.1 shows an overview of a general-purpose microprocessor. The following discussion references this diagram.

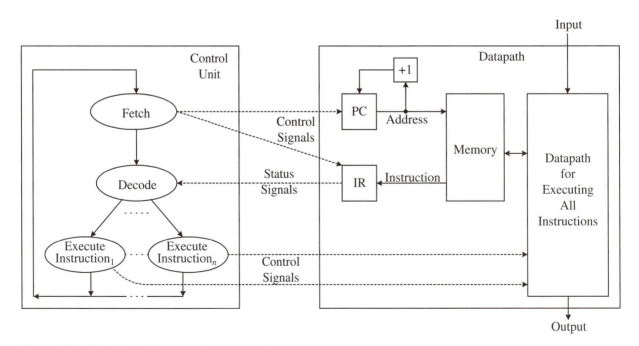

Figure 12.1   Overview of a general-purpose microprocessor.

In designing a CPU, we must first define its instruction set and how the instructions are encoded and executed. We need to answer questions such as:

- How many instructions do we want?
- What are the instructions?
- What binary encoding (normally referred to as the **operation code** or **opcode**) do we assign to each of the instructions?
- How many bits do we use to encode an instruction?

Once we have decided on the instruction set, we can proceed to designing a datapath that can execute all of the instructions in the instruction set. In this step, we are creating a custom datapath, so we need to answer questions such as:

- What functional units do we need?
- How many registers do we need?
- Do we use a single register file or separate registers?
- How are the different units connected together?

Creating the datapath for a general-purpose microprocessor is exactly the same as creating the datapath for a dedicated microprocessor. However, in addition to being able to perform all of the instructions in the instruction set, there are other data operations and registers that must be included in the datapath for the general-purpose microprocessor. These data operations and registers deal with how the general-purpose microprocessor fetches the instructions from memory and executes them. In particular, there is a **program counter (*PC*)** that contains the memory location of where the next instruction is stored. There is also an **instruction register (*IR*)** for storing the instruction being fetched from the memory. Each time an instruction is fetched from a memory location pointed to by the *PC*, the *PC* normally must be incremented to the next memory location for the next instruction. Alternatively, if the instruction is a jump instruction, the *PC* must be loaded with a new memory address instead.

The control unit for a general-purpose microprocessor basically cycles through three main steps, usually referred to as the **instruction cycle**:

Step 1 fetches an instruction

Step 2 decodes the instruction

Step 3 executes the instruction

Each step is executed in one state of the finite state machine. For Step 3, each instruction usually is executed in one clock cycle, although some memory access instructions may require two or more clock cycles to complete. Hence, they may require several states for correct timing.

For fetching the instruction in Step 1, the control unit simply reads the memory location specified by the *PC* and copies the content of that location into the instruction register (*IR*). The *PC* is then incremented by 1 (assuming that each instruction occupies one memory location). For decoding the instruction in Step 2, the control unit extracts the opcode bits from the instruction register and determines what the current instruction is by jumping to the state that is assigned for executing that instruction. Once in that particular state, the control unit performs Step 3 by simply asserting the appropriate control signals for controlling the datapath to execute that instruction.

Instructions for the program usually are stored in external memory, so in addition to the CPU, there is external memory that is connected to the CPU via an address bus and a data bus. Hence, Step 1 (fetch an instruction) usually involves the control unit setting up a memory address on the address bus and telling the external memory to output the instruction from that memory location onto the data bus. The control unit then reads the instruction from the data bus. To keep our design simple, instead of having external memory, we will include the memory as part of the datapath. This way, we do not have to worry about the handshaking and timing issues involved for accessing external memory.

● ● ● ● ● ● ● ● ● ● ● ● ● ● ● ● ●

12.2 The EC-1 General-Purpose Microprocessor

This first version of the EC[1] computer is extremely small and very limited as to what it can do, and therefore, its general-purpose microprocessor is very "E-Cee" to design manually. In order to keep the manual design of the microprocessor manageable, we have to keep the number of variables small. Since these variables determine the number of states and input signals for the finite state machine, these factors have to be kept to the bare minimum. Nevertheless, the building of this computer demonstrates how a general-purpose microprocessor is designed and how the different components are put together. After this exercise, you quickly will appreciate the power of designing with VHDL at a higher abstraction level and the use of an automatic synthesizer.

We will now design the general-purpose microprocessor for our EC-1 computer. After which, we will interface this microprocessor to external I/Os, and implement the complete computer using the FLEX FPGA chip on the UP2 development board to make it into a real-working general-purpose computer. Using the few instructions available, we will then write a program to execute on the EC-1 and see that it actually works!

12.2.1 Instruction Set

The instructions that our EC-1 general-purpose microprocessor can execute and the corresponding encodings are defined in Figure 12.2. The *Instruction* column shows the syntax and mnemonic to use for the instruction when writing a program in assembly language. The *Encoding* column shows the binary encoding defined for the instruction, and the *Operation* column shows the operation of the instruction.

---

[1] "EC" is the acronym for Enoch's computer.

Instruction	Encoding	Operation	Comment
IN A	011 ×××××	$A \leftarrow Input$	Input to A
OUT A	100 ×××××	$Output \leftarrow A$	Output from A
DEC A	101 ×××××	$A \leftarrow A - 1$	Decrement A
JNZ address	110 × aaaa	IF (A != 0) THEN PC = aaaa	Jump to address if A is not zero
HALT	111 ×××××	$Halt$	Halt execution

Notations:

    A = accumulator

    PC = program counter

    aaaa = four bits for specifying a memory address

    × = don't-cares

Figure 12.2   Instruction set for the EC-1.

As we can see from Figure 12.2, our little computer's instruction set has only five instructions. To encode five instructions, the operation code (opcode) will require three bits—giving us eight different combinations. As shown in the *Encoding* column, the first three most significant bits are the opcode given to an instruction. For example, the opcode for the IN A instruction is 011, the opcode for OUT A is 100, and so on. The three encodings, 000, 001, and 010, are not defined and so can be used as a "no-operation" (NOP) instruction. Since the width of each instruction is fixed at 8 bits, the last 5 bits are not used by all of the instructions, except for the JNZ (Jump Not Zero) instruction. Normally, for a more extensive instruction set, these extra bits are used as operand bits to specify what registers or other resources to use. In our case, only the JNZ instruction uses the last 4 bits, designated as aaaa, to specify an address in the memory to jump to.

The IN A instruction inputs an 8-bit value from the data input port, *Input*, and stores it into the **accumulator (A)**. The accumulator is an 8-bit register for performing data operations. The OUT A instruction copies the content of the accumulator to the output port, *Output*. For the EC-1, the content of the accumulator is always available at the output port, so this OUT A instruction really is not necessary. It is included just because a program should have an output instruction. The DEC A instruction decrements the content of A by 1 and stores the result back into A. The JNZ (Jump Not Zero) instruction tests to see if the value in A is equal to 0 or not. If A is equal to 0, then nothing is done. If A is not equal to 0, then the last 4 bits (aaaa) of the instruction is loaded into the PC. When this value is loaded into the PC, we essentially are performing a jump to this new memory address, since the value stored in the PC is the location for the next fetch operation. Finally, the HALT instruction halts the CPU by having the control unit stay in the *Halt* state indefinitely until reset.

12.2.2 Datapath

Having defined the instruction set for the EC-1 general-purpose microprocessor, we are now ready to design the custom datapath that will execute all of the operations as defined by all of the instructions. The custom datapath for the EC-1 is shown in Figure 12.3.

The datapath can be viewed as having three separate parts: (1) for performing the instruction cycle operations of fetching an instruction and incrementing or loading the PC, (2) the memory, and (3) for performing the data operations for all of the instructions in the instruction set.

The portion of the datapath for performing the instruction cycle operations basically contains the instruction register (IR) and the program counter (PC). The bit width of the instructions determine the size of the IR; whereas, the number of addressable memory locations determines the size of the PC. For this datapath, we want a memory with 16 locations, each being 8-bits wide, so we need a 4-bit ($2^4 = 16$) address. Hence, the PC is 4-bits wide, and the IR is 8-bits wide. A 4-bit increment unit is used to increment the PC by 1. The PC needs to be loaded with either the result of the increment unit or the address from the JNZ instruction. A 2-to-1 multiplexer is used for this purpose. One input of the multiplexer is from the increment unit, and the other input is from the four least significant bits of the IR, IR_{3-0}.

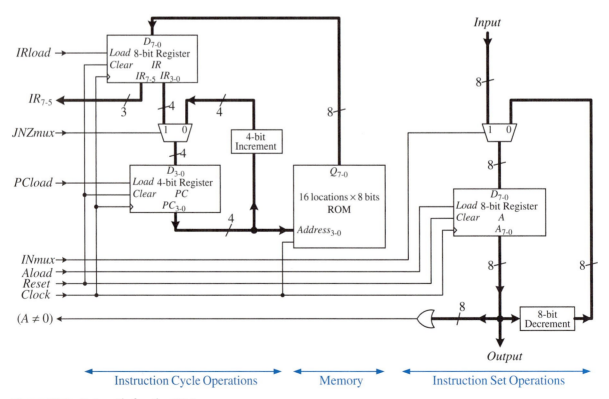

Figure 12.3 Datapath for the EC-1.

As mentioned in Section 12.1, to keep our design simple, instead of having external memory, we have included the memory as part of the datapath. In this design, the memory is a 16 locations $\times$ 8-bits wide read-only memory (ROM). Since the instruction set does not have an instruction that writes to memory, we only need a read-only memory. The output of the PC is connected directly to the 4-bit memory address lines, because the memory location always is determined by the content of the PC. The 8-bit memory output, Q_{7-0}, is connected to the input of the *IR* for executing the instruction fetch operation (Step 1 of the instruction cycle).

The portion of the datapath for performing the instruction set operations includes the 8-bit accumulator *A* and an 8-bit decrement unit. A 2-to-1 multiplexer is used to select the input to the accumulator. For the IN A instruction, the input to the accumulator is from the data input port, *Input*; whereas for the DEC A instruction, the input is from the output of the decrement unit, which performs the decrement of *A*. The output of the accumulator is connected directly to the data output port, *Output*, hence the OUT A instruction does not require any specific datapath actions. Furthermore, with this direct connection, it is equivalent to always performing the OUT A instruction. The JNZ instruction requires an 8-input OR gate connected to the output of the accumulator to test for the condition ($A \neq 0$). The actual operation required by the JNZ instruction is to load the *PC* with the four least significant bits of the *IR*. The HALT instruction also does not require any specific datapath actions.

The control word for this custom datapath has five control signals: *IRload*, *PCload*, *INmux*, *Aload*, and *JNZmux*. The datapath provides one status signal, ($A \neq 0$), to the control unit. The control words for executing the instruction cycle operations and the instruction set operations are discussed in the next section.

12.2.3 Control Unit

The state diagram for the control unit is shown in Figure 12.4(a), and the actions that are executed, specifically the control signals that are asserted in each state, are shown in (d). States for executing the instructions are given the same name as the instruction mnemonics. The first *Start* state, 000, serves as the initial reset state. No action is performed in this state. In addition, this *Start* state provides one extra clock cycle for instructions that require an extra clock cycle to complete its operation. Although, those instructions that do not require this extra clock cycle to complete its operation technically should go back to the *Fetch* state, however, we have made all instructions go back to the *Start* state so that the next-state table and the excitation equations are simpler.

Of the five instructions, only the JNZ instruction requires an extra clock cycle to complete its operation. This is because the *PC* must be loaded with a new address value if the condition is tested true. This new address value, however, is loaded into the *PC* at the beginning of the next clock cycle. So, if we have the FSM go to the *Fetch* state in the next clock cycle, then the *IR* will be loaded with the memory from the old address and not from the new address. However, by making the FSM go back to the *Start* state, the *PC* will be updated with the new address in this state, and the memory will be accessed during the next *Fetch* state from the new address.

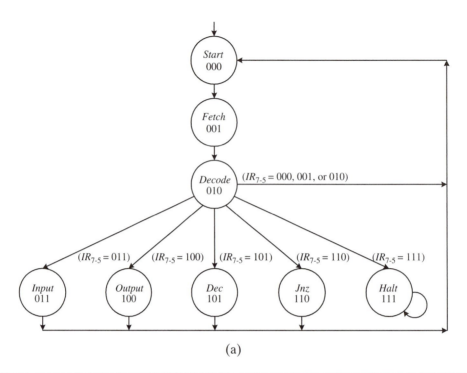

(a)

Current State	Next State (Implementation)							
$Q_2 Q_1 Q_0$	$Q_{2next} Q_{1next} Q_{0next}\ (D_2 D_1 D_0)$							
	IR_7, IR_6, IR_5							
	000	001	010	011	100	101	110	111
	NOP	NOP	NOP	INPUT	OUTPUT	DEC	JNZ	HALT
000 *Start*	001	001	001	001	001	001	001	001
001 *Fetch*	010	010	010	010	010	010	010	010
010 *Decode*	000	000	000	011	100	101	110	111
011 *Input*	000	000	000	000	000	000	000	000
100 *Output*	000	000	000	000	000	000	000	000
101 *Dec*	000	000	000	000	000	000	000	000
110 *Jnz*	000	000	000	000	000	000	000	000
111 *Halt*	111	111	111	111	111	111	111	111

(b)

Figure 12.4 Control unit for the EC-1: (a) state diagram; (b) next-state and implementation table; (c) excitation equations; (d) control words and output table; (e) output equations; (f) circuit. *(continued on next page)*

$$D_2 = Q_2 Q_1 Q_0 + Q_2' Q_1 Q_0' I R_7$$

$$D_1 = Q_2 Q_1 Q_0 + Q_2' Q_1 Q_0' (I R_6 I R_5 + I R_7 I R_6) + Q_2' Q_1' Q_0$$

$$D_0 = Q_2 Q_1 Q_0 + Q_2' Q_1 Q_0' (I R_6 I R_5 + I R_7 I R_5) + Q_2' Q_1' Q_0'$$

(c)

Control Word	State $Q_2 Q_1 Q_0$	IRload	PCload	INmux	Aload	JNZmux	Halt
0	000 *Start*	0	0	0	0	0	0
1	001 *Fetch*	1	1	0	0	0	0
2	010 *Decode*	0	0	0	0	0	0
3	011 *Input*	0	0	1	1	0	0
4	100 *Output*	0	0	0	0	0	0
5	101 *Dec*	0	0	0	1	0	0
6	110 *Jnz*	0	IF $(A \neq 0)$ THEN 1 ELSE 0	0	0	1	0
7	111 *Halt*	0	0	0	0	0	1

(d)

$$IRload = Q_2' Q_1' Q_0$$

$$PCload = Q_2' Q_1' Q_0 + Q_2 Q_1 Q_0' \; (A \neq 0)$$

$$INmux = Q_2' Q_1 Q_0$$

$$Aload = Q_2' Q_1 Q_0 + Q_2 Q_1' Q_0$$

$$JNZmux = Q_2 Q_1 Q_0'$$

$$Halt = Q_2 Q_1 Q_0$$

(e)

Figure 12.4   Control unit for the EC-1: (a) state diagram; (b) next-state and implementation table; (c) excitation equations; (d) control words and output table; (e) output equations; (f) circuit.
(continued on next page)

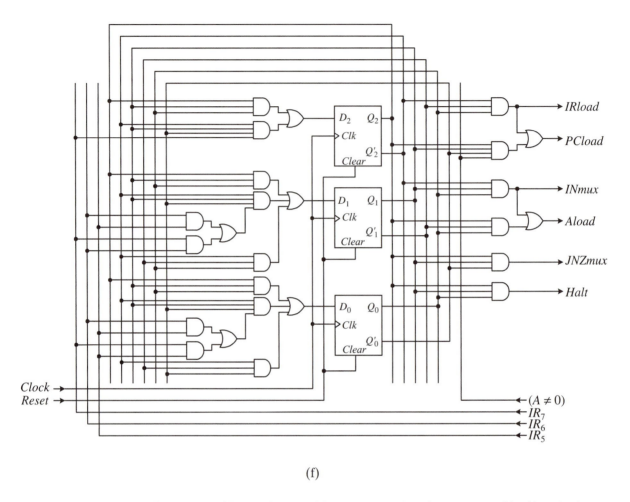

(f)

Figure 12.4 Control unit for the EC-1: (a) state diagram; (b) next-state and implementation table; (c) excitation equations; (d) control words and output table; (e) output equations; (f) circuit.

From the *Start* state, the control unit goes to the *Fetch* state unconditionally. In the *Fetch* state, the *IR* is loaded with the memory content from the location specified by the *PC* by asserting the *IRload* signal. Furthermore, the *PC* is incremented by 1, and the result is loaded back to the *PC* by asserting the *PCload* signal. The *Decode* state tests the three most significant bits of the *IR*, IR_{7-5}, and goes to the corresponding state as encoded by the 3-bit opcode for executing the instruction.

In the five execute states corresponding to the five instructions, the appropriate control signals for the datapath are asserted to execute that instruction. For example, the IN A instruction requires setting the *INmux* signal to a 1 for the input multiplexer, and setting the *Aload* signal to a 1 to load the input value into *A*. Notice that, in order for the input instruction to read in the correct value, the input value must be set up first before resetting the CPU. Furthermore, since the *Input* state does not wait for an *Enter* key signal, only one value can be read in, even if there are multiple input statements. The DEC A instruction requires setting *INmux* to 0 and *Aload* to 1, so the output from the decrement unit is routed back to the accumulator and gets loaded in.

The JNZ instruction asserts the *JNZmux* signal to route the four address bits from the *IR*, IR_{3-0}, to the *PC*. Whether the PC actually gets loaded with this new address depends on the condition of the $(A \neq 0)$ status signal. Hence, the *PCload* control signal is asserted only if $(A \neq 0)$ is a 1. By asserting the *PCload* signal conditionally depending on the status signal $(A \neq 0)$, the state diagram will require one less state, thus making the finite state machine smaller, and making it into a Mealy FSM. Otherwise, the FSM will need two states for the JNZ instruction: one state for asserting the *PCload* signal when $(A \neq 0)$ is true, and one state for de-asserting the *PCload* signal when $(A \neq 0)$ is false.

Once the FSM enters the *Halt* state, it unconditionally loops back to the *Halt* state, giving the impression that the CPU has halted.

The next-state and implementation table for the state diagram and the three excitation equations, as derived from the implementation table, are shown in Figure 12.4(b) and (c), respectively. With eight states, three D flip-flops are used for the implementation of the control unit circuit. Notice that the derivation of the excitation equations is fairly straightforward, since most of the entries in the table contain 0's. Only the *Decode* state row contains different values. The output equations shown in Figure 12.4(e) are derived directly from the output table in (d).

Finally, we can derive the circuit for the control unit based on the excitation equations and the output equations. The complete control unit circuit for the EC-1 general-purpose microprocessor is shown in Figure 12.4(f).

12.2.4   Complete Circuit

The complete circuit for the EC-1 general-purpose microprocessor is constructed by connecting the datapath from Figure 12.3 and the control unit from Figure 12.4(f) together using the designated control and status signals, as shown in Figure 12.5.

12.2.5   Sample Program

Dedicated microprocessors, as discussed in the previous chapter, have the algorithm built right into the circuit of the microprocessor. General-purpose microprocessors, on the other hand, do not have an algorithm built into it. It is designed only to execute program instructions fetched from the memory. Hence, in order to test out the EC-1 computer, we need to write a program using the instructions available in the instruction set.

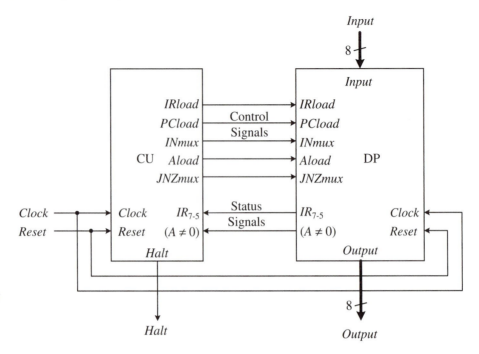

Figure 12.5
Complete circuit for the EC-1 general-purpose microprocessor.

There are only five instructions in the instruction set defined for the EC-1, as shown in Figure 12.2. For our sample program, we will use these five instructions to write a program to input a number and then to count down from this input number to 0. The program is shown in Figure 12.6(a).

Since we do not have a compiler for the EC-1, we need to manually compile this program by hand. The binary executable code for this program is shown in Figure 12.6(b). The binary code is obtained by replacing each instruction with its corresponding 3-bit opcode, as listed in Figure 12.2, followed by 5 bits for the operand. All of the instructions, except for the JNZ instruction, do not use these five operand bits, so either a 0 or a 1 can be used. From Figure 12.2, we find that the opcode for the IN A instruction is 011; therefore, the encoding for this first instruction is 01100000. Similarly, the opcode for the OUT A instruction is 100; therefore, the encoding used is 10000000.

For the JNZ instruction, the 4 least significant bits represent the memory address to jump to if the condition is true. In the example, we are assuming that the first instruction, IN A, is stored in memory location 0000. Since the JNZ instruction jumps to the second instruction, OUT A, which will be stored in memory location 0001, therefore, the four address bits for the JNZ instruction are 0001. The opcode for the JNZ instruction is 110; hence, the encoding for the complete JNZ instruction is 11000001.

```
        IN A          -- input a value into the A register
loop:   OUT A         -- output the value from the A register
        DEC A         -- decrement A by one
        JNZ loop      -- go back to loop if A is not zero
        HALT          -- halt
```

(a)

```
Memory      Instruction
Address     Encoding
--------    ----------
0000        01100000;   -- IN A
0001        10000000;   -- OUT A
0010        10100000;   -- DEC A
0011        11000001;   -- JNZ loop
0100        11111111;   -- HALT
```

Figure 12.6
Countdown program
to run on the EC-1:
(a) assembly code;
(b) binary executable
code.

(b)

Normally, the program instructions are stored in memory that is external to the CPU, and the computer (with the help of the operating system) will provide means to independently load the instructions into the memory. However, to keep our design simple, we have included the memory as part of the CPU inside the datapath. Furthermore, we do not have an operating system for loading the instructions into the memory separately. Therefore, our program must be "loaded" into the memory before the synthesis of the datapath and the microprocessor. The memory circuit that we have used is from MAX+plus II's LPM component library. Information on the usage of this component can be obtained from the Help menu in the MAX+plus II software. This memory is initialized with the content of the text file named PROGRAM.MIF. Therefore, to load our memory with the program, we must enter the binary encoding of the program in this text file and then re-synthesize the microprocessor circuit.

The content of this PROGRAM.MIF file for the countdown program is shown in Figure 12.7. Texts after the two hyphens (--) are comments, and all of the capitalized words are keywords.

Remember to re-synthesize your computer every time you make changes to the PROGRAM.MIF file. For this purpose, you may want to check the Processing | Smart Recompile option under the Compiler menu. With this option checked, the compiler will not have to re-synthesize the entire computer circuit.

12.2.6 Simulation

Having typed in the countdown program into the PROGRAM.MIF file and re-synthesized the computer, we can now run a simulation of the microprocessor executing the

```
-- Content of the ROM memory in the file PROGRAM.MIF

DEPTH = 16;          -- Number of memory locations: 4-bit address
WIDTH = 8;           -- Data Width of memory: 8-bit data

ADDRESS_RADIX = BIN;           -- Specifies the address values are in binary
                               -- Other valid radixes are HEX, DEC, OCT, BIN
DATA_RADIX = BIN;              -- Specifies the data values are in binary

-- Specify memory content.
-- Format of each memory location is
--      address : data

CONTENT
  BEGIN

  [0000..1111]     :     00000000;   -- Initialize locations range 0-F to 0
-- Program to countdown from an input to 0

  0000 : 01100000;   -- IN A
  0001 : 10000000;   -- OUT A
  0010 : 10100000;   -- DEC A
  0011 : 11000001;   -- JNZ 0001
  0100 : 11111111;   -- HALT

END;
```

Figure 12.7
The PROGRAM.MIF file containing the countdown program to run on the EC-1.

program. A sample simulation of the microprocessor is in Figure 12.8, showing the countdown from the input 3 on the *Output* signal. This complete microprocessor circuit is on the accompanying CD-ROM in the file MP.GDF and can be synthesized and simulated.

12.2.7 Hardware Implementation

A complete computer, according to the Von Neumann model, includes not only the microprocessor but also the memory, input and output devices. So far, we have only constructed the general-purpose microprocessor with the built-in memory for the EC-1, as shown in Figure 12.5. The final step in building the complete EC-1 computer is to add the input and output devices. Figure 12.9 shows the interface between the microprocessor and the input and output devices on the UP2 development board. The input simply consists of eight DIP switches, and the output is the two 7-segment LED displays. Since the microprocessor outputs an 8-bit binary number, we need

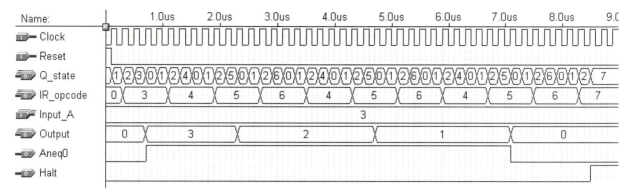

Figure 12.8 A sample simulation trace of the countdown program running on the EC-1 starting at the input 3.

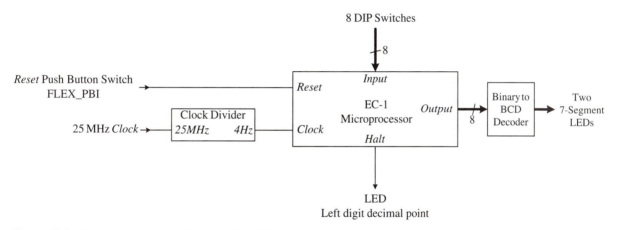

Figure 12.9 Hardware implementation of the EC-1 computer.

an 8-bit binary to 2-digit BCD (binary coded decimal) decoder, so that we can see the 8-bit binary number as two decimal digits on the two 7-segment LED displays. A single LED is used to show when the microprocessor has halted. A push button switch is used as the *Reset* key. Finally, the on-board 25 MHz clock is slowed down with a clock divider circuit. The reason for using the slower clock speed is so that we can see some intermediate results on the display.

This complete hardware implementation circuit is in the file UP2FLEX.GDF in the appropriate directory on the accompanying CD-ROM. You can use this file to download this complete EC-1 computer onto the UP2 development board and see the countdown on the 7-segment display from any given input provided by the eight DIP switches. The FLEX_PB1 push-button switch is the *Reset* key. The output is shown on the two 7-segment LEDs as a decimal number. The decimal point on the right digit is lit if the number is in the 100's. The decimal point on the left digit is the *Halt* light and is lit when the program terminates with the HALT instruction.

● ● ● ● ● ● ● ● ● ● ● ● ● ● ●

12.3 The EC-2 General-Purpose Microprocessor

For our next example, we will design the general-purpose microprocessor for a second version of the EC computer, the EC-2.

12.3.1 Instruction Set

The instruction set for the EC-2 general-purpose microprocessor has eight instructions, as shown in Figure 12.10. The reason for keeping this number at eight is so that we can still use only 3 bits to encode them.

The LOAD instruction loads the content of the memory at the specified address into the accumulator, A. The address is specified by the five least significant bits of the instruction. The STORE instruction is similar to the LOAD instruction, except that it stores the value in A to the memory at the specified address. The ADD and SUB

Instruction	Encoding	Operation	Comment
LOAD A, address	000 aaaaa	$A \leftarrow M[aaaaa]$	Load A with content of memory location aaaaa
STORE A, address	001 aaaaa	$M[aaaaa] \leftarrow A$	Store A into memory location aaaaa
ADD A, address	010 aaaaa	$A \leftarrow A + M[aaaaa]$	Add A with $M[aaaaa]$ and store the result back into A
SUB A, address	011 aaaaa	$A \leftarrow A - M[aaaaa]$	Subtract A with $M[aaaaa]$ and store result back into A
IN A	100 $\times\times\times\times\times$	$A \leftarrow Input$	Input to A
JZ address	101 aaaaa	IF $(A = 0)$ THEN $PC = aaaaa$	Jump to address if A is zero
JPOS address	110 aaaaa	IF $(A \geq 0)$ THEN $PC = aaaaa$	Jump to address if A is a positive number
HALT	111 $\times\times\times\times\times$	Halt	Halt execution

Notations:

    A = accumulator.

    M = memory.

    PC = program counter.

    aaaaa = five bits for specifying a memory address.

    $\times$ = don't-cares.

Figure 12.10   Instruction set for the EC-2.

instructions, respectively, add and subtract the content of A with the content in a memory location and store the result back into A. The IN instruction inputs a value from the data input port, *Input*, and stores it into A. The JZ instruction loads the PC with the specified address if A is zero. Loading the PC with a new address simply causes the CPU to jump to this new memory location. The JPOS instruction loads the PC with the specified address if A is a positive number. The value in A is taken as a two's complement signed number, so a positive number is one where the most significant bit of the number is a 0. Finally, the HALT instruction halts the CPU.

12.3.2 Datapath

The custom datapath for the EC-2 is shown in Figure 12.11. The portion of the datapath for performing the instruction cycle operations is very similar to that of the previous EC-1 with the instruction register (IR), the program counter (PC), and

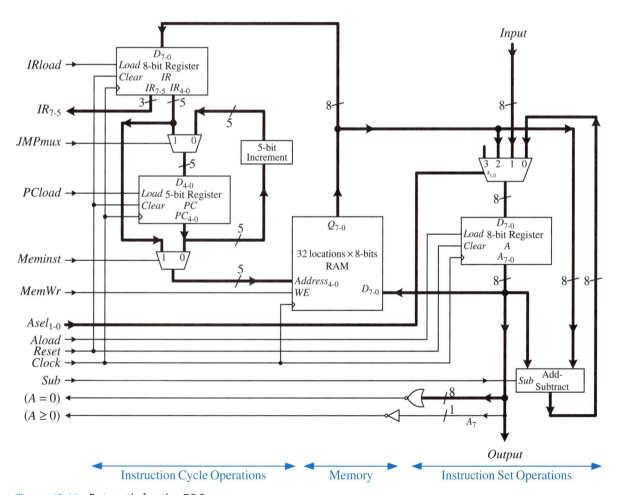

Figure 12.11 Datapath for the EC-2.

the increment unit for incrementing the PC. The minor differences between the two are in the size of the PC and the increment unit. In this second design, we want a memory with 32 locations; hence, the address and, therefore, the size of the PC and the increment unit must all be 5-bits wide.

The main modification to this portion of the datapath is the addition of a second 2-to-1 multiplexer that is connected between the output of the PC and the memory address input. One input of this multiplexer comes from the PC, and the other input comes from the five least significant bits of the IR, IR_{4-0}. The reason for this is because there are now two different types of operations that can access memory. The first is still for the fetch operation, where the memory address is given by the content of the PC. The second type is for the four instructions, LOAD, STORE, ADD, and SUB, where they use the memory as an operand. Hence, the memory address for these four instructions is given by the five least significant bits of the IR, IR_{4-0}. The select signal for this multiplexer is *Meminst*.

The memory size for the EC-2 is increased to 32 locations, thus requiring five address bits. The memory is still included as part of the datapath rather than as an independent external unit to the CPU. In order to accommodate the STORE instruction for storing the value of A into the memory, we need to use a RAM instead of the previous ROM. To realize this operation, the output of the accumulator A is connected to the memory data input, D_{7-0}. The signal *MemWr*, when asserted, causes the memory to write the value from register A into the location specified by the address in the instruction.

The output of the memory at Q_{7-0} is connected to both the input of the IR and to the input of the accumulator, A, through a 4-to-1 multiplexer. The connection to the IR is for the fetch operation just like in the EC-1 design. The connection to the accumulator is for performing the LOAD instruction, where the content of the memory is loaded into A. Since the memory is only one source among two other sources that is loaded into A, the multiplexer is needed.

The portion of the datapath for performing the instruction set operations includes the 8-bit accumulator A, an 8-bit adder-subtractor unit, and a 4-to-1 multiplexer. The adder-subtractor unit performs the ADD and SUB instructions. The *Sub* signal, when asserted, selects the subtraction operation, and when de-asserted selects the addition operation. The 4-to-1 multiplexer allows the accumulator input to come from one of three sources. For the ADD and SUB instructions, the A input comes from the output of the adder-subtractor unit. For the IN instruction, the A input comes from the data input port, *Input*. For the LOAD instruction, the A input comes from the output of the memory, Q_{7-0}. The selection of this multiplexer is through the two signal lines, $Asel_{1-0}$. The fourth input of the multiplexer is not used.

Similar to the previous EC-1 design, the output port is connected directly to the output of the accumulator, A. Therefore, the value of the accumulator is always available at the output port, and no specific output instruction is necessary to output the value in A.

For the two conditional jump instructions, JZ and JPOS, the datapath provides the two status signals, $(A = 0)$ and $(A \geq 0)$, respectively, that are generated from two comparators. The $(A = 0)$ status signal outputs a 1 if the value in A is a 0, hence an 8-input NOR gate is used. The $(A \geq 0)$ status signal outputs a 1 if the value in A, which is treated as a two's complement signed number, is a positive number. Since for a two's complement signed number, a leading 0 means positive and a leading 1 means negative, hence, $(A \geq 0)$ is simply the negated value of A_7 (the most significant bit of A).

The control word for this custom datapath has nine control signals, *IRload*, *JMPmux*, *PCload*, *Meminst*, *MemWr*, $Asel_{1-0}$, *Aload*, and *Sub*. The datapath provides two status signals, $(A = 0)$ and $(A \geq 0)$, to the control unit. The control words for executing the instruction cycle operations and the instruction set operations are discussed in the next section.

12.3.3 Control Unit

The state diagram for the control unit is shown in Figure 12.12(a), and the actions that are executed, specifically the control signals that are asserted in each state, are shown in (d). States for executing the instructions are given the same name as the instruction mnemonics. The first three states, *Start*, *Fetch*, and *Decode*, serve the same purpose as in the previous EC-1's control unit. The *Decode* state for this second

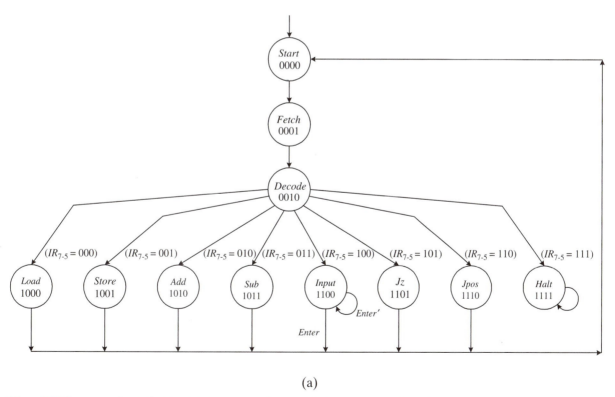

(a)

Figure 12.12   Control unit for the EC-2: (a) state diagram; (b) next-state and implementation table; (c) excitation equations; (d) control words and output table; (e) output equations; (f) circuit. *(continued on next page)*

Current State $Q_3 Q_2 Q_1 Q_0$	Next State (Implementation) $Q_{3\text{next}} Q_{2\text{next}} Q_{1\text{next}} Q_{0\text{next}} \; (D_3 \; D_2 \; D_1 \; D_0)$									
	IR_7, IR_6, IR_5								Enter	
	000	001	010	011	100	101	110	111	0	1
	LOAD	STORE	ADD	SUB	INPUT	JZ	JPOS	HALT		
0000 *Start*	0001	0001	0001	0001	0001	0001	0001	0001		
0001 *Fetch*	0010	0010	0010	0010	0010	0010	0010	0010		
0010 *Decode*	1000	1001	1010	1011	1100	1101	1110	1111		
1000 *Load*	0000	0000	0000	0000	0000	0000	0000	0000		
1001 *Store*	0000	0000	0000	0000	0000	0000	0000	0000		
1010 *Add*	0000	0000	0000	0000	0000	0000	0000	0000		
1011 *Sub*	0000	0000	0000	0000	0000	0000	0000	0000		
1100 *Input*									1100	0000
1101 *Jz*	0000	0000	0000	0000	0000	0000	0000	0000		
1110 *Jpos*	0000	0000	0000	0000	0000	0000	0000	0000		
1111 *Halt*	1111	1111	1111	1111	1111	1111	1111	1111		

(b)

$$D_3 = Q_3' Q_2' Q_1 Q_0' + Q_3 Q_2 Q_1' Q_0' Enter' + Q_3 Q_2 Q_1 Q_0$$

$$D_2 = Q_3' Q_2' Q_1 Q_0' IR_7 + Q_3 Q_2 Q_1' Q_0' Enter' + Q_3 Q_2 Q_1 Q_0$$

$$D_1 = Q_3' Q_2' Q_1' Q_0 + Q_3' Q_2' Q_1 Q_0' IR_6 + Q_3 Q_2 Q_1 Q_0$$

$$D_0 = Q_3' Q_2' Q_1' Q_0' + Q_3' Q_2' Q_1 Q_0' IR_5 + Q_3 Q_2 Q_1 Q_0$$

(c)

Figure 12.12 Control unit for the EC-2: (a) state diagram; (b) next-state and implementation table; (c) excitation equations; (d) control words and output table; (e) output equations; (f) circuit. *(continued on next page)*

State $Q_3Q_2Q_1Q_0$	IRload	JMPmux	PCload	Meminst	MemWr	$Asel_{1-0}$	Aload	Sub	Halt
0000 Start	0	0	0	0	0	00	0	0	0
0001 Fetch	1	0	1	0	0	00	0	0	0
0010 Decode	0	0	0	1	0	00	0	0	0
1000 Load	0	0	0	0	0	10	1	0	0
1001 Store	0	0	0	1	1	00	0	0	0
1010 Add	0	0	0	0	0	00	1	0	0
1011 Sub	0	0	0	0	0	00	1	1	0
1100 Input	0	0	0	0	0	01	1	0	0
1101 Jz	0	1	$(A = 0)$	0	0	00	0	0	0
1110 Jpos	0	1	$(A \geq 0)$	0	0	00	0	0	0
1111 Halt	0	0	0	0	0	00	0	0	1

(d)

$$IRload = Q_3'Q_2'Q_1'Q_0$$

$$JMPmux = Q_3Q_2Q_1'Q_0 + Q_3Q_2Q_1Q_0'$$

$$PCload = Q_3'Q_2'Q_1'Q_0 + Q_3Q_2Q_1'Q_0(A = 0) + Q_3Q_2Q_1Q_0'(A \geq 0)$$

$$Meminst = Q_3'Q_2'Q_1Q_0' + Q_3Q_2'Q_1'Q_0$$

$$MemWr = Q_3Q_2'Q_1'Q_0$$

$$Asel_1 = Q_3Q_2'Q_1'Q_0'$$

$$Asel_0 = Q_3Q_2Q_1'Q_0'$$

$$Aload = Q_3Q_1'Q_0' + Q_3Q_2'Q_1$$

$$Sub = Q_3Q_2'Q_1Q_0$$

$$Halt = Q_3Q_2Q_1Q_0$$

(e)

Figure 12.12 Control unit for the EC-2: (a) state diagram; (b) next-state and implementation table; (c) excitation equations; (d) control words and output table; (e) output equations; (f) circuit. *(continued on next page)*

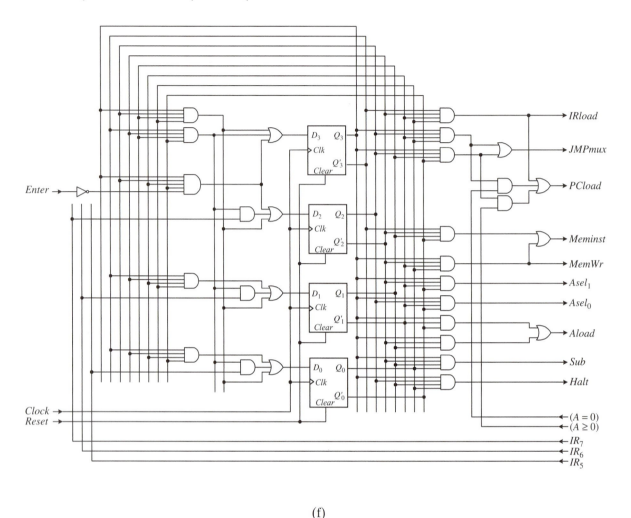

(f)

Figure 12.12 Control unit for the EC-2: (a) state diagram; (b) next-state and implementation table; (c) excitation equations; (d) control words and output table; (e) output equations; (f) circuit.

design, however, needs to decode eight opcodes by branching to eight different states for executing the corresponding eight instructions. Like before, the decoding of the opcodes depends on the three most significant bits of the IR.

A very important timing issue for this control unit has to do with the memory accesses of the four instructions, LOAD, STORE, ADD, and SUB. The problem here is that only after fetching these instructions will the address of the memory location for these instructions be available. Furthermore, only after decoding the instruction will the control unit know that the memory needs to be read. If we change the memory address during the *Execute* state, the memory will not have enough time to output the value for the instruction to operate on.

Normally, for instructions requiring a memory access for one of its operands, an extra memory read state will be inserted between the *Decode* state and the *Execute* state. This way, the memory will have one clock cycle to output the data for the instruction to operate on in the following clock cycle. This, of course, is assuming that the memory requires only one clock cycle for a read operation. If the memory is slower, then more states must be inserted in between.

To minimize the number of states in our design, we have used the *Decode* state to also perform the memory read. This way, when the control unit gets to the *Execute* state, the memory will already have the data ready. Whether the data from the memory actually is used or not will depend on the instruction being executed. If the instruction does not require the data from the memory, it is simply ignored. On the other hand, if the instruction needs the data, then the data is there and ready to be used. This solution works in this design because it does not conflict with the operations for the rest of the instructions in our instruction set. The memory read operation performed in the *Decode* state is accomplished by asserting the *Meminst* signal from this state. Looking at the output table in Figure 12.12(d), this is reflected by the 1 under the *Meminst* column for the *Decode* state.

The actual execution of each instruction is accomplished by asserting the correct control signals to control the operation of the datapath. This is shown by the assignments made for the respective rows in the output table in Figure 12.12(d). At this point, you should be able to understand why each assignment is made by looking at the operation of the datapath. For example, for the LOAD instruction, the $Asel_1$ signal needs to be asserted and the $Asel_0$ signal needs to be de-asserted in order to select input 2 of the multiplexer so that the output from the memory can pass to the input of the accumulator, A. The actual loading of A is done by asserting the *Aload* signal. To perform the STORE instruction, the memory address is taken from the *IR* by asserting *Meminst*. The writing into memory takes place when *MemWr* is asserted.

The *Input* state for this state diagram waits for the *Enter* key signal before looping back to the *Start* state. In so doing, we can read in several values correctly by having multiple input statements in the program. Notice that after the *Enter* signal is asserted, there is no state that waits for the *Enter* signal to be de-asserted (i.e., for the Enter key to be released). Hence, the input device must resolve this issue by outputting exactly one clock pulse each time the Enter key is pressed. This is accomplished at the computer circuit level by using a one-shot circuit.

The next-state and implementation table for the state diagram and the three excitation equations, as derived from the implementation table, are shown in Figure 12.12(b) and (c), respectively. To keep the table reasonable small, all of the possible combinations of the input signals are not listed. All of the states, except for the *Input* state, depend only on the three *IR* bits, IR_{7-5}; whereas, the *Input* state depends only on the *Enter* signal. The blank entries in the table, therefore, can be viewed as having all 0's. With 11 states, four D flip-flops are used for the implementation of the control unit circuit. The output equations shown in Figure 12.12(e) are derived directly from the output table in (d).

Finally, we can derive the circuit for the control unit based on the excitation equations and the output equations. The complete control unit circuit for the EC-2 general-purpose microprocessor is shown in Figure 12.12(f).

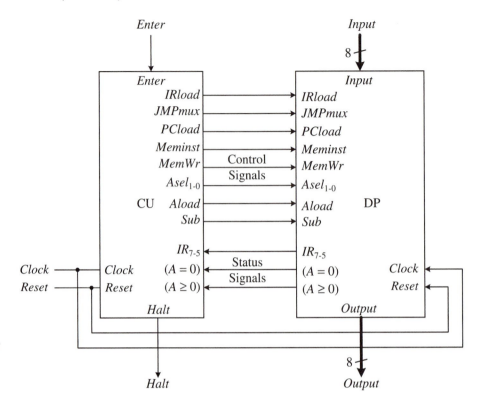

Figure 12.13
Complete circuit for the EC-2 general-purpose microprocessor.

12.3.4 Complete Circuit

The complete circuit for the EC-2 general-purpose microprocessor is constructed by connecting the datapath from Figure 12.11 and the control unit from Figure 12.12(f) together using the designated control and status signals as shown in Figure 12.13.

12.3.5 Sample Program

The memory for our computer is initialized with the content of the text file PROGRAM.MIF. A sample file is shown in Figure 12.14. The file contains three programs: GCD calculates the greatest common divisor of two input numbers; SUM evaluates the sum of all of the numbers between an input n and 1; and COUNT displays the count from input n down to 0. Only the last program (COUNT) listed in the file is executed. To try out another program, move the code for that program to the end of the file, then recompile the entire microprocessor circuit (i.e., the UP2FLEX.GDF file as explained in the next section) and download the new code to the UP2 board.

```
-- Content of the RAM memory in the file PROGRAM.MIF

DEPTH = 32;        -- Depth of memory: 5-bit address
WIDTH = 8;         -- Width of memory: 8-bit data

ADDRESS_RADIX = BIN;    -- All values in binary (HEX, DEC, OCT, BIN)
DATA_RADIX = BIN;

-- Opcodes for the EC-2
-- 000 = LOAD A,aaaaa
-- 001 = STORE A,aaaaa
-- 010 = ADD A,aaaaa
-- 011 = SUB A,aaaaa
-- 100 = IN A
-- 101 = JZ aaaaa
-- 110 = JPOS aaaaa
-- 111 = HALT

-- Specify the memory content
-- Format of each memory location is
--     address : data

CONTENT
  BEGIN
  [00000..11111] : 00000000;   -- Initialize locations range 00-1F to 0

  --------------------------------------------------------
  -- There are three programs listed below: GCD, SUM, and COUNT
  -- Only the program listed last is ran.
  -- To try out a different program, move the code for the program you
  -- want to the end of the list, re-compile, and download to the FPGA

  --------------------------------------------------------
  -- GCD
  -- Program to calculate the GCD of two numbers, x and y

  00000 : 10000000;   -- IN A
  00001 : 00111110;   -- STORE A,x
  00010 : 10000000;   -- IN A
  00011 : 00111111;   -- STORE A,y

  00100 : 00011110;   -- loop: LOAD A,x -- x=y?
  00101 : 01111111;   -- SUB A,y
  00110 : 10110000;   -- JZ out       -- x=y
  00111 : 11001100;   -- JPOS xgty     -- x>y
```

(continued on next page)

Figure 12.14
The PROGRAM.MIF file containing three programs for the EC-2. Only the last program, COUNT, is executed.

```
01000 : 00011111;   -- LOAD A,y      -- y>x
01001 : 01111110;   -- SUB A,x       -- y-x
01010 : 00111111;   -- STORE A,y
01011 : 11000100;   -- JPOS loop

01100 : 00011110;   -- xgty: LOAD A,x -- x>y
01101 : 01111111;   -- SUB A,y        -- x-y
01110 : 00111110;   -- STORE A,x
01111 : 11000100;   -- JPOS loop

10000 : 00011110;   -- LOAD A,x
10001 : 11111111;   -- HALT

11110 : 00000000;   -- storage for variable x
11111 : 00000000;   -- storage for variable y


-----------------------------------------------------------

-- SUM
-- Program to sum n downto 1

00000 : 00011101;   -- LOAD A,one     -- zero sum by doing 1-1
00001 : 01111101;   -- SUB A,one
00010 : 00111110;   -- STORE A,sum

00011 : 10000000;   -- IN A
00100 : 00111111;   -- STORE A,n

00101 : 00011111;   -- loop: LOAD A,n -- n + sum
00110 : 01011110;   -- ADD A,sum
00111 : 00111110;   -- STORE A,sum

01000 : 00011111;   -- LOAD A,n       -- decrement A
01001 : 01111101;   -- SUB A,one
01010 : 00111111;   -- STORE A,n

01011 : 10101101;   -- JZ out
01100 : 11000101;   -- JPOS loop
01101 : 00011110;   -- out: LOAD A,sum
01110 : 11111111;   -- HALT

11101 : 00000001;   -- storage for the constant 1
11110 : 00000000;   -- storage for variable sum
11111 : 00000000;   -- storage for variable n
```

Figure 12.14
The PROGRAM.MIF file containing three programs for the EC-2. Only the last program, COUNT, is executed.
(continued on next page)

```
----------------------------------------------------------
-- COUNT
-- Program to countdown from input n to 0

  00000 : 10000000;    -- IN A
  00001 : 01111111;    -- SUB A,11111
  00010 : 10100100;    -- JZ 00100
  00011 : 11000001;    -- JPOS 00001
  00100 : 11111111;    -- HALT
  11111 : 00000001;    -- storage for the constant 1

END;
```

Figure 12.14
The PROGRAM.MIF file containing three programs for the EC-2. Only the last program, COUNT, is executed.

12.3.6   Hardware Implementation

Figure 12.15 shows the interface between the EC-2 microprocessor and the input and output devices on the UP2 development board. The input simply consists of eight DIP switches, and the output is the two 7-segment LED displays. Since the microprocessor outputs an 8-bit binary number, we need an 8-bit binary to two-digit BCD (binary coded decimal) decoder so that we can see the 8-bit binary number as two decimal digits on the two 7-segment LED displays. A single LED is used to show when the microprocessor has halted. A push button switch is used as the *Reset* key, and a second push button switch is used as the *Enter* key. A one-shot circuit is used to generate only one clock pulse for each keypress of the *Enter* key. Finally, the on board 25 MHz clock is slowed down with a clock divider circuit. The reason for using the slower clock speed is so that we can see some intermediate results on the display.

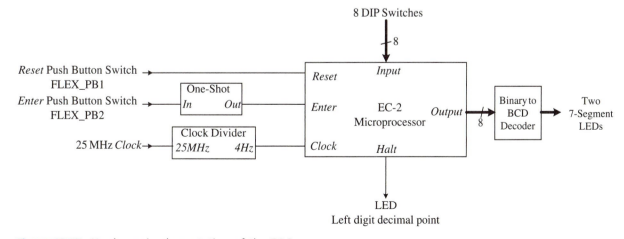

Figure 12.15 Hardware implementation of the EC-2.

This complete hardware implementation circuit is in the file UP2FLEX.GDF in the appropriate directory on the accompanying CD-ROM. You can use this file to download this complete EC-2 computer onto the UP2 development board, and see your very own microprocessor running a program. The COUNT program that is listed at the end of the PROGRAM.MIF file is executed. The FLEX_PB1 push button switch is the *Reset* key. The FLEX_PB2 push button switch is the *Enter* key. The eight DIP switches are for the data input port. The output is shown on the two 7-segment LEDs as a decimal number. The decimal point on the right digit is lit if the number is in the 100's. The decimal point on the left digit is the Halt light and is lit when the program terminates with the HALT instruction.

12.4 VHDL for General-Purpose Microprocessors

This section presents the VHDL code for the EC-2 general-purpose microprocessor using both the structural and the behavioral levels. First, in Section 12.4.1, we use the VHDL structural level to define the microprocessor based on the FSM+D model, as designed manually in Section 12.3. In order to use this method, the microprocessor first must be designed manually. Remember that in practice, we really do not want to use this method. This exercise is just to show how tedious and complicated it is to do it this way, especially for a more realistic microprocessor.

Section 12.4.2 uses the VHDL behavioral level to define the EC-2 microprocessor based on the FSMD model. Using this method, we only need to manually derive the state diagram. Using the VHDL template for defining an FSM (from Section 7.6), we can translate the state diagram to the behavioral VHDL code. Comparing this behavioral code with the structural code, we can see how much easier it is to construct a microprocessor circuit using the behavioral method and quickly appreciate the power of the VHDL synthesizer.

12.4.1 Structural FSM+D

Figure 12.16 shows the structural VHDL code of the datapath for the EC-2 general-purpose microprocessor. This code is based on the datapath circuit as derived manually from Section 12.3.2. The memory used is from MAX+plus II's LPM component library. Figure 12.17 shows the code for the control unit as derived manually from Section 12.3.3. This control unit code follows very closely the template for writing an FSM sequential circuit, as discussed in Section 7.6. Figure 12.18 shows the structural code for the complete EC-2 general-purpose microprocessor.

```vhdl
LIBRARY IEEE;
USE IEEE.STD_LOGIC _1164.ALL;

LIBRARY lpm;                        -- for memory
USE lpm.lpm_components.ALL;

ENTITY dp IS PORT (
   Clock, Clear: IN STD_LOGIC;
   -- datapath input
   Input: IN STD_LOGIC_VECTOR(7 DOWNTO 0);
   -- control signals
   IRload, JMPmux, PCload, MemInst, MemWr: IN STD_LOGIC;
   ASel: IN STD_LOGIC_VECTOR(1 DOWNTO 0);
   Aload, Sub: IN STD_LOGIC;
   -- status signals
   IR: OUT STD_LOGIC_VECTOR(7 DOWNTO 5);
   Aeq0, Apos: OUT STD_LOGIC;
   -- datapath output
   Output: OUT STD_LOGIC_VECTOR(7 DOWNTO 0));
END dp;

ARCHITECTURE Structural OF dp IS
   COMPONENT reg
   GENERIC (size: INTEGER := 4); -- the actual size is defined in the instantiation GENERIC MAP
   PORT (
      Clock, Clear, Load: IN STD_LOGIC;
      D: IN STD_LOGIC_VECTOR(size-1 DOWNTO 0);
      Q: OUT STD_LOGIC_VECTOR(size-1 DOWNTO 0));
   END COMPONENT;


   COMPONENT increment
   GENERIC (size: INTEGER := 8); -- default number of bits
   PORT (
      A: IN STD_LOGIC_VECTOR(size-1 DOWNTO 0);
      F: OUT STD_LOGIC_VECTOR(size-1 DOWNTO 0));
   END COMPONENT;


   COMPONENT mux2
   GENERIC (size: INTEGER := 8);                    -- default size
   PORT (
      S: IN STD_LOGIC;                              -- select line
      D1, D0: IN STD_LOGIC_VECTOR(size-1 DOWNTO 0);  -- data bus input
      Y: OUT STD_LOGIC_VECTOR(size-1 DOWNTO 0));     -- data bus output
   END COMPONENT;
```

Figure 12.16   VHDL code for the datapath of the EC-2.
(continued on next page)

```
COMPONENT mux4
GENERIC (size: INTEGER := 8);                      -- default size
PORT (
   S: IN STD_LOGIC_VECTOR(1 DOWNTO 0);             -- select line
   D3, D2, D1, D0: IN STD_LOGIC_VECTOR(size-1 DOWNTO 0); -- data bus input
   Y: OUT STD_LOGIC_VECTOR(size-1 DOWNTO 0));      -- data bus output
END COMPONENT;

COMPONENT addsub
GENERIC (n: INTEGER :=4);       -- default number of bits = 4
PORT (S: IN std_logic;          -- select subtract signal
   A: IN STD_LOGIC_VECTOR(n-1 DOWNTO 0);
   B: IN STD_LOGIC_VECTOR(n-1 DOWNTO 0);
   F: OUT STD_LOGIC_VECTOR(n-1 DOWNTO 0);
   unsigned_overflow: OUT STD_LOGIC;
   signed_overflow: OUT STD_LOGIC);
END COMPONENT;

SIGNAL dp_IR, dp_RAMQ: STD_LOGIC_VECTOR(7 DOWNTO 0);
SIGNAL dp_JMPmux, dp_PC, dp_increment, dp_meminst : STD_LOGIC_VECTOR(4 DOWNTO 0);
SIGNAL dp_Amux, dp_addsub, dp_A: STD_LOGIC_VECTOR(7 DOWNTO 0);

BEGIN
   -- doing structural modeling for the datapath here
   -- IR
   U0: reg GENERIC MAP(8) PORT MAP(Clock, Clear, IRload, dp_RAMQ, dp_IR);
   IR <= dp_IR(7 DOWNTO 5);

   -- JMPmux
   U1: mux2 GENERIC MAP(5) PORT MAP(JMPmux,dp_IR(4 DOWNTO 0),dp_increment,dp_JMPmux);

   -- PC
   U2: reg GENERIC MAP(5) PORT MAP(Clock, Clear, PCload, dp_JMPmux, dp_PC);

   -- Meminst
   U3: mux2 GENERIC MAP(5) PORT MAP(Meminst,dp_IR(4 DOWNTO 0),dp_PC,dp_meminst);

   -- increment
   U4: increment GENERIC MAP(5) PORT MAP(dp_PC,dp_increment);

   -- memory
   U5: lpm_ram_dq
      GENERIC MAP (
         lpm_widthad => 5,
```

Figure 12.16 VHDL code for the datapath of the EC-2.
(continued on next page)

```
              lpm_outdata => "UNREGISTERED",
--            lpm_indata => "UNREGISTERED",
--            lpm_address_control => "UNREGISTERED",
              lpm_file => "program.mif",  -- fill ram with content of file program.mif
              lpm_width => 8)

        PORT MAP (
            data => dp_A,
            address => dp_meminst,
            we => MemWr,
            inclock => Clock,
            q => dp_RAMQ);


    -- A input mux
    U6: mux4 GENERIC MAP(8) PORT MAP (Asel,dp_RAMQ,dp_RAMQ,Input,dp_addsub,dp_Amux);


    -- Accumulator
    U7: reg GENERIC MAP(8) PORT MAP(Clock, Clear, Aload, dp_Amux, dp_A);


    -- Adder-subtractor
    U8: addsub GENERIC MAP(8) PORT MAP(Sub,dp_A,dp_RAMQ,dp_addsub,OPEN,OPEN);


    Aeq0 <= '1' WHEN dp_A = "00000000" ELSE '0';    -- (A = 0)
    Apos <= NOT dp_A(7);                             -- (A >= 0)
    Output <= dp_A;
END Structural;
```

Figure 12.16   VHDL code for the datapath of the EC-2.


```
LIBRARY IEEE;
USE IEEE.STD_LOGIC_1164.ALL;

ENTITY cu IS PORT (
    Clock, Reset : IN STD_LOGIC;
    -- control input
    Enter: IN STD_LOGIC;
    -- control signals
    IRload, JMPmux, PCload, MemInst, MemWr: OUT STD_LOGIC;
    Asel: OUT STD_LOGIC_VECTOR(1 DOWNTO 0);
    Aload, Sub: OUT STD_LOGIC;
```

Figure 12.17   VHDL code for the control unit of the EC-2.
(continued on next page)

```
   -- status signals
   IR: IN STD_LOGIC_VECTOR(7 DOWNTO 5);
   Aeq0, Apos: IN STD_LOGIC;
   -- control outputs
   Halt: OUT STD_LOGIC);
END cu;

ARCHITECTURE FSM OF cu IS
   TYPE state_type IS (s_start,s_fetch,s_decode,s_load,s_store,s_add,s_sub,s_in,s_jz,s_jpos,s_halt);
   SIGNAL state: state_type;
BEGIN

   next_state_logic: PROCESS(Reset, Clock)
   BEGIN
     IF(Reset = '1') THEN
           state <= s_start;
     ELSIF(Clock'EVENT AND Clock = '1') THEN
        CASE state IS
        WHEN s_start => -- reset
           state <= s_fetch;
        WHEN s_fetch =>
           state <= s_decode;
        WHEN s_decode =>
           CASE IR IS
              WHEN "000" => state <= s_load;
              WHEN "001" => state <= s_store;
              WHEN "010" => state <= s_add;
              WHEN "011" => state <= s_sub;
              WHEN "100" => state <= s_in;
              WHEN "101" => state <= s_jz;
              WHEN "110" => state <= s_jpos;
              WHEN "111" => state <= s_halt;
              WHEN OTHERS => state <= s_halt;
           END CASE;

        WHEN s_load =>
           state <= s_start;
        WHEN s_store =>
           state <= s_start;
        WHEN s_add =>
           state <= s_start;
        WHEN s_sub =>
           state <= s_start;
```

Figure 12.17 VHDL code for the control unit of the EC-2.
(continued on next page)

```vhdl
      WHEN s_in =>
         IF (Enter = '0') THEN   -- wait for the Enter key for inputs
            state <= s_in;
         ELSE
            state <= s_start;
         END IF;
      WHEN s_jz =>
         state <= s_start;
      WHEN s_jpos =>
         state <= s_start;
      WHEN s_halt =>
         state <= s_halt;
      WHEN OTHERS =>
         state <= s_start;
      END CASE;
   END IF;
END PROCESS;

output_logic: PROCESS(state)
BEGIN
   CASE state IS
   WHEN s_fetch =>
      IRload <= '1';     -- load IR
      JMPmux <= '0';
      PCload <= '1';     -- increment PC
      Meminst <= '0';
      MemWr <= '0';
      Asel <= "00";
      Aload <= '0';
      Sub <= '0';
      Halt <= '0';
   WHEN s_decode =>      -- also set up for memory access
      IRload <= '0';
      JMPmux <= '0';
      PCload <= '0';
      Meminst <= '1';    -- pass IR address to memory
      MemWr <= '0';
      Asel <= "00";
      Aload <= '0';
      Sub <= '0';
      Halt <= '0';
```

Figure 12.17   VHDL code for the control unit of the EC-2.
(continued on next page)

```
WHEN s_load =>
   IRload <= '0';
   JMPmux <= '0';
   PCload <= '0';
   Meminst <= '1';
   MemWr <= '0';
   Asel <= "10";      -- pass memory to A
   Aload <= '1';      -- load A
   Sub <= '0';
   Halt <= '0';
WHEN s_store =>
   IRload <= '0';
   JMPmux <= '0';
   PCload <= '0';
   Meminst <= '1';    -- pass IR address to memory
   MemWr <= '1';      -- store A to memory
   Asel <= "00";
   Aload <= '0';
   Sub <= '0';
   Halt <= '0';
WHEN s_add =>
   IRload <= '0';
   JMPmux <= '0';
   PCload <= '0';
   Meminst <= '1';
   MemWr <= '0';
   Asel <= "00";      -- pass add/sub unit to A
   Aload <= '1';      -- load A
   Sub <= '0';        -- select add
   Halt <= '0';
WHEN s_sub =>
   IRload <= '0';
   JMPmux <= '0';
   PCload <= '0';
   Meminst <= '1';
   MemWr <= '0';
   Asel <= "00";      -- pass add/sub unit to A
   Aload <= '1';      -- load A
   Sub <= '1';        -- select subtract
   Halt <= '0';
```

Figure 12.17 VHDL code for the control unit of the EC-2.
(continued on next page)

```
WHEN s_in =>
   IRload <= '0';
   JMPmux <= '0';
   PCload <= '0';
   Meminst <= '0';
   MemWr <= '0';
   Asel <= "01";      -- pass input to A
   Aload <= '1';      -- load A
   Sub <= '0';
   Halt <= '0';
WHEN s_jz =>
   IRload <= '0';
   JMPmux <= '1';     -- pass IR address to PC
   PCload <= Aeq0;    -- load PC if condition is true
   Meminst <= '0';
   MemWr <= '0';
   Asel <= "00";
   Aload <= '0';
   Sub <= '0';
   Halt <= '0';
WHEN s_jpos =>
   IRload <= '0';
   JMPmux <= '1';     -- pass IR address to PC
   PCload <= Apos;    -- load PC if condition is true
   Meminst <= '0';
   MemWr <= '0';
   Asel <= "00";
   Aload <= '0';
   Sub <= '0';
   Halt <= '0';
WHEN s_halt =>
   IRload <= '0';
   JMPmux <= '0';
   PCload <= '0';
   Meminst <= '0';
   MemWr <= '0';
   Asel <= "00";
   Aload <= '0';
   Sub <= '0';
   Halt <= '1';
```

Figure 12.17   VHDL code for the control unit of the EC-2.
(continued on next page)

```
      WHEN OTHERS =>
         IRload <= '0';
         JMPmux <= '0';
         PCload <= '0';
         Meminst <= '0';
         MemWr <= '0';
         Asel <= "00";
         Aload <= '0';
         Sub <= '0';
         Halt <= '0';
      END CASE;
   END PROCESS;
END FSM;
```

Figure 12.17 VHDL code for the control unit of the EC-2.

```
LIBRARY IEEE;
USE IEEE.STD_LOGIC_1164.ALL;

ENTITY mp IS PORT (
   Clock, Reset: IN STD_LOGIC;
   Enter: IN STD_LOGIC;
   Input: IN STD_LOGIC_VECTOR(7 DOWNTO 0);
   Output: OUT STD_LOGIC_VECTOR(7 DOWNTO 0);
   Halt: OUT STD_LOGIC);
END mp;

ARCHITECTURE Structural OF mp IS
   COMPONENT cu PORT (
      Clock, Reset : IN STD_LOGIC;
      -- control input
      Enter: IN STD_LOGIC;
      -- control signals
      IRload, JMPmux, PCload, MemInst, MemWr: OUT STD_LOGIC;
      Asel: OUT STD_LOGIC_VECTOR(1 DOWNTO 0);
      Aload, Sub: OUT STD_LOGIC;
      -- status signals
      IR: IN STD_LOGIC_VECTOR(7 DOWNTO 5);
      Aeq0, Apos: IN STD_LOGIC;
      -- control outputs
      Halt: OUT STD_LOGIC);
      END COMPONENT;
```

Figure 12.18 Structural VHDL code for the complete EC-2 general-purpose microprocessor.
(continued on next page)

```vhdl
COMPONENT dp PORT (
   Clock, Clear: IN STD_LOGIC;
   -- datapath input
   Input: IN STD_LOGIC_VECTOR(7 DOWNTO 0);
   -- control signals
   IRload, JMPmux, PCload, MemInst, MemWr: IN STD_LOGIC;
   ASel: IN STD_LOGIC_VECTOR(1 DOWNTO 0);
   Aload, Sub: IN STD_LOGIC;
   -- status signals
   IR: OUT STD_LOGIC_VECTOR(7 DOWNTO 5);
   Aeq0, Apos: OUT STD_LOGIC;
   -- datapath output
   Output: OUT STD_LOGIC_VECTOR(7 DOWNTO 0));
END COMPONENT;

-- control signals
SIGNAL mp_IRload, mp_JMPmux, mp_PCload, mp_MemInst, mp_MemWr: STD_LOGIC;
SIGNAL mp_Asel: STD_LOGIC_VECTOR(1 DOWNTO 0);
SIGNAL mp_Aload, mp_Sub: STD_LOGIC;
-- status signals
SIGNAL mp_IR: STD_LOGIC_VECTOR(7 DOWNTO 5);
SIGNAL mp_Aeq0, mp_Apos: STD_LOGIC;

BEGIN
   -- doing structural modeling for the microprocessor here
   U0: cu PORT MAP (
      Clock, Reset,
      -- control input
      Enter,
      -- control signals
      mp_IRload, mp_JMPmux, mp_PCload, mp_MemInst, mp_MemWr,
      mp_Asel,
      mp_Aload, mp_Sub,
      -- status signals
      mp_IR,
      mp_Aeq0, mp_Apos,
      -- control outputs
      Halt);

   U1: dp PORT MAP (
      Clock, Reset,
      -- datapath input
      Input,
```

Figure 12.18   Structural VHDL code for the complete EC-2 general-purpose microprocessor.
(continued on next page)

```
        -- control signals
        mp_IRload, mp_JMPmux, mp_PCload, mp_MemInst, mp_MemWr,
        mp_Asel,
        mp_Aload, mp_Sub,
        -- status signals
        mp_IR,
        mp_Aeq0, mp_Apos,
        -- datapath output
        Output);


END Structural;
```

Figure 12.18 Structural VHDL code for the complete EC-2 general-purpose microprocessor.

12.4.2 Behavioral FSMD

Figure 12.19 shows the complete behavioral FSMD VHDL code for the EC-2 general-purpose microprocessor. The memory used is from the MAX+plus II's LPM component library. The three registers, *IR*, *PC*, and *A*, are declared as signals in the architecture section. Recall from Section 6.13, that if a signal is not assigned a value from all conditional paths, then it is synthesized as a register. The process block is structured just like a regular FSM following the state diagram from Figure 12.12(a). The *Decode* state uses a CASE statement to check the opcode, which is the first three bits of the *IR*. From there, the FSM jumps to the state for executing the corresponding instruction. For the STORE instruction, an extra state is needed to de-assert the *MemWr* signal before changing the memory address back to the location of the next instruction.

```
-- EC-2 Behavioral FSMD description
LIBRARY IEEE;
USE IEEE.STD_LOGIC_1164.ALL;
USE IEEE.STD_LOGIC_ARITH.ALL;
USE IEEE.STD_LOGIC_UNSIGNED.ALL;

LIBRARY lpm;
USE lpm.lpm_components.ALL;

ENTITY mp IS PORT (
    clock, reset: IN STD_LOGIC;
    enter: IN STD_LOGIC;
```

Figure 12.19 Behavioral VHDL code for the complete EC-2 general-purpose microprocessor. *(continued on next page)*

```vhdl
    -- data input
    input: IN STD_LOGIC_VECTOR(7 DOWNTO 0);
    -- data output
    output: OUT STD_LOGIC_VECTOR(7 DOWNTO 0);
    -- control outputs
    halt: OUT STD_LOGIC);
END mp;

ARCHITECTURE FSMD OF mp IS
    TYPE state_type IS(s_start,s_fetch,s_decode,s_load,s_store,s_store2,
                s_add,s_sub,s_input,s_jz,s_jpos,s_halt);
    SIGNAL state: state_type;                          -- states
    SIGNAL IR: STD_LOGIC_VECTOR(7 DOWNTO 0);           -- Instruction register
    SIGNAL PC: STD_LOGIC_VECTOR(4 DOWNTO 0);           -- Program counter
    SIGNAL A: STD_LOGIC_VECTOR(7 DOWNTO 0);            -- Accumulator
    SIGNAL memory_address: STD_LOGIC_VECTOR(4 DOWNTO 0); -- memory address
    SIGNAL memory_data: STD_LOGIC_VECTOR(7 DOWNTO 0);    -- memory data input
    SIGNAL MemWr: STD_LOGIC;

BEGIN
    memory: lpm_ram_dq        -- 32 locations x 8 bits wide asynchronous memory
      GENERIC MAP (
        lpm_widthad => 5,
        lpm_outdata => "UNREGISTERED",
        lpm_indata  => "UNREGISTERED",
        lpm_address_control => "UNREGISTERED",
        lpm_file => "program.mif", -- fill ram with content of file program.mif
        lpm_width => 8)
      PORT MAP (
        data        => A,
        address     => memory_address,
        we          => MemWr,
        q           => memory_data);

    PROCESS(clock,reset)
    BEGIN
      IF(reset = '1') THEN
        PC <= "00000";
        IR <= "00000000";
        A  <= "00000000";
        MemWr <= '0';
        halt  <= '0';
        state <= s_start;
      ELSIF(clock'EVENT AND clock = '1') THEN
```

Figure 12.19   Behavioral VHDL code for the complete EC-2 general-purpose microprocessor. *(continued on next page)*

```
CASE state IS
WHEN s_start => -- reset, start
   memory_address <= PC;
   state <= s_fetch;
WHEN s_fetch => -- fetch
   IR <= memory_data;
   PC <= PC + 1;
   state <= s_decode;
WHEN s_decode => -- decode
   -- memory access using last 5 bits of IR
   memory_address <= IR(4 DOWNTO 0);
   CASE IR(7 DOWNTO 5) IS      -- decode first 3 bits of IR as opcode
      WHEN "000" => state <= s_load;
      WHEN "001" => state <= s_store;
      WHEN "010" => state <= s_add;
      WHEN "011" => state <= s_sub;
      WHEN "100" => state <= s_input;
      WHEN "101" => state <= s_jz;
      WHEN "110" => state <= s_jpos;
      WHEN "111" => state <= s_halt;
      WHEN OTHERS => state <= s_start;
   END CASE;
WHEN s_load =>                 -- load A from memory
   A <= memory_data;
   state <= s_start;
WHEN s_store =>               -- store A to memory
   MemWr <= '1';
   state <= s_store2;
WHEN s_store2 =>             -- need an extra state to de-assert MemWr
   MemWr <= '0';             -- before changing the memory address
   state <= s_start;
WHEN s_add =>                -- add
   A <= A + memory_data;
   state <= s_start;
WHEN s_sub =>               -- subtract
   A <= A - memory_data;
   state <= s_start;
WHEN s_input =>
   A <= input;
   IF (Enter = '0') THEN     -- wait for Enter key
      state <= s_input;
   ELSE
      state <= s_start;
   END IF;
```

Figure 12.19 Behavioral VHDL code for the complete EC-2 general-purpose microprocessor. *(continued on next page)*

```
        WHEN s_jz =>
            IF (A = 0) THEN              -- jump if A is 0
                PC <= IR(4 DOWNTO 0);
            END IF;
            state <= s_start;
        WHEN s_jpos =>
            IF (A(7) = '0') THEN         -- jump if MSB(A) is 0
                PC <= IR(4 DOWNTO 0);
            END IF;
            state <= s_start;
        WHEN s_halt =>
            halt <= '1';
            state <= s_halt;
        WHEN OTHERS =>
            state <= s_halt;
        END CASE;
    END IF;
  END PROCESS;

  output <= A;        -- send value of Accumulator to the output
END FSMD;
```

Figure 12.19   Behavioral VHDL code for the complete EC-2 general-purpose microprocessor.

● ● ● ● ● ● ● ● ● ● ● ● ● ● ⋯⋯

12.5 Summary Checklist

- ■ Instruction set
- ■ Operation code (opcode)
- ■ Instruction register (*IR*)
- ■ Program counter (*PC*)
- ■ Instruction cycle
- ■ Be able to derive the datapath circuit for a given instruction set
- ■ Be able to derive the state diagram for a given instruction set
- ■ Be able to derive the control unit circuit for a given instruction set
- ■ Be able to derive the complete general-purpose microprocessor circuit
- ■ Be able to implement the complete computer using the general-purpose microprocessor developed
- ■ Be able to write programs in machine language using the implemented instruction set
- ■ Be able to run the programs on the implemented computer
- ■ Be able to write behavioral VHDL codes for general-purpose microprocessors

● ● ● ● ● ● ● ● ● ● ● ● ● ● ● ● ●

12.6 Problems

P12.1. Manually redesign the EC-1 microprocessor to accommodate each of the following changes. The changes are to be done separately.

(a) Modify the OUTPUT instruction so that the output port outputs a value only when the OUTPUT instruction is executed.

(b) Modify the INPUT instruction so that it will wait for an external *Enter* key signal before continuing to the next instruction.

(c) Add an extra INC instruction, using the opcode 000, to the EC-1 instruction set. The INC instruction increments the accumulator.

(d) Add an extra LOAD instruction, using the opcode 001, to the EC-1 instruction set. The LOAD instruction loads the accumulator with the content of memory location aaaa, where aaaa are the least four significant bits of the instruction encoding.

P12.2. Write the behavioral VHDL code for the EC-1 microprocessor.

P12.3. Rewrite the behavioral VHDL code for the EC-1 microprocessor with each of the changes from Problem P12.1.

P12.4. Write and run the following programs on the EC-2 computer:

(a) Input two numbers, and output the sum of these two numbers.

(b) Input two numbers, and output the larger of the two numbers.

(c) Input two numbers, and output the product of these two numbers.

(d) Keep inputting numbers until a 0. Output the total number of numbers entered.

(e) Keep inputting numbers until a 0. Output the sum of these numbers.

(f) Keep inputting numbers until a 0. Output the largest of these numbers.

(g) Keep inputting numbers until a 0. Output the largest and second largest of these numbers.

(h) Input three numbers, and output these numbers in ascending order.

P12.5. Manually redesign the EC-2 microprocessor to accommodate each of the following changes. The changes are to be done separately.

(a) Replace the SUB instruction in the EC-2 instruction set with an LSHIFT instruction. The LSHIFT instruction shifts the content of the accumulator left by one bit. The result of the shift operation is written back into the accumulator.

(b) Replace the SUB instruction in the EC-2 instruction set with an OUTPUT instruction. The OUTPUT instruction outputs the content of the accumulator to the output port. The output port should not show anything when the OUTPUT instruction is not being executed.

(c) Add an extra LSHIFT instruction to the EC-2 instruction set. The LSHIFT instruction is defined in part (a) above.

(d) Add an extra OUTPUT instruction to the EC-2 instruction set. The OUTPUT instruction is defined in part (b) above.

P12.6. Rewrite the behavioral VHDL code for the EC-2 microprocessor with each of the changes from Problem P12.5.

P12.7. Given the instruction set as defined in Figure P12.7, manually design a datapath that can realize this instruction set.

Instruction	Encoding	Operation	Comment
Data Movement Instructions			
LDA A,rrr	0001 0rrr	$A \leftarrow R[rrr]$	Load accumulator from register
STA rrr,A	0010 0rrr	$R[rrr] \leftarrow A$	Load register from accumulator
LDM A,aaaaaa	0011 0000 00 aaaaaa	$A \leftarrow M[aaaaaa]$	Load accumulator from memory
STM aaaaaa,A	0100 0000 00 aaaaaa	$M[aaaaaa] \leftarrow A$	Load memory from accumulator
LDI A,iiiiiiii	0101 0000 iiiiiiii	$A \leftarrow iiiiiiii$	Load accumulator with immediate value (iiiiiiii is a signed number)
Jump Instructions			
JMP absolute	0110 0000 00 aaaaaa	$PC = aaaaaa$	Absolute unconditional jump
JMPR relative	0110 smmm	IF (smmm != 0) THEN IF (s == 0) THEN $PC = PC + mmm$ ELSE $PC = PC - mmm$	Relative unconditional jump (smmm is in sign and magnitude format)
JZ absolute	0111 0000 00 aaaaaa	IF (A == 0) THEN $PC = aaaaaa$	Absolute jump if A is zero
JZR relative	0111 smmm	IF (A == 0 AND smmm != 0) THEN IF (s == 0) THEN $PC = PC + mmm$ ELSE $PC = PC - mmm$	Relative jump if A is zero (smmm is in sign and magnitude format)
JNZ absolute	1000 0000 00 aaaaaa	IF (A != 0) THEN $PC = aaaaaa$	Absolute jump if A is not zero
JNZR relative	1000 smmm	IF (A != 0 AND smmm != 0) THEN IF (s == 0) THEN $PC = PC + mmm$ ELSE $PC = PC - mmm$	Relative jump if A is not zero (smmm is in sign and magnitude format)
JP absolute	1001 0000 00 aaaaaa	IF (A == positive) THEN $PC = aaaaaa$	Absolute jump if A is positive
JPR relative	1001 smmm	IF (A == positive AND smmm != 0) THEN IF (s == 0) THEN $PC = PC + mmm$ ELSE $PC = PC - mmm$	Relative jump if A is positive (smmm is in sign and magnitude format)

Figure P12.7
(continued on next page)

Instruction	Encoding	Operation	Comment

Arithmetic and Logical Instructions

Instruction	Encoding	Operation	Comment
AND A,rrr	1010 0rrr	$A \leftarrow A \text{ AND } R[\text{rrr}]$	Accumulator AND register
OR A,rrr	1011 0rrr	$A \leftarrow A \text{ OR } R[\text{rrr}]$	Accumulator OR register
ADD A,rrr	1100 0rrr	$A \leftarrow A + R[\text{rrr}]$	Accumulator + register
SUB A,rrr	1101 0rrr	$A \leftarrow A - R[\text{rrr}]$	Accumulator − register
NOT A	1110 0000	$A \leftarrow \text{NOT } A$	Invert accumulator
INC A	1110 0001	$A \leftarrow A + 1$	Increment accumulator
DEC A	1110 0010	$A \leftarrow A - 1$	Decrement accumulator
SHFL A	1110 0011	$A \leftarrow A << 1$	Shift accumulator left
SHFR A	1110 0100	$A \leftarrow A >> 1$	Shift accumulator right
ROTR A	1110 0101	$A \leftarrow \text{Rotate\_right}(A)$	Rotate accumulator right

Input/Output and Miscellaneous

Instruction	Encoding	Operation	Comment
In A	1111 0000	$A \leftarrow Input$	Input to accumulator
Out A	1111 0001	$Output \leftarrow A$	Output from accumulator
HALT	1111 0010	Halt	Halt execution
NOP	0000 0000	No operation	No operation

Notations:

A = accumulator

R = general register

M = memory

PC = program counter

rrr = three bits for specifying the general register number (0–7)

aaaaaa = six bits for specifying the memory address

iiiiiiii = an 8-bit signed number

smmm = four bits for specifying the relative jump displacement in sign and magnitude format. The most significant bit(s) determines whether to jump forward or backward (0 = forward, 1 = backward). The last three bits (mmm) specify the number of locations to increment or decrement from the current PC location.

Figure P12.7

P12.8. Write the behavioral VHDL code for a microprocessor that can execute the instructions in the instruction set defined in Problem P12.7.

Schematic Entry—Tutorial 1

The MAX+plus II software and the UP2 development board provide all of the necessary tools for implementing and trying out all of the examples, including building the final general-purpose microprocessor, discussed in this book. The MAX+plus II software offers a completely integrated development tool and easy-to-use graphical-user interface for the design, and synthesis of digital logic circuits. Together with the UP2 development board, these circuits can be implemented on a programmable logic device (PLD) chip. After downloading the circuit netlist to the PLD, you can see the actual operation of these circuits in the hardware.

A Student Edition version of the MAX+plus II software is included on the accompanying CD-ROM and can also be downloaded from the Altera website found at www.altera.com. The optional UP2 development board can be purchased directly from Altera. The full User Guide for using the UP2 board is on the CD-ROM, and can also be downloaded from the Altera website. This tutorial assumes that you are familiar with the Windows environment, and that the MAX+plus II software has already been installed on your computer. Instructions for the installation of the MAX+plus II software can be found on the CD-ROM. You also must obtain a license file from the Altera website in order for the software to function correctly. Be careful that you obtain the Student Edition license, and not the Baseline Edition license.

The MAX+plus II development software provides for both schematic and text entry of a circuit design. The Schematic Editor is used to enter a schematic drawing of a circuit. Using the Schematic Editor, logic symbols for the circuit can be inserted and connected together using the drawing tools. The Text Editor is used to enter VHDL or Verilog code for describing a circuit.

This tutorial provides a step-by-step instruction for the schematic entry, synthesis, and simulation of an 8-bit 2-to-1 multiplexer circuit. Tutorial 3 (Appendix C) will show how a circuit can be downloaded to the PLD on the UP2 development board so that you actually can see this circuit executed in the hardware.

● ● ● ● ● ● ● ● ● ● ● ● ● ● ●

A.1 Getting Started

A.1.1 Preparing a Folder for the Project

Each circuit design in MAX+plus II is called a project. Each project should be placed in its own folder, since the synthesizer creates many associated working files for a project. Using Windows File Manager, create a new folder for your new project.

● For this tutorial, create a folder called 2x8mux in the root directory of the C drive.

A.1.2 Starting MAX+plus II

After the successful installation of the MAX+plus II software, there should be a link for the program under the Start button. Click on this link to start the program. You should see the MAX+plus II Manager window, as shown in Figure A.1.

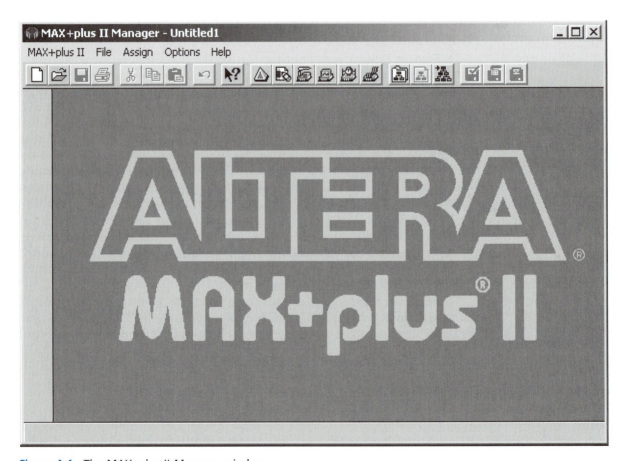

Figure A.1 The MAX+plus II Manager window.

Figure A.2   The MAX+plus II development software toolbar.

Figure A.2 shows the toolbar for accessing the main development tools. The buttons from left to right are:

- *Hierarchy display*—to show the design files used in the current project
- *Floorplan Editor*—to map the I/O signals from the circuit to the pins on the PLD chip
- *Compiler (synthesizer)*—to synthesize the circuit to its netlist
- *Simulator*—to perform circuit simulation
- *Timing analyzer*—to perform circuit timing analysis
- *Programmer*—to program the circuit to the PLD chip
- *Open existing or new project*—to select a new project
- *Change project name to current filename*—to use the current file as the new project
- *Open top-level design file*—to open the top-level design file for the current project

In the MAX+plus II software, different commands in the menus are available when different windows are activated. This might cause some confusion at first. If you cannot find a particular command from the menu, make sure that the correct window for that command is the active window.

A.1.3   Starting the Graphic Editor

From the **Manager** window, select **Max+plus II | Graphic Editor**. You should see the **Graphic Editor** window similar to the one shown in Figure A.3. Any circuit diagram can be drawn in this **Graphic Editor** window.

- Alternatively, you can select **File | New** from the **Manager** window menu. Select **Graphic Editor** file using the extension .gdf, and click OK.

● ● ● ● ● ● ● ● ● ● ● ● ● ● ● ● ●

A.2 Using the Graphic Editor

A.2.1   Drawing Tools

In Figure A.3, the tools for drawing a circuit in the Graphic Editor are shown in the toolbar on the left side. There are the standard text writing tool, line drawing tools for making connections between logic symbols, and zooming tools. The main tool that you will use is the multifunction pointer tool. This multifunction pointer allows you to perform many different operations depending on the context in which it is used. Two

Mulitifunction Pointer Tool ⟶

Text Tool ⟶

Line/Connection Tools ⟶

Zooming ⟶

Fit Drawing in Window ⟶

Connection Dot ⟶

Turns On Rubberbanding ⟶

Turns Off Rubberbanding ⟶

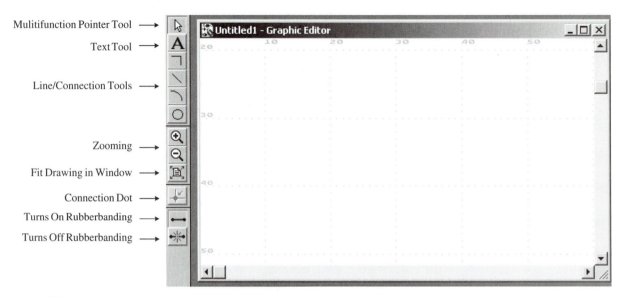

Figure A.3 The Graphic Editor window with the graphics toolbar on the left.

main operations performed by this multifunction pointer are selecting objects and making connections between logic symbols. The connection dot tool either makes or deletes a connection point between two crossing lines. Finally, the two rubberbanding buttons turn on or off the rubberbanding function. When rubberbanding is turned on, connection lines are adjusted automatically when symbols are moved from one location to another. When rubberbanding is turned off, moving a symbol will not affect the lines connected to it.

A.2.2 Inserting Logic Symbols

1. To insert a logic symbol, first select the multifunction pointer tool, and then double-click the pointer on an empty spot in the Graphic Editor window. You should see the Enter Symbol window, as shown in Figure A.4.

    Available symbol libraries are listed in the Symbol Libraries selection box. These libraries include the standard primitive gates, standard combinational and sequential components, and your own logic symbols located in the current project directory.

    All of the primitive logic gates, latches, flip-flops, and input and output connectors that we need are in the primitive library: ...\prim. Your own circuits that you want to reuse in building larger circuits are in the directory that they are stored in.

2. Double-click on the **prim** library to see a list of logic symbols available in that library. A list of logic symbols is shown in the Symbol Files selection box. The logic symbols are sorted in alphabetical order. To narrow down the list, you can type the first few letters of the symbol name followed by an asterisk in the Symbol Name text box. You need to either press Enter or click on the OK button to update the list. For example, typing a*, and pressing Enter will produce a list of all of the symbols whose name starts with the letter "a."

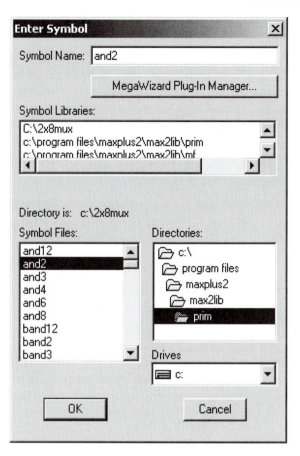

Figure A.4
The Enter Symbol selector window.

3. Double-click on the logic symbol name that you want in order to insert that symbol into the Graphic Editor. If you just select the symbol with a single-click, then you will also have to click on the **OK** button.

   For this tutorial, insert the following symbols:

   - A 2-input AND gate (and2)

   - A 2-input OR gate (or2)

   - A NOT gate (not)

   - An input signal connector (input)

   - An output signal connector (output)

   A unique number is given to each instance of a symbol and is written at the lower-left corner of the symbol. This number is used only as a reference number in the output netlist and report files. These numbers may be different from those in the examples.

A.2.3 Selecting, Moving, Copying, and Deleting Logic Symbols

- To select a logic symbol in the Graphic Editor, simply single-click on the symbol using the multifunction pointer tool. You can also select multiple symbols by holding down the **Shift** key while you select the symbols. An alternative method is to trace a rectangle with the multifunction tool around the objects that you want to select. All objects inside the rectangle will be selected.

- To move a symbol, simply drag the symbol.

- To copy a symbol, first select it and then perform the Copy and Paste operations. An alternative method is to hold down the **Ctrl** key while you drag the symbol.

- To delete a symbol, first select it and then press the **Delete** key.

- To rotate a symbol, right-click on the symbol, select **Rotate** from the drop-down menu and select the angle to rotate the symbol.

Perform the following operations for this tutorial:

1. Make a copy of the 2-input AND gate

2. Make two more copies of the input signal connector

3. Position the symbols to look like Figure A.5

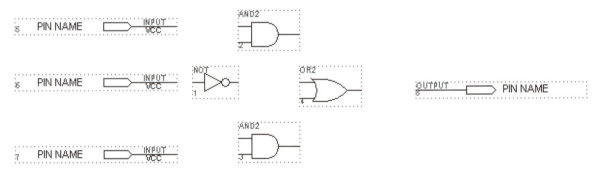

Figure A.5 Symbol placements for the 2-to-1 multiplexer circuit.

A.2.4 Making and Naming Connections

- To make a connection between two connection points, use the multifunction pointer tool and drag from one connection point to the second connection point. Notice that, when you position the multifunction pointer tool to a connection point, the arrow pointer changes to a crosshair.

- To change the direction of a connection line while dragging the line, simply release and press the mouse button again and then continue to drag the connection line.

- You can also make a connection between two connection points by moving a symbol so that its connection point touches the other connection point.
- If you want to make a connection line that does not start from a connection point, you will need to use either the straight line drawing tool or the vertical and horizontal line drawing tool instead of the multifunction pointer tool.
- Once a connection is made to a symbol, you can move the symbol to another location, and the connection line is adjusted automatically if the rubberbanding function is turned on. However, if the rubberbanding function is turned off, the connection will be broken if the symbol is moved.
- To make a connection between two lines that cross each other, you need to use the multifunction pointer tool to select the junction point (i.e., the point where the two lines cross) and then press the connection tool button, as shown in Figure A.6. The connection tool button is enabled only after you have selected a junction point.

Figure A.6 Making or deleting a connection point with the connection tool button.

- To remove a connection point, select (single-click) the point and then press the connection tool button.
- To make a bus connection (for grouping two or more lines together), first draw a regular connection line, then right-click on the line, select **Line Style**, and select the thicker, solid line style from the pop-up menu.
- A bus must also have a name and a width associated with it. Select the bus line at the point where you want to place the name and then type in the name and the width for the bus. For example, data[7..0] is an 8-bit bus with the name data, as shown in Figure A.7.

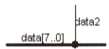

Figure A.7 A single connection line to an 8-bit bus with the name data.

- To change the name, just double-click on the name and edit it.
- To connect one line to a bus, connect a single line to the bus, and then give it the same name as the bus with the line index appended to it. For example, data2, is bit two of the data bus. as shown in Figure A.7.
- To check whether a name is attached correctly to a line, select the line, and the name that is attached to the line will also be selected.
- To name an input or output connector, select its name label by double-clicking it and then type in the new name. Pressing the **Enter** key will move the text entry cursor to the name label for the symbol below the current symbol.

- A bus line connected to an input or output connector must have the same name as the connector.

Perform the following operations for this tutorial:

1. Name the three input connectors d0[7..0], s, and d1[7..0], as shown in Figure A.8
2. Name the output connector y[7..0], as shown in Figure A.8
3. Connect and name the five bus lines d0[7..0], d1[7..0], and0[7..0], and1[7..0], and y[7..0], as shown in Figure A.8
4. Connect the single lines from the input connector s to the inverter and to the two AND gates, as shown in Figure A.8.

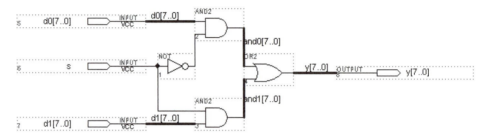

Figure A.8   Connections and names for the 2-to-1 multiplexer circuit.

A.2.5   Selecting, Moving and Deleting Connection Lines

- To select a straight connection line, just single-click on it.
- To select an entire connection line with horizontal and vertical segments, double-click on it.
- To select a portion of a line, trace a rectangle with the multifunction tool around the segment.
- After a line is selected, it can be moved by dragging.
- After a line is selected, it can be deleted by pressing the **Delete** key.

A.3   Specifying the Top-Level File and Project

A.3.1   Saving the Schematic Drawing

1. From the **Graphic Editor** menu, select **File | Save**. Select the 2x8mux directory that you created on the C drive in Section A.1.1. Type in the filename 2x8mux. The extension should be .gdf (for graphic design file).
2. Click **OK**.

A.3.2   Specifying the Project

To use the schematic drawing file saved in Section A.3.1 as the top-level project file, select File | Project | Set Project to Current File from the Manager window menu, or simply click on the icon 🔳 .

- You can open any graphic design file (with the extension .gdf) using the Manager menu command File | Open, and then select File | Project | Set Project to Current File, or click on the icon 🔳 to make that particular file the top-level project file.

● ● ● ● ● ● ● ● ● ● ● ● ● ● ● ●
A.4   Synthesis for Functional Simulation

1. From the Manager window menu, select MAX+plus II | Compiler, or click on the icon 🔳 to bring up the Compiler window.

2. From the Compiler window menu (that is, with the Compiler window selected as the active window), select Processing | Functional SNF Extractor so that a check mark appears next to it. To actually see whether there is a check mark or not, you need to select the Processing menu again. The Compiler window for functional extraction is shown in Figure A.9.

3. Click on the Start button to start the synthesis. You will then see the progress of the synthesis.

4. At the end of the synthesis, if there are no syntax errors, you will see a message window saying that the compilation was successful. Click OK to close the message window.

5. If there are errors, go back to Section A.2.4 and double check your circuit with the one in Figure A.8.

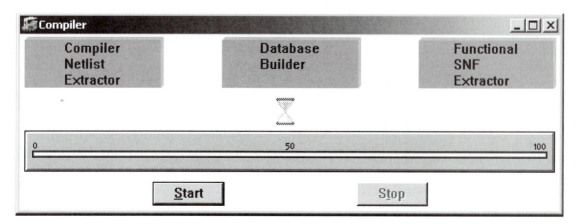

Figure A.9   Compiler window for functional extraction.

· · · · · · · · · · · · · · ·

A.5 Circuit Simulation

A.5.1 Selecting Input Test Signals

1. Before you can simulate the design, you need to create test vectors for specifying what the input values are. From the **Manager** window menu, select MAX+plus II | Waveform Editor.

2. From the **Waveform Editor** window menu, select **Node | Enter Nodes from SNF**. You can also right-click under the **Name** section in the **Waveform Editor** window and select **Enter Nodes from SNF** from the pop-up menu. You will see something similar to the **Enter Nodes from SNF** window shown in Figure A.10.

3. Click on the **List** button in the **Enter Nodes from SNF** window, and a list of available nodes and groups will be displayed in the **Available Nodes & Groups** box.

4. Select the signals that you want to see in the simulation trace. The signals that we want are: s(I), d1[7..0] (I), d0[7..0] (I), and y[7..0] (O). The letters I and O in parenthesis next to each signal denote whether the signal is an input or output signal, respectively. Note that the signal name such as y7 is bit seven of the bus named y, and d16 is bit six of the bus named d1. Multiple nodes can be selected by holding down the **Ctrl** or **Shift** key while clicking on the signal names.

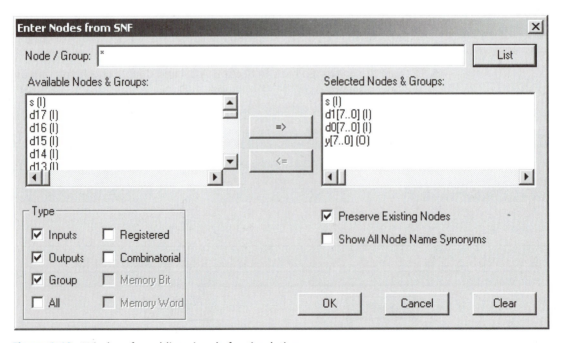

Figure A.10 Window for adding signals for simulation.

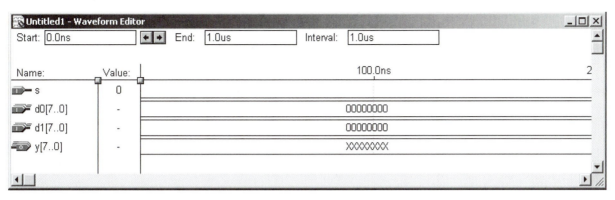

Figure A.11   Waveform Editor window for simulation.

5. After selecting the signals, click on the => button to move the selected signals to the **Selected Nodes & Groups** box.

6. Repeat Steps 4 and 5 until all of the signals that you want to see in the simulation are moved to the **Selected Nodes & Groups** box.

7. Click on **OK** when you are finished. The selected signals will now be inserted in the **Waveform Editor** window similar to Figure A.11.

A.5.2 Customizing the Waveform Editor

- You can rearrange the signals in the Waveform Editor by dragging the signal icons like up or down.

- To delete a signal in the Waveform Editor, just select the signal by clicking on its name and press the **Delete** key.

- For signals that are composed of a group of bits (such as the data input d1), you can separate them into individual bits or change the radix for the displayed value by first selecting that signal and then right-click the mouse. A drop-down menu appears. Select **Ungroup** to separate the bits. To regroup them, select the bits you want to group and then right-click the mouse. A drop-down menu appears. Select **Enter Group**. Type in a group name, and select the **Decimal** radix for the display.

1. We want to simulate for 500 ns. To change the simulation end time, select **File | End Time** from the **Waveform Editor** window menu.

2. In the **End Time** window, type in 500ns, and click **OK**.

3. To fit the entire simulation time range inside the window, select **View | Fit in Window** from the **Waveform Editor** window menu, or click on the icon in the toolbar on the left. Your **Waveform Editor** window should now look like the one in Figure A.12.

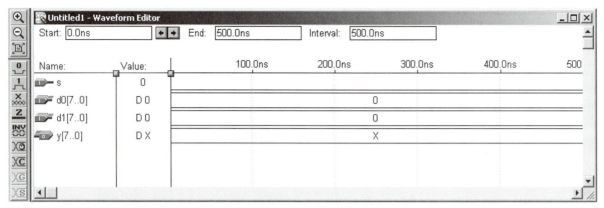

Figure A.12 Waveform Editor window after changing the value radix and fitting the entire time range inside the window. Notice also the toolbar buttons on the left.

A.5.3 Assigning Values to the Input Signals

The next thing is to assign values to all of the input signals.

1. Using the multifunction pointer tool, drag from time 200ns to 400ns for the s signal only, as shown in Figure A.13.

2. Click on the icon ⬚ in the toolbar on the left to set the signal in the selected range to a logical 1 value.

3. Drag from time 0ns to 100ns for the d0 signal only, as shown in Figure A.13.

4. Click on the icon ⬚ in the toolbar on the left. Type in the value 5 and click OK to set the value for the d0 bus signal to decimal 5.

5. Repeat Steps 3 and 4 for the remaining input values for d0 and d1, as shown in Figure A.13.

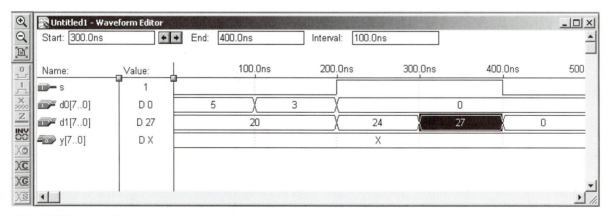

Figure A.13 Changing the input signal values.

A.5.4   Saving the Waveform File

1. Save the **Waveform Editor** window by selecting **File | Save**. The **Save As** window appears. Notice that the default file name is the same as the top-level entity name, and the extension is .scf. For this example, the name is 2x8mux.scf.

2. Click on **OK** to save the file.

A.5.5   Starting the Simulator

1. We are now ready to simulate the design. From the **Manager** window menu, select **MAX+plus II | Simulator**, or click on the icon 🔧 to bring up the **Simulator** window.

   - You can also save the waveform file and start the simulator in one step by selecting from the **Manager** window menu **File | Project | Save & Simulate**.

2. The **Simulator** window as shown in Figure A.14 is displayed. Make sure that the **Simulation Input** filename is 2x8mux.scf. This is the same name as your top-level entity.

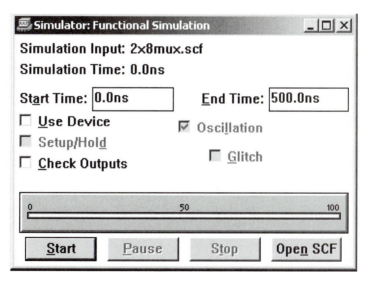

Figure A.14   Simulator window for the multiplexer design.

3. Click on the **Start** button and watch the progress of the simulation.

4. At the end of the simulation, if there are no errors, you will see a message window saying that the simulation was successful. Click **OK** to close the message window.

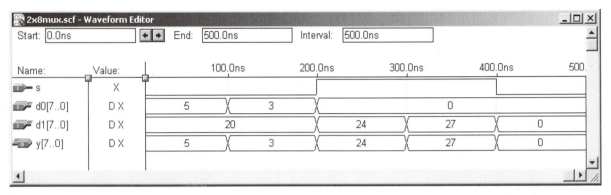

Figure A.15 Resulting waveform after the simulation.

5. Click on the **Open SCF** button in the **Simulator** window to bring up the Waveform Editor with the resulting simulation waveforms. The simulation result is shown in Figure A.15. The signal y is the multiplexer output.

   • Notice that when s is a 0, the y output follows the d0 input, and when s is a 1, the y output follows the d1 input.

6. You can change the input signal values of s, d0, and d1 to something different and run the simulation again.

• • • • • • • • • • • • • • • •

A.6 Creating and Using the Logic Symbol

If you want to use this circuit as part of another circuit, you need to create a logic symbol for this circuit.

1. To create a logic symbol for the current active circuit diagram that is in the **Graphic Editor** window, from the **Graphic Editor** menu, select **File | Create Default Symbol**. The name of this logic symbol is the same as the name of the current active circuit diagram in the Graphic Editor, but with the extension .sym.

   • If your **Graphic Editor** window is closed, you can click on the icon 🖧 to open the top project circuit diagram again.

2. You can view and edit the logic symbol by selecting **File | Edit Symbol**. The placements of the input and output signals can be moved to different locations by dragging them. The size of the symbol can also be changed by dragging the edges of the symbol rectangle.

3. To use this circuit in another project, you need to copy this .sym file and the corresponding .gdf circuit file to the other project's directory. This new symbol name will now show up in the **Enter Symbol** window like the one shown in Figure A.4.

VHDL Entry—Tutorial 2

This tutorial provides a step-by-step instruction for the VHDL entry, synthesis, and simulation of a 4-bit binary counter circuit. However, no knowledge of VHDL is required to follow this tutorial. Tutorial 3 in Appendix C will show how this circuit can be downloaded to the PLD on the UP2 development board so that you can actually see this circuit executing in the hardware.

This tutorial is very similar to Tutorial 1 in Appendix A for schematic entry. The main difference is in using the Text Editor rather than the Graphic Editor. The procedures for project creation, synthesis, and simulation are the same in both cases. Even if you do not intend to write VHDL code, you should go through this tutorial so that you can continue on to Tutorial 3 in Appendix C and learn how to download a circuit to the PLD on the UP2 development board.

● ● ● ● ● ● ● ● ● ● ● ● ● ● ● ● ●

B.1 Getting Started

B.1.1 Preparing a Folder for the Project

1. Each circuit design in MAX+plus II is called a project. Each project should be placed in its own folder, since the synthesizer creates many associated working files for a project. Using Windows File Manager, create a new folder for your new project. This tutorial uses the folder called counter created in the root directory of the C drive.

2. The VHDL source code for the counter circuit can be found on the accompanying CD-ROM in the file counter.vhd located in the directory <CD-ROM drive>:\VHDL Examples\Appendix B VHDL Entry Tutorial 2\Source. Using Windows File Manager, copy this file to the new folder c:\counter that you created in Step 1.

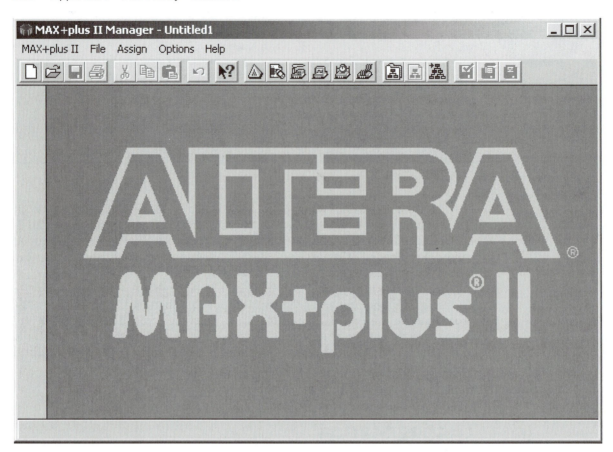

Figure B.1 The MAX+plus II Manager window.

B.1.2 Starting MAX+plus II

After the successful installation of the MAX+plus II software, there should be a link for the program under the **Start** button. Click on this link to start the program. You should see the **MAX+plus II Manager** window, as shown in Figure B.1.

Figure B.2 shows the toolbar for accessing the main development tools. The buttons from left to right are:

- *Hierarchy display*—to show the design files used in the current project
- *Floorplan Editor*—to map the I/O signals from the circuit to the pins on the PLD chip

Figure B.2 The MAX+plus II development software toolbar.

- *Compiler (synthesizer)*—to synthesize the circuit to its netlist
- *Simulator*—to perform circuit simulation
- *Timing analyzer*—to perform circuit timing analysis
- *Programmer*—to program the circuit to the PLD chip
- *Open existing or new project*—to select a new project
- *Change project name to current filename*—to use the current file as the new project
- *Open top-level design file*—to open the top-level design file for the current project

In the MAX+plus II software, different commands in the menus are available when different windows are activated. This might cause some confusion at first. If you cannot find a particular command from the menu, make sure that the correct window is the active window.

B.1.3   Creating a Project

1. From the Manager window menu, select File | Project | Name, or simply click on the icon [icon]. You should see the Project Name window similar to the one shown in Figure B.3.

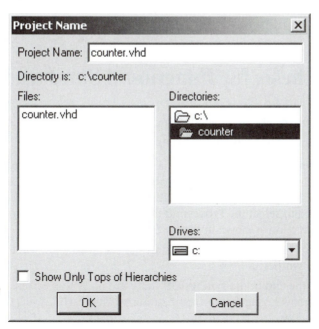

Figure B.3
Project Name window for creating a new project.

2. Select the C drive from the **Drives** drop-down list.

3. Move to the `counter` directory on the C drive and double-click on it. You should see the file `counter.vhd` listed in the **Files** box.

4. Select the file `counter.vhd`. The filename will be copied to the **Project Name** text field.

5. Click **OK**. The **MAX+plus II Manager** window title should now show `c:\counter\counter`.

- Alternatively, you can open any VHDL entity source file (with the extension `.vhd`) using the **Manager** menu command **File | Open**. With the VHDL entity source file in the active **Text Editor** window, select **File | Project | Set Project to Current File**, or click on the icon ![icon] to make that particular file the top-level project file. Note that the name of the file must be the same as the name of the entity that is in this file.

B.1.4 Editing the VHDL Source Code

- From the **Manager** window menu, select **File | Hierarchy Project Top**, or click on the icon ![icon] to open the VHDL source code for the counter. Notice that the entity name for this circuit is also `counter`. The top entity name for the project must be the same as the project name and the file name.

- You can use the Text Editor to modify the code if necessary. For this tutorial, we will not make any modifications, so you can close the **Editor** window.

- If you need to create a new VHDL source file, select **File | New** from the **Manager** window menu. Select **Text Editor file**, and click **OK**.

● ● ● ● ● ● ● ● ● ● ● ● ● ● ●
B.2 Synthesis for Functional Simulation

1. From the **Manager** window menu, select **MAX+plus II | Compiler**, or click on the icon ![icon] to bring up the **Compiler** window.

2. From the **Compiler** window menu (that is, with the **Compiler** window selected as the active window), select **Processing | Functional SNF Extractor** so that a check mark appears next to it. To actually see whether there is a check mark or not, you need to select the **Processing** menu again. The **Compiler** window for functional extraction is shown in Figure B.4.

3. Click on the **Start** button to start the synthesis. You will then see the progress of the synthesis.

4. At the end of the synthesis, if there are no syntax errors, you will see a message window saying that the compilation was successful. Click **OK** to close the message window.

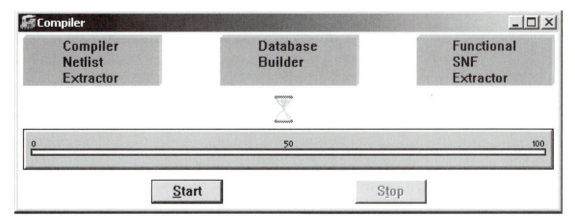

Figure B.4   Compiler window for functional extraction.

● ● ● ● ● ● ● ● ● ● ● ● ● ● ● ● ●

B.3 Circuit Simulation

B.3.1 Selecting Input Test Signals

1. Before you can simulate the design, you need to create test vectors for specifying what the input values are. From the **Manager** window menu, select MAX+plus II | Waveform Editor.

2. From the **Waveform Editor** window menu, select **Node | Enter Nodes from SNF**. You can also right-click under the **Name** section in the **Waveform Editor** window, and select **Enter Nodes from SNF** from the pop-up menu. You will see something similar to the **Enter Nodes from SNF** window shown in Figure B.5.

3. Click on the **List** button in the **Enter Nodes from SNF** window, and a list of available nodes and groups will be displayed in the **Available Nodes & Groups** box.

4. Select the signals that you want to see in the simulation trace. The signals that we want are: ResetN (I), Clock (I), and Q (O). Be careful that it is Q, and not Q0, Q1, Q2, or Q3. The letters I and O in parenthesis next to each signal denote whether the signal is an input or output signal, respectively. Note that the signal name, such as Q3, is bit three of the bus named Q. Multiple nodes can be selected by holding down the **Ctrl** or **Shift** key while clicking on the signal names.

5. After selecting the signals, click on the => button to move the selected signals to the **Selected Nodes & Groups** box.

6. Repeat Steps 4 and 5 until all of the signals that you want to see in the simulation are moved to the **Selected Nodes & Groups** box.

7. Click on **OK** when you are finished. The selected signals will now be inserted in the **Waveform Editor** window similar to Figure B.6.

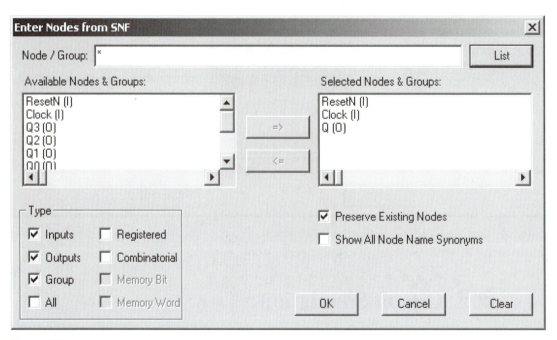

Figure B.5 Windows for adding signals for simulation.

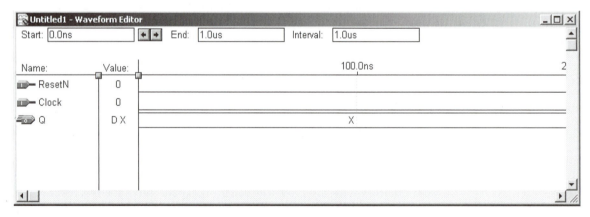

Figure B.6 Waveform Editor window for simulation.

B.3.2 Customizing the Waveform Editor

1. You can rearrange the signals in the Waveform Editor by dragging the signal icons like up or down. Drag the Clock signal to the very top of the list.

● To delete a signal in the Waveform Editor, just select the signal by clicking on its name and press the **Delete** key.

- For signals that are composed of a group of bits (such as the counter output Q), you can separate them into individual bits or change the radix for the displayed value by first selecting that signal and then right-click the mouse. A drop-down menu appears. Select Ungroup to separate the bits. To regroup them, select the bits you want to group and then right-click the mouse. A drop-down menu appears. Select Enter Group. Type in a group name, and select the radix you want for the display.

2. We want to simulate for 2 microseconds. To change the simulation end time, select File | End Time from the Waveform Editor window menu.

3. In the End Time window, type in 2us, and click OK.

4. To fit the entire simulation time range inside the window, select View | Fit in Window from the Waveform Editor window menu, or click on the icon ▣ in the toolbar on the left. Your Waveform Editor window should now look like the one in Figure B.7.

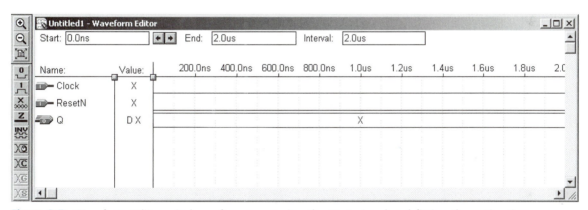

Figure B.7 Waveform Editor window after rearranging the Clock signal and fitting the entire time range inside. Notice also the toolbar buttons on the left.

B.3.3 Assigning Values to the Input Signals

The next thing is to assign values to all of the input signals.

1. Select the Clock signal by clicking on the signal name.

2. Click on the icon ⊠ in the toolbar on the left to define the Clock signal.

3. Click on OK to set the clock pulse.

4. Select the ResetN signal. The ResetN signal is active-low, (i.e., a 0 value will enable the signal).

5. Click on the icon ⊐ in the toolbar on the left to set the signal to a 1 value.

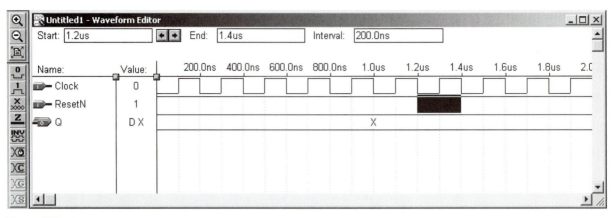

Figure B.8   Changing the *Reset* signal value between times 1.2 μs and 1.4 μs.

6. Drag from time 1.2us to 1.4us for the ResetN signal only, as shown in Figure B.8.

7. Click on the icon in the toolbar on the left to set the signal in this selected time range to a 0 value.

B.3.4 Saving the Waveform File

1. Save the **Waveform Editor** window by selecting **File | Save**. The **Save As** window appears. Notice that the default file name is the same as the top-level entity name, and the extension is .scf. For this example, the name is counter.scf.

2. Click on **OK** to save the file.

B.3.5 Starting the Simulator

1. We are now ready to simulate the design. From the **Manager** window menu, select **MAX+plus II | Simulator**, or click on the icon to bring up the **Simulator** window.

   ● You can also save the waveform file and start the simulator in one step by selecting from the **Manager** window menu **File | Project | Save & Simulate**.

2. The **Simulator** window, as shown in Figure B.9, is displayed. Make sure that the **Simulation Input** filename is counter.scf. This is the same name as your top-level entity.

3. Click on the **Start** button and watch the progress of the simulation.

4. At the end of the simulation, if there are no errors, you will see a message window saying that the simulation was successful. Click **OK** to close the message window.

Simulator: Functional Simulation —□×

Simulation Input: counter.scf

Simulation Time: 0.0ns

Start Time: 0.0ns   **End Time:** 2.0us

☐ **Use Device**   ☑ **Oscillation**
☐ **Setup/Hold**
☐ **Check Outputs**   ☐ **Glitch**

0   50   100

Start   **Pause**   **Stop**   **Open SCF**

Figure B.9
Simulator window
for the counter
design.

5. Click on the **Open SCF** button in the **Simulator** window to bring up the Waveform Editor with the resulting simulation waveforms. The simulation result is shown in Figure B.10. The signal Q is the counter output.

Notice that when ResetN is de-asserted (value 1), Q increments at the rising edge of each clock cycle. At 600 ns, the count is at 3. When ResetN is asserted at 1.2 μs, Q is immediately reset to 0. When ResetN is de-asserted again at 1.4 μs, the count starts again at the next rising clock edge.

6. You can change the ResetN signal values to something different and run the simulation again.

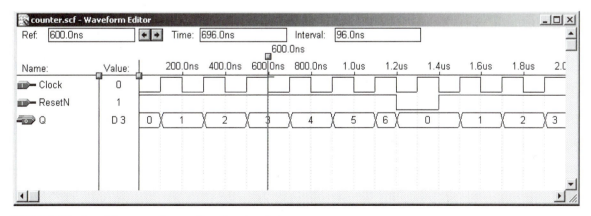

Figure B.10 Resulting waveform after the simulation.

UP2 Programming—Tutorial 3

Regardless of whether the design file is a VHDL source file or a schematic drawing, the procedures for synthesis, simulation, and PLD programming are the same. In fact, a project can contain design files of both VHDL codes and schematic drawings. In this tutorial, we will program and test a 4-bit up-counter circuit on the UP2 development board. You can start with either the VHDL source code or the schematic drawing for this counter circuit. In either case, the procedure and the final result will be the same.

In Tutorial 2 (Appendix B), you saw how a VHDL description of a 4-bit up-counter circuit is synthesized and simulated in MAX+plus II. Test values for the input signals *Clock* and *Clear* were setup manually in the simulator. In order for the synthesized circuit to operate in the hardware, these input signals must be provided for by the hardware. For example, the *Clear* signal must be connected to an input switch, and a clock generator is needed for the *Clock* signal. Furthermore, the counter output signal *Q* must be connected to LEDs in order for you to see that the counter really works.

In this tutorial, we will expand on the 4-bit up-counter circuit by adding a clock divider, and a 7-segment decoder. The UP2 development board already has a built-in clock source running at a frequency of 25 MHz. The clock-divider circuit simply divides this clock speed down to approximately 1 Hz so that you can see the counting. The 7-segment decoder converts the 4-bit counter output to drive a 7-segment LED display. A top-level module called up2flex is used to connect these three components (clockdiv, counter, and decoder) together to form one complete circuit. This circuit is then downloaded to the FLEX chip on the UP2 development board, and after applying power, you actually can see the count being displayed on the 7-segment LED. Alternatively, you can use the top-level module up2max for programming the MAX chip instead.

The schematic for this complete counter circuit is shown in Figure C.1. The switches and LEDs on the UP2 board are all active-low. This is why the inverters are needed, and the decoder outputs are all designated with an *N* to denote active-low. The eight *Vcc* lines are used optionally to turn off the second 7-segment LED.

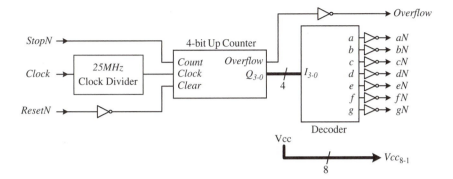

Figure C.1
Complete counter circuit for implementing on the UP2 board.

C.1 Getting Started

C.1.1 Preparing a Folder for the Project

1. Use Windows File Manager to create a new folder for this project. This tutorial uses a folder called up2 in the root directory on the C drive. If you want to work with the schematic drawings for the counter, go to Step 2 to copy the schematic drawing source files. If you want to work with the VHDL codes for the counter, go to Step 3 to copy the VHDL source files.

2. The schematic drawings for the four components, up2flex, clockdiv, counter, and decoder, are located on the accompanying CD-ROM in the four files up2flex.gdf, clockdiv.gdf, counter.gdf, and decoder.gdf in the directory <CD-ROM drive>:\Schematic Examples\Appendix C UP2 Tutorial 3\Source. In addition to these four files, there is a fifth file, ha.gdf, that is used by the counter. Using Windows File Manager, copy all of the files in this folder to the new folder c:\up2 that you have created in Step 1.

   • All of the schematic drawing files have the extension .gdf, which stands for graphic design file. The Graphic Editor is used to view and edit the graphic design files.

   • The file up2flex.gdf is for programming the FLEX chip. The file up2max.gdf is for programming the MAX chip.

3. The VHDL source code for the four entities, up2flex, clockdiv, counter, and decoder, are located on the accompanying CD-ROM in the four files up2flex.vhd, clockdiv.vhd, counter.vhd, and decoder.vhd in the directory <CD-ROM drive>:\VHDL Examples\Appendix C UP2 Tutorial3\Source. Using Windows File Manager, copy all the files in this folder to the new folder c:\up2 that you have created in Step 1.

   • All of the VHDL code files have the extension .vhd. The Text Editor is used to view and edit the VHDL source files.

   • The file up2flex.vhd is for programming the FLEX chip. The file up2max.vhd is for programming the MAX chip.

4. Start MAX+plus II if it is not already started. If there are windows in MAX+plus II that are opened from a previous session, you can close them.

C.1.2 Creating a Project

1. From the **Manager** window menu, select **File | Project | Name**, or simply click on the icon 🔳.

2. Select the C drive from the **Drives** dropdown list.

3. Select the up2 directory on the C drive. You should see the file up2flex.gdf (for schematic drawing) or up2flex.vhd (for VHDL code) listed in the **Files** box.

4. Select this up2flex file. The filename will be copied to the **Project Name** text field.

5. Click **OK**. The **MAX+plus II Manager** window title should now show c:\up2\up2flex.

   - The top-level file up2flex.gdf or up2flex.vhd is for programming the FLEX chip on the UP2 board. If you want to use the MAX chip, you need to use the file up2max.gdf or up2max.vhd instead. All subsequent references to up2flex should then be changed to up2max. The FLEX chip has a much larger capacity, but it is volatile (that is, a circuit that is programmed on it will remain only as long as power is applied to it). The MAX chip, on the other hand, is smaller but it is nonvolatile.

C.1.3 Viewing the Source File

1. From the **Manager** window menu, select **File | Hierarchy Project Top**, or click on the icon 🔲 to open the top-level source file up2flex.gdf or up2flex.vhd.

   - The Text Editor is used for viewing and editing the VHDL code. Notice that the entity name for this circuit is up2flex. The top entity name for the project must be the same as the project name and the file name.
   - The Graphic Editor is used for viewing and editing the graphic design file.

2. Use **File | Open** from the **Manager** window menu, or click on the icon 📂 to open the other source files for viewing or to make changes. For this tutorial, we do not need to make any modifications.

● ● ● ● ● ● ● ● ● ● ● ● ● ● ● ● ●

C.2 Synthesis for Programming the PLD

C.2.1 Selecting the Target Device

Since we want to download the circuit to a PLD, we need to specify what the target device is.

1. Open the **Device** selection window by selecting **Assign | Device** from the **Manager** window menu, as shown in Figure C.2.

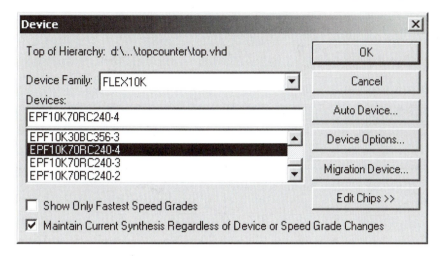

Figure C.2
Selecting the FLEX
10K70RC240-4PLD
chip.

2. Remove the check mark from **Show Only Fastest Speed Grades**.

3. In the **Device Family** dropdown box, select FLEX10K. This is the larger one of the two PLD devices that is on the UP2 board.

   ● Select MAX7000S if you want to use the smaller PLD device on the UP2 board.

4. In the **Devices** list, select EPF10K70RC240-4.

   ● Select EPM7128SLC84-7 if you want to use the smaller PLD device on the UP2 board.

5. Click on **OK**.

C.2.2   Synthesis

1. From the **Manager** window menu, select MAX+plus II | Compiler, or click on the icon 🖻 to bring up the **Compiler** window.

2. From the **Compiler** window menu (that is, with the **Compiler** window selected as the active window), select **Processing** and make sure that there is no check mark next to **Functional SNF Extractor**. If there is, then select it to remove the check mark. The **Compiler** window for full synthesis should look like Figure C.3.

3. From the **Compiler** window menu, select **Processing | Smart Recompile**. With this option turned on, if you change any pin assignments later on and recompile, the compiler does not have to perform a full synthesis.

4. Click on the **Start** button to start the synthesis. You will then see the progress of the synthesis.

5. At the end of the synthesis, if there are no errors, you will see a message window saying that the compilation was successful. You can ignore the warnings, if any, for this project. Click **OK** to close the message window.

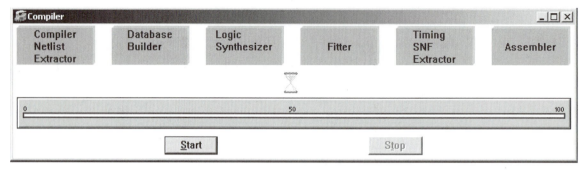

Figure C.3 Compiler window for full synthesis.

C.3 **Circuit Simulation**

The following steps for circuit simulation in this section are only necessary if you want to perform a simulation of the circuit. In practice, it is advisable to simulate the circuit to make sure that it is correct before implementing it on a PLD. For this tutorial, you can skip this step and go directly to Section C.4 for programming the PLD.

1. From the **Manager** window menu, select **MAX+plus II | Waveform Editor**.

2. From the **Waveform Editor** window menu, select **Node | Enter Nodes from SNF**. You can also right-click under the **Name** section in the **Waveform Editor** window and select **Enter Nodes from SNF** from the pop-up menu.

3. Click on the **List** button, and a list of available nodes and groups will be displayed in the **Available Nodes & Groups** box. You should see something similar to Figure C.4.

4. Select the signals that you want to see in the simulation trace and then click on the => button. The signals that we want are: StopN, ResetN, Clock, Overflow, aN, bN, cN, dN, eN, fN, and gN. The signals aN to gN are the signals for driving the seven LEDs on the 7-segment LED. After clicking on the => button, the selected signals will be moved to the **Selected Nodes & Groups** box.

   For this particular circuit, you may have a slight problem with the simulation, because the signal Clock is assumed to be running at 25 MHz, and the clockdiv component divides the clock down from 25 MHz to 1 Hz. So to even see a few counts, you will need the simulation end time to be very large. To make this simulation work, you need to remove the clockdiv component from the circuit and synthesize the circuit again. Remember that this is only necessary if you want to perform the simulation. You need to have this clockdiv component in order to see the counting on the UP2 board.

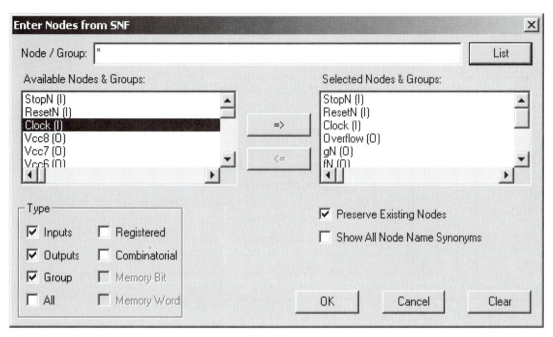

Figure C.4   Window for adding signals for simulation.

5. Click on **OK** when you are finished. The selected signals will now be inserted in the **Waveform Editor** window.

6. Select **File | End Time** from the **Waveform Editor** window menu and type in 3.5us to set the simulation end time to 3.5 μs.

7. Assign values to the Clock signal using the button.

8. Assign a logic 1 value to both the StopN and ResetN signals using the button.

9. Save the **Waveform Editor** window file as up2flex.scf.

10. From the **Manager** window menu, select **MAX+plus II | Simulator**, or click on the icon to bring up the **Simulator** window, and then click on the **Start** button to start the simulation of the design.

11. After the simulation has terminated, click on the **Open SCF** button to view the simulation result in the **Waveform Editor** window, as shown in Figure C.5. At 200 ns, the signals aN to gN are showing the count for 1 with bN and cN being 0 and the rest being 1. For these seven signals, a 0 turns on the LED and a 1 turns it off, as shown in Figure C.6.

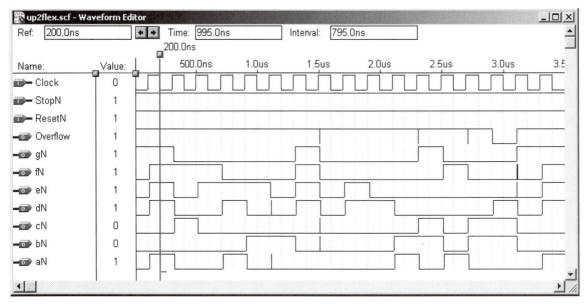

Figure C.5 Resulting waveform after the simulation for 3.5 μs.

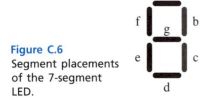

Figure C.6
Segment placements
of the 7-segment
LED.

C.4 Mapping the I/O Pins with the Floorplan Editor

1. Since we want to implement the ciucuit on a PLD, we need to map the I/O signals to the actual pins on the PLD. From the **Manager** window menu, select **MAX+plus II | Floorplan Editor** or click on the icon to bring up the **Floorplan Editor** window.

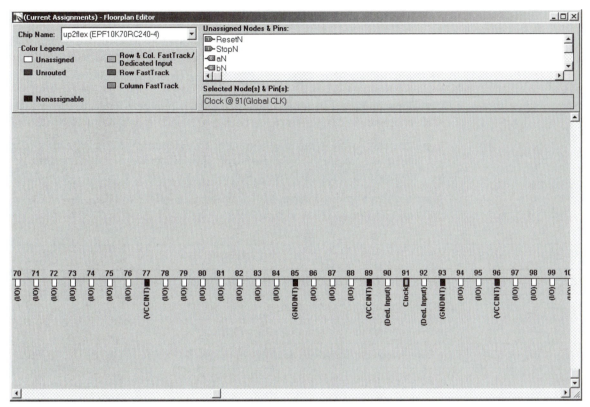

Figure C.7 The Floorplan Editor showing a portion of the FLEX 10K70RC240-4 chip pins and the signal names for the circuit.

2. From the Floorplan Editor window menu, select Layout | Device View. You should see a physical layout of the pins for the selected PLD chip, as shown in Figure C.7 for the FLEX chip. You may have to scroll the window or zoom out to see the pins.

3. From the Floorplan Editor window menu, select Layout | Current Assignments Floorplan, or click on the icon ▨ in the toolbar on the left.

All of the pins are labeled with a pin number and the name of the signal that is assigned to it. Pins either in blue or red color are already assigned to a signal. In Figure C.7, the signal Clock is already assigned to pin 91. All the pins in white are currently unassigned and available to use. Signals from the circuit that have not yet been assigned to any pins are listed in the top-right box labeled Unassigned Nodes & Pins.

4. To assign a signal to a pin, simply drag the icon ▣ or ▣ next to the signal from the **Unassigned Nodes & Pins** list box to one of the white pins in the floorplan. The pin color will change when a signal is assigned to it. Make sure that you drag the icon ▣ or ▣ and not the signal name.

- After making a signal assignment, you can change it by dragging the signal from one pin to another pin or back to the **Unassigned Nodes & Pins** list box.

- You can delete a signal assignment by selecting the pin with that signal and then pressing the **Delete** key. This will move the signal back to the **Unassigned Nodes & Pins** list box.

- Perform the following signal-to-pin assignments if you have selected to use the FLEX chip from Section C.2.1:

Signal	Pin Number	Comment
Clock	91	Pin 91 is connected to the built-in 25-MHz clock source
ResetN	28	Pin 28 is connected to push-button switch FLEX_PB1
StopN	29	Pin 29 is connected to push-button switch FLEX_PB2
aN	6	Pin 6 is connected to segment a on digit 1 of the FLEX 7-segment LED
bN	7	Pin 7 is connected to segment b on digit 1 of the FLEX 7-segment LED
cN	8	Pin 8 is connected to segment c on digit 1 of the FLEX 7-segment LED
dN	9	Pin 9 is connected to segment d on digit 1 of the FLEX 7-segment LED
eN	11	Pin 11 is connected to segment e on digit 1 of the FLEX 7-segment LED
fN	12	Pin 12 is connected to segment f on digit 1 of the FLEX 7-segment LED
gN	13	Pin 13 is connected to segment g on digit 1 of the FLEX 7-segment LED
Overflow	14	Pin 14 is connected to the decimal point on digit 1 of the FLEX 7-segment LED
Vcc1	17	Optional assignment to turn off segment a on digit 2 of the FLEX 7-segment LED
Vcc2	18	Optional assignment to turn off segment b on digit 2 of the FLEX 7-segment LED
Vcc3	19	Optional assignment to turn off segment c on digit 2 of the FLEX 7-segment LED
Vcc4	20	Optional assignment to turn off segment d on digit 2 of the FLEX 7-segment LED
Vcc5	21	Optional assignment to turn off segment e on digit 2 of the FLEX 7-segment LED
Vcc6	23	Optional assignment to turn off segment f on digit 2 of the FLEX 7-segment LED
Vcc7	24	Optional assignment to turn off segment g on digit 2 of the FLEX 7-segment LED
Vcc8	25	Optional assignment to turn off the decimal point on digit 2 of the FLEX 7-segment LED

● Perform the following signal-to-pin assignments if you have selected to use the MAX chip from Section C.2.1:

Signal	Pin Number	Comment
Clock	83	Pin 83 is connected to the built-in 25-MHz clock source
ResetN	1	Connect a hook-up wire between Pin 1 and the push-button switch MAX_PB1
StopN	2	Connect a hook-up wire between Pin 2 and the push-button switch MAX_PB2
aN	58	Pin 58 is connected to segment a on digit 1 of the MAX 7-segment LED
bN	60	Pin 60 is connected to segment b on digit 1 of the MAX 7-segment LED
cN	61	Pin 61 is connected to segment c on digit 1 of the MAX 7-segment LED
dN	63	Pin 63 is connected to segment d on digit 1 of the MAX 7-segment LED
eN	64	Pin 64 is connected to segment e on digit 1 of the MAX 7-segment LED
fN	65	Pin 65 is connected to segment f on digit 1 of the MAX 7-segment LED
gN	67	Pin 67 is connected to segment g on digit 1 of the MAX 7-segment LED
Overflow	68	Pin 68 is connected to the decimal point on digit 1 of the MAX 7-segment LED
Vcc1	69	Optional assignment to turn off segment a on digit 2 of the MAX 7-segment LED
Vcc2	70	Optional assignment to turn off segment b on digit 2 of the MAX 7-segment LED
Vcc3	73	Optional assignment to turn off segment c on digit 2 of the MAX 7-segment LED
Vcc4	74	Optional assignment to turn off segment d on digit 2 of the MAX 7-segment LED
Vcc5	76	Optional assignment to turn off segment e on digit 2 of the MAX 7-segment LED
Vcc6	75	Optional assignment to turn off segment f on digit 2 of the MAX 7-segment LED
Vcc7	77	Optional assignment to turn off segment g on digit 2 of the MAX 7-segment LED
Vcc8	79	Optional assignment to turn off the decimal point on digit 2 of the MAX 7-segment LED

● ● ● ● ● ● ● ● ● ● ● ● ● ● ●

C.5   Fitting the Netlist and Pins to the PLD

We need to compile the counter circuit again for fitting the netlist and pins to the selected PLD. If you have selected **Processing | Smart Recompile** in Step 3 of Section C.2.2, then this step would be faster, since the compiler does not have to resynthesize the circuit.

1. Bring up the Compiler window again by selecting MAX+plus II | Compiler from the Manager window menu or click on the icon 📇 .

2. Click on the Start button to start the compilation.

3. At the end of the synthesis, if there are no syntax errors, you will see a message window saying that the compilation was successful. Click OK to close the message window.

C.6 Hardware Setup

C.6.1 Installing the ByteBlaster Driver

1. Windows XP, 2000, and NT users must also install the ByteBlaster driver in order to program the PLD chip on the UP2 board. Perform the following steps to install the driver:

   - Get a command prompt window from Windows with Start | Run, and type in cmd.

   - Change to the \maxplus2\drivers\i386 directory under the directory where you have installed the MAX+plus II program.

   - Enter the command bblpt /i <Enter> at the command prompt to install the driver.

   Refer to the ByteBlaster installation instructions in the UP2 User Guide on the accompanying CD-ROM for more information.

2. From the Manager window menu, select Max+plus II | Programmer or click on the icon 👍 to bring up the Programmer window.

3. From the Programmer window menu, select Options | Hardware Setup.

4. Select ByteBlaster(MV) from the Hardware Type dropdown list. If this is not available, then go back to Step 1 to install the ByteBlaster device driver.

5. Select the correct parallel port (usually LPT1).

6. Click OK.

C.6.2 Jumper Settings

The four JTAG jumpers, TDI, TDO, DEVICE, and BOARD on the UP2 board must be set correctly, depending on which PLD chip you want to use.

- Use the following settings for the FLEX chip:

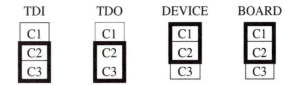

- Use the following settings for the MAX chip:

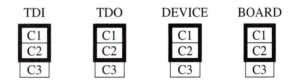

Refer to Figure C.8 for the location of these four jumpers.

C.6.3   Hardware Connections

1. Attach the ByteBlaster parallel cable directly to the PC's parallel port and the JTAG_IN connector on the UP2 development board. Refer to Figure C.8 for the location of the JTAG_IN connector.
2. Plug in the 9 V power supply.

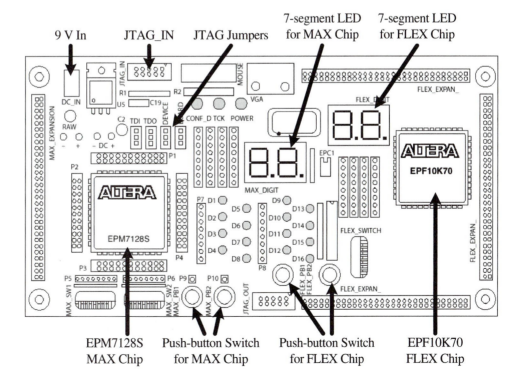

Figure C.8
UP2 development
board layout.

● ● ● ● ● ● ● ● ● ● ● ● ● ● ● ●

C.7 Programming the PLD

1. Select MAX+plus II | Programmer from the Manager window menu, or click on the icon 🐾 to bring up the Programmer window.

● If you have selected the FLEX chip from Section C.2.1, then the window will look like the one in Figure C.9. Make sure that the file name shown in the window is up2flex, and the device is EPF10K70RC240-4. If it is not correct, then you have either selected the wrong project file from Section C.1.2, selected the wrong chip from Section C.2.1, or set the wrong jumpers from Section C.6.2.

● If you have selected the MAX chip from Section C.2.1, then the window will look like the one in Figure C.10. Make sure that the file name shown in the window is up2max, and the device is EPM7128SLC84-7. If it is not correct, then you have either selected the wrong project file from Section C.1.2, selected the wrong chip from Section C.2.1, or set the wrong jumpers from Section C.6.2.

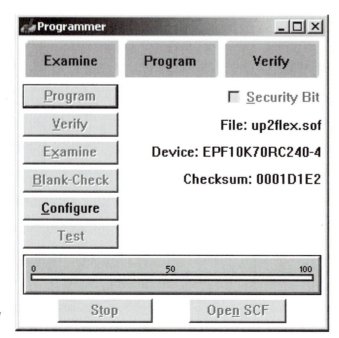

Figure C.9
Programmer window for the FLEX chip.

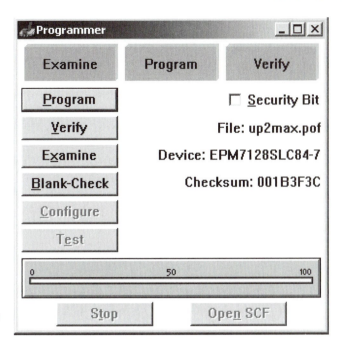

Figure C.10
Programmer window
for the MAX chip.

2. Click on the Configure button to start the configuration (programming) of the FLEX chip.

- Click on the Program button to start the programming of the MAX chip.

3. At the end of the programming, you will see a message window saying that the programming is successful. Click OK to close the message window.

- If the programming is unsuccessful, the most likely reason is because you have either selected the wrong project file from Section C.1.2, selected the wrong chip from Section C.2.1, or have set the wrong jumpers from Section C.6.2.

• • • • • • • • • • • • • • •

C.8 Testing the Hardware

1. If you have programmed the FLEX chip, the push-button FLEX_PB1 is the Reset button, and the FLEX_PB2 is the Stop button.

- If you have programmed the MAX chip, you need to first connect a hook-up wire between Pin 1 on the MAX dual-row female header strip P1 and the push-button MAX_PB1. This push-button acts as the **Reset** button. Connect another hook-up wire between Pin 2 on the MAX dual-row female header strip P1 and the push-button MAX_PB2. This push-button acts as the **Stop** button. Refer to Section C.9.3 for the location of Pins 1 and 2 on the header strip, and Section C.9.4 for the MAX_PB1 and MAX_PB2 push-button switches.

- You may want to make all of the jumper connections in Section C.9 now, since all of the circuits implemented on the MAX chip require these connections.

2. You should see the counter counts on the 7-segment LED. When you press the push-button FLEX_PB1 (if you are using the FLEX chip) or MAX_PB1 (if you are using the MAX chip), the counter will reset to zero. When you press FLEX_PB2 (or MAX_PB2), the counter will stop.

● ● ● ● ● ● ● ● ● ● ● ● ● ● ● ●

C.9 MAX7000S EPM7128SLC84-7 Summary

The UP2 development board contains two PLDs: an EPM7128SLC84-7 and an EPF10K70RC240-4. The EPM7128S has a capacity of 2,500 gates and the EPF10K70 has a capacity of 70,000 gates. The board provides switches, LEDs, and connectors for prototyping digital circuits. Figure C.11 shows a picture and physical layout of the UP2 development board.

This section provides a summary of the resources available for the MAX EPM7128S PLD and the jumper connections for using them. All of the LEDs and switches are active-low. In other words, a 0 turns on a LED, and a 0 is generated when a switch is pressed or in the on position. Refer to the UP2 User Guide on the accompanying CD-ROM for a detailed description and usage of the board.

C.9.1 JTAG Jumper Settings

The four JTAG jumpers, TDI, TDO, DEVICE, and BOARD, on the UP2 development board must be set according to Figure C.12 in order to program the EPM7128S PLD.

C.9.2 Prototyping Resources for Use

- Female headers for signal pins
- Two momentary push-button switches
- Two octal, dual in-line package (DIP) switches
- 16 LEDs
- Dual-digit 7-segment LED display

(a)

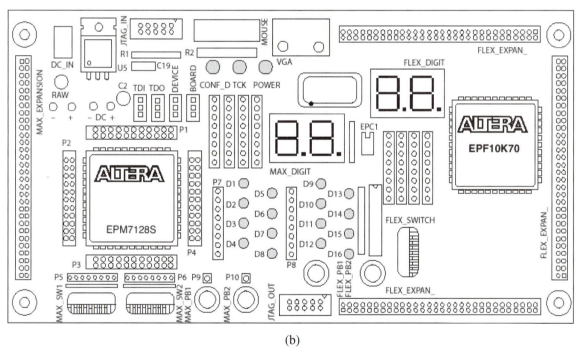

(b)

Figure C.11   UP2 development board: (a) picture; (b) layout.

TDI	TDO	DEVICE	BOARD
C1	C1	C1	C1
C2	C2	C2	C2
C3	C3	C3	C3

Figure C.12
JTAG jumper settings
for the EPM7128S
device.

- On-board oscillator (25.175 MHz)
- Expansion port with 42 I/O pins and the dedicated global CLR, OE1, and OE2/GCLK2 pins

C.9.3 General Pin Assignments

The signal pins on the EPM7128S are connected to four dual-row female header strips surrounding the chip. The pin numbers for the EPM7128S device are printed on the board and summarized in Figure C.13. An "×" indicates an unconnected pin, and a crossed-out pin number indicates that the pin is non-assignable.

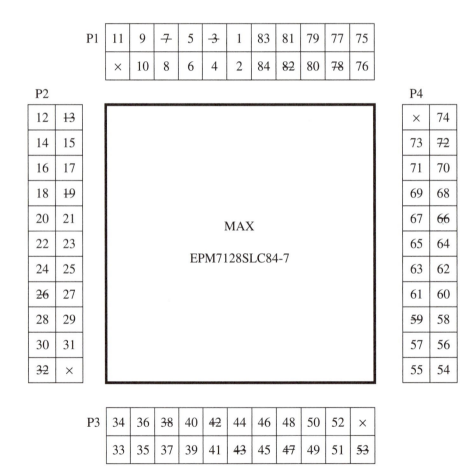

Figure C.13
Connection pins for the EPM7128S device.

C.9.4 Two Push-Button Switches

- The MAX_PB1 and MAX_PB2 are two push-button switches that provide active-low signals (i.e., they output a 0 when pressed).

MAX_PB Switch	Pin
MAX_PB1	1
MAX_PB2	2

- Connections to these two push-buttons are made by inserting one end of the hook-up wire into the female header associated with it.
- All of the circuits for programming the MAX chip on the accompanying CD-ROM map the two push-button switches to the header pins on the MAX chip, as shown in Figure C.14. You may want to make these connections now using hook-up wires and leave them connected this way for all of the exercises.

C.9.5   16 DIP Switches

- The MAX_SW1 and MAX_SW2 are two dual in-line package (DIP) switches that provide active-low signals (i.e., they output a 0 when set to on).
- Connections to these DIP switches are made by inserting one end of the hook-up wire into the female header associated with it.
- All of the circuits for programming the MAX chip on the accompanying CD-ROM map the 16 DIP switches to the header pins on the MAX chip, as shown in Figure C.15. You may want to make these connections now using hook-up wires and leave them connected this way for all of the exercises.

MAX_SW1 Switch	Pin
MAX_SW1 Switch 1	33
MAX_SW1 Switch 2	34
MAX_SW1 Switch 3	35
MAX_SW1 Switch 4	36
MAX_SW1 Switch 5	37
MAX_SW1 Switch 6	39
MAX_SW1 Switch 7	40
MAX_SW1 Switch 8	41

MAX_SW2 Switch	Pin
MAX_SW2 Switch 1	44
MAX_SW2 Switch 2	45
MAX_SW2 Switch 3	46
MAX_SW2 Switch 4	48
MAX_SW2 Switch 5	49
MAX_SW2 Switch 6	50
MAX_SW2 Switch 7	51
MAX_SW2 Switch 8	52

LEDs	Pin
D1	11
D2	12
D3	15
D4	16
D5	17
D6	18
D7	20
D8	21

LEDs	Pin
D9	22
D10	24
D11	25
D12	27
D13	28
D14	29
D15	30
D16	31

Figure C.16
Hook-up wire connections for the 16 MAX LEDs.

C.9.6 16 LEDs

- Each LED is turned on with a logic 0.
- Connections to the 16 LEDs are made by inserting one end of the hook-up wire into the female header associated with that LED.
- All of the circuits for programming the MAX chip on the accompanying CD-ROM map the 16 LEDs to the header pins on the MAX chip, as shown in Figure C.16. You may want to make these connections now using hook-up wires and leave them connected this way for all of the exercises.

C.9.7 7-Segment LEDs

- Each LED segment is turned on with a logic 0.
- The dual-digit 7-segment display LEDs are connected permanently to the pins on the MAX chip, as show in Figure C.17.

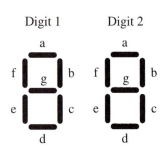

Segment	Digit 1 Pins	Digit 2 Pins
a	58	69
b	60	70
c	61	73
d	63	74
e	64	76
f	65	75
g	67	77
Decimal Point	68	79

Figure C.17   Connections for the two MAX 7-segment LEDs.

C.9.8 Clock

A 25.175-MHz clock signal is connected permanently to Pin 83 on the MAX chip.

● ● ● ● ● ● ● ● ● ● ● ● ● ● ⋯⋯

C.10 **FLEX10K EPF10K70RC240-4 Summary**

This section provides a summary of the resources available for the FLEX EPF10K70 PLD and the pin connections for using them. All of the LEDs and switches are active-low. In other words, a 0 turns on a LED, and a 0 is generated when a switch is pressed or in the on position. Refer to the UP2 User Guide on the accompanying CD-ROM for a detailed description and usage of the board.

C.10.1 **JTAG Jumper Settings**

The four JTAG jumpers, TDI, TDO, DEVICE, and BOARD, on the UP2 development board must be set according to Figure C.18 in order to program the EPF10K70 PLD.

Figure C.18
JTAG jumper settings for the EPF10K70 device.

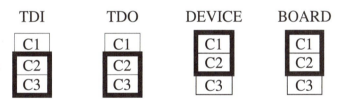

C.10.2 **Prototyping Resources for Use**

- Two momentary push-button switches
- One octal DIP switch
- Dual-digit 7-segment display
- On-board oscillator (25.175 MHz)
- PS/2 mouse or keyboard port
- VGA port
- Three expansion ports, each with 42 I/O pins and seven global pins

C.10.3 **Two Push-Button Switches**

- The FLEX_PB1 and FLEX_PB2 are two push-button switches that provide active-low signals (i.e., they output a 0 when pressed).

- They are connected permanently to the pins on the FLEX chip, as shown in Figure C.19.

Name	Pin
FLEX_PB1	28
FLEX_PB2	29

C.10.4 8 DIP Switches

- The FLEX_SW1 is a dual in-line package (DIP) switch that provides active-low signals (i.e., they output a 0 when set to on).

- They are connected permanently to the pins on the FLEX chip, as shown in Figure C.20.

Name	Pin
FLEX_SWITCH 1	41
FLEX_SWITCH 2	40
FLEX_SWITCH 3	39
FLEX_SWITCH 4	38
FLEX_SWITCH 5	36
FLEX_SWITCH 6	35
FLEX_SWITCH 7	34
FLEX_SWITCH 8	33

C.10.5 7-Segment LEDs

- Each LED segment is turned on with a logic 0.

- The dual-digit 7-segment display LEDs are connected permanently to the pins on the FLEX chip, as show in Figure C.21.

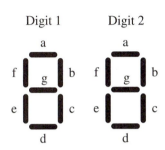

Segment	Digit 1 Pins	Digit 2 Pins
a	6	17
b	7	18
c	8	19
d	9	20
e	11	21
f	12	23
g	13	24
Decimal Point	14	25

Figure C.21   Connections for the two FLEX 7-segment LEDs.

C.10.6   Clock

A 25.175-MHz clock signal is connected permanently to Pin 91 on the FLEX chip.

C.10.7   PS/2 Port

- A 6-pin mini-DIN connector is available to receive data from either a PS/2 mouse or a PS/2 keyboard.
- The connector is connected permanently to the FLEX chip, as shown in Figure C.22.

Mouse/Keyboard Signal	Mini-DIN Pin	FLEX Pin
MOUSE_CLK, KEYBOARD_CLK	1	30
MOUSE_DATA, KEYBOARD_DATA	3	31
VCC	5	–
GND	2	–

Figure C.22   Connections for the PS/2 port.

C.10.8 **VGA Port**

- The VGA interface allows the FLEX device to control an external monitor.
- The D-sub VGA connector is connected permanently to the FLEX chip, as shown in Figure C.23.

VGA Signal	Mini-DIN Pin	FLEX Pin
RED	1	236
GREEN	2	237
BLUE	3	238
GND	6, 7, 8, 10, 11	–
HORIZ_SYNC	13	240
VERT_SYNC	14	239
No connect	4, 5, 9, 15	–

Figure C.23 Connections for the VGA interface.

VHDL Summary

VHDL is a hardware description language for modeling digital circuits that can range from the simple connection of gates to complex systems. VHDL is an acronym for VHSIC Hardware Description Language, and VHSIC in turn is an acronym for Very High Speed Integrated Circuits. This appendix gives a brief summary of the basic VHDL elements and its syntax. Many advanced features of the language are omitted. Interested readers should refer to other references for detailed coverage.

● ● ● ● ● ● ● ● ● ● ● ● ● ●···

D.1 Basic Language Elements

D.1.1 Comments

Comments are preceded by two consecutive hyphens (--) and are terminated at the end of the line.

Example:

```
-- This is a comment
```

D.1.2 Identifiers

VHDL identifier syntax:

- A sequence of one or more uppercase letters, lowercase letters, digits, and the underscore
- Upper and lowercase letters are treated the same (i.e., case insensitive)
- The first character must be a letter
- The last character cannot be the underscore
- Two underscores cannot be together

D.1.3 Data Objects

There are three kinds of data objects: signals, variables, and constants.

- The data object SIGNAL represents logic signals on a wire in the circuit. A signal does not have memory; thus, if the source of the signal is removed, the signal will not have a value.

- A VARIABLE object remembers its content and is used for computations in a behavioral model.

- A CONSTANT object must be initialized with a value when declared, and this value cannot be changed.

Example:

```
SIGNAL x: BIT;
VARIABLE y: INTEGER;
CONSTANT one: STD_LOGIC_VECTOR (3 DOWNTO 0) := "0001";
```

D.1.4 Data Types

BIT and BIT_VECTOR

The BIT and BIT_VECTOR types are predefined in VHDL. Objects of these types can have the values '0' or '1.' The BIT_VECTOR type is simply a vector of type BIT. A vector with all bits having the same value can be obtained using the OTHERS keyword.

Example:

```
SIGNAL x: BIT;
SIGNAL y: BIT_VECTOR (7 DOWNTO 0);
x <= '1';
y <= "00000010";
y <= (OTHERS => '0'); -- same as "00000000"
```

STD_LOGIC and STD_LOGIC_VECTOR

The STD_LOGIC and STD_LOGIC_VECTOR types provide more values than the BIT type for modeling a real circuit more accurately. Objects of these types can have the following values.

'0' normal 0

'1' normal 1

'Z' high impedance[1]

'-' don't-care[2]

'L' weak 0[2]

---

[1] Must use uppercase. This is only a MAX+plus II restriction.

[2] MAX+plus II only supports the values 0, 1, Z, and X.

'H'   weak 1^2

'U'   uninitialized^2

'X'   unknown^1

'W'   weak unknown^2

The STD_LOGIC and STD_LOGIC_VECTOR types are not predefined, and so the following two library statements must be included in order to use these types.

```
LIBRARY IEEE;
USE IEEE.STD_LOGIC_1164.ALL;
```

If objects of type STD_LOGIC_VECTOR are to be used as binary numbers in arithmetic manipulations, then either one of the following two USE statements must also be included

```
USE IEEE.STD_LOGIC_SIGNED.ALL;
```

for signed number arithmetic, or

```
USE IEEE.STD_LOGIC_UNSIGNED.ALL;
```

for unsigned number arithmetic. A vector with all bits having the same value can be obtained using the OTHERS keyword, as shown in the next example.

Example:

```
LIBRARY IEEE;
USE IEEE.STD_LOGIC_1164.ALL;

SIGNAL x: STD_LOGIC;
SIGNAL y: STD_LOGIC_VECTOR(7 DOWNTO 0);

x <= 'Z';
y <= "0000001Z";
y <= (OTHERS => '0'); -- same as "00000000"
```

INTEGER

The predefined INTEGER type defines binary number objects for use with arithmetic operators. By default, an INTEGER signal uses 32 bits to represent a signed number. Integers using fewer bits can also be declared with the RANGE keyword.

Example:

```
SIGNAL x: INTEGER;
SIGNAL y: INTEGER RANGE -64 to 64;
```

BOOLEAN

The predefined BOOLEAN type defines objects having the two values TRUE and FALSE.

Example:

```
SIGNAL x: BOOLEAN;
```

Enumeration TYPE

An enumeration type allows the user to specify the values that the data object can have.

Syntax:

TYPE identifier IS (value1, value2, ...);

Example:

```
TYPE state_type IS (S1, S2, S3);
SIGNAL state: state_type;
state <= S1;
```

ARRAY

The ARRAY type groups single data objects of the same type together into a one-dimensional or multi-dimensional array.

Syntax:

TYPE identifier IS ARRAY (range) OF type;

Example:

```
TYPE byte IS ARRAY (7 DOWNTO 0) OF BIT;
TYPE memory_type IS ARRAY (1 TO 128) OF byte;
SIGNAL memory: memory_type;
memory(3) <= "00101101";
```

SUBTYPE

A SUBTYPE is a subset of a type, that is, a type with a range constraint.

Syntax:

SUBTYPE identifier IS type RANGE range;

Example:

```
SUBTYPE integer4 IS INTEGER RANGE -8 TO 7;
SUBTYPE cell IS STD_LOGIC_VECTOR(3 DOWNTO 0);
TYPE memArray IS ARRAY(0 TO 15) OF cell;
```

Some standard subtypes include:

- NATURAL—an integer in the range 0 to INTEGER'HIGH
- POSITIVE—an integer in the range 1 to INTEGER'HIGH

D.1.5   Data Operators

The VHDL built-in operators are listed in Figure D.1.

Logical Operators	Operation	Example
AND	AND	n <= a AND b
OR	OR	n <= a OR b
NOT	NOT	n <= NOT a
NAND	NAND	n <= a NAND b
NOR	NOR	n <= a NOR b
XOR	XOR	n <= a XOR b
XNOR	XNOR	n <= a XNOR b
Arithmetic Operators	Operation	Example
+	Addition	n <= a + b
−	Subtraction	n <= a − b
*	Multiplication (integer or floating point)	n <= a * b
/[1]	Division (integer or floating point)	n <= a / b
MOD[2]	Modulus (integer)	n <= a MOD b
REM[1]	Remainder (integer)	n <= a REM b
**	Exponentiation	n <= a ** 2
&	Concatenation	n <= 'a'& 'b'
ABS	Absolute	
Relational Operators	Operation	Example
=	Equal	IF (n = 10) THEN
/ =	Not equal	IF (n / = 10) THEN
<	Less than	IF (n < 10) THEN
<=	Less than or equal	IF (n <= 10) THEN
>	Greater than	IF (n > 10) THEN
>=	Greater than or equal	IF (n >= 10) THEN
Shift Operators	Operation	Example
SLL	Shift left logical	n <= "1001010"SLL 2
SRL	Shift right logical	n <= "1001010"SRL 1
SLA	Shift left arithmetic	n <= "1001010"SLA 2
SRA	Shift right arithmetic	n <= "1001010"SRA 1
ROL	Rotate left	n <= "1001010"ROL 2
ROR	Rotate right	n <= "1001010"ROR 3

[1] Can only divide by a power of 2. This is only a MAX+plus II restriction.

[2] Not supported by MAX+ plus II.

Figure D.1   VHDL built-in data operators.

D.1.6 ENTITY

An ENTITY declaration declares the external or user interface of the module similar to the declaration of a function. It specifies the name of the entity and its interface. The interface consists of the signals to be passed into the entity or out from it using the two keywords IN and OUT, respectively.

Syntax:

```
ENTITY entity-name IS
   PORT (list-of-port-names-and-types);
END entity-name;
```

Example:

```
LIBRARY IEEE;
USE IEEE.STD_LOGIC_1164.ALL;

ENTITY Siren IS PORT (
   M:  IN STD_LOGIC;
   D:  IN STD_LOGIC;
   V:  IN STD_LOGIC;
   S:  OUT STD_LOGIC);
END Siren;
```

D.1.7 ARCHITECTURE

The ARCHITECTURE body defines the actual implementation of the functionality of the entity. This is similar to the definition or implementation of a function. The syntax for the architecture varies, depending on the model (dataflow, behavioral, or structural) you use.

Syntax: Dataflow model

```
ARCHITECTURE architecture-name OF entity-name IS
   signal-declarations;
BEGIN
   concurrent-statements;
END architecture-name;
```

The concurrent statements are executed concurrently.

Example:

```
ARCHITECTURE Siren_Dataflow OF Siren IS
   SIGNAL term_1: STD_LOGIC;
BEGIN
   term_1 <= D OR V;
   S <= term_1 AND M;
END Siren_Dataflow;
```

Syntax: Behavioral model

```
ARCHITECTURE architecture-name OF entity-name IS
   signal-declarations;
   function-definitions;
   procedure-definitions;
BEGIN
   PROCESS-blocks;
   concurrent-statements;
END architecture-name;
```

Statements within the PROCESS block are executed sequentially. However, the PROCESS block itself is a concurrent statement.

Example:

```
ARCHITECTURE Siren_Behavioral OF Siren IS
   SIGNAL term_1: STD_LOGIC;
BEGIN
   PROCESS (D, V, M)
   BEGIN
      term_1 <= D OR V;
      S <= term_1 AND M;
   END PROCESS;
END Siren_Behavioral;
```

Syntax: Structural model

```
ARCHITECTURE architecture-name OF entity-name IS
   component-declarations;
   signal-declarations;
BEGIN
   instance-name: PORT MAP statements;
   concurrent-statements;
END architecture-name;
```

For each component declaration used, there must be a corresponding entity and architecture for that component. The PORT MAP statements are concurrent statements.

Example:

```
ARCHITECTURE Siren_Structural OF Siren IS
   COMPONENT myOR PORT (
      in1, in2: IN STD_LOGIC;
      out1: OUT STD_LOGIC);
   END COMPONENT;
   SIGNAL term1: STD_LOGIC;
BEGIN
   U0: myOR PORT MAP (D, V, term1);
   S <= term1 AND M;
END Siren_Structural;
```

D.1.8 GENERIC

Generics allow information to be passed into an entity so that, for example, the size of a vector in the PORT list does not have to be known until elaboration time. Generics of an entity are declared with the GENERIC keyword before the PORT list declaration for the entity. An identifier that is declared as GENERIC is a constant that only can be read. The identifier then can be used in the entity declaration and its corresponding architectures wherever a constant is expected.

Syntax: In an ENTITY *declaration*

```
ENTITY entity-name IS
GENERIC (identifier: type);          --with no default value
...
```

or

```
ENTITY entity-name IS
GENERIC (identifier: type := constant);  --with a default value
                                           given by the constant
...
```

Example:

```
ENTITY Adder IS
-- declares the generic identifier n having a default value 4
GENERIC (n: INTEGER := 4);
PORT (
   -- the vector size is 3 downto 0 since n is 4
   A, B: IN STD_LOGIC_VECTOR(n-1 DOWNTO 0);
   Cout: OUT STD_LOGIC;
   SUM:  OUT STD_LOGIC_VECTOR(n-1 DOWNTO 0));
   S:    OUT STD_LOGIC);
END Adder;
```

The value for a generic constant can also be specified in a component declaration or a component instantiation statement.

Syntax: In a COMPONENT *declaration*

```
COMPONENT component-name
   GENERIC (identifier: type := constant);  --with an optional value given
      PORT (list-of-port-names-and-types);        by the constant
END COMPONENT;
```

Syntax: In a COMPONENT *instantiation*

```
label: component-name GENERIC MAP (constant)
                 PORT MAP (association-list);
```

Example:

```
ARCHITECTURE ...
  COMPONENT mux2 IS
    -- declares the generic identifier n having a default value 4
    GENERIC (n: INTEGER := 4);
    PORT (
       S: IN STD_LOGIC;                            -- select line
       D1, D0: IN STD_LOGIC_VECTOR(n-1 DOWNTO 0);  -- data bus input
       Y: OUT STD_LOGIC_VECTOR(n-1 DOWNTO 0));     -- data bus output
  END COMPONENT;
  ...
BEGIN
  U0: mux2 GENERIC MAP (8) PORT MAP (mux_select, A, B, mux_out);
  -- change vector to size 8
  ...
```

D.1.9 PACKAGE

A PACKAGE provides a mechanism to group together and share declarations that are used by several entity units. A package itself includes a declaration and, optionally, a body. The PACKAGE declaration and body usually are stored together in a separate file from the rest of the design units. The file name given for this file must be the same as the package name. In order for the complete design to synthesize correctly using MAX+plus II, you must first synthesize the package as a separate unit. After that, you can synthesize the unit that uses that package.

PACKAGE Declaration and Body

The PACKAGE declaration contains declarations that may be shared between different entity units. It provides the interface, that is, items that are visible to the other entity units. The optional PACKAGE BODY contains the implementations of the functions and procedures that are declared in the PACKAGE declaration.

Syntax: PACKAGE *declaration*

```
PACKAGE package-name IS
  type-declarations;
  subtype-declarations;
  signal-declarations;
  variable-declarations;
  constant-declarations;
  component-declarations;
  function-declarations;
  procedure-declarations;
END package-name;
```

Syntax: PACKAGE BODY *declaration*

```
PACKAGE BODY package-name IS
   function-definitions;    --for functions declared in the package declaration
   procedure-definitions;  --for procedures declared in the package declaration
END package-name;
```

Example:

```
LIBRARY IEEE;
USE IEEE.STD_LOGIC_1164.ALL;

PACKAGE my_package IS
  SUBTYPE bit4 IS STD_LOGIC_VECTOR(3 DOWNTO 0);
  FUNCTION Shiftright (input: IN bit4) RETURN bit4; -- declare a function
  SIGNAL mysignal: bit4;  -- a global signal
END my_package;

PACKAGE BODY my_package IS
  -- implementation of the Shiftright function
  FUNCTION Shiftright (input: IN bit4) RETURN bit4 IS
  BEGIN
     RETURN '0' & input(3 DOWNTO 1);
  END shiftright;
END my_package;
```

Using a PACKAGE

To use a package, you simply include a LIBRARY and USE statement for that package.
Before synthesizing the module that uses the package, you need to first synthesize
the package by itself as a top-level entity.
Syntax:

```
LIBRARY WORK;
USE WORK.package-name.ALL;
```

Example:

```
LIBRARY WORK;
USE WORK.my_package.ALL;

ENTITY test_package IS PORT (
   x: IN bit4;
   z: OUT bit4);
END test_package;

ARCHITECTURE Behavioral OF test_package IS
BEGIN
   mysignal <= x;
   z <= Shiftright(mysignal);
END Behavioral;
```

● ● ● ● ● ● ● ● ● ● ● ● ● ● ● ● ● ●

D.2 Dataflow Model—Concurrent Statements

Concurrent statements used in the dataflow model are executed concurrently. Hence, the ordering of these statements does not affect the resulting output.

D.2.1 Concurrent Signal Assignment

The concurrent signal assignment statement assigns a value or the result of evaluating an expression to a signal. This statement is executed whenever a signal in its expression changes value. However, the actual assignment of the value to the signal takes place after a certain delay and not instantaneously as for variable assignments. The expression can be any logical or arithmetical expressions.

Syntax:

```
signal <= expression;
```

Example:

```
y <= '1';
z <= y AND (NOT x);
```

A vector with all bits having the same value can be obtained using the OTHERS keyword as shown below

```
SIGNAL x: STD_LOGIC_VECTOR(7 DOWNTO 0);
x <= (OTHERS => '0');   -- 8-bit vector of 0, same as "00000000"
```

D.2.2 Conditional Signal Assignment

The conditional signal assignment statement selects one of several different values to assign to a signal based on different conditions. This statement is executed whenever a signal in any one of the value or condition changes.

Syntax:

```
signal <= value1 WHEN condition ELSE
          value2 WHEN condition ELSE
          . . .
          value3;
```

Example:

```
z <= in0 WHEN sel = "00" ELSE
     in1 WHEN sel = "01" ELSE
     in2 WHEN sel = "10" ELSE
     in3;
```

D.2.3   Selected Signal Assignment

The selected signal assignment statement selects one of several different values to assign to a signal based on the value of a select expression. All possible choices for the expression must be given. The keyword OTHERS can be used to denote all remaining choices. This statement is executed whenever a signal in the expression or any one of the value changes.

Syntax:

```
WITH expression SELECT
  signal <= value1 WHEN choice1,
            value2 WHEN choice2 | choice3,
            . . .
            value4 WHEN OTHERS;
```

In the above syntax, if *expression* is equal to *choice1*, then *value1* is assigned to *signal*. Otherwise, if *expression* is equal to *choice2* or *choice3*, then *value2* is assigned to *signal*. If *expression* does not match any of the above choices, then *value4* in the optional WHEN OTHERS clause is assigned to *signal*.

Example:

```
WITH sel SELECT
  z <=  in0 WHEN "00",
        in1 WHEN "01",
        in2 WHEN "10",
        in3 WHEN OTHERS;
```

D.2.4   Dataflow Model Sample

```
-- outputs a 1 if the 4-bit input is a prime number, 0 otherwise
LIBRARY IEEE;
USE IEEE.STD_LOGIC_1164.ALL;

ENTITY Prime IS PORT (
   number: IN STD_LOGIC_VECTOR(3 DOWNTO 0);
   yes: OUT STD_LOGIC);
END Prime;

ARCHITECTURE Prime_Dataflow OF Prime IS
BEGIN
   WITH number SELECT
      yes <=  '1' WHEN "0001" | "0010",
              '1' WHEN "0011" | "0101" | "0111" | "1011" | "1101",
              '0' WHEN OTHERS;
END Prime_Dataflow;
```

● ● ● ● ● ● ● ● ● ● ● ● ● ● ● ●

D.3 Behavioral Model—Sequential Statements

The behavioral model allows statements to be executed sequentially just like in a regular computer program. Sequential statements include many of the standard constructs, such as variable assignments, IF-THEN-ELSE statements, and loops.

D.3.1 PROCESS

The PROCESS block contains statements that are executed sequentially. However, the PROCESS statement itself is a concurrent statement. Multiple process blocks in an architecture will be executed simultaneously. These process blocks can be combined together with other concurrent statements.

Syntax:

```
process-name: PROCESS (sensitivity-list)
   variable-declarations;
BEGIN
   sequential-statements;
END PROCESS process-name;
```

The sensitivity list is a comma-separated list of signals, which the process is sensitive to. In other words, whenever a signal in the list changes value, the process will be executed (i.e., all the statements in the sequential order listed). After the last statement has been executed, the process will be suspended until the next time that a signal in the sensitivity list changes value before it is executed again.

Example:

```
Siren: PROCESS (D, V, M)
BEGIN
   term_1 <= D OR V;
   S <= term_1 AND M;
END PROCESS;
```

D.3.2 Sequential Signal Assignment

The sequential signal assignment statement assigns a value to a signal. This statement is just like its concurrent counterpart, except that it is executed sequentially (i.e., only when execution reaches it).

Syntax:

```
signal <= expression;
```

Example:

```
y <= '1';
z <= y AND (NOT x);
```

D.3.3 **Variable Assignment**

The variable assignment statement assigns a value or the result of evaluating an expression to a variable. The value always is assigned to the variable instantaneously whenever this statement is executed.

Variables are only declared within a process block.

Syntax:

```
signal := expression;
```

Example:

```
y := '1';
yn := NOT y;
```

D.3.4 **WAIT**

When a process has a sensitivity list, the process always suspends after executing the last statement. An alternative to using a sensitivity list to suspend a process is to use a WAIT statement, which must also be the first statement in a process.[3]

Syntax:[4]

```
WAIT UNTIL condition;
```

Example:

```
-- suspend until a rising clock edge
WAIT UNTIL clock'EVENT AND clock = '1';
```

D.3.5 **IF-THEN-ELSE**

Syntax:

```
IF condition THEN
   sequential-statements1;
ELSE
   sequential-statements2;
END IF;
```

or

```
IF condition1 THEN
   sequential-statements1;
ELSIF condition2 THEN
   sequential-statements2;
   ...
ELSE
   sequential-statements3;
END IF;
```

---

[3] This is only a MAX+plus II restriction.

[4] There are three different formats of the WAIT statement, however, MAX+plus II only supports one.

Example:

```
IF (count /= 10) THEN      -- not equal
   count := count + 1;
ELSE
   count := 0;
END IF;
```

D.3.6   CASE

Syntax:

```
CASE expression IS
   WHEN choices => sequential-statements;
   WHEN choices => sequential-statements;

   . . .
   WHEN OTHERS => sequential-statements;
END CASE;
```

Example:

```
CASE sel IS
   WHEN "00" => z <= in0;
   WHEN "01" => z <= in1;
   WHEN "10" => z <= in2;
   WHEN OTHERS => z <= in3;
END CASE;
```

D.3.7   NULL

The NULL statement does not perform any actions.

Syntax:

```
NULL;
```

D.3.8   FOR

Syntax:

```
FOR identifier IN start [TO | DOWNTO] stop LOOP
   sequential-statements;
END LOOP;
```

Loop statements must have locally static bounds.[5] The identifier is implicitly declared, so no explicit declaration of the variable is needed.

---

[5] This is only a MAX+plus II restriction.

Example:

```
sum := 0;
FOR count IN 1 TO 10 LOOP
   sum := sum + count;
END LOOP;
```

D.3.9 WHILE

Syntax:[6]

```
WHILE condition LOOP
  sequential-statements;
END LOOP;
```

D.3.10 LOOP

Syntax:[6]

```
LOOP
  sequential-statements;
  EXIT WHEN condition;
END LOOP;
```

D.3.11 EXIT

The EXIT[6] statement can only be used inside a loop. It causes execution to jump out of the innermost loop and usually is used in conjunction with the LOOP statement.

Syntax:

```
EXIT WHEN condition;
```

D.3.12 NEXT

The NEXT statement can be used only inside a loop. It causes execution to skip to the end of the current iteration and continue with the beginning of the next iteration. It usually is used in conjunction with the FOR statement.

Syntax:

```
NEXT WHEN condition;
```

Example:

```
sum := 0;
FOR count IN 1 TO 10 LOOP
   NEXT WHEN count = 3;
   sum := sum + count;
END LOOP;
```

---

[6] Not supported by MAX+ plus II.

D.3.13   FUNCTION

Syntax: Function declaration

> FUNCTION function-name (parameter-list) RETURN return-type;

Syntax: Function definition

> FUNCTION function-name (parameter-list) RETURN return-type IS
> BEGIN
>   sequential-statements;
> END function-name;

Syntax: Function call

> function-name (actuals);

Parameters in the parameter list can be either signals or variables of mode IN only.

Example:

```
LIBRARY IEEE;
USE IEEE.STD_LOGIC_1164.ALL;

ENTITY test_function IS PORT (
   x: IN STD_LOGIC_VECTOR(3 DOWNTO 0);
   z: OUT STD_LOGIC_VECTOR(3 DOWNTO 0));
END test_function;

ARCHITECTURE Behavioral OF test_function IS

   SUBTYPE bit4 IS STD_LOGIC_VECTOR(3 DOWNTO 0);

   FUNCTION Shiftright (input: IN bit4) RETURN bit4 IS
   BEGIN
      RETURN '0' & input(3 DOWNTO 1);
   END shiftright;

   SIGNAL mysignal: bit4;

BEGIN
   PROCESS
   BEGIN
      mysignal <= x;
      z <= Shiftright(mysignal);
   END PROCESS;
END Behavioral;
```

D.3.14 PROCEDURE

Syntax: Procedure declaration

PROCEDURE procedure-name (parameter-list);

Syntax: Procedure definition

PROCEDURE procedure-name (parameter-list) IS
BEGIN
  sequential-statements;
END procedure-name;

Syntax: Procedure call

procedure-name (actuals);

Parameters in the parameter-list are variables of modes IN, OUT, or INOUT.

Example:

```
LIBRARY IEEE;
USE IEEE.STD_LOGIC_1164.ALL;

ENTITY test_procedure IS PORT (
    x: IN STD_LOGIC_VECTOR(3 DOWNTO 0);
    z: OUT STD_LOGIC_VECTOR(3 DOWNTO 0));
END test_procedure;

ARCHITECTURE Behavioral OF test_procedure IS

   SUBTYPE bit4 IS STD_LOGIC_VECTOR(3 DOWNTO 0);

   PROCEDURE Shiftright (input: IN bit4; output: OUT bit4) IS
   BEGIN
      output := '0' & input(3 DOWNTO 1);
   END shiftright;

BEGIN
   PROCESS
      VARIABLE mysignal: bit4;
   BEGIN
      Shiftright(x, mysignal);
      z <= mysignal;
   END PROCESS;
END Behavioral;
```

D.3.15 **Behavioral Model Sample**

```
LIBRARY IEEE;
USE IEEE.STD_LOGIC_1164.ALL;

ENTITY bcd IS PORT (
    I: IN STD_LOGIC_VECTOR(3 DOWNTO 0);
    Segs: OUT STD_LOGIC_VECTOR(1 TO 7));
END bcd;

ARCHITECTURE Behavioral OF bcd IS
BEGIN
    PROCESS(I)
    BEGIN
        CASE I IS
        WHEN "0000" => Segs <= "1111110";
        WHEN "0001" => Segs <= "0110000";
        WHEN "0010" => Segs <= "1101101";
        WHEN "0011" => Segs <= "1111001";
        WHEN "0100" => Segs <= "0110011";
        WHEN "0101" => Segs <= "1011011";
        WHEN "0110" => Segs <= "1011111";
        WHEN "0111" => Segs <= "1110000";
        WHEN "1000" => Segs <= "1111111";
        WHEN "1001" => Segs <= "1110011";
        WHEN OTHERS => Segs <= "0000000";
        END CASE;
    END PROCESS;
END Behavioral;
```

● ● ● ● ● ● ● ● ● ● ● ● ● ● · · · ·

D.4 Structural Model—Concurrent Statements

The structural model allows the manual connection of several components together using signals. All components used must first be defined with their respective ENTITY and ARCHITECTURE sections, which can be in the same file or can be in separate files.

In the topmost module, each component used in the netlist is first declared using the COMPONENT statement. The declared components are then instantiated with the actual components in the circuit using the PORT MAP statement. SIGNALS are then used to connect the components together according to the netlist.

D.4.1 **COMPONENT Declaration**

The COMPONENT declaration statement declares the name and the interface of a component that is used in the circuit description. For each COMPONENT declaration used, there must be a corresponding ENTITY and ARCHITECTURE for that component. The declaration name and the interface must match exactly the name and interface that is specified in the ENTITY section for that component.

Syntax:

```
COMPONENT component-name IS
   PORT (list-of-port-names-and-types);
END COMPONENT;
```

or

```
COMPONENT component-name IS
   GENERIC (identifier: type := constant);
   PORT (list-of-port-names-and-types);
END COMPONENT;
```

Example:

```
COMPONENT half_adder IS PORT (
   xi, yi, cin: IN STD_LOGIC;
   cout, si: OUT STD_LOGIC);
END COMPONENT;
```

D.4.2 **PORT MAP**

The PORT MAP statement instantiates a declared component with an actual component in the circuit by specifying how the connections to this instance of the component are to be made.

Syntax:

```
label: component-name PORT MAP (association-list);
```

or

```
label: component-name GENERIC MAP (constant)
                    PORT MAP (association-list);
```

The association list can be specified using either the positional or named method.

Example: Positional association

```
SIGNAL x0, x1, y0, y1, c0, c1, c2, s0, s1: STD_LOGIC;
U1: half_adder PORT MAP (x0, y0, c0, c1, s0);
U2: half_adder PORT MAP (x1, y1, c1, c2, s1);
```

Example: Named association

```
SIGNAL x0, x1, y0, y1, c0, c1, c2, s0, s1: STD_LOGIC;
U1: half_adder PORT MAP (cout=>c1, si=>s0, cin=>c0, xi=>x0, yi=>y0);
U2: half_adder PORT MAP (cin=>c1, xi=>x1, yi=>y1, cout=>c2, si=>s1);
```

D.4.3   OPEN

The OPEN keyword is used in the PORT MAP association list to signify that a particular output port is not connected or used. It cannot be used for an input port.

Example:

```
U1: half_adder PORT MAP (x0, y0, c0, OPEN, s0);
```

D.4.4   GENERATE

The GENERATE statement works like a macro expansion. It provides a simple way to duplicate similar components.

Syntax:

```
label: FOR identifier IN start [TO | DOWNTO] stop GENERATE
    port-map-statements;
END GENERATE label;
```

Example:

```
-- using a FOR-GENERATE statement to generate four instances of the full adder
-- component for a 4-bit adder
LIBRARY IEEE;
USE IEEE.STD_LOGIC_1164.ALL;

ENTITY Adder4 IS PORT (
    Cin: IN STD_LOGIC;
    A, B: IN STD_LOGIC_VECTOR(3 DOWNTO 0);
    Cout: OUT STD_LOGIC;
    SUM: OUT STD_LOGIC_VECTOR(3 DOWNTO 0));
END Adder4;

ARCHITECTURE Structural OF Adder4 IS
    COMPONENT FA PORT (
        ci, xi, yi: IN STD_LOGIC;
        co, si: OUT STD_LOGIC);
    END COMPONENT;

    SIGNAL Carryv: STD_LOGIC_VECTOR(4 DOWNTO 0);

BEGIN
    Carryv(0) <= Cin;
    Adder: FOR k IN 3 DOWNTO 0 GENERATE
        FullAdder: FA PORT MAP (Carryv(k), A(k), B(k), Carryv(k+1), SUM(k));
        END GENERATE Adder;
    Cout <= Carryv(4);
END Structural;
```

D.4.5   **Structural Model Sample**

This example is based on the following circuit:

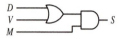

```
-- declare and define the 2-input OR gate
LIBRARY IEEE;
USE IEEE.STD_LOGIC_1164.ALL;

ENTITY myOR IS PORT (
   in1, in2: IN STD_LOGIC;
   out1: OUT STD_LOGIC);
END myOR;

ARCHITECTURE OR_Dataflow OF myOR IS
BEGIN
   out1 <= in1 OR in2;     -- performs the OR operation
END OR_Dataflow;

-- declare and define the 2-input AND gate
LIBRARY IEEE;
USE IEEE.STD_LOGIC_1164.ALL;

ENTITY myAND IS PORT (
   in1, in2: IN STD_LOGIC;
   out1: OUT STD_LOGIC);
END myAND;

ARCHITECTURE AND_Dataflow OF myAND IS
BEGIN
   out1 <= in1 AND in2;     -- performs the AND operation
END AND_Dataflow;

-- topmost module for the siren
LIBRARY IEEE;
USE IEEE.STD_LOGIC_1164.ALL;

ENTITY Siren IS PORT (
      M: IN STD_LOGIC;
      D: IN STD_LOGIC;
      V: IN STD_LOGIC;
      S: OUT STD_LOGIC);
END Siren;
```

```
ARCHITECTURE Siren_Structural OF Siren IS
   -- declaration of the needed OR gate
   COMPONENT myOR PORT (
      in1, in2: IN STD_LOGIC;
      out1: OUT STD_LOGIC);
   END COMPONENT;

   -- declaration of the needed AND gate
   COMPONENT myAND PORT (
      in1, in2: IN STD_LOGIC;
      out1: OUT STD_LOGIC);
   END COMPONENT;

   -- signal for connecting the output of the OR gate
   -- with the input to the AND gate
   SIGNAL term1: STD_LOGIC;

BEGIN
   U0: myOR PORT MAP (D, V, term1);
   U1: myAND PORT MAP (term1, M, S);
END Siren_Structural;
```

● ● ● ● ● ● ● ● ● ● ● ● ● ● ⋯⋯

D.5 Conversion Routines

D.5.1   CONV_INTEGER()

The CONV_INTEGER() routine converts a STD_LOGIC_VECTOR type to an INTEGER type. Its use requires the inclusion of the following library.

```
LIBRARY IEEE;
USE IEEE.STD_LOGIC_UNSIGNED.ALL;
```

Syntax:

CONV_INTEGER(std_logic_vector)

Example:

```
LIBRARY IEEE;
USE IEEE.STD_LOGIC_UNSIGNED.ALL;

SIGNAL four_bit: STD_LOGIC_VECTOR(3 DOWNTO 0);
SIGNAL n: INTEGER;

n := CONV_INTEGER(four_bit);
```

D.5.2 **CONV_STD_LOGIC_VECTOR(,)**

The CONV_STD_LOGIC_VECTOR() routine converts an INTEGER type to a STD_LOGIC_VECTOR type. Its use requires the inclusion of the following library.

```
LIBRARY IEEE;
USE IEEE.STD_LOGIC_ARITH.ALL;
```

Syntax:

CONV_STD_LOGIC_VECTOR (integer, number_of_bits)

Example:

```
LIBRARY IEEE;
USE IEEE.STD_LOGIC_ARITH.ALL;

SIGNAL four_bit: STD_LOGIC_VECTOR(3 DOWNTO 0);
SIGNAL n: INTEGER;

four_bit := CONV_STD_LOGIC_VECTOR(n, 4);
```

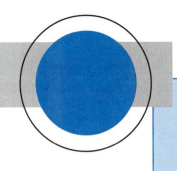

Index